CW00923838

THE OXFORD HANDBOOK OF SOUND STUDIES

THE OXFORD HANDBOOK OF SOUND STUDIES

Edited by
Trevor Pinch and
Karin Bijsterveld

OXFORD
UNIVERSITY PRESS

Oxford University Press, Inc., publishes works that further
Oxford University's objective of excellence
in research, scholarship, and education.

Oxford New York
Auckland Cape Town Dar es Salaam Hong Kong Karachi
Kuala Lumpur Madrid Melbourne Mexico City Nairobi
New Delhi Shanghai Taipei Toronto

With offices in
Argentina Austria Brazil Chile Czech Republic France Greece
Guatemala Hungary Italy Japan Poland Portugal Singapore
South Korea Switzerland Thailand Turkey Ukraine Vietnam

Published by Oxford University Press, Inc.
198 Madison Avenue, New York, New York 10016
www.oup.com

Library of Congress Cataloging-in-Publication Data

The Oxford handbook of sound studies / edited by Trevor Pinch, Karin Bijsterveld.
p. cm.
Includes bibliographical references and index.
ISBN 978-0-19-538894-7 (alk. paper)
1. Sound. 2. Sounds. 3. Noise. I. Pinch, T. J. (Trevor J.) II. Bijsterveld, Karin, 1961-
QC225.15.O93 2011
534—dc22 2010046062

1 3 5 7 9 8 6 4 2
Printed in the United States of America
on acid-free paper

Acknowledgments

We would like to thank all of the contributors to this volume not only for their hard intellectual work in preparing their chapters and for their forbearance in meeting our many deadlines but also for their detailed comments on each other's chapters made during the Sound Studies Conference held at Maastricht, November 21–22, 2009. We are also grateful to the Research Program Science, Technology, and Society of the Faculty of Arts and Sciences at Maastricht University for its generous funding and hosting of the conference. We could not have organized the meeting without the excellent assistance provided by Lidwien Hollanders and Sabine Kuipers. We also owe a deep debt of gratitude to our editor, Norman Hirschy, for his patience, perspicacity, and excellent advice in helping us with every phase of the production of this volume. We would also like to thank Norm for attending the Sound Studies Conference and for providing detailed comments on all of the chapters. In addition, we thank the various anonymous referees who at different stages reviewed this collection. Finally, we express our thanks to Ton Brouwers for correcting the English of the chapters written by nonnative speakers.

Contents

Contributors

Mitchell Akiyama, McGill University, Canada

Katherine Athanasiades, Ford Foundation, New York

Karin Bijsterveld, Maastricht University, the Netherlands

Hans-Joachim Braun, Helmut Schmidt University, Hamburg

Michael Bull, University of Sussex, United Kingdom

Joeri Bruyninckx, Maastricht University, the Netherlands

Eefje Cleophas, Maastricht University, the Netherlands

Tia DeNora, University of Exeter, United Kingdom

Andreas Fickers, Maastricht University, the Netherlands

Rayvon Fouché, University of Illinois at Urbana–Champaign

Trever Hagen, University of Exeter, United Kingdom

Mark Grimshaw, University of Bolton, United Kingdom

Stefan Helmreich, Massachusetts Institute of Technology

Myles Jackson, Polytechnic Institute and Gallatin School, New York University

Mark Katz, University of North Carolina–Chapel Hill

Stefan Krebs, Maastricht University, the Netherlands

Julia Kursell, Max Planck Institute for the History of Science, Berlin

Mara Mills, New York University

Cyrus Mody, Rice University

Trevor Pinch, Cornell University

Tom Rice, University of Exeter, United Kingdom

Hillel Schwartz, independent scholar

Mark M. Smith, University of South Carolina

Jonathan Sterne, McGill University, Canada

Alexandra Supper, Maastricht University, the Netherlands

Timothy D. Taylor, University of California–Los Angeles

William Whittington, University of Southern California

About the Companion Website: www.oup.com/us/ohss

Since this Oxford Handbook deals with sound and a book is an overwhelmingly visual medium, we have taken advantage of the Web to post sound samples of some of the sonic material considered by contributors. At the companion website, authors of many of the individual chapters have posted pertinent examples that can be listened to by following the links at the website. Occasionally there are visuals, such as videos, which accompany the sound samples; one of the major themes of our handbook is that the visual and sonic often work together. Enjoy what you hear (and see) and, if you find more relevant examples, do not hesitate to let us know: the beauty of the Web is that material can be continually updated there.

Trevor Pinch and Karin Bijsterveld

THE OXFORD HANDBOOK OF SOUND STUDIES

NEW KEYS TO THE WORLD OF SOUND

TREVOR PINCH AND
KARIN BIJSTERVELD

1. Introduction

We were sitting in the Tudor Arms, Stockholm, recently voted as "the best British pub in the world"—outside of Great Britain. With oak-paneled walls, wooden beams, nooks and crannies, warm beer, and special British culinary delights such as steak and mushroom pie, the Tudor Arms is a perfect emulation of a pub. We were with two friends, Thomas and Otto, one Swedish, one German. The discussion was getting heated; the topic was the SoundEar.

The SoundEar is a special device found in many kindergartens in Scandinavia. It is a box that displays an outline ear with green glowing LEDs that can be attached to a classroom wall. The SoundEar measures sound.[1] When the noise in the classroom is at or below a predetermined level, the ear glows green. When the noise reaches 10 decibels above this level, the inner part of the ear glows amber; if the noise continues at this level or above for longer than ten seconds, the ear will glow red. The kids have been warned: Be quiet!

Otto had encountered this device when his child first entered Swedish kindergarten. He had noticed that the kids had shut up as soon as the ear turned amber. For Otto, who was used to rowdy German kids and their teachers' use of more traditional means of keeping order, it was a telling example of the difference between Swedish and German society. In public places the Swedes liked to be "civilized." Even on the train home to the suburbs of Stockholm, where he lived, he had observed older school kids monitoring the behavior of younger kids and telling them to pipe down if they were too loud. Thomas quietly asserted that Otto was missing the point. The SoundEar was a safety device that had been introduced into all Swedish schools by the caring state to protect the hearing of school employees. This attention to noise in the workplace was actually a sign of Swedish progressiveness. Of course, it might be a means of discipline, but Thomas pointed out that this is what hard hats at a building site do and surely Otto wouldn't want more head injuries. Otto wasn't buying this argument, however. He claimed that the kids at first took no notice of the SoundEar and that their natural inclination was just to talk louder as their excitement rose. Thomas conceded that the SoundEar sometimes didn't work properly or was switched off in the realization that teachers sometimes just couldn't stop the kids from being kids. As Otto and Thomas argued, we realized that the pub had become almost silent; everyone was looking our way. The Tudor Arms was a perfect emulation of a British pub except for one thing. In Britain the noise gets louder as the evening wears on and more and more beer is consumed. In Sweden that crescendo never comes!

We wondered, was this the future of pubs? Once the pub would have been full of cigarette smoke; now pubs everywhere (except for Berlin, as Otto proudly pointed out) are smoke free. Will new laws and policies eventually be enacted to control sound in public spaces? Will one day machines like the SoundEar glow everywhere? And the SoundEar *is* starting to glow in other places. The largest market for sales is hospitals in the United States. Patients who are coming out of anesthesia are especially sensitive to sound. Furthermore, as Hillel Schwartz points out in this volume, hospitals, which were once islands of silence, are getting noisier and noisier, and one of the major contributors to the increased cacophony are technologies, whether the humming of ventilators or the beeping of monitoring equipment. The SoundEar helps maintain a quiet environment for certain patients. Researchers are also exploring the use of the SoundEar during surgery: It has been claimed that a percentage of surgical error arises from unwanted noise in the operating room.

This true story[2] of our encounter with the SoundEar captures some of the main themes in this book. Modernity has brought about developments in science, technology, and medicine and at the same time increasingly new ways of producing, storing, and reproducing sound. Sound is no longer just sound; it has become technologically produced and mediated sound. This allows it to be more easily transformed, or "transduced," the term used by some of the authors in this book. Transduction turns sound into something accessible to other senses. The SoundEar turns sound into sight. Of course, we should not forget (as Jonathan Sterne and Mitchell Akiyama point out in their chapter), that the human ear itself is a transducer, turning sound into vibrations that then become nerve impulses that are registered in the brain. Yet now that technologies of transduction are everywhere, we would like to foreground their appropriation and consequences in science, society, and culture as important topics for study.

Sound is no longer produced only by humans and nature, for machines roar everywhere and technologies not only measure sound in a myriad of new ways but also produce and emulate sounds, such as in video games and movies. New sources of sound such as the ubiquitous iPod and cell phones demonstrate that sounds have become personal and mobile. New sounds—never heard before, such as the sound of industrialization, the gasoline engine of the automobile, and electronic sounds—have entered what Murray Schafer has so felicitously called "the soundscape" (Schafer 1994/1977).

Sounds can be captured in new ways, such as with the parabolic microphone used to record bird sound, and stored in new ways, such as with the first such device, the recently rediscovered phonoautograph, and the better-known phonograph, which is used not only for music but also as a new scientific tool for the science of ethnology. Sounds that could not be heard before can now be heard for the first time, such as the sound of atoms, which lies at the heart of the newly invented scanning probe microscope, one of the key instruments of the new field of nanotechnology. Digital technologies today provide ever-new ways of storing, manipulating, and transferring sound and music. The recording of music has been transformed (and is still transforming) what it means to make and experience music. Sounds can be reproduced in new sorts of places, such as under water. The establishment of a whole new science of sound—acoustics—has led to new instruments, new ways of measuring, conceptualizing, and controlling sound, and new contributions to music. With the development of the new field of acoustical architecture, buildings and rooms can be specially designed for their sound qualities (Thompson 2002). We interact with sound in new ways. Scientists turn to sonification—innovative technologies and techniques for rendering scientific data into sound.

However, new ways of interacting with sound are also part of everyday life. Consider, for instance, the humble radio dial, which for many people at the dawn of radio opened up a global world accessible only by sound. One of the more significant advances in the history of medicine has come through sound. The ability to listen to the body via the stethoscope provides doctors a powerful new means of diagnosis (Lachmund 1994, 1999). In modern terms, the stethoscope enables them to do an "audification" of bodily phenomena. Sound continues to transform medicine through technologies such as ultrasound (Casper 1998). Novel forms of hearing aid that directly stimulate the brain mean that deaf people can almost miraculously start to hear again.

The consequences of all of this are vast. Sound becomes more materially mediated in a whole host of newfangled ways. Sound becomes more "thinglike"—a commodity to be bought and sold on iTunes, a thing to be worn, as with personal stereos. Sound becomes a means in itself to sell and market goods. Sound can not only be listened to but also be measured, regulated, and controlled (Thompson 2002; Bijsterveld 2008). It can even become an important part of political dissent as with subversive listening to the radio and making of tapes in former Communist countries, which Trevor Hagen and Tia DeNora discuss in this volume.

For us, science, technology, and medicine are the keys to unlock these new worlds of sound. Science, technology, and medicine do not only—intentionally or unintentionally—create novel sources of sound but also provide us with innovative tools for using sound and theories about it. Which sounds have been produced, captured, stored, and transferred by science, technology, and medicine? By which means? How have society and culture appropriated these sounds and means, and how have scientists, engineers, and doctors themselves listened to the objects, machines, and bodies they study—with or without the help of sonic equipment?

These questions have informed our decision to organize the twenty-three chapters that make up this handbook around the different sorts of places where sound is experienced. Some of these are the exemplary sites of science, technology, and medicine, such as the laboratory, the test site, the design studio, and the clinic. Such sites are populated by experts, test subjects, and carefully monitored patients and objects. The other sites we have chosen—the workshop, the field, the home—are much more open and accessible to nonexperts. By stressing this we do *not* mean to say that "society and culture" are not to be found in labs and clinics, unlike in fields and homes, but that different sites have different gatekeepers and that this affects how sounds are listened to, talked about, and given meaning. In the earlier sections these sites tend to be well-defined circumscribed environments, but in the later sections, particularly sections VI and VII, we encounter more diffuse environments. Section I deals with how machine sounds are encountered on the shop floor and at the test site. Section II covers what we call the field, including the ocean, and Section III covers the laboratory. Section IV takes the clinic as its location, and Section V the design studio. Section VI focuses mainly upon the home but also includes karaoke bars, clubs, concert halls, and Internet cafés. Section VII is the most diffuse in terms of location as it focuses upon digital storage of sound, and, of course, as such devices become ever smaller they can be played back in a myriad of places. In addition, as the titles of our sections suggest, we explore how at these sites sounds have been reworked (literally and metaphorically), staged, spoken for, edited, consumed, and moved from one place to another. We stress, however, that this ordering does not preclude alternative routes through our volume. You can work through the chapters in order or dip in and out on your own self-directed journeys through sound.

2. Now That Sound Is in the Air

Our focus on science, technology, and medicine as the keys to unlock the worlds of sound unsurprisingly helped to attract authors with a background in the history and sociology of technology and science, as well as in science and technology studies (STS). Nonetheless, we could not have made this book without contributions from scholars who are working on the history of the senses, cultural history, anthropology of medicine and the body, media studies, film studies, game studies, and musicology.

At the same time, we do not claim that this book fully covers "sound studies." In 2004 we defined *sound studies* as "an emerging interdisciplinary area that studies the material production and consumption of music, sound, noise, and silence and how these have changed throughout history and within different societies" (Pinch and Bijsterveld 2004: 636). Although we still endorse this definition, the word *emerging* may now be slightly too modest. Sound studies has become a vibrant new interdisciplinary field with many different, yet often overlapping, strands. Among the areas involved are acoustic ecology, sound and soundscape design, anthropology of the senses, history of everyday life, environmental history, cultural geography, urban studies, auditory culture, art studies, musicology, ethnomusicology, literary studies, and STS. New fields of study always come with competing definitions of what should be studied, and sound studies is no exception. Each strand conceptualizes its topic and thus the reality it constructs as its proper subject in slightly different ways.

Acoustic ecology has its roots in the environmental concerns of the late 1960s and the 1970s. In those years Raymond Murray Schafer, Barry Truax, and Hildegard Westerkamp established their World Soundscape Project at Simon Fraser University, Vancouver. Their approach was and still is a highly original mix of raising environmental awareness of "our sonic environment and the tuning of the world," mapping past and present soundscapes, endorsing psychoacoustic research that reaches beyond the individual, and contributing to a higher-quality sonic environment through composition and sound design (Schafer 1967, 1969, 1977, 1994/1977; Truax 1978, 2001; Westerkamp 1974). Their allies and heirs are to be found among today's innumerable projects that record and play the soundscapes of contemporary cities or intervene in such soundscapes by adding sound art. In addition, their work has been cited and paraphrased so often that a notion like soundscape as sonic environment, area of aural study, or composition is now the sole focus of publications that trace its roots, discuss its merits and shortcomings, and elaborate on alternative definitions (for an excellent overview see Kelman 2010).

Closely related to acoustic ecology are the emerging fields of soundscape design and sound design. While *soundscape* design is dominated by architects and sound artists (Arkette 2004; Augoyard and Torgue 2005; Martin 1994; LaBelle and Roden 1999), *sound* design is largely the domain of engineers, designers, and marketers (Langemaier 1993; Özcan Vieira 2008). Such sound designers measure the sound-level emission of a consumer good, attempt to curtail or silence its noise, add specifically designed sound to it, study the perception of these sounds, and design the sound of media products such as films or video games (Collins 2008; Grimshaw and Whittington, this volume). Acoustic ecology and soundscape and sound design together have been the feeder disciplines for imaginative interdisciplinary sound studies programs. The sound studies research and education activities coordinated by Berlin cultural theorist Holger Schulze are an example of this new alignment. In his work the theory and history of auditory culture, which draws on acoustic ecology and historical anthropology, are combined with experimental sound design—including sound art—sonic branding, and audio production. It explicitly incorporates a "new materialism" that studies sound as both technical-physical emanation and artistic-aesthetical imagination, and as a "tangible and rich subject of our experience, feelings and thoughts" (Schulze 2008, 11; Spehr 2009). This strand of research thus overlaps with more positivistic approaches to sound studies. Sonic interaction design, for instance, focuses on the "exploitation of sound as one of the principal channels conveying information, meaning, and aesthetic/emotional qualities in interactive contexts." It combines an interest in interaction design with the auditory display of information, "sound modeling," sound perception, cognition and emotion, and sound and music computing.[3]

Without the interactive media available today, such initiatives would not have been possible, and it is therefore not surprising that the field of media studies has made a significant contribution to sound studies. Yet, while sonic interaction design starts out in perception psychology and the communication sciences, most of this work in media studies is firmly rooted in the humanities, qualitative sociology, and cultural studies. This is true for the more historically oriented works on sound and media technologies such as the phonograph, gramophone, telephone, tape recorder, and radio (Badenoch 2008; Douglas 1999; Fickers 1998; Fischer 1992; Gitelman 1999, 2003; Haring 2008; Katz 2004; Morton 2000; Schiffer 1991; Weber 2008). It is also true for the contributions by sociologists and media theorists on the meaning and use of both these older and the more contemporary sound, media, and communication technologies such as the Walkman, iPod, cell phone, and film (Agar 2003; Alderman 2001; Birdsall and Enns 2008; Bull 2000, 2007; Goggin 2006; Lastra 2000; Sterne 2003; Wurtzler 2007). Indeed, several of these authors, who often combine an interest in media studies with the history of technology, are included in the present volume.

The World Soundscape Project has been rightfully credited for coining the word *soundscape* and putting sound studies on the map. One should not forget, however, that without the silent assistance of long-standing scholarly traditions such as the anthropology of the senses, the Annales school of history, cultural history, environmental history, and musicology, the study of sound and soundscapes might have had a much more problematic reception in the humanities (see also Smith 2010). The studies of the smells and sounds of everyday French life by Annales historian Alain Corbin not only brought the most intangible aspects of our past to the attention of historians and others but also contributed to the notion that modes of listening and other sensory experiences are as historically variable as many other aspects of life (Corbin 1986/1982, 1995, 1999). Cultural histories of reading, speech, and sound sensitized our knowledge about who was allowed (or not allowed) to speak or to make sounds and when and where this occurred (Bailey 1996; Burke 1993; Cockayne 2007; Johnson 1995; Payer 2006; Picker 2003; Rath 2003; B. Smith 1999; M. Smith 2000, 2001, 2002, 2003, 2004; Rée 1999). Environmental historians did a similar yet more governance-oriented job by tracking down the rise of noise abatement regulation (Coates 2005; Smilor 1980; Saul 1996a, 1996b), while cultural geography and urban studies have informed us about the spatial distribution of noise and its political implications (Rodaway 1994; Revill 2000). Studies by anthropologists and historians on the hierarchy, meaning, and employment of the senses in different societies substantiated arguments for cultural differences in sensory experience (Classen 1997; Parr 2001; Stoller 1989). With a journal (*The Senses and Society*) and introductory volumes (Howes 2005; Jütte 2005; M. Smith 2007), sensory studies is now an established field in its own right. These introductory volumes include studies of "auditory culture" (Bull and Back 2003) and "hearing cultures" (Erlmann 2004) and feed into museum exhibits on the history and ethnography of sound and noise (Gonseth 2010; Smith, this volume).

The fields of art studies, musicology, and ethnomusicology have recently widened their scope to include some sound studies. This probably reflects the shifting interest among students from classical music to popular music and from the skills necessary for playing traditional musical instruments to those involved in electronic and digital recording, as well as sampling, composing, and consuming music (DeNora 2000; Kahn 1992, 1999; Kraft 1996; Kelly 2009; Lysloff and Gay 2003; Perlman 2004; Greene and Porcello 2005; Théberge 1997; Schmidt-Horning 2004; Stadler 2010; Chanan 1995; Taylor 2001). Perhaps as a sign of the times, the Department of Music and Center for Ethnomusicology at Columbia University in 2009 organized a conference, "Listening In/Feeding Back: Listening and the Circulation of Sound Media," which brought together scholars from sound studies and the more traditional fields of musicology and ethnomusicology. Moreover, the rise of recording has stirred a subspecialty within musicology that has been coined *phonomusicology*: the study of recorded music (Cottrell 2010, 15; Philip 2004; Day 2000).

Close neighbors of ethnomusicology and phonomusicology are new musicology and radical musicology, which have fully digested cultural studies approaches (Rodgers 2010). Often such work also refers to more literary approaches to sound. Literary studies examine the ways that literary writers textualize sound, hearing, and listening, express the sounds of their times, imbue sound with meaning, and evoke noise (for an overview see Bernhart 2008). Some of their publications center on textualized sound in experimental poetics, poetry performances, and poetry (Morris 1997). Others analyze references to specific forms of hearing and listening in literary texts, such as eavesdropping (Gaylin 2002) or mishearing (Connor 2009)—although Connor's humorous talk on mishearing reaches far beyond literature. Highly insightful are the studies that unravel expressions of sound in Shakespearean theater (Folkerth 2002) and in twentieth-century plays (Meszaros 2005), as well as textualizations of noise in late nineteenth- and twentieth-century literary texts. Michael Cowan (2006), for instance, has written wonderfully on Rainer Maria Rilke's resistance to worldly noise. and Philipp Schweighauser (2006) intelligently unravels the noises of naturalist, modernist, and postmodernist American literature in his "history of literary acoustics."

Textualization of sound is again a transformation of one sensory experience into another. This and the rise of sensory studies might lead to the question whether one sense—hearing—deserves its own field and indeed its own handbook, but this would be to misconstrue the senses involved in sound studies. Although sensory studies in general has contributed to the rise of sound studies and sound studies is clearly part of sensory studies, we feel that sound studies, especially when taken in the direction of science, technology, and medicine, raises a far wider range of issues. We have already pointed out that as sound becomes more materially embedded, it becomes mediated by many senses, including seeing and touch. It is exactly this relationship between the material embedding and multisensory mediation of modern sound that intrigues us and has led to our key questions.

Sound studies thus is a flourishing interdisciplinary area with several overlapping disciplines and a range of methods that touch upon the fields of acoustic ecology, sound design, urban studies, cultural geography, media and communication studies, cultural studies, the history and anthropology of the senses, the history and sociology of music, and literary studies. We are not the first authors to tie in some of these wider concerns with the specific fields of history and sociology of science and technology and STS. Hans-Joachim Braun, a historian of technology, has long been an advocate of studying music and sound technologies within the context of the history of technology and has organized many sessions on such themes at international meetings of the history of technology. This development led to Braun's important edited collection, *"I Sing the Body Electric": Music and Technology in the 20th Century* (Braun 2000, 2002). Emily Thompson, another scholar whose roots lie in the history of technology, has shown in her authoritative book, *The Soundscape of Modernity* (Thompson 2002), how the science and practice of acoustical architecture developed in parallel with twentieth-century concerns and changes in technology and how sound was experienced. Jonathan Sterne, a scholar from media studies, draws upon STS in his important book, *The Audible Past*, which deals with how, in the twentieth century, listening changed in concert with technologies such as the telegraph and the stethoscope (Sterne 2003).

Our own interest in sound studies should also be mentioned here. One of us, Trevor Pinch, stumbled into the field in the course of writing his book on the history of the Moog electronic music synthesizer (Pinch and Trocco 2002). Pinch, a professor of science and technology studies who has a background in the sociology of science and technology, was initially interested in linking STS to music and held sessions called "STS Faces the Music" at an international science studies conference at Bielefeld, Germany, in 1996, in which Karin Bijsterveld also participated. Eventually Pinch teamed up with Bijsterveld to pursue this theme in research on musical instruments (Pinch and Bijsterveld 2003) and with a special issue of *Social Studies of Science*: "Sound Studies: New Technologies and Music," published in 2004. Pinch first taught a course in sound studies at Cornell University in 2004, while Bijsterveld started a course called Sound Technologies and Cultural Practices at Maastricht University the same year. She has a background in history, musicology, and STS and has contributed to the history of noise with *Mechanical Sound: Technology, Culture, and Public Problems of Noise in the Twentieth Century* (Bijsterveld 2008), our knowledge of the everyday use of audio technologies with *Sound Souvenirs: Audio Technologies, Memory, and Cultural Practices* (Bijsterveld and Van Dijck 2009), and our understanding of acoustic cocooning in the car (Bijsterveld 2010).

3. Listening for Knowledge

The new audio technologies discussed in this book, such as the phonograph, gramophone, radio, cassette player, and digital audio equipment, have not only drawn on earlier instruments used in science, engineering, and medicine but also contributed to the processes of knowledge creation in these fields (Brady 1999; Kursell 2008; Stangl 2000). Such contributions have never been self-evident, however.

If there is any field that is associated with seeing rather than with hearing, it is science. Scholars who emphasize the visual bias in Western culture even point to science as their favorite example. Because doing research seems impossible without using images, graphs, and diagrams, science is—in their view—a visual endeavor par excellence. Historians and sociologists of science have recently corrected this claim by showing how senses other than seeing, including listening, have been significant in the development of knowledge, notably in the laboratory. They stress that scientific work involves more than visual observation. The introduction of measurement devices that merely seem to require the reading of results and thus seeing has not ruled out the deployment of the scientists' other senses. On the contrary, scientific work in experimental settings often calls for bodily skills, one of which is listening. We'll come back to these views later. The world of science itself, however, still considers listening a less objective entrance into knowledge production than seeing. Why?

One of the aims of our book is to offer readers a better understanding of this contested position of sonic skills—by which we mean listening skills *and* the skills needed to employ the tools for listening—in knowledge production. Several contributors to this handbook focus specifically on the role of sound and listening in science, technology, and medicine. This includes fields that have acoustic phenomena as their exclusive topic and domains that examine other subjects with the help of sound. Most of these contributions focus on the period from the 1920s on, thus taking the rise of recording technologies into account. Our readers can find these contributions in the first four sections of our book, which focus on listening on shop floors and at test sites, in the field, in labs, and in clinics, respectively. While Stefan Krebs, Eefje Cleophas, and Karin Bijsterveld unravel the listening practices of engineers and mechanics, Joeri Bruynincks, Julia Kursell, Myles Jackson, Cyrus Mody, and Alexandra Supper discuss those of scientists, and Tom Rice and Mara Mills those of physicians. Moreover, Stefan Helmreich and some of the authors just mentioned also show what artists take from scientists when dealing with sound.

Three questions in particular are of concern here.[4] The *first* is how scientists, engineers, and physicians have employed their ears in making sense of what they have studied. How did they listen to the objects, machines, and bodies they studied? What tools did they use? And how did they acquire their sonic skills? The *second* question is how such listening practices, regardless of whether or not mediated by recording and amplification technologies, have generated new scientific knowledge, technological designs, and medical devices. In short, did listening elicit new questions and findings in these fields, and if so, what kind of listening? The final and most crucial question is why listening has nonetheless remained contested and still lacks the same legitimization given to other means of knowing. In other words, under which conditions have sonic skills been accepted as an "objective" means of inquiry alongside visual ones in science, engineering, and medicine? In some cases, why have visual skills partially replaced sonic skills? Why, for instance, is "sonification," that is, the auditory (in contrast to the visual) display of scientific data still deeply contested today?

These questions have almost naturally evolved from a particular set of issues predominating the historiography of the senses and STS for quite some time. To many scholars, as we have already suggested, science is the most telling example of the visual orientation of Western culture (Attali 1985; Berendt 1985). To the indignation of such scholars, even acousticians illustrate their work with slides and charts rather than with sounds (Schafer 1994/1977). Others claim that the numerous visual metaphors for knowing in Western languages ("Yes, I see") proves the visual bias of Western culture (Tyler 1984). Comparisons with cultures that have epistemologies based on alternative sensory orientations, such as auditory ones, rhetorically underline this argument (Feld 2003).

For many years, the discussion boiled down to the issue of *when*, exactly, Western culture had turned from privileging hearing to preferring sight. Did it occur with the rise of print (Bailey 1996)? Yet Constance Classen has claimed that the standard ranking of the senses in Western culture, in which sight occupies the highest position, preceded print culture by many centuries (Classen 1997). Furthermore, other scholars have stressed the continuing significance of sound for people's everyday spatial and symbolic orientation in modern times (Corbin 1999, Bull and Back 2003). As Jonathan Sterne has clarified, the search for an overall shift in the sensory selectivity of the West unjustly assumed that the history of the senses should be "a zero-sum game, where the dominance of one sense by necessity leads to the decline of another sense" (Sterne 2003, 16). Consequently, many academics today argue instead for a "perceptual equilibrium" that has been present since at least the later medieval period (Woolf 2004).

For a long time, however, the idea that *modern science* is a visual affair remained undisputed. The big picture was that profound changes in natural philosophy in the sixteenth and seventeenth centuries—including the new reliance on observation and experiment as legitimate sources of knowledge, as well as the growing importance of print as a vehicle of academic communication—had given rise to a situation in which scientific knowledge increasingly came to be expressed in visual terms. "Came to be expressed" happened to be an important qualification, though. The empirical turn in STS moved scholars away from formulating universal criteria that demarcated rational science from irrational beliefs and shifted their focus from *overidealized* conceptions of science to the everyday *practices* of science.

This change did not result in debunking the significance of visual routines in science per se. On the contrary, the new approach fine-tuned the understanding of the historical processes and sociological mechanisms that had made visualizations the most widely accepted forms of scientific *representation*. Yet the empirical turn also created a heightened awareness of the contingencies in the skills used to *acquire* data, both the visual skills and those related to the other senses. As to the issue of representation, sociologist of science Bruno Latour showed how inscriptions such as tables and diagrams, because they were both immutable and mobile, effectively circulated locally acquired data across geographically widespread networks of knowledge (Latour 1986, 1987). Moreover, publications at the intersection of the history of science, media studies, and art history added insights into the shifting forms of visual display and the educational entertainment that preceded the spread of visual literacy (Stafford 1994; Tufte 1997).

At the very same time, scholars in STS began studying what actually happened in the laboratory *before* scientists sent out their inscriptions. These highlighted the subtle interconnections between the experimenter's body, the use of scientific instruments, and the outcomes of laboratory work. Ethnographic lab studies revealed that experimental skills involve so much tacit knowledge that the exact replication of experiments is, in fact, extremely difficult to attain and could lead to an infinite "experimenters' regress" (Collins 1985). Collins based the notion of tacit knowledge upon the work of Michael Polanyi (1967), considering it "an embodied kind of know-how irreducible to symbolic terms" (Mody 2005, 176). In addition, historians of science produced studies of the intricate relationships between research instruments and the content of the knowledge generated (Hankins and Silverman 1995; Schaffer 1999). A more nuanced account of visual instruments, for instance, led to the idea of an "externalized retina" that structures the cognitions of scientists (Lynch 1990). Similar in approach, studies of nonverbal "visual thinking" and materially located "situated actions" in engineering acquired salience (Ferguson 1992; Suchman 1987).

Closely in line with this type of research and echoing the rise of the history and anthropology of the senses, the study of the role of senses *other than seeing* in science, engineering, and medicine gained momentum. We have already mentioned Lachmund's work (1994, 1999) on the rise of the stethoscope and auditory skills in nineteenth-century medicine and how this related to knowledge of pulmonary diseases (see also Duffin 1998). Others showed how technicians and engineers listened to photocopiers (Orr 1996) and cars (Borg 2007) to detect flaws in the mechanisms, as well as how in materials science the sounds of laboratory instruments and data informed researchers of the quality and content of the experiments (Mody 2005).

Such work does *not* imply that all scientific work draws on the senses. Many sciences depend on the use of mathematics and logics that rely less directly on sensory skills. Moreover, many historians of science accept the claims of Lorraine Daston and Peter Galison (1992, 2007) on the rise of "mechanical objectivity." As they have shown, the moral of mechanical objectivity in science gradually gained significance, reaching its zenith in the 1920s. This moral dismissed the human mind and body as trustworthy witnesses of natural phenomena in favor of the registration of such phenomena by machines. Yet as Lissa Roberts (1995, 519) has argued, eighteenth-century chemists publicly sidelined touch, hearing, smell, and taste and in fact transformed "their bodies into appendages of a machine."

To understand the role of the senses in knowledge dynamics, it is thus useful to distinguish between the different *sites* at which scientists, engineers, and physicians perform their work: in behind-the-doors laboratories and out-of-sight fields, in the factories and hospitals, where professional engineers and doctors examine individual machines and bodies, as well as at the conferences at which they present their results in scientific papers. These different sites are all represented in the chapters that follow.

Across these sites of knowledge production, different *modes* of listening are involved, such as monitory, diagnostic, exploratory, and synthetic listening. *Monitory listening* is the kind of auditory surveillance that scientists, engineers, and physicians employ in order to check the proper functioning of instruments, machines, and patients' bodies (Mody 2005; Bijsterveld 2006). *Diagnostic listening* refers to the mode of listening that physicians apply to identify pathologies when using a stethoscope (Lachmund 1994, 1999; Rice and Coltart 2006) and that engineers use to detect the origin of calculation mistakes in computers by amplifying their sound (Alberts 2000, 2003). Whereas *monitory* listening is used to determine *whether* something is wrong, diagnostic listening reveals *what* is wrong. *Exploratory listening* is listening to discover new phenomena. Susan Douglas (1999) used it to refer to the way in which early amateur radio operators searched the ether for radio stations. The use of sound recording technologies by ornithologists to identify new bird species is a similar form of exploratory listening. *Synthetic listening*, finally, focuses on the understanding of polyphonic patterns of sound, such as in the sonification of scientific data (Dayé and de Campo 2006).

All of these *modes of listening* require specific skills. These are not only skills for understanding what exactly one is listening to but also technical skills linked to the listening devices used, such as positioning a stethoscope properly on a patient's body or effectively handling a magnetic tape recorder. Even musical skills may be required as, for instance, when engineers single out the different "melodies" and "rhythms" of humming computers. This explains why we prefer to speak of *sonic skills* rather than auditory skills.

The chapters that follow have produced highly contextualized insights into the fundamental issue of sensory selectivity in the production and validation of scientific knowledge and technological design. Rather than proclaiming a victory of the visual in science, pleading for the emancipation of hearing at the expense of seeing, or defending a perceptual equilibrium, the chapters *investigate* when, how, and under what conditions the ear has contributed to knowledge dynamics in tandem with or instead of the eye. All of the authors have their own particular take on these issues, colored by the scholarly traditions they draw from. We do not want to give away all of their conclusions. We do, however, want to *flag* a few remarkable issues that return in many of the chapters, albeit under different headings.

The first issue concerns the contexts in which an interest in sound leads to the acquisition of knowledge of objects, machines, bodies, and environments—what

Steven Feld has called "acoustemology" (Feld 2003) and Tom Porcello has retuned into "technostemology" for technologically mediated forms of acoustemology (Greene and Porcello 2005). Apart from the availability of instruments that have enabled or assisted listening, there seems to be a clear relationship with military activities—detecting "the enemy" with the help of sound—or, in contrast, the wish to leave such a military focus behind, such as in reconversion activities (Mody, this volume). Another relevant context involves efforts to aid those who are blind or deaf (Supper and Mills, this volume). Moreover, the epistemological *status* of sound and the attitude toward acoustic instruments or practices seem related to the dominance of particular sensory-related crafts and instruments within certain disciplines or medical practices, such as the microscope in biology or the stethoscope in medicine (Mody and Rice, this volume).

Another intriguing issue is that in their attempts to make sound "speak" to them and to inform them about the topics they study, scientists, engineers, physicians, and test subjects struggle with the verbalization or description of sound. To compare sounds—whether those of car engines, birds, hearts, sonar, or sonified data—and to delineate even the tiniest differences, knowledge makers need a sound vocabulary and a sound language—in short, a way to talk about sound. In doing so, they not only use dictionary words for sound, such as "roaring," "whistling," or "crackling" sounds, but also draw upon analogies with the sounds of musical instruments, the human voice, and everyday artifacts. Creating a sound vocabulary is a prerequisite for making a "sound index," "sound codification," or "sound taxonomy": a way of distinguishing and classifying sounds. Subsequently, the taxonomy enables "sound mapping": the assignment of sounds to the information they represent, as Stefan Krebs explains in his chapter.

He and other authors nicely illustrate the various modes of listening in medical, engineering, and scientific contexts, such as monitory listening or sonic surveillance (Rice, this volume), diagnostic listening (Krebs and Rice, this volume), and exploratory listening (Fickers, this volume). The empirical evidence in these chapters also implies, however, that it is helpful to distinguish conceptually between the *intent* behind the listening, such as in our initial definition of monitory or diagnostic listening, and the more perceptual or epistemological characteristics of listening, such as in sound mapping.

Also remarkable are the many ways in which the sonic and the visual interfere with one another. Some of the chapters here describe instances of conversion or "transduction" of sonic information into visual information; we have already alluded to these phenomena. Examples are recording sound by way of music notation and onomatopoeia or with the help of instruments such as the kymograph or the sound spectrograph. More important, these chapters also explain why conversion took place and what effects resulted. Other chapters illustrate how even the listening experience itself has been visualized, such as through the design of the radio dial (Fickers, this volume) or how visual conceptualizations of sound, such as the "close up," entered the world of sound recording in science (Bruyninckx, this volume).

Finally, we want to share with our readers the observation that tinkering, repairing, and do-it-yourself practices in research and art seem to be sociologically related to the tendency to adapt audification, sonification, and careful listening, to—in other words—some sort of sensory sophistication and sensory flexibility. There is a link in this respect between Krebs's mechanics, Mody's acoustic microscope people, Jackson's, Helmreich's, and Supper's tinkerer-composers, and Mill's technophile cochlear-implant test subjects, which they share with Marc Perlman's audiophiles (2004) and Kristen Haring's amateur radio operators (hams) (2008).

4. New Sources and Means of Capturing, Storing, and Reproducing Sound

This book covers many new and old sources of sound, some intentional and some unintentional. Science, technology, and medicine themselves constitute one of the biggest new sources of *unintentional* sounds. The machines of American industrialization such as the new weaving machines used in Lowell, Massachusetts (Smith, this volume), or the industrial machines found in Nazi Germany, the former DDR, and West Germany (Braun, this volume) were not deliberately designed to be noisy—although extra noise may on occasion also be symbolic of power.[5] It just so happens that enormous engines or industrial machinery housed in enclosed spaces such as factories produce sound as a byproduct of their operation. Even in hospitals, medical technology is a source of often unintentional sound, whether it is the buzz of the ventilators of life-support systems or the beeps of more routine technologies such as digital thermometers (Schwartz, this volume). This is in contrast to, say, the *intentional* sounds of music, which, via radio and newly developed loudspeakers, was piped into factories in the twentieth century as part of the "music while you work" movement also described by Braun (this volume).

There is of course nothing "natural" about machines having to produce noise (Bijsterveld 2008). Once we understand the adverse health consequences of these noisy machines, we can design them to be quieter. The lack of naturalness of machine noise is nicely demonstrated by the sound engineering of automobiles (Cleophas and Bijsterveld, this volume), where a car company such as Porsche will employ acoustic engineers precisely to engineer the "throaty" roar of the Porsche engine to make it more appealing to its customer base —a sound that interestingly appeals differently depending upon the gender and nationality of consumers.

Unintentional sounds produced in new sorts of locations, such as factories, hospitals, and laboratories, can lead to the reconfiguration of space. One of the scenes Braun evokes is the workers testing some of the loudest diesel engines found in the Rostock diesel factory in the DDR in a special soundproofed room that they have constructed because of the health-threatening aspects of the sounds. Sound and acoustic architecture can together subtly change space in new designs of buildings and concert halls (Thompson 2002). Urban space can also be reconfigured by mobile listeners with the use of personal stereos and iPods (Bull, this volume). Geographical space too, can be dramatically changed as with the case of the early radio dial, which was for many listeners their gateway to different cities, regions, and countries (Fickers, this volume).

Science, technology, and medicine are themselves important sources of innovation for new sonic technologies and instruments. Sounds that have never been heard before can be listened to through new devices and pieces of equipment such as the stethoscope, which we have already mentioned. Interestingly, in the early history of the automobile a form of stethoscope was also developed specially for listening to car engines (Krebs, this volume). New sound reproduction equipment such as hydrophonic headphones and special loudspeakers enabled people to listen to sounds under water, thus facilitating a whole new form of art (Helmreich, this volume). New medical devices such as cochlear implants have helped deaf people or those with a severe hearing impairment almost miraculously to hear again (Mills, this volume).

New ways of producing and capturing sounds also change the way we mediate sound. For instance, in the early days of radio, listeners' experience of geography was mediated by the tactile feel and visual organization of the radio dial (Fickers, this volume). The new parabolic microphone allowed bird sound to be captured but required a truckload of equipment to accompany it, thus mediating the sorts of "natural" environments where bird sounds could be recorded (Bruyninckx, this volume). With new instruments and techniques, new sorts of skills and tacit knowledge can also emerge (Rice, Krebs, and Kursell, this volume) and may involve new regimes of training, as well as the new types of listening skills, which we discussed earlier.

Science, technology, and medicine offer new ways of transforming or "transducing" sound. The transformation of sound to another material medium enables it to be more easily stored and transported, such as with phonograph cylinders, player piano rolls, vinyl records, tapes, cassettes, compact discs, and later a vast array of digital storage and transmission media. This is an important point for sound studies to digest and lies at the core of the STS approach to sound.

The strength of science studies is in its dealing with materiality, the senses, culture, and politics within the same analytical register. Special pieces of equipment that render the external world into visual traces that are in turn reproducible and transportable are known within STS as "inscription devices" (Latour and Woolgar 1979). In sound studies we are dealing with an even wider range of transformations or conversions between *different senses* and *different media* (for instance, the underwater sounds described by Helmreich). Scholars in this volume employ terms such as *transduction* (Helmreich) or *conversion* (Mody) to describe such processes. Indeed, from the perspective of sound studies it is possible to see the "inscription device" as a rather limited, purely visual form of a much more general class of devices for transforming the external world into reproducible and transportable sensory traces.

The increasing movement toward sonification has produced a range of new ways of rendering scientific data into the sensory realm (Supper, Sterne and Akiyama, this volume). The transformation to another media or material also produces the possibility of new forms of sound storage such as the phonoautograph (Sterne and Akiyama, this volume), the phonograph (Kursell and Katz, this volume), the player piano (the punched paper holes can even be thought of as an early means of digital storage) (Katz, this volume), and the current digital revolution aligned to computers, software, and the Internet in the ways that sound and music are produced, stored, and consumed (Grimshaw, Whittington, Katz, Fouché, Pinch and Athanasiades, and Bull, this volume). Sometimes new inscription devices, such as the kymograph,[6] invented by German physiologist Karl Ludwig in the 1840s, can be combined with new sound technologies such as the phonograph to provide new tools for scientific investigation (Kursell, this volume). Another example is the combination of the new parabolic microphone for recording bird sound with the audio spectrogram to advance scientific studies of birdsong (Bruyninckx, this volume)

With these new ways of transforming, storing, and reproducing sound come interesting conceptual issues about how the "transformed" sounds are experienced. For instance, with the development of the phonograph, new sorts of musical idioms became popular because of phonograph recording techniques, such as overcoming the sonic limitations of the bullhorn microphone, by singers "crooning" or excessive vibrato on a violin (Katz 2004). Indeed, such musical effects, rather than being seen as artifacts of the recording process, started to sound "real," and performers began including them in their live performances—a process known as *phonorealism*. With the digitization of sound, new sorts of sonic experiences have become possible, such as the "hyperrealism" of sound and the "immersion" of the listener in the sonic experience of modern movies and video games (Whittington and Grimshaw, this volume). Novel technologies, such as digital samplers and synthesizers, enable sounds to be specifically "designed" to enhance visual stimuli in movies (Whittington, this volume) or to sell commercial products (Taylor, this volume). Technology, it can be argued, is creating original sorts of phenomenal experiences, such as with cochlear implants—a foretaste of what some futurists claim will be life increasingly emulating video games (Mills, this volume). Psychological states are more and more linked directly to sounds, whether these states are the audiotopia sought after by iPod listeners (Bull, this volume) or involve the monitoring of EEGs linked to sound sources, which has been proposed as a new way of making video games even more sonically immersive (Grimshaw, this volume). However, the significance of these transformations between the senses is even deeper for the field of sound studies. As Sterne and Akiyama argue, as the senses themselves become ever more technologically mediated, they become more "plastic," and this challenges standard accounts of individual senses that one typically encounters in histories, anthropologies, and cultural studies.

The new technologies of sound have had a dramatic impact on the arts. The invention of the electronic music synthesizer has been described as one of the major musical transformations of the twentieth century (Pinch and Trocco 2002). The boundaries between what is natural and what is synthetic in music has always been contested. For example, are the keys on a flute a "mechanical" interference or an integral part of a "natural" instrument (Pinch and Bijsterveld 2003)? What makes a musical instrument an instrument as opposed to a machine or piece of technology (Pinch and Trocco 2002)? What counts as music and what counts as noise have, of course, been fiercely contested (Kahn 1999). Unintentional sounds such as the sound of wind made by an Aeolian harp or the sounds heard in John Cage's famous 4′33″ challenge our notion of music. New sounds, such as that made by a scratch on a vinyl record, can be a source of innovation for a whole new genre of music such as in hip-hop (Katz and Fouché, this volume). New instruments developed in the lab, such as the metronome, siren, and tuning fork, can become sources of musical inspiration that lead to new compositions (Jackson, this volume), as can the instruments taken over from war-time sonar used to make underwater music (Helmreich, this volume). Digital software enables users to transform their personal computers into recording studios and leads to new genres of music such as "remixing" and "mash-ups" (Katz, Pinch, and Athanasiades, this volume). Sometimes the new sonic technologies and the sorts of sound they produce lead to desires to acquire and experience earlier ways of producing and listening to music, such as the "technostalgia" involved in the turn to vintage electronic instruments (Pinch and Reineker 2009) and the desire for authentic vinyl scratching practices in hip-hop (Fouché and Katz, this volume).

The border crossings between the arts and science documented in this book reveal again that essentialist definitions of instruments and their proper domains are unnecessarily restrictive. "Follow the instruments" is the methodological heuristic heard in this volume (Pinch and Bijsterveld 2004). Instruments developed in science and engineering can become part of the arts and the commercial world and vice versa (Kursell and Jackson, this volume). The lack of one meaning of scientific results and technologies is referred to as "interpretative flexibility" within STS (Collins 1985; Shapin and Schaffer 1985; Latour 1987; Bijker, Hughes, and Pinch 1987). The role of users is crucial here as well. Users, such as Grand Wizard Theodore, who is reputed to have invented the hip-hop vinyl scratch, have in effect redesigned the technology of the record player for an entirely new use in music making (Eglash et al. 2004). Within STS we describe this as the "interpretive flexibility" of technology once more appearing but in this case in the context of use (Kline and Pinch 1996; Oudshoorn and Pinch 2003). This book is full of such instances, including the remarkable example of one the earliest methods of *magnitizdat* found in Eastern Europe during the Soviet era, documented by Trever Hagen and Tia DeNora (this volume). The emulsion on discarded X-rays from the 1950s in the Soviet Union provided a material that could be engraved as a record. The production and distribution of these homemade records became known as *roentgenizdat*, or "playing the bones." This case and the hip-hop one add a political valence to these user practices.

Running like a bass ostinato throughout the transformations, travels, and terminations undergone by sound in this volume are indeed the politics of sound. A common struggle is which professional group gets authorized to use which sound technologies and where (Krebs, Katz, Supper, and Rice, this volume). The politics of sound crops up in the status of the amateur in music (Katz, Pinch, and Athanasiades, this volume), the struggles over racialized and gender identities (Smith, Braun, Schwartz, Fouché, Pinch, and Athansiades, this volume), and the sorts of subversive listening and music practices encountered during the Cold War in Eastern Europe (Hagen and DeNora, this volume). Earlier struggles over noise abatement (Thompson 2002; Bijsterveld 2008; Braun, this volume) and new struggles over the possible damage to hearing from new devices such as earbuds (Bull and Schwartz, this volume) remind us that sound, as in the story of the SoundEar, is part of the lived politics of everyday life.

5. Chapter Summaries

One of the most enduring sources of unintended sound is natural phenomena, such as thunderstorms, waterfalls, the ocean, volcanoes, and the sounds made by the creatures that inhabit the natural world. This concern is central to Mark Smith's opening chapter of the Shop Floor Section, where he reconsiders the "machine in the garden" thesis of industrialization in the United States, made famous by Leo Marx. How was the sound of nature experienced, and how did it contrast with the new sorts of industrial sounds found in factories? Furthermore, how did the sounds of the new industrial soundscape contrast with the sound of another means of production—the more silent and agrarian use of slavery in the South?

What machine sounds mean to listeners and how they respond to them crucially depend upon historical context, as Smith shows in his chapter. Hans-Joachim Braun writes about a very different period, Germany in the Nazi era and the "two Germanies" after the Second World War. One of the most intriguing questions he discusses is how the differing German political regimes and their dissimilar commitments to the rights and safety of workers have actually dealt with industrial noise.

The sound of the early automobile engine in Germany from the 1930s to the 1950s is the topic of Stefan Krebs's chapter. Here a community of listeners arises who specialize in diagnosing a car's problems from the sound of its engine, gearbox, and so on. This community turns out to be transient. First, it consists of the car owners themselves, who listen to the car—a form of listening that Krebs calls "listening by driving," but later (by the 1950s) it is the specialized garages and auto mechanics who do most of the listening—a practice that over time largely vanishes in favor of other techniques of diagnosing automobile problems. Again, issues of "diagnostic listening" are important, but Krebs also offers a new theoretical addition with Michel Foucault's notion of a "dispositive" to link these listening practices and their demise in the wake of wider power and discourse issues in society at large.

Cleophas and Bijsterveld write mainly about car sound and how European manufacturers in the 1990s tested what consumers wanted in that regard. They focus largely on the interior of cars and how even things such as the "crackle" of leather upholstery should come with the "right sound" for a particular make or model. The tests carried out reveal that laypeople's vocabulary of sound and their listening skills do not always match those of expert engineers. They extend STS approaches toward understanding testing to this new sonic environment. Negotiating a common shared meaning—the outcome of a test—is even harder in the context of different European cultures. Moreover, Cleophas and Bijsterveld draw upon Gerhard Schulze's notion of *Erlebnisgesellschaft*, or "experience-driven society," to explain why the car industry began caring so much about sound design at all. Schulze argues that since many products have been perfected, sensory experience plays an ever-greater role in their marketing and sales.

The sound of nature—in this case, birds—is a central feature of Joeri Bruyninckx's chapter on how the recording of birdsong has evolved and become more scientific since the late 1920s. He focuses upon ornithologists in the United Kingdom and the United States and shows how special microphones and devices for turning sound into visual traces, such as the vibralyzer, the oscillograph, and the audio spectrograph (taken over from wartime uses of sonar to search for submarines), have played a crucial role. Often trips into the wild to record the sounds of birds were stymied by the large mobile studios needed to house the recording equipment of the day. Part of his story is the role played by people in the movie industry, who were often the first to develop new sound technologies for places such as the well-known Cornell Lab of Ornithology. Interestingly, later on in another very different sonic context we find scientists returning the compliment by contributing to the movie industry. Bill Whittington writes in his chapter about how Cornell computer scientists in collaboration with companies such as Pixar are developing new ways to digitally realize sounds for movies. Although for the purpose of organization we have divided this book into sections such as the Laboratory, Field, and Studio, the actors studied do not always respect or follow such categories. As Bruyninckx points out, although we have placed his chapter in the field section, the field, lab, and studio become "blurred spaces" in his study. Perhaps as this is a sounds study book we could better say that the different sites (as in the recording studio) suffer "leakage."

The sound of nature does not feature much in other chapters although the medium of water is an important theme in Stefan Helmreich's chapter on underwater music. Helmreich shows how the explorers of underwater music learned much from military underwater technologies such as sonar and the development of hydrophonic headphones. New technologies of loudspeakers are also required to produce underwater sounds. There is fluid movement between the sciences and the arts. Helmreich is one of the first authors in the handbook to raise the issue of transduction—sound that is usually experienced in air must be experienced in a totally different medium, water, and this requires special devices: "transducers." What does it mean to "transduce" or transform sounds from one sonic medium or sense to another, and what is lost and gained in the process?

Julia Kursell shows in her chapter how the phonograph and the kymograph were used by Carl Stumpf at the Berlin Institute of Psychology in the early twentieth century as a tool for the scientific investigation of sound and hearing. Stumpf explored the ways in which these new instruments could be used for the study of language. For instance, he recorded and replayed sounds, such as human vowels, at different speeds. In 1908 Stumpf also established a phonograph archive that contained examples of musical use from all over the world as part of an ethnologically inspired endeavor. At the same time, the archive provided a basis for the new field science of comparative musicology—which today we would call ethnomusicology. Kursell uses German media theorist Friedrich Kittler to rethink what is involved in the standardization of a media technology.

Myles Jackson also reminds us that sound travels between the arts and science in his investigation of several devices developed in the nineteenth century by scientists and acousticians interested in better understanding musical phenomena such as pitch. He documents the invention of the tuning fork, the siren, and the portable chronometer (the latter became the metronome) and the different uses they acquired later in musical composition and performances in the twentieth century (the tuning fork and metronome, he notes, also became important instruments in nineteenth-century physiology). This traveling across disciplines and indeed between the sciences and the arts in general makes following these acoustic and sound technologies even more salient. As these instruments get used in new ways in new contexts by new groups of users, we start to hear a richer story of how human creativity and invention occur. While bricolage and tinkering are going on, their sounds are calling out to be listened to. What is at one moment a scientific instrument can later become a new musical instrument. Similarly, a musical instrument such as the phonograph can become part of a scientific laboratory, as Julia Kursell details. The world of sound has for too long been walled off—it is an important part of the means and methods on which the world of human ingenuity and creativity thrive.

At the start of his chapter, Cyrus Mody again draws attention to the issue of the transformation of sound by the community of probe microscopists, who developed many of the new sorts of instruments that are crucial to nanotechnology. Mody uses the term *synesthetic conversion*, meaning conversion and reconversion between any of the senses, and points out that probe microscopists were attempting to engage the haptic dimensions of the practice, as well as the audio and the visual. Mody's chapter is largely about the wider context in which this community of probe microscopists emerges at specific locations, such as Stanford University during the 1960s. He shows how this group was also embroiled in the "reconversion" of American academic research from military funding and applications to civilian funding and an orientation to "human problems." Mody thus shows how the specific technical concerns and practices of the scientists he studies were linked to the wider social and political context in which they worked.

The transformation of scientific data into auditory information is part of a wider process known as *sonification*. In her chapter, which concludes the section on the laboratory, Alexandra Supper looks at the recent history of this emergent field. Researchers in the field must not only find a way to reduce the residual bias toward the visual but also convincingly legitimize the various sonification techniques, often linked to music and the arts,to make this a discrete field in its own right. No one describes "visualization" as a separate field of science, so what warrants a field of "sonification"? The search for the "killer application" that Supper documents is further testimony to its current striving for professional status. Supper follows sonification through different contexts in great detail by studying its discursive formations and strategic demeanor and its attempts to professionalize. Part of the novelty of her work lies in her focus upon conference presentations by scientists—an area neglected in much of the history and sociology of science.

Hospital sounds are discussed in the chapter by Hillel Schwartz. Drawing upon an astounding range of sources, including literary greats such as Homer and Kafka, he details the history of the silent hospital, starting with Florence Nightingale's experiences in the Crimean war. Schwartz considers hospital architecture and the ever-increasing encroachment of noise in parallel with the history of another audio technology—the earplug. He traces its use particularly through the First World War and up to the noise-canceling headphones of today. It is the conjunction of the technology of personal sound (and its prevention) and the institutional sound of the hospital that gives Schwartz's chapter its tension. Just as in the social sciences, where we wrestle with a concern for both individuals qua individuals and wider social formations such as groups and institutions, so, too, with sound technologies: Is sound an institutional problem to be treated by reforming hospitals, or is it up to individuals to wear sound devices to protect themselves? The wearing of individual devices, of course, soon gets coupled with issues of stigmata and gender (the "manliness" or not of such devices) as Schwartz details and which Braun and Smith in the very different context of industrial noise also note.

Another important source of sound is the human body. The process known as *auscultation*, which is the topic of Tom Rice's chapter, concerns the sounds of internal organs such as the heart and how they can be listened to with a new device, the stethoscope. Interestingly, Rice was drawn to this topic while carrying out a project similar to Schwartz's, namely a study of how patients in a modern hospital experienced sound. Rice's study is one of the most anthropological in the volume as he details his own "ears-on" experiences as he learned auscultation from a practitioner in a leading London teaching hospital. We have already mentioned diagnostic listening skills, and this chapter explores them in detail. It discusses issues such as how the sound of a heartbeat can help medical practitioners make a diagnosis, how stable the categories it produces are, and how reliable the technique itself is, especially when other medical contexts (such as that in the United States) play down auscultation in favor of heart monitoring by echocardiography (which uses another sonic technology, ultrasound, to produce a sonograph of the heart). Rice not only is concerned with how the listening skills are acquired and passed on but also notices how listening to the body requires medical practitioners to acquire special bodily skills and postures as they learn to position their bodies in the best way to employ this specialized form of listening. He thus engages with the discussion on tacit skills in this volume.

Another chapter in the section on clinics discusses a new device that utilizes sound technology, the cochlear implant, which enables deaf people or those with a severe hearing impairment to hear again. This controversial topic—some members of the Deaf community welcome cochlear implants, whereas others reject them as a violation of "Deaf culture"—has been treated within STS before where there is a growing literature on disability studies (Blume 2010; Haraway 2008). Mara Mills describes the cochlear implant as the most common neural-computer interface in the world with more than two hundred thousand users. She sets her chapter within the context of neuroenhancement and its futurist discourse and offers a new history of cochlear-implant development by focusing upon the previously neglected topic of the subjects actually used in the research. By treating the subjects as "users," she shows how their concerns sometimes get incorporated and sometimes ignored in the design of these devices, which must also accommodate economic concerns and the pressures of the wider medical establishment. One thus gains a sense of the new forms of subjectivity that these devices can enable. Mills intriguingly quotes media theorist Vilém Flusser, who found that his own deafness and a hearing aid gave him more control as to what he chose to hear— a form of "ear lid" that enabled him to better navigate his immersion in the world's noises and voices and permitted him to "see through" the programming of his own hearing aid and thus both "see" and hear better.

Immersion is an important theme in the handbook and no more so than in the section on the design studio. Often work on the phenomenology of sound, such as Don Idhe's important early investigations (Idhe 1976, 2007), starts by pointing out the difference between listening and seeing. The visual field appears as a form of screen surrounded by darkness in front of our eyes, but we experience sound all around us—in short without the aid of "ear lids" we are immersed in sound. The issue of sound immersion particularly comes to the fore in the design of video games, which is the theme of Mark Grimshaw's chapter. Grimshaw, who is himself a game player and designer, shows how, in video games, the players have enormous control over where in the game the character they operate is at any moment; in addition, the sounds must respond appropriately (for instance, movement must generate the sound of footsteps). Furthermore, these sounds may convey important information to the player, such as footsteps coming from behind, signaling a threat. Because of the vast range of sounds that need to be generated in real time, the sonic experience in video games is also much more contingent upon the sound-generating technology available. Grimshaw's chapter shows how technological constraint, the users' own imaginative experiences, the designers' goals, and the visual material offered to the user all work together to produce a constructed immersive experience.

Grimshaw ends his chapter by pointing to an as yet unrealized possibility in games whereby sound itself might respond to the affective state of the player as measured by psychophysical sensors such as an EEG. In such a scenario one could imagine a frightening sound being made even scarier because the player did not yet seem to be frightened enough! This more complete enhanced immersive experience bears an uncanny similarity to the futurist scenarios Mills discusses in her chapter, whereby brain implants and neuroenhancers will, according to some futurists, make ordinary life appear more and more like a video game.

In contrast to Grimshaw's chapter on video design, where the perspective of the player and the designer form the core of the analysis, Bill Whittington, a film scholar, considers the emergence of a whole new way of designing sound for film and a specialized company (Pixar) associated with it. Pixar, of course, is famous for having developed the computer-generated animation methods that replaced the extremely labor-intensive way that cartoons used to be made. Whittington shows that part of the story of this revolution in film production is a sound story. Film sound underwent a digital transformation of its own with the introduction of new sound formats and the rise of the sound design movement, which involved new sound technologies such as synthesizers and samplers. Indeed, Whittington argues, the technological innovations and new aesthetics that Pixar developed, including the use of sound techniques from live action movies, gained credibility in part because of sound. Whittington focuses mainly on the early animated short films that Pixar made. Important is the development of the technique and the aesthetic of what he calls "hyperrealism," where designers created stylized constructions that accessed familiar sound events yet recombined these effects to create new sound impressions. For instance, the sound of a dog barking in *Toy Story* was made by combining the sound of a barking dog with that of a tiger. Whittington unpacks the cinematic codes embedded in the new methods and traces the relationships among the complicated networks of production companies, sound houses, and designers in the San Francisco Bay area, which preceded Pixar and led to its eventual success. With the current success of 3D productions such as *Avatar*, Hollywood is still undergoing dramatic changes, and, as Whittington notes, the increasing turn to a "spectacle of sensation" means that the immersion experience of film is drawing ever closer to that of the video game technologies discussed by Mark Grimshaw in his chapter.

In the early 1960s, commercial electronic music was first becoming realizable. Tim Taylor, an ethnomusicologist, in the final chapter in the section on the design studio documents the early use of electronic sounds, such as those of the newly invented Moog synthesizer, to sell products such as beer and coffee. Taylor focuses on two of the best known U.S. commercial musicians of the period, Raymond Scott and Eric Siday, and dissects in great musicological detail their most famous ads. He contrasts Scott's and Siday's early efforts with those of a 1980s' composer, Suzanne Ciani, who was famous for a Coke ad realized on her Buchla synthesizer, which evokes the kind of hyperrealism Whittington discusses. In the ad, the sound of Coke being poured is much more evocative than the sound of "the real thing."

One can speculate that just as for the 1950s, when movies and ads led to the growing acceptance of electronic sounds, hyperrealism as a form of listening—perhaps a new sonic skill—was conveyed via the popular medium of ads. Taylor's chapter is interesting when read beside the earlier chapter by Cleophas and Bijsterveld on selling car sounds. In their case the car sounds help sell the product, and in Taylor's case the hyperreal sound of the product sells the product.

In the opening chapter on the section on the home, Andreas Fickers tells us about early radio and in particular the role played by the iconic radio dial. For many people who grew up in the 1950s (as one of the editors of this book did), the dial glowing on the radio on the mantelpiece provided the first window into a global world—a world where listeners in Europe learned for the first time of the importance of, for instance, Hilversum, Holland (always featured prominently on such dials). Fickers details technical developments in early radio whereby it became possible to easily tune in stations—a precondition for the radio dial. However, the weight of his chapter is on spectrum-allocation issues that the International Broadcasting Union mediated in the 1920 and 1930s as different national radio interests negotiated nothing less than the layout of the dial. Fickers thus bridges the detailed listening habits in the home and the wider world of regulatory authorities to provide an intriguing account of the material, institutional, and symbolic aspects of this crucial feature of a single audio technology.

Radio and the elicit tuning-in of stations feature in Trever Hagen and Tia DeNora's chapter on the role of sound technologies in Hungary and Czechoslovakia during the 1960s and 1970s as growing opposition to the official Communist regimes and their state-sanctioned music practices developed. We learn fascinating details about this world, such as that the sound of the Soviets' attempts to jam Western stations was referred to by listeners in Czechoslovakia as "Stalin's bagpipes." They use anthropologist Victor Turner's notion of a liminal space (defined by Turner as "betwixt and between") to describe the sorts of nonofficial practice in which their subjects engage. The listening and performances that Hagen and DeNora document often take place in private spaces away from the prying eyes of the state. We learn how technologies such as radio, records, and cassette tapes were all important in the development of new forms of subversive practice and also how the sound technologies were in turn shaped by these practices. In Hagen and DeNora's chapter the twist is that these subversive listening and musical practices help coproduce a sort of unofficial antistate as users learn how to develop a "disposition" to challenge the official state.

Mark Katz shows the continued importance of what might be described as "amateurs." He starts by examining the use of the phonograph and the player piano and puts paid to the myth that the invention of these devices killed off amateur music making. Indeed, the story he tells is the converse: These new devices actually facilitated and encouraged amateur music making, and manufacturers often responded to these new users by modifying the devices to facilitate more new uses. Katz shows that technologies such as phonographs and the player piano employed devices to adjust tone, tempo, and loudness, along with the performance, so that the listener was far from being passive. He notes that, years before the digital medium arose, there was a lively tradition of home recording with the phonograph, with dubbing and the making of what we would today call "mash-ups." Other amateur uses discussed by Katz are the development of karaoke in Japan (one of the few sections of this handbook to deal with music in Asia), the ways users adapt the video game "Guitar Hero," and, as mentioned already, the birth of hip-hop.

In the concluding chapter to the section on the home, Trevor Pinch and Katherine Athanasiades examine a specific form of digital musical community that has arisen with the advent of the Internet. These amateur musicians, who form a worldwide community, post their musical compositions to a special website, where their music can in turn be downloaded and reviewed by other users and which offers a chart position for every piece of music posted. Pinch and Athansiades show how online reputations at this site are established with the help of yet another form of transduction, one between sounds and music and the words and symbols that populate online ranking systems. They also show that, although the website does open up new possibilities for musical identities and forms of collaboration, many of the identities (including gendered identities) and processes found there actually bear more resonance to existing offline musical practices and identities.

The last section of the handbook contains three chapters that deal with the new possibilities of moving sound that digital storage offers. However, Rayvon Fouché's chapter, which opens this section, starts off in the analog world. The origin story of the vinyl record "scratch" is well known. At first an unwanted and annoying aspect of vinyl record listening, in the hands of a skillful DJ, "scratching" became part of the new music making of hip-hop. Special tools and technologies and eventually digital scratching systems evolved as Fouché discusses. He points to the crucial role played by hybrid digital-vinyl systems, which allows DJs to control digital music with specially encoded vinyl records and turntables, thus preserving some of the authentic craft of analog scratching while seeming to embrace the latest digital advances. Rather than having to schlep heavy vinyl records around, DJs can now have all their music conveniently nearby, stored on a digital computer. The practices of digital scratching and the use of vinyl have, however, remained in tension within the hip-hop community. Fouché's chapter is one of the few in the volume where issues of race and ethnicity and audio technology are interconnected. He documents how Grandmaster Flash played an important and often-neglected role in outlining the technical specifications for one of the mixers used by DJs.

The sound of the urban soundscape forms the background to Michael Bull's chapter on iPod listeners. The iPod (along with other MP3 players and cell phones) is one of the most pervasive new sound technologies ever introduced. As Bull (2000, 2007) has shown, personal stereos have profound consequences not only for how users' psychological states can be mediated, modulated, and controlled but also for how they experience their external environment, including space and social interaction. In this chapter Bull argues that users must navigate between an extreme state of enjoyment, which he calls audiotopia, and the opposite state, audio toxicity—where iPod users lose all sense of their grounding (or, as Bull calls it, "tethering") in the surrounding social world and even risk physical damage to their hearing. Bull draws upon a set of theoretical concerns from cultural studies to show how internal experiences and mental states are coproduced or coconstructed with particular new audio technologies.

The final chapter in the volume, which is by Jonathan Sterne and Mitchell Akiyama and takes Scott's phonoautograph as its topic, raises a fascinating set of issues having to do with old ways of storing sound and modern digital methods of retrieving and recovering it. The phonoautograph received worldwide attention in 2008 after scientists managed to play back some of the songs recorded by Scott in 1860, thus producing what was claimed to be the "world's oldest recording." Sterne and Akiyama note that Scott rather belittled Edison's phonograph as it merely played back sound. The phonoautograph was a new way to *write* sound—in other words, to render sound into a visual form. The chapter thus further explores the topic of sonification, taken up earlier by Supper. Sterne and Akiyama develop the cultural studies idea of "articulation," which they describe as an approach that is more attentive to issues of power than other constructivist approaches such as the social construction of technology or actor network theory. Their conclusion is a dramatic one for sound studies. The examples of sonification they discuss demonstrate that data intended for one sense can be readily transformed into another, and they claim this shows the increasing plasticity of the senses. Their goal here is to "lay bare the degree to which the senses themselves are articulated into different cultural, technological, and epistemic formations." In short, according to Sterne and Akiyama, it is no longer possible to treat particular technologies or cultural forms as predestined for or determined by a single sense. In a way, the topic of a sensory history, of anthropology, or of cultural studies has according to them been dissolved. This offers a radical challenge for the future!

NOTES

1 The device is made by the SoundShip Company in Copenhagen and has won a Danish Design Council award. It has an interesting history: Before becoming adapted for classroom purposes, the device was first made with the idea of protecting musicians from damaging their ears.

2 There actually was such an incident experienced by one of the editors (Trevor Pinch). The story has been slightly simplified for our purposes here. We are grateful to Otto Sibum for drawing our attention to the SoundEar; he and Thomas Kaiserfeld have graciously allowed us to use their friendly disagreement in this book.

3 Http://www.cost-sid.org/ (accessed Sept. 2, 2010).

4 These questions also guide the Maastricht research program "Sonic Skills: Sound and Listening in Science, Technology, and Medicine, 1920s–Now," coordinated by Karin Bijsterveld and funded by a Dutch NWO VICI Award.

5 For instance, American-made vacuum cleaners are reputed to be designed to be louder than similar machines in Europe and Asia because Americans associate a loud cleaner with more suction power. See John Seabrook, "How to Make It," *New Yorker*, Sept. 20, 2010, 66–73.

6 The kymograph consisted of a rotating drum covered with paper, over which a stylus moved back and forth recording a physiological variable such as blood pressure.

REFERENCES

Agar, Jon. *Constant Touch: A Global History of the Mobile Phone*. Cambridge: Icon, 2003.

Alberts, Gerard. "Computergeluiden." In *Informatica & Samenleving*, ed. Gerard Alberts and Ruud van Dael, 7–9. Nijmegen: Katholieke Universiteit Nijmegen, 2000.

———. "Een halve eeuw computers in Nederland." *Nieuwe Wiskrant* 22 (2003): 17–23.

Alderman, John. *Sonic Boom: Napster, MP3, and the New Pioneers of Music*. Cambridge, Mass.: Perseus, 2001.

Arkette, Sophie. "Sounds like City." *Theory, Culture, and Society* 21 (2004): 159.

Attali, Jacques. *Noise: The Political Economy of Music*. Manchester: Manchester University Press, 1985.

Augoyard, Jean-François, and Henry Torgue, eds. *Sonic Experience: A Guide to Everyday Sounds*. Montreal: McGill-Queen's University Press, 2005.

Badenoch, Alexander. *Voices in Ruins: West German Radio across the 1945 Divide*. Basingstoke, UK: Palgrave MacMillan, 2008.

Bailey, Peter. "Breaking the Sound Barrier: A Historian Listens to Noise." *Body and Society* 2(2) (1996): 49–66.

Berendt, Joachim-Ernst. *Das dritte Ohr: Vom Hören der Welt*. Reinbek bei Hamburg: Rowohlt, 1985.

Bernhart, Toni. "Stadt hören: Auditive Wahrnehmung in Berlin Alexanderplatz von Alfred Döblin." *Zeitschrift für Literaturwissenschaft und Linguistik* 38(149) (2008): 51–67.

Bijker, Wiebe, E., Thomas P. Hughes, and Trevor J. Pinch, eds. *The Social Construction of Technological Systems: New Directions in the Sociology and History of Technology*. Cambridge, Mass.: MIT Press, 1987.

Bijsterveld, Karin. "Acoustic Cocooning: How the Car Became a Place to Unwind." *The Senses and Society* 5 (July 2010): 189–211.

———. "Listening to Machines: Industrial Noise, Hearing Loss, and the Cultural Meaning of Sound." *Interdisciplinary Science Reviews* 31(4) (2006): 323–37.

———. *Mechanical Sound: Technology, Culture, and Public Problems of Noise in the Twentieth Century*. Cambridge, Mass.: MIT Press, 2008.

———, and José van Dijck, eds. *Sound Souvenirs: Audio Technologies, Memory, and Cultural Practices*. Amsterdam: Amsterdam University Press, 2009.

Birdsall, Carolyn, and Anthony Enns. *Sonic Mediations: Body, Sound, Technology*. Newcastle: Cambridge Scholars Publishing, 2008.

Blume, Stuart. *The Artificial Ear: Cochlear Implants and the Culture of Deafness*. New Brunswick, N.J.: Rutgers University Press, 2010.

Borg, Kevin. *Auto Mechanics: Technology and Expertise in Twentieth-Century America*. Baltimore: Johns Hopkins University Press, 2007.

Brady, Erika. *A Spiral Way: How the Phonograph Changed Ethnography*. Jackson: University Press of Mississippi, 1999.

Braun, Hans-Joachim, ed. 2000. *"I Sing the Body Electric": Music and Technology in the 20th Century*. Hofheim, Germany: Wolke, 2000. Reprint, Baltimore: Johns Hopkins University Press, 2002.

Bull, Michael. *Sound Moves: iPod Culture and Urban Experience*. New York: Routledge, 2007.

———. *Sounding Out the City: Personal Stereos and the Management of Everyday Life.* New York: Berg, 2000.

———, and Les Back, eds. *The Auditory Culture Reader*. Oxford: Berg, 2003.

Burke, Peter. "Notes for a Social History of Silence in Early Modern Europe." In *The Art of Conversation*, 123–41. Ithaca, N.Y.: Cornell University Press, 1993.

Casper, Monica. The Making of the Unborn Patient: A Social Anatomy of Fetal Surgery. New Brunswick, NJ: Rutgers University Press, 1998.

Chanan, Michael. *Repeated Takes: A Short History of Recording and Its Effects on Music.* New York: Verso, 1995.

Classen, Constance. "Foundations for an Anthropology of the Senses." *International Social Science Journal* 49 (1997): 401–12.

Coates, Peter A. "The Strange Stillness of the Past: Toward an Environmental History of Sound and Noise." *Environmental History* 10(4) (2005): 636–65.

Cockayne, Emily. *Hubbub: Filth, Noise, & Stench in England, 1600–1770*. New Haven, Conn.: Yale University Press, 2007.

Collins, Harry M. *Changing Order: Replication and Induction in Scientific Practice*. London: Sage, 1985.

———. "Tacit Knowledge, Trust, and the Q of Sapphire." *Social Studies of Science* 31 (2001): 71–86.

Collins, Karen. *Game Sound: An Introduction to the History, Theory, and Practice of Video Game Music and Sound Design*. Cambridge, Mass.: MIT Press, 2008.

Connor, Steven. "Earslips: Of Mishearings and Mondegreens"; http://www.stevenconnor.com/earslips/ (accessed October 10, 2010).

Corbin, Alain. *The Foul and the Fragrant: Odor and the French Social Imagination*. Cambridge, Mass.: Harvard University Press, 1986 (originally published in 1982 as Le miasme et la jonquille: l'odorat et l'imaginaire social, XVIIIe–XIXe siècles, Paris: Aubier Montaigne).

———. *Time, Desire, and Horror: Towards a History of the Senses*. Cambridge, UK: Polity, 1995.

———. *Village Bells: Sound and Meaning in the Nineteenth-Century French Countryside*. London: Macmillan, 1999.

Cottrell, Stephen. "The Rise and Rise of Phonomusicology." In *Recorded Music: Performance, Culture, and Technology*, ed. Amanda Bayley, 15–36. New York: Cambridge University Press, 2010.

Cowan, Michael. "Imagining Modernity through the Ear." *Arcadia* 41(1) (2006): 124–46.

Daston, Lorraine, and Peter Galison. "The Image of Objectivity." *Representations* 10(40) (1992): 81–128.

———. *Objectivity*. New York: Zone, 2007.

Day, Timothy. *A Century of Recorded Music: Listening to Musical History*. New Haven, Conn.: Yale University Press, 2000.

Dayé, Christian, and Alberto de Campo. "Sounds Sequential: Sonification in the Social Sciences." *Interdisciplinary Science Reviews* 31(4) (2006): 349–64.

DeNora, Tia. *Music in Everyday Life*. New York: Cambridge University Press, 2000.

Douglas, Susan J. *Listening In: Radio and the American Imagination, from Amos 'n' Andy and Edward R. Murrow to Wolfman Jack and Howard Stern*. New York: Times Books, 1999.

Duffin, Jacalyn. *To See with a Better Eye: A Life of R. T. H. Laennec*. Princeton, N.J.: Princeton University Press, 1998.

Eglash, Ron, Jennifer L. Croissant, Giovanna Di Chiro, and Rayvon Fouché, eds. *Appropriating Technology: Vernacular Science and Cultural Invention*. Minneapolis: University of Minnesota Press, 2004.

Erlmann, Veit, ed. *Hearing Cultures: Essays on Sound, Listening, and Modernity*. New York: Berg, 2004.
Feld, Steven. "A Rainforest Acoustemology." In *The Auditory Culture Reader*, ed. Michael Bull and Les Back, 223–39. Oxford: Berg, 2003.
Ferguson, Eugene S. *Engineering and the Mind's Eye*. Cambridge, Mass.: MIT Press, 1992.
Fickers, Andreas. *"Der Transistor" als technisches und kulturelles Phaenomen: Die Transistorisierung der Radio- und Fernseheempfaenger in der deutschen Rundfunkindustrie 1955 bis 1965*. Bassum, Germany: Verlag für Geschichte der Naturwissenschaften und der Technik, 1998.
Fisher, Claude S. *America Calling: A Social History of the Telephone to 1940*. Berkeley: University of California Press, 1992.
Folkerth, Wes. *The Sound of Shakespeare*. New York: Routledge, 2002.
Galison, Peter. "Computer Simulations and the Trading Zone." In *The Disunity of Science: Boundaries, Contexts, and Power*, ed. P. Galison, and D. J. Stump, 118–57. Stanford: Stanford University Press, 1996.
Garfinkel, Harold. *Studies in Ethnomethodology*. Englewood Cliffs, N.J.: Prentice Hall, 1967.
Gaylin, Ann. *Eavesdropping in the Novel from Austen to Proust*. New York: Cambridge University Press, 2002.
Gitelman, Lisa. *Scripts, Grooves, and Writing Machines: Representing Technology in the Edison Era*. Stanford: Stanford University Press, 1999.
———, and Geoffrey B. Pingree, eds. *New Media 1740–1915*. Cambridge, Mass.: MIT Press, 2003.
Goggin, Gerard. *Cell Phone Culture: Mobile Technology in Everyday Life*. London: Routledge, 2006.
Gonseth, Marc-Olivier, Yann Laville, and Gregoire Mayor, eds. *Bruits*. Neuchâtel: Musée d'ethnographie, 2010.
Greene, Paul D., and Thomas Porcello, eds. *Wired for Sound: Engineering and Technologies in Sonic Cultures*. Middletown, Conn.: Wesleyan University Press, 2005.
Hankins, Thomas L., and Robert J. Silverman. *Instruments and the Imagination*. Princeton, N.J.: Princeton University Press, 1995.
Haraway, Donna. *When Species Meet*. Minneapolis: University of Minnesota Press, 2008.
Haring, Kristen. *Ham Radio's Technical Culture*. Cambridge, Mass.: MIT Press, 2008.
Howes, David. *Empire of the Senses: The Sensual Culture Reader*. Oxford: Berg, 2005.
Ihde, Don. *Listening and Voice: A Phenomenology of Sound*. Athens: Ohio University Press, 1976.
———. *Listening and Voice: Phenomenologies of Sound*. Albany: State University of New York Press, 2007.
Johnson, James H. *Listening in Paris*. Berkeley: University of California Press, 1995.
Jütte, Robert. *A History of the Senses: From Antiquity to Cyberspace*. Malden, Mass.: Polity, 2005.
Kahn, Douglas. *Noise, Water, Meat: A History of Sound in the Arts*. Cambridge Mass.: MIT Press, 1999.
———, ed. *Wireless Imagination: Sound, Radio, and the Avant-Garde*. Cambridge, Mass.: MIT Press, 1992.
Katz, Mark. *Capturing Sound: How Technology Has Changed Music*. Los Angeles: University of California Press, 2004.
Kelman, Ari Y. "Rethinking the Soundscape: A Critical Genealogy of a Key Term in Sound Studies." *The Senses and Society* 5 (July 2010): 212–34.

Kelly, Caleb. *Cracked Media: The Sound of Malfunction*. Cambridge, Mass.: MIT Press, 2009.

Kline, Ronald, and Trevor Pinch. "Users as Agents of Technological Change: The Social Construction of the Automobile in the Rural United States." *Technology and Culture* 37 (1996): 763–95.

Kraft, James. *Stage to Studio: Musicians and the Sound Revolution, 1890–1950*. Baltimore: Johns Hopkins University Press, 1996.

Kursell, Julia, ed. *Sounds of Science—Schall im Labor (1800–1930)*. Berlin: Max-Planck-Institut für Wissenschaftsgeschichte, 2008.

LaBelle, Brandon, and Steve Roden, eds. *Site of Sound: Of Architecture and the Ear*. Los Angeles: Errant Bodies, 1999.

Lachmund, Jens. *Der abgehorchte Körper: Zur historischen Soziologie der medizinischen Untersuchung*. Opladen: Westdeutscher Verlag, 1994.

———. "Making Sense of Sound: Auscultation and Lung Sound Codification in Nineteenth-Century French and German Medicine." *Science, Technology, and Human Values* 24(4) (1999): 419–50.

Langenmaier, Arnica-Verena, ed. *Der Klang der Dinge: Akustik—eine Aufgabe des Design*. Munich: Design Zentrum/Verlag Silke Schreiber, 1993.

Lastra, James. *Sound Technology and the American Cinema: Perception, Representation, Modernity*. New York: Columbia University Press, 2000.

Latour, Bruno. *Science in Action: How to Follow Scientists and Engineers through Society*. Milton Keynes, UK: Open University Press, 1987.

———. "Visualisation and Cognition: Thinking with Eyes and Hands." *Knowledge and Society: Studies in the Sociology of Culture Past and Present* 6 (1986): 1–40.

———, and Steve Woolgar. *Laboratory Life: The Social Construction of Scientific Facts*. London: Sage, 1979.

Lynch, Michael. "The Externalized Retina: Selection and Mathematization in the Visual Documentation of Objects in the Life Sciences." In *Representation in Scientific Practice*, ed. M. Lynch and S. Woolgar, 153–86. Cambridge, Mass.: MIT Press, 1990.

Lysloff, René T. A., and Leslie C. Gay. *Music and Technoculture*. Middletown, Conn.: Wesleyan Press, 2003.

Martin, Elizabeth. *Architecture as a Translation of Music*. New York: Princeton Architectural Press, 1994.

Meszaros, Beth. "Infernal Sound Cues: Aural Geographies and the Politics of Noise." *Modern Drama* 48(1) (2005): 118–31.

Mody, Cyrus. "The Sounds of Science: Listening to Laboratory Practice." *Science, Technology, and Human Values* 30 (2005): 175–98.

Morris, Adalaide, ed. *Sound States: Innovative Poetics and Acoustical Technologies*. Chapel Hill and London: The University of North Carolina Press, 1997.

Morton, David. *Off the Record: The Technology and Culture of Sound Recording in America*. New Brunswick, N.J.: Rutgers University Press, 2000.

Orr, Julian E. *Talking about Machines: An Ethnography of a Modern Job*. Ithaca, N.Y.: Cornell University Press, 1996.

Oudshoorn, Nelly, and Trevor Pinch. *How Users Matter: The Coconstruction of Users and Technologies*. Cambridge Mass.: MIT Press, 2003.

Özcan Vieira, Elif. *Product Sounds: Fundamentals and Applications*. PhD diss. Delft: Delft University Press, 2008.

Parr, Joy. "Notes for a More Sensuous History of Twentieth-Century Canada: The Timely, the Tacit, and the Material Body." *Canadian Historical Review* 82(4) (2001): 720–45.

Payer, Peter. "Vom Geräuch zum Lärm: Zur Geschichte des Hörens im 19. und frühen 20. Jahrhundert." In *Der Aufstand des Ohrs—die neue Lust am Hören*, ed. Volker Bernius, Peter Kemper, and Regina Oehler, 105–19. Göttingen: Vandenhoeck and Ruprecht, 2006.

Perlman, Marc. "Golden Ears and Meter Readers: The Contest for Epistemic Authority in Audiophilia." *Social Studies of Science* 34 (2004): 783–807.

Philip, Robert. *Performing Music in the Age of Recording*. New Haven, Conn.: Yale University Press, 2004.

Picker, John M. *Victorian Soundscapes*. New York: Oxford University Press, 2003.

Pinch, Trevor, and Karin Bijsterveld. Introduction to "Sound Studies: New Technologies and Music." Special issue, *Social Studies of Science* 34(5) (2004): 635–48.

———. "Should One Applaud? Breaches and Boundaries in the Reception of New Technology in Music." *Technology and Culture* 44 (2003): 536–59.

Pinch, Trevor, and David Reinecker. "Technostalgia: How Old Gear Lives on in New Music." In *Sound Souvenirs: Audio Technologies, Memory, and Cultural Practices*, ed. Karin Bijsterveld and José van Dijck, 152–66. Amsterdam: Amsterdam University Press, 2009.

Pinch, Trevor, and Frank Trocco. *Analog Days: The Invention and Impact of the Moog Synthesizer*. Cambridge, Mass.: Harvard University Press, 2002.

Polanyi, Michael. *The Tacit Dimension*. New York: Anchor, 1967.

Porcello, Thomas. "Speaking of Sound: Language and the Professionalization of Sound Recording Engineers." *Social Studies of Science* 34(5) (2004): 733–58.

Rath, Richard Cullen. *How Early America Sounded*. Ithaca, N.Y.: Cornell University Press, 2003.

Rée, Jonathan. *I See a Voice: A Philosophical History of Language, Deafness, and the Senses*. London: Flamingo, 1999.

Revill, G. "Music and the Politics of Sound: Nationalism, Citizenship, and Auditory Space." *Environment and Planning* 18 (October 2000): 597–613.

Rice, Tom, and John Coltart. "Getting a Sense of Listening: An Anthropological Perspective on Auscultation." *British Journal of Cardiology* 13(1) (2006): 56–57.

Roberts, Lissa. "The Death of the Sensuous Chemist. The 'New' Chemistry and the Transformation of Sensuous Technology." *Studies in the History and Philosophy of Science* 4 (1995): 503–29.

Rodaway, Paul. *Sensuous Geographies: Body, Sense, and Place*. New York: Routledge, 1994.

Rodgers, Tara. *Pink Noises: Women on Electronic Music and Sound*. Durham, N.C.: Duke University Press, 2010.

Saul, Klaus. "'Kein Zeitalter seit Erschaffung der Welt hat so viel und so ungeheuerlichen Lärm gemacht . . .'—Lärmquellen, Lärmbekämpfung, und Antilärmbewegung im Deutschen Kaiserreich." In *Umweltgeschichte—Methoden, Themen, Potentiale: Tagung des Hamburger Arbeitskreises für Umweltgeschichte, Hamburg 1994*, ed. Günther Bayerl, Norman Fuchsloch, and Torsten Meyer, 187–217. Münster: Waxmann, 1996a.

———. "Wider die 'Lärmpest': Lärmkritik und Lärmbekämpfung im Deutschen Kaiserreich." In *Macht Stadt krank? Vom Umgang mit Gesundheit und Krankheit*, ed. Dittmar Machule, Olaf Mischer, and Arnold Sywottek, 151–92. Hamburg: Dölling and Galitz, 1996b.

Schafer, R. Murray. *Ear Cleaning: Notes for an Experimental Music Course*. Toronto: Berandol Music, 1967.

———. *European Sound Diary*. Burnaby, Canada: Simon Fraser University Press, 1977.

———. *The New Soundscape: A Handbook for the Modern Music Teacher*. Toronto: Clark and Cruickshank, 1969.

———. *The Soundscape: Our Sonic Environment and the Tuning of the World*. Rochester, Vt.: Destiny, 1994 (originally published in 1977 as *The Tuning of the World*, New York: Knopf).

Schaffer, Simon. "Late Victorian Metrology and Its Instrumentation: A Manufactory of Ohms." In *The Science Studies Reader*, ed. M. Biagioli, 457–78. London: Routledge, 1999.

Schiffer, Michael B. *The Portable Radio in American Life*. Tucson: University of Arizona Press, 1991.

Schmidt-Horning, Susan. "Engineering the Performance: Recording Engineers, Tacit Knowledge, and the Art of Controlling Sound." *Social Studies of Science* 34 (2004): 703–31.

Schulze, Holger, ed. *Sound Studies: Traditione—Methoden—Desiderate: Eine Einführung*. Bielefeld, Germany: Transcript, 2008.

Schweighauser, Philipp. *The Noises of American Literature, 1890–1985: Toward a History of Literary Acoustics*. Gainesville: University Press of Florida Press, 2006.

Shapin, Steven, and Simon Schaffer. *Leviathan and the Air-Pump: Hobbes, Boyle, and the Experimental Life*. Princeton, N.J.: Princeton University Press, 1985.

Smilor, Raymond Wesley. "Toward an Environmental Perspective: The Anti-Noise Campaign, 1893–1932." In *Pollution and Reform in American Cities, 1870–1930*, ed. Martin V. Melosi, 135–51. Austin: University of Texas Press, 1980.

Smith, Bruce R. *The Acoustic World of Early Modern England: Attending to the O-Factor*. Chicago: University of Chicago Press, 1999.

Smith, Mark M. "Echoes in Print: Method and Causation in Aural History." *Journal of the Historical Society* 2 (Summer/Fall 2002): 317–336.

———. "Futures of Hearing Pasts." Lecture, Hearing Modern History: Auditory Cultures in the 19th and 20th Century, 9th Blankensee Colloquium, Free University of Berlin, June 17–19, 2010.

———. *Hearing History: A Reader*. Athens: University of Georgia Press, 2004.

———. *Listening to Nineteenth-Century America*. Chapel Hill: University of North Carolina Press, 2001.

———. "Listening to the Heard Worlds of Antebellum America." *Journal of the Historical Society* 1 (Spring 2000): 65–99.

———. "Making Sense of Social History." *Journal of Social History* 37(1) (2003): 165–86.

———. *Sensing the Past: Seeing, Hearing, Smelling, Tasting, and Touching in History*. Berkeley: University of California Press, 2007.

Spehr, Georg, ed. *Funktionale Klänge: Hörbare Daten, klingende Geräte, und gestaltete Hörerfahrungen*. Bielefeld, Germany: Transcript, 2009.

Stadler, Gustavus, ed. Special issue, *Social Text* 102 (Spring 2010).

Stafford, Barbara M. *Artful Science: Enlightenment Entertainment and the Eclipse of Visual Education*. Cambridge, Mass.: MIT Press, 1994.

Stangl, Burkhard. *Ethnologie im Ohr: Die Wirkungsgeschichte des Phonographen*. Vienna: WUV Universitätsverlag, 2000.

Sterne, Jonathan. The Audible Past: Cultural Origins of Sound Reproduction. Durham, N.C.: Duke University Press, 2003.

Stoller, Paul. *The Taste of Ethnographic Things: The Senses in Anthropology*. Philadelphia: University of Pennsylvania Press, 1989.

Suchman, Lucy. *Plans and Situated Actions: The Problem of Human-Machine Communication*. New York: Cambridge University Press, 1987.
Taylor, Timothy. *Strange Sounds: Music, Technology, and Culture*. New York: Routledge, 2001.
Théberge, Paul. *Any Sound You Can Imagine: Making Music/Consuming Technology*. Hanover, N.H.: Wesleyan University Press, 1997.
Thompson, Emily. *The Soundscape of Modernity: Architectural Acoustics 1900–1933*. Cambridge, Mass.: MIT Press, 2002.
Tyler, Stephen A. "The Vision Quest in the West, or What the Mind's Eye Sees." *Journal of Anthropological Research* 40 (1984): 23–40.
Truax, Barry. *Acoustic Communication*, 2nd ed., Norwood, N.J.: Ablex, 2001.
———, ed. *The World Soundscape Project's Handbook for Acoustic Ecology*. Vancouver, British Columbia: ARC, 1978.
Tufte, Edward R. *Visual Explanations: Images and Quantities, Evidence and Narrative*. Cheshire, Conn.: Graphics, 1997.
Weber, Heike. *Das Versprechen mobiler Freiheit: Zur Kultur- und Technikgeschichte von Kofferradio, Walkman, und Handy*. Bielefeld, Germany: Transcript, 2008.
Westerkamp, Hildegard. "Soundwalking." *Sound Heritage* 3(4) (1974) (unpaginated); http://www.sfu.ca/ westerka/writings%20page/articles%20pages/soundwalking.html (accessed October 8, 2010).
Woolf, Daniel R. "Hearing Renaissance England." In *Hearing History: A Reader*, ed. Mark M. Smith, 112–35. Athens: University of Georgia Press, 2004.
Wurtzler, Steve J. *Electric Sounds: Technological Change and the Rise of Corporate Mass Media*. New York: Columbia University Press, 2007.

SECTION I

REWORKING MACHINE SOUND: SHOP FLOORS AND TEST SITES

CHAPTER 1

THE GARDEN IN THE MACHINE: LISTENING TO EARLY AMERICAN INDUSTRIALIZATION

MARK M. SMITH

INTRODUCTION

Go to http://www.lowell.com/museums.boott-cotton-mills-museum, "Your Guide to Historic Lowell, Massachusetts," and you, as a virtual visitor, will be greeted with the following: "One entire floor of the Boott Cotton Mills Museum shows visitors exactly how a working mill actually looks. The floor is called the 'Weave Room,' and it is filled with industrial-grade looms, running at top speed, allowing visitors to feel the buzz of a working mill. This is one of the largest industrial history exhibits in the nation, and it is probably one of the loudest—it's not called the 'roar of industry' for nothing!"[1]

Loudest. Roar. Buzz. These are the key words that people generally, historians included, deploy to capture and render simple and digestible what was, in fact, a very complicated, situated, and highly contextualized process: industrialization in the United States. On the one hand, of course, historians of aurality should be pleased that their preferred way of excavating the past—listening—occupies such a prominent position in historical understanding. As a great deal of work has shown, historians and general historical consciousness is prone, for a variety of reasons, to read and understand the past through the eyes. Certainly, histories of industrialization are often indebted to vision—the visual rise of the factories, the slicing of erstwhile rural landscapes by impossibly tall chimney stacks, the visual power of the smoke, sparks, and sheer scale of the early factories eliciting plenty of visually indexed primary evidence. Nevertheless, of all of the principal developments usually associated with the coming of modernity in the nineteenth century, industrialization was understood, especially by contemporaries but also by some historians, as very much an aural affair.

Yet the way in which historians have chosen to frame the aural history of industrialization, especially in the context of early nineteenth-century America, is arguably quite misleading; it is easily understood, interpretatively portable and exportable (especially to living museums and their customers), certainly, but is it really accurate? Does it convey the full texture and, critically, the context in which people, especially those on the shop floor, heard and listened? In this chapter I suggest first that historians in the United States have tended to ignore the aural history of early industrialization (most commentaries, themselves quite rare, tend to latch on to the sounds and noises of large-scale, mature industrial America, mainly in the late nineteenth century, when industry was becoming a behemoth and, as such, occupied a more obvious place in the soundscape). I also maintain that they have tended to slip too easily—and misleadingly—into a binary understanding of the transition to industrialism, which makes artificial distinctions between a "quiet" countryside and a "loud" factory shop floor that would not always have been apparent to the people who, in fact, experienced the transition. I stress the preeminent importance of context in that transition. Factory workers in Lowell, Massachusetts—the first industrial laborers in the United States—not only understood their laboring experience in terms of aurality but, critically, also mediated that aural experience within a larger context. That context involved increasing sectional tensions between a free-wage labor North and a slave South, Romanticism, and the imperatives of a factory-based paternalism that focused very much on the gender of the Lowell workforce.

It is something of a commonplace among historians of aurality—including me—to argue that hearing is qualitatively different from seeing because we have eyelids and no ear lids. For this reason, so the argument goes, sound has a more transgressive quality than sight. While this may be true, the insight tends to raise questions. How, for example, do people who find themselves quite suddenly immersed into qualitatively and quantitatively new acoustic environments—such as with the first generation of female factory workers in the United States—incorporate the new, invariably loud sounds into their worldviews and soundscapes? What were the precise mechanisms that mediated, in other words, the transition from rural life to industrial life? Sound, noise, silence—habits of listening—were central to that transition, and, in fact, the prevailing cultural New England norms of the time concerning the meaning and experience of sound were central to effecting what was, for the most part, a smooth transition to industrialism for the first generation of factory workers. In other words, the cultural backdrop of religious, natural, and sectional sound helped grease the transition to industrial sound so that the sounds of the shop floor were deemed just that: sound, at worst noise that was necessary, but rarely simple, intolerable noise. Context, in other words, matters a great deal when trying to listen to the aural experience of the first industrial workers in the United States (Smith 2007: 41–57).

The Machine in the American Garden

What we might call the conventional aural history of industrialization in the antebellum United States rests on a basic and enduring interpretive binary that stresses that the preindustrial quietude of pastoral America became overwhelmed by ever-louder industrialism. Quiet countryside versus loud factory is certainly an intuitively attractive way to think of the transition not least because it fits, almost too neatly, something of a whiggish teleological mandate. It also happens to bear the imprimatur of some key historians, most notably Leo Marx, whose seminal 1964 study, *The Machine in the Garden*, has influenced an entire generation of historians in the United States and, in fact, has impacted the work of scholars of hearing. For Marx, the shriek of the locomotive especially functioned as an actual and symbolic disruption of the pastoral ideal by positioning the pastoral and the industrial in tension with the sounds of the factory, which were ruthlessly penetrating and disrupting the countryside. For Marx, the pastoral increasingly functioned as an escape from the ravages—aural included—of modernity (Marx 1964; Bijsterveld 2008, 50–51).

While Marx's metaphor works very well for transgressive technologies, such as the railroad, which literally pierced the countryside, it not only is less effective for explaining the experiential aspects of work in the first mills but also fails to take full stock of just how braided pastoral sounds were with machine sounds in those factories. In other words, it cannot take account of that very first wave of industrialization, a wave in which the country—in the form of rural women—came to the factory. Capturing the complexity of the sounds of the first factory shop floors in the United States requires a careful ear, one attuned to the larger context in which industrialization occurred and that takes the words and life experiences of Lowell workers seriously and arrives at an understanding of the way that pastoral sounds were cobbled onto and braided with factory sounds.

I use the term *braided* deliberately. No process—even one as iconic and as sonic as early industrialization—sweeps clean. Everyday life—lived experience—tends to foster attitudes that import elements of our memory in an effort to manage the experience of the present and cope with anticipated futures. In this sense, transitions are just that: processes, lived and woven. They are temporal fabrics, an ever-evolving cloth, replete with patterns, contours, and textures understood by contemporaries but whose threads are sometimes difficult for the historian to discern. Here, listening to how early factory workers understood their experience helps reveal, in unusual relief, those threads, the process of the weaving, that daily braiding. In other words, the heard worlds of women factory workers grant us unusual access to the complexity of their experience. Braiding, then, functions as both apt metaphor and descriptor.

Now, to what extent the experience of female workers in Lowell was exceptional or unusual is a matter worthy of future research, and I can only gesture toward answers here. To some extent, all instances of early, first-wave industrialization, regardless of location, involved transitions from agriculture to industry with machines making their way into gardens of ancient stock. Braiding, metaphorical and literal, was probably inherent to the process and sometimes by design. English steam peppered nineteenth-century Prussian gardens, at once elaborating national ideals but also giving voice to the technological sublime; everywhere in the industrializing world, railroads snaked through the countryside; and the first generation of factories in England and the United States drew from riparian sources, nature's oil greasing and powering people's machines (Roberts 2000, 55–58; Wise 1999). It is also the case that even in maturing industrial societies—as was the case in late nineteenth- and early twentieth-century America—rural eastern European immigrants found themselves thrust into powerfully and disconcertingly large and cacophonous factories. However, as (albeit limited and largely literary) evidence suggests, these second-generation industrial laborers, unlike their early nineteenth-century Lowell counterparts, found their pastoralist and bucolic metaphors challenged by the sheer scale, loudness, and alienating noises of some of the United States' largest industrial enterprises (Sinclair 1906, 4, 29, 237). The Lowell workers, by contrast, entered factories at once smaller and, very probably, quieter. Certainly, these factories were noisy, and many Lowell workers commented on the matter, often describing the noise as oppressive and unhealthy. Critically, though, they did so in a cultural, economic, political, and social context that helped them braid their rural backgrounds with a new industrial reality.

Here I wish to make more of the importance of gender to the shaping of the experience and reception of factory noise. The paternalism underwriting America's first factory system was very much informed by the fact that the workers were women, and these imperatives led factory managers to go feminize the workplace somewhat both visually and, in the form of what they had the girls read, wrote, and learned, aurally. Many of these women were reasonably well educated, could read, and could draw on literary representations—as well as personal experience—to help frame, mediate, and represent what they actually heard on the Lowell shop floor. For this reason, they held, in mind's eye and ear, literary representations of sound, noise, and silence and were able to weave their actual experience in Lowell with their lived and read experience, thereby creating new, braided narratives that helped them make sense of their world.

None of this is to say that the machine did not intrude upon the American garden; it certainly did. However, missing from the narrative is the way in which the garden became incorporated into the machine and, in the process, tattooed some of the most important acoustic signatures onto the first wave of industrial life. Plainly, sounds did function as a metric of industrialization in the United States and at Lowell in particular. However, there is a great deal more to be made of that process. Rural sounds and factory sounds certainly did compete but not as much as the machine-in-the-garden metaphor leads us to believe. Instead, the way Lowell workers listened and the meaning they attached to the sounds around them suggest that the sounds of industrialization were not necessarily alienating or jarring (Bijsterveld 2008, 78–79).

To be sure, not all historians of the debate about the transition to capitalism in the United States have embraced the country as quiet, industry as noisy dichotomy. Robert A. Gross in an important article on the coming of agricultural capitalism in rural New England made the important point that the presence of the market revolution in rural areas—and the trains and roads and canals that facilitated it—tended to up a preexisting volume rather than change it qualitatively. The "new system of agricultural capitalism" could be heard in "the steady chopping of the ax; the bustle of men spading up meadows, hauling gravel, and raking hay; the clanging of milk pails." Urban demand for agricultural goods in the 1840s merely intensified these sounds and, in this way, built on to an already existing rural soundscape. These were old sounds, sounds built upon by the distant demands of urbanizing and industrializing America. For the most part, according to Gross, the sounds of the countryside in the antebellum period remained similar in register and meaning, and the principal change was an increase in frequency (Gross 1982, 54).

Historians of other regions in the early industrial North have offered helpful remarks on the highly stuttered transition from countryside to factory. Anthony F. C. Wallace's influential "account of the coming of the machines, the making of a new way of life in the mill hamlets, the triumph of evangelical capitalists over socialists and infidels, and the transformation of the workers into Christian soldiers in a cotton-manufacturing district in Pennsylvania in the years before and during the Civil War," as the subtitle of his study of Rockdale puts it, uses sound to counterpoint preindustrial and industrial Rockdale. Wallace titles his first chapter "Sweet, Quiet Rockdale," capturing, in Wallace's ear, preindustrial (pre-1850s) Rockdale, a place where "There was as yet no jarring sound of locomotive engines and shrieking whistles." Nascent industrialization was also tranquil. "The mills themselves," he goes on, "powered only by water, whispered and grunted softly; the looms clattered behind windows closed to keep moisture in the air; even when the workers were summoned, it was by the bell in the cupola and not by a steam whistle." He concludes: "The machine was in the garden, to be sure, but it was a machine that had grown almost organically in its niche" (Wallace 1978, 4).

Yet the larger interpretive matrix established by Leo Marx has tended to predominate basic thinking on the history of industrialization in the New England and other northern states. And there is some legitimacy to that influence. In some important ways, the machine did, in fact, enter the American garden. This is most obvious with the railroad. Numerous sources from the 1830s to the eve of the Civil War chart the puncturing sound of the train on the putative silence and quiet of the countryside. Spectators—ear witnesses—were "surprised to observe what a sudden change was made in the scene by the departure of the cars. A moment before, all had been noise, tumult and confusion. But when the sound of the engine died away in the distance, they found themselves left in a scene of almost entire silence and solitude" (Abbott 1843, 98). Here was the familiar trope of industrial capitalism in the form of the railroad puncturing rural quietude.

The machine also became increasingly intrusive, so much so that the garden was smothered. Urban industrialization quieted the sounds of largely preindustrial, rural production. Rural New England workers who grew flax and processed it in their homes had their artisan form of craft production eclipsed by the massive factories at Waltham and Lowell. By the 1860s, the craft industry had died, and its death was heard. As one contemporary in the 1860s remarked, "So complete has been the change that few persons under . . . thirty years of age, have ever . . . heard the buzzing of the flaxwheel" (quoted in Rivard 2002, 33). Indeed, in both the United States and Great Britain, the impact of centralized factory production and the concomitant extraction of quasi-industrial production from rural households was heard by an older generation, "grand-mothers" included, who, as early as the 1830s, saw "misery and ruin close at hand, because the sound of the spinning-wheel and the loom is no longer heard in all our farm-houses" (Anon. 1832, 221). Little wonder, then, that the experience of the first factory workers at Lowell has been wrapped around the same binary by historians. Transported from supposedly serene, tranquil New England farms by hirers sent out from the mills to recruit them, they, in the words of more than one historian, "arrived in Lowell, heard the racket of the mills," and returned to their "tranquil" farmsteads (Moran 2002, 1, 43).

Part of the disruption of the pastoral ideal resided in the collateral sounds produced by the transition, most notably the sound of urban class conflict. Paul Faler's *Mechanics and Manufacturers*, a detailed study of the industrial revolution in the shoemaking industry in Lynn, Massachusetts, describes how, in the late 1850s, Lynn's city officials, operating from a bourgeoisie reformist perspective, charged workers with "rowdyism" and equated social order and "quietness" with discipline and obedience. Booze and lateness caused social noise, threatening, by extension, social "tranquility." In other words, this was the sound of class conflict—which culminated in a strike by the shoemakers in 1860. For Faler, attentiveness to the role of sound and perceived noise is actually quite important in trying to capture the contested nature of class conflict, the formation of class consciousness, and the larger process of industrialization. For example, he points out how, in 1828, Lynn city officials passed a series of ordinances designed to prohibit what they saw—and heard—as objectionable. Bylaws prohibited, in Faler's words, "insulting language and "making tumultuous noises." Faler ponders these ordinances: Perhaps industrialization and accompanying urbanization in Lynn in the 1820s had increased noise. Nonetheless, he concludes: "It is more likely though that there had been little or no increase in . . . coarse language, or noise in the streets but that these practices had become objectionable by their incompatibility with the values" of the urban bourgeoisie (Faler 1981, 210, 118, 116–17).

However, is the machine-in-the garden framework necessarily helpful or even accurate for capturing the full range of the experience of industrialization in New England? In the case of the establishment and the growth of the very first factories in Lowell, Massachusetts, I think it less so. In fact, and as I argue, the experience of the first factory operatives at Lowell was not just about the auditory history of the machine domineering the garden but also had a great deal to do with how the larger context in which the Lowell workers existed led the garden into the machine.

Lowell: The Garden in the Machine

Although it is well known, a brief social history of Lowell and especially of the living and working conditions in the factories is instructive not least because it helps provide the context necessary for understanding the ways in which the workers braided their experience with the countryside with their new factory lives. Context—both specific and general—is key for determining how and why people heard as they did—for determining the meaning contemporaries attached to particular sounds. A quick architecture of the Lowell women's lives allows us access to that context.

The "first efforts to promote manufactures in this place, were made in 1813," courtesy of the War of 1812, Thomas Jefferson's Embargo Act, and efforts by Americans to wean themselves from British manufacturing. Beginnings were modest, as most beginnings are. Initial efforts focused on woolen textile production and the acquisition of water rights to expand factory production. In the 1820s, more substantial mills dedicated to cotton textiles emerged, and the register of expansion can almost be heard in the statistics. In 1820, the population of Lowell was roughly 200 souls; in 1828, it stood at 3,532; in 1833, 12,363. By 1840, the population was 20,981. By the 1830s, the city had ten textile corporations that ran thirty-two mills. By 1840, the factories employed a total of about eight thousand workers, and the majority of them were women aged sixteen to thirty-five (H. F. 1843, 146–47).

The "Lowell system" was a deliberate attempt to blend large-scale industrial manufacturing with the educational and moral improvement of its female workforce. The average female operative was twenty-four years of age; some arrived at Lowell with mothers or sisters; most were hired on annual contracts; and, on average, the women stayed four years at Lowell. At work, the newer, younger arrivals were paired with older and more experienced female workers, who trained them on the job and also in the ways of Lowell. Male overseers held authority in the shop or room, each of which averaged eighty female workers. They worked from five in the morning until seven at night, averaging seventy-three hours a week. After work, the women were required to sleep and live in a series of female-only boarding houses located on the mill grounds. They were expected to attend church, behave in a moral fashion, and attend lectures sponsored by the Lowell factory owners (Dublin 1975, 1979).

In other words, industrial paternalism underwrote Lowell from beginning to end. The labor force—on average about three-quarters female—was designed to be temporary. Indeed, rural women were recruited by men who told tales of high wages and good, moral living conditions. Part of the reasoning here was to avoid the development of "ignorance" and "depravity" among workers, which, courtesy of the English (and, to some extent, the French) experience, was deeply associated with industrialization in New Englanders' minds. "As long as our mills are wrought by operatives from the country, or from the common schools of Lowell, they will not be filled with a depraved and ignorant class." That a "great preponderance" of the workforce was composed "of a youthful female population" animated efforts to institute habits and training designed to sidestep perceived depravity and also make the young women into good, conscientious workers. Lowell girls were, therefore, encouraged to become involved in religious and charitable societies, read and write, and improve their moral character while also learning the ropes of industrial labor and earning what was considered a good wage. When all was said and done, supporters of this system believed that, viewed as a whole, Lowell "might almost lead observers to believe that our hard-working, matter-of-fact city had been transformed to fairy land" (H. F. 1843, 147–49).

In stressing the Lowell experience and the specific context of paternalism, I take my cue from Herbert Gutman, an eminent U.S. labor historian. In his seminal 1973 article, "Work, Culture, and Society in Industrializing America, 1815–1919," Gutman examined the rate, nature, and intensity of laborers' resistance to the coming of the industrial order. Like E. P. Thompson before him and Jonathan Prude after him, Gutman stressed the process of industrialization and placed the workers' experience at center stage. Different generations of American workers, explained Gutman, accommodated or resisted industrialization depending on a number of factors, including the scale of manufacturing and the policies of industrial capitalist and factory managers. For our purposes, it is important to note that Gutman elected to examine the Lowell mills as typical of early industrialization. As Gutman makes clear, not all was roses. When Lowell began to rationalize production after 1840, Lowell women engaged increasingly in collective protest and were eventually replaced by Irish immigrants. He is also careful to point out that *The Lowell Offering*, a magazine written by the Lowell women themselves (although it was funded and produced by Lowell's factory owners), is a helpful source for detecting criticism of the factory, if read carefully. "Historians have dismissed it too handily," says Gutman. It was much more than a factory-owner mouthpiece. Careful reading reveals expressed dissatisfactions embedded in prose and poetry. Beyond their criticism of work speed-ups and decreasing real wages, Gutman detects in these writings the women's "attachment to nature . . . the concern of persons working machines in a society still predominantly 'a garden.' " Certainly, the women's expectation that they would work at Lowell only temporarily (they had no intentions of becoming a permanent proletariat) eased their transition to factory life and helps account for the relative absence of protest, according to Gutman. However, that relatively smooth transition also had a lot to do with the ways that the factory owners and the workers themselves braided pastoral concerns with factory life. Sound and modalities of listening occupied an important place in that transition (Gutman 1973, 552–53).

Of course, there is no doubt that the Lowell workers found aspects of the mills' soundscapes alienating, deadening, and disruptive. For the Lowell women, there were certainly acoustic hurdles. One was simply vocal. Young women fresh from the countryside heard their own accents, their "up-country" twangs accentuated now that they spoke in the context of more voices. Some found the competing brogues mildly disconcerting, while many attempted to modulate their dialect to better fit in with the workforce (in Eisler 1977, 46).

Also, as Gutman suggested, the Lowell authors were not unknown to vein their miscellaneous offerings with subtle critiques of their industrial world. Occasionally—for they were relatively rare—such critique came in the form of detangling the sounds of nature from the noise of industry, setting the two as antagonistic and irreconcilable. "How beautiful, and yet how striking, is the thought, that the most wonderful operations of nature are effected without the noise and tumult attendant on the works of man," mused S. J. H. in 1842. Human "vanity," "pomp, and show" were mere "noise" compared to "Nature's plan, And silent work" (S. J. H. 1843, 102).

Other mill girls expressed their angst in poetry:

> And amidst the clashing noise and din
> Of the ever beating loom,
> Stood a fair young girl with throbbing brow,
> Working her way to the tomb. (quoted in Moran 2002, 42)

Indeed, the alienating effect of industrialization was also tackled by contemporary labor reformers through an appeal to sound. In 1846, factory operatives in Philadelphia urged the adoption of a ten-hour working day by depicting the enervated, exploited worker as a near-dead automaton, "listening in despondency to hear, through the clattering of machinery, the hour of his delivery from toil" (quoted in Wallace 1978, 389).

Beyond this, though, there are very particular reasons that the sounds of Lowell are better understood as being consonant with the sounds of rural New England. There is little doubt that female workers in Lowell expected their new environments to be qualitatively and quantitatively different from their villages, but their experience upon arrival in the factory town suggested that the factory boosters and the rumors had exaggerated that difference. Take the experience of "Susan," a typical Lowell girl. She traveled from her rural village in New Hampshire and, soon after arriving in Lowell, wrote home telling her family her impressions of the place. "I waited, one day, to see the cars come in from Boston," wrote Susan. She was, though, underimpressed: "They moved, as you know, very swiftly, but not so much like a 'streak of lightning' as I anticipated." Moreover, she mused, "If all country girls are like me their first impressions of a city are far below their previous conceptions, and they think there is more difference than there really is. . . . I see that the difference is more apparent than real" (in Eisler 1977, 49). Indeed, it should be remembered that industrialization and industrial progress did not always equate with louder machines. Quieter leather belts replaced noisy gearing beginning in the 1820s in some factories in Massachusetts and Rhode Island, for example (Rivard 2002, 58).

Many of the workers mediated Lowell through a pastoral idiom simply because that idiom was the prevailing one at the time. The mill women's ability to braid the soundscapes of rural life and the sounds of the shop floor and thereby make sense of their transition was due, in no small part, to other highly valued cultural assumptions about the importance of nature and God, a movement known as Romanticism. While certainly a national experience, Romanticism's roots were planted firmly in New England, and it emerged in the context of the long arc of the American Enlightenment. A complicated and highly varied movement, Romanticism tended to read nature and God into the world partly in reaction to excessive materialism and industrialization and partly in reaction to what was perceived as a failed Calvinism. While it is certainly true that some of the most active literary advocates of Romanticism—most notably the Transcendentalist Henry David Thoreau—grimaced when they heard industrialization, even they were not above hearing nature in modernity. Thoreau heard the telegraph as "faint music in the air like an Aeolian harp" and understood "the whistle of the steam-engine" as "arousing a country to its progress" (quoted in Smith 2001, 127). Beyond Thoreau, Romanticism was broadly shared, and the Lowell women were not only a part of it but integral to it as well. Page after page of *The Lowell Offering* shows that the Lowell women believed God and nature were ubiquitous. "His name is written indelibly upon every particle of created matter," the factory floor included. "[N]oisy" grandeur" could be heard in nature, especially in rivers and waterfalls, just as could the "soft music" of "whispering winds" and other "silent accents" (Annaline 1842, 32–33).

Other Lowell workers understood the sounds, the smells, and the experience of working on the shop floor in distinctly religious and natural terms:

> In the mills, we are not so far from god and nature, as many persons might suppose. . . . A large and beautiful variety of plants is placed around the walls of the rooms, giving them more the appearance of a flower garden than a workshop. It is there we inhale the sweet perfume of the rose, the lily, and geranium; and, with them, send the sweet incense of sincere gratitude to the bountiful Giver of these rich blessings." (in Eisler 1977, 64)

The Lowell Offering is replete with examples of mill women braiding the sounds of rural life with the registers of industrial work. The Lowell women came to understand the noises of the factory—which were very loud and, superficially at least, wholly alien to the soundscapes of rural new England—in the context of remembered pastoral sounds. Some suggested that the soundscape of "The Power of Industry" was not altogether different from the heard worlds of life back on the farm. "The hens' constant cackling" was indicative of "the Spirit of Industry," their product, "*two* eggs *a* day." Village life when "Every body was busy" was heard in "an incessant cackle" (Kate 1842, 27). "[W]hen I went out at night," wrote one Lowell worker in 1844 after a day in the cacophonous mill, "the sound of the mill was in

LOWELL OFFERING

December, 1845.

"Is Saul also among the prophets?"

A REPOSITORY
OF ORIGINAL ARTICLES, WRITTEN BY
"FACTORY GIRLS."

LOWELL: MISSES CURTIS & FARLEY.
BOSTON: JORDAN & WILEY, 121
Washington street.
1845.

Entered according to Act of Congress, in the year 1845, in the Clerk's Office of the District Court of the District of Massachusetts.

LOWELL OFFERING, TITLE PAGE FROM 1845
COLLECTION BY WORKING WOMEN

Figure 1.1 *[Title page], The Lowell Offering (December, 1845).*

my ears." She compared the clanking of metal, the rush of belts, and the slamming of the looms to her aural past. The sounds of the mill were "as of crickets, frogs, and jewsharps, all mingled together in strange discord." Possibly suffering from tinnitus (the author's reference to crickets is especially suggestive), the register was certainly different, and the sounds of the factory lacked nature's harmony, to be sure. Still, the referents were pastoral as, in fact, they had to be. More than that, though, the pastoral metaphor functioned to ease the transition to factory labor. "Susan" went on to explain that after a day in the mill, "it seemed as though cotton-wool was in my ears, but now I do not mind at all. You know that people learn to sleep with the thunder of Niagara in their ears, and a cotton mill is no worse, though you wonder that we do not have to hold our breath in such a noise." Susan reached into the loudest sound of rural New England—the roar of Niagara—and made sense of her new aural environment on the factory floor (in Eisler 1977, 51–52).

Even the dreaded clanging of the factory bell was rendered comparable to farm life. In their literary offerings, the working women of Lowell played out the tension between farm and factory life and tried to make the connection seamless. " 'What difference does it make,' said I, 'whether you shall be awakened by a bell, or the noisy bustle of a farm-house? For, you know, farmers are generally up as early in the morning as we are obliged to rise.' " Enter the counterargument, given by "Ellen." " 'But then,' said Ellen, 'country people have none of the clattering of machinery constantly dinning in their ears.' " " 'True,' I replied, 'but they have what is worse—and that is, a dull, lifeless silence all around them. The [hens] may cackle sometimes, and the geese gabble, and the pigs squeal' " (in Eisler, 1977, 162).

There were good structural, physical reasons behind the Lowell workers' incorporation of industrial noise into pastoralism. That mill workers should have understood the sound of the factory in the idiom of nature's sounds is hardly surprising because early—and, in fact, some later—New England's mills were literally connected to and dependent on waterpower. The machine was literally in the riparian garden. In April 1843, "H. F." offered a description of Lowell in which nature and industry played handmaidens. Water figured prominently in the braiding. "The city of Lowell stands upon the Merrimack river; upon a point of land, formed by the Concord river, at its confluence with the Merrimack, and a bend in that river, from which its direction is at a right angle with its former course. It is intersected by many canals, the principal of which is the Pawtucket . . . thus forming an island of the city; it being *entirely surrounded by water*." Certainly, the making of Lowell was a noisy affair with "Five hundred men . . . constantly employed in digging and blasting" to carve out the canals, but the effort was worthwhile: The labor had effected a quite "sudden . . . transition from the monotony of a quiet village to the hurry and bustle of a manufacturing city" (H. F. 1843, 145, 147; see also Rivard 2002, 55–58; Lewis 1844, 242). However, the specific technology underwriting early New England mills—waterpower—helped in the mill women's articulation of rural, natural sounds with the sounds of factory life. "Directly below my window passes the combination of nature, and human invention, forming a canal," wrote one Lowell worker in 1842. Water powered this "American Manchester," and the customary and traditional sound of rushing torrents could be heard in the mills, less in terms of the sound of the torrents themselves but more in the noise of the machinery powered by that same water (M. T. 1842, 57). While Lowell was louder than it had once been, its volume and soundscape were not necessarily in tension with its audible past.

Paternalist imperatives led factory owners to bring the garden to the machine. According to Susan in 1844, her experience on the factory floor was hardly pleasant—the hours were long, the warm rooms could be stifling, and the constant standing while tending the loom "makes my feet ache and swell." Yet, taken as a whole, the experience was located at an intellectual, emotional, and even physical intersection between nature and machine. Susan considered the factory floor "Very pleasant . . . light, spacious, and clean . . . the machinery so brightly polished or nicely painted." Space and light: both reminiscent of the countryside. Moreover, "The plants in the windows, or on the overseer's bench or desk, gave a pleasant aspect to things." The garden was literally in the machine room, water and all: "The dressing-rooms are very neat, and the frames move with a gentle undulating motion which is really graceful," not unlike the undulations of water (in Eisler 1977, 51–52).

But what was the value of the garden in the machine if the machine was so loud it damaged the intellect? Factory operatives writing in *The Lowell Offering* also went to some lengths to deflect the criticism that the noise of factory production was necessarily inimitable to the life of the mind, a charge that had a deep genealogy and one that, as John Picker has shown, was championed by intellectuals at the end of the nineteenth century (Picker, 1999/2000; Agar, 2002; Baron, 1982). However, the factory women of Lowell maintained not only that the *vita contemplativa* was unsuppressed by factory noise but also that the soundscape of industrial labor could actually benefit thinking. Reject solitude, moved a *Lowell Offering* writer in 1843, "for we are social creatures, creatures of sympathy." Thinking, godliness, and spirituality were not achieved alone in nature, surrounded by "the low fluttering breeze . . .the faint murmur of the brook, the loud roar of the cataract—or even the rich music of the birds—these are nothing more than unmeaning, senseless sounds," sounds that, experienced alone, fall on a "senseless ear." Sound in social context was far more productive of social "harmony," populating social space "where the current of thought could have free circulation" (Adeline 1843, 162–63).

Neither did entering the social world—even if it was the world of the factory floor—detract from the value of thinking. Indeed, factory noise could be overcome and even help. "Strangers appear to think that the noise which affects them so severely and unpleasantly, must cause the same sensibilities in all who hear it," read a *Lowell Offering* editorial. "They forget the power of habit, the benumbing influence of every constant action of sensation. Those who live within hearing the dash of Niagara, become insensible to its deafening roar; and, amid the clatter of wheels, bands, and spindles, the still small voice within may be as plainly heard as in the chamber's solitude, or when beneath a midnight sky." "In truth," the writer continued, "the factory is rather favorable than otherwise to reflection. We become unconscious of the machinery's din, and it completely deadens every other sound." So much so that "the best articles in the Offering we know to have been composed in the mill" (Editorial 1843, 164).

As one worker explained: "In the sweet June weather I would lean far out the window, and try not to hear the unceasing clash of sound inside. . . . I discovered, too, that I could so accustom myself to the noise that it became like a silence to me. And I defied the machinery to make me its slave. Its incessant discords could not drown out the music of my thoughts if I would let them fly high enough" (quoted in Moran 2002, 21). Noise certainly precluded chatter among the operatives: "But, aside from the talking, where can you find a more pleasant place for contemplation?" Labor focused the mind, the division of labor freed it to think, and the noise did not intrude (in Eisler 1977, 63). More than that, managing the noise in order to think was imperative: "While at her work, the clattering and rumbling around her prevent any other noise from attracting her attention, and she *must think*, or her life would be dull indeed" (in Eisler 1977, 66). This was the lived experience of the first generation of factory operatives, one in which sounds of the factory floor, although objectively loud, discordant, and noisy, were rendered manageable through prevailing New England idioms stressing nature, God, and mind.

The Sectional Sounds of Lowell

If Lowell sounded like God and nature and empowered the intellect, another factor made Lowell's noise not only manageable but even attractive. That was the sound of sectionalism and, more generally, the soundscape of a particular mode of production (that modes of production have particular soundscapes is clear from other work; see Braun, this volume, Chapter 2). Although they did not phrase it in such formal terms, antebellum Americans, North and South, listened to their respective modes of production, processing, mediating, constructing, and shaping their respective identities through what they heard, literally and figuratively, about one another. The mode of production in the antebellum South—one premised on the use of enslaved labor—had to northern ears increasingly accustomed to hearing a distinctly free-wage-labor society its own particular—and worrying—soundscape. Northern ears heard the voice of democracy, the hum of industry, and the hum of economic progress in their own society; in southern slavery, they heard disturbing quiet, the silence of economic backwardness, and the muffling of slave voices (Smith 2001).

Lowell especially was the vanguard of northern modernity and stood in stark visual—and aural—contrast to southern slavery. It was the great experiment of the early American republic, northern style. The Lowell women existed within an emerging capitalist mode of production that fashioned itself in part acoustically, adopted its own specific aural metaphors, and came to counterpoint itself to an at-once silent but scream-ridden slave South that refused to industrialize on the free-wage labor terms being sponsored in Lowell.

It was this aural counterpoint against a silent, quiet, and presumably economically backward, largely rural, and precapitalist slave South that led northern factory owners to hear not din but progress and harmony in their industrial ventures (see also Braun, this volume, Chapter 2). In some important respects, the tenor of factory life was heard because its counterpoint—silence—was too reminiscent of stagnation and, especially, death, all of which were associated with southern slavery (J. S. W. 1842, 106; Smith 2001, 150–71).

Many observers made this point, including the abolitionist Ebenezer Davies while steaming along the Kentucky-Ohio border in the 1840s. Davies let his ears do the telling. To his left stood slavery, a place half asleep, "slaves loitering in half-desert fields." To the north, from Ohio, "a confused hum is heard which proclaims the presence of industry . . . and man appears to be in the enjoyment of that wealth and contentment which are the reward of [free] labour." Others agreed that from the South "the sounds of the steam engine and the manufactory rarely falls [*sic*] upon his ears" (Davies 1849, 177; Anon. 1850, 46).

For their part, northern sponsors of industrial capitalism stressed the silent role of capital in the creation of what they styled the "hum of industry" (Smith 2001, 129). Paternalists such as William D. Haley, pastor of the First Congregational Church of Alton, Illinois, offered a "Series of Lectures to Workingmen, Mechanics, and Apprentices" in 1855 and warned them of the dangers of labor unrest by pointing to aurality. Without capital, maintained Haley, quoting "one Dr. Boardman," "All the businesses which now rest on a credit basis would cease. Not a hammer would be heard in the ship-yards. The silence of death would replace the intolerable but productive clatter of the foundries and machine shops." You might not enjoy the noise of industrial capitalism, Haley told the workers, but you will loathe the silence of economic recession (Haley 1855, 31). Moreover, the sound of industry (as opposed to its noise) was a fixed register with similar meaning regardless of country. In his celebratory *Memoir of Samuel Slater*, George White offered an account of "the machinery in French mills," which echoed, quite precisely, the conventions used to tout American progress. The "perpetual din" was powerful testimony to France's economic progress (White 1836, 341).

But boosters were careful not to go too far. As much as they wanted to avoid the silent perils of slavery, they also proved anxious to fall short of what they understood as excessive industrialization. From the virtue of the hum of industry, industrial capitalists pivoted seamlessly to the argument that workers nonetheless were afforded quiet when they needed it, especially when it came to the quiet necessary for moral instruction. In fact, American boosters nationalized the association, claiming that their paternalism, their American system, was superior to that of England's especially. Their mill houses and Sunday schools were "quiet fortresses" where "the operatives with their families pass the tranquil tenor of their lives" (White 1836, 173, 223).

Even in the factories, capitalists made the case that while the sound of machinery was inherently jarring, workers earned aural respite. Sometimes they even generated it and thereby fed the paternalist conceits of managers. Samuel Slater wistfully recalled the following about a factory floor in Falls River, Massachusetts:

> I shall always recollect with pleasure one little incident in one of the weaving rooms of the manufactory, where the noise was very distracting, arising from a vast number of looms going at once. The machinery suddenly stopped, and a strain of music arose simultaneously from every part of the room, in such perfect concord that I at first thought it a chime of bells. My conductor smiled when I asked him if it was not, and pointed to the girls, who each kept their station until they had sung the tune through. (White 1836, 235)

In this way, industrial capitalists tried to establish, through aural metaphor, what they wanted most: a mode of production stressing the harmony of labor and capital. Sound was central to that connection, and this is what they meant by the "hum of industry": the sound of compliant labor and factories, the former refraining from introducing the noise of social dislocation and labor unrest into the soundscape of economic progress. This is why in Lowell he could hear "the peaceful hum of an industrious population, whose movements are regulated like clockwork" (White 1836, 47, 238; Moran 2002, 15).

Conclusion

While we should remain skeptical of capitalists' claims concerning the putative harmony of labor and early industrialization, it is nevertheless true that workers accepted Lowell's noise precisely because they were the sounds of freedom and not those of slavery and because they were framed within a larger matrix of paternalism and Romanticism. This is precisely what Lucy Larcom, perhaps the most famous of the young women of Lowell's factories, meant when she wrote that, having returned from the countryside to the mill, "I found that I enjoyed even the familiar, unremitting clatter of the mill, because it indicated that something was going on" (in Larcom 2007, 88). There was, to be sure, class conflict in the North, but workers—at least the earliest generation—and capitalists agreed on their preferred soundscape, and there was impressive unity in that aurality.

Certainly the context in which the factory women of Lowell were made workers was one in which a narrative of sound-as-progress prevailed, courtesy (at least in part) of capitalists who themselves were making the transition from rural to urban, industrial society. By pointing out that Lowell workers mediated their transition from rural to factory life through the aural idioms of religion, nature, and the emerging capitalist mode of production is not to render them mere dupes to the principal authors who described and promoted these developments. Rather, it is to suggest that they were active participants in the same processes, their voices contributing in important ways to those transitions. If we listen to them listening, we begin to get a deeper, more complex understanding of the role that sound and noise played in the early years of American industrialization. The transition to industrialism did not necessarily come with a bang, no matter how noisy, alien, and new the sounds of the shop floor were at first. Rather, it is more profitable and telling to understand why and how the first generation of American factory workers managed to mediate the new sounds and turn noises into registers they found palatable and consonant.

The value of attending to context in our efforts to deploy and capture the histories of sound and listening, broadly construed, should also be understood in terms that reach beyond specific historiographies. As historians of the auditory world begin increasingly to accept invitations by their colleagues in museums and historic preservation to help them use sound to educate the general public, we will need to think carefully about the ways in which we can be helpful but reliable interpreters of the heard past. After all, the curatorial tendency is to attempt to re-create the sounds of buildings and factories in an effort to generate public interest and heighten education. And this is certainly laudable. Nonetheless, as we enter these conversations and advisory roles, we should do our best to stress the importance of context to the history of sound. And therein is the challenge. Context changes everything, and it is extremely difficult to convey its preeminent value to museum patrons. The context in which the sounds are being produced—even if those sounds remain the same—alters meaning and is likely not only unrecoverable but possibly beyond communication in mere aural form as well. Historic preservations and museum curators will need to provide context lest visitors to, say, the Lowell mills walk away with the impression that the mills were simply loud. Or roaring. Or buzz ridden. Yes, they certainly were that. All the same, leaving the description and experience at that does enormous violence to the way that the actual workers understood, mediated, and came to manage those same sounds (Smith 2007; Matthews 2007). If historical acoustemology is to fulfill its promise outside of the academy, we will need to think creatively within its walls.

NOTE

1 http://www.lowell.com/museums.boott-cotton-mills-museum. This chapter benefits materially from insights offered by Michael Bull, Karin Bijsterveld, Myles Jackson, Norm Hirschy, and fellow contributors to the volume. A special thanks to David Prior for invaluable help with the research.

REFERENCES

Abbott, Jacob. *Marco Paul's Travels and Adventures in the Pursuit of Knowledge on the Erie Canal*. Boston: Harrington Carter, 1843.

Adeline. "Solitude." *Lowell Offering* (April 1843): 162–63.

Agar, Jon. "Bodies, Machines, and Noise." In *Bodies/Machines*, ed. Iwan Rhys Morus, 197–220. Oxford: Berg, 2002.

Annaline. "Evidence of Design in Nature." *Lowell Offering* (November 1842): 32–33.

Anon. "Art. III—The Southern States." *De Bow's Review* (January 1850).

Anon. "Effects of Machinery." *North American Review* 34 (1832).

Baron, Lawrence. "Noise and Degeneration: Theodore Lessing's Crusade for Quiet." *Journal of Contemporary History* 17 (January 1982): 165–78.

Bijsterveld, Karin. *Mechanical Sound: Technology, Culture, and the Public Problems of Noise in the Twentieth Century*. Cambridge, Mass.: MIT Press, 2008.

Cott, Nancy F., ed. *Root of Bitterness: Documents of the Social History of American Women*. New York: Dutton, 1972.

Davies, Ebenezer. *American Scenes and Christian Slavery: A Recent Tour of Four Thousand Miles in the United States*. London: Snow, 1849.

Dublin, Thomas. *Women at Work: The Transformation of Work and Community in Lowell, Massachusetts, 1826–1860*. New York: Columbia University Press, 1979.

———. "Women, Work, and Protest in the Early Lowell Mills: 'The Oppressing Hand of Avarice Would Enslave Us.' " *Labor History* 16 (1975): 99–116.

Editorial. *Lowell Offering* (April 1843): 164.

Eisler, Benita, ed. *The Lowell Offering: Writings by New England Mill Women (1840–1845)*. New York: Harper and Row, 1977.

Faler, Paul G. *Mechanics and Manufacturers in the Early Industrial Revolution: Lynn, Massachusetts, 1780–1860*. Albany: State University of New York Press, 1981.

Gross, Robert A. "Culture and Cultivation: Agriculture and Society in Thoreau's Concord." *Journal of American History* 69 (June 1982): 42–61.

Gutman, Herbert G. "Work, Culture, and Society in Industrializing America, 1815–1819." *American Historical Review* 78 (June 1973): 531–88.

H. F. "Lowell." *Lowell Offering* (April 1843): 146–47.

Haley, William D. *Words for the Workers; in a Series of Lectures to Workingmen, Mechanics, and Apprentices*. Boston: Crosby, Nichols, 1855.

J. S. W. "The Village Burial." *Lowell Offering* (December 1842): 106.

Kate. "Aunt Letty; or, the Useful." *Lowell Offering* (November 1842): 27.

Lewis, Alonzo. *The History of Lynn, including Nahant*. Boston: Dickinson, 1844.

M. T. "The Prospect from My Window in the Mill." *Lowell Offering* (December 1842): 57.

Marx, Leo. *The Machine in the Garden: Technology and the Pastoral Ideal in America*. New York: Oxford University Press, 1964.

Matthews, Anne. "If Walls Could Talk." *Preservation: The Magazine of the National Trust for Historic Preservation* 59 (November/December 2007): 34–37.

Moran, William. *The Belles of New England: The Women of the Textile Mills and the Families Whose Wealth They Wove*. New York: St. Martin's, 2002.

Picker, John M. "The Soundproof Study: Victorian Professionals, Work Space, and Urban Noise." *Victorian Studies* 42 (Spring 1999/2000): 427–53.

Rivard, Paul A. *A New Order of Things: How the Textile Industry Transformed New England*. Hanover, N.H.: University Press of New England, 2002.

Roberts, Lissa. "Water, Steam, and Change: The Roles of Land Drainage, Water Supplies, and Garden Fountains in the Early Development of the Steam Engine." *Endeavour* 24 (June 2000): 55–58.

S. J. H. "The Silent Expressions of Nature." *Lowell Offering* (February 1843): 102.

Sinclair, Upton. *The Jungle*. New York: Doubleday, Page, 1906.

Smith, Mark M. *Listening to Nineteenth-Century America*. Chapel Hill: University of North Carolina Press, 2001.

———. "Producing Sense, Consuming Sense, Making Sense: Perils and Prospects for Sensory History." *Journal of Social History* 40 (Summer 2007): 841–58.

———. *Sensing the Past: Seeing, Hearing, Smelling, Touching, and Tasting in History*. Berkeley: University of California Press, 2008.

Wallace, Anthony F. C. *Rockdale: The growth of an American village in the early Industrial Revolution. An account of the coming of the machines, the making of a new way of life in the mill hamlets, the triumph of evangelical capitalists over socialists and infidels, and the transformation of the workers into Christian soldiers in a cotton-manufacturing district in Pennsylvania in the years before and during the Civil War*. New York: Knopf, 1978.

White, George S. *Memoir of Samuel Slater, the Father of American Manufactures. Connected with a History of the Rise and Progress of the Cotton Manufacture in England and America*. Philadelphia, 1836.

Wise, M. Norton. "Architectures for Steam." In *The Architecture of Science*, ed. Peter Galison and Emily Thompson, 107–40. Cambridge: MIT Press, 1999.

CHAPTER 2

TURNING A DEAF EAR? INDUSTRIAL NOISE AND NOISE CONTROL IN GERMANY SINCE THE 1920S

HANS-JOACHIM BRAUN

INTRODUCTION

The title of this chapter, "Turning a Deaf Ear?" sounds catchy, of course, and is rather ambiguous.[1] In dealing with industrial noise and noise abatement in the twentieth century, however, this title also serves as a sort of leitmotiv because, at least before the 1970s, industrial noise abatement was not very popular. Employers and employees turned a deaf ear to this noise not only in a metaphorical sense but also in a literal sense because many employees were already deaf or hard of hearing because of long-term exposure to industrial noise.

Although today almost everyone in the industrialized world will identify noise as a major problem, it is surprising that so little has been published on its history. Whereas several publications treat the history of "environmental" or "public noise" and noise-abatement measures, particularly as regards means of transport such as automobiles and airplanes, historians have largely ignored the issue of industrial noise. Several case studies discuss various countries and periods, but the only comprehensive survey on this topic is Bijsterveld (2008). In addition, several smaller surveys cover different aspects of the topic such as measuring noise (Thompson 2002; Bijsterveld 2008) and noise legislation (Wiethaup 1967; Koch 2002). Aside from Bijsterveld's work and a study by Mark M. Smith (2001) on the nineteenth century, however, there is a conspicuous lack of information on what happened on the shop floor. To what extent, for instance, did employers or workers take advice from doctors and industrial psychologists seriously? How did specific recommendations by insurance organizations and governmental regulations influence noise concerns?

This chapter focuses on industrial noise and noise abatement in three different regimes: National Socialism during the Third Reich, socialism in the German Democratic Republic, and (Western) democracy in the Federal Republic of Germany. It is mainly based on contemporary journal articles, on articles in company journals, and on interviews. My argument reveals that the political economy within these regimes played a large role in how they dealt with unwanted sound. Questions addressed include the following: To what extent did ideology influence or even determine legislation on noise abatement? What was the relationship between theory and practice? Was employees' resistance to hearing-protection devices completely irrational, or did it make sense? Part of the argument explores the use of music on the shop floor in the three German cases, also as compared to such practice in Great Britain.

The phenomenon of work-related noise dates back at least as far as Roman antiquity. A major way of tackling this problem was zoning. For example, copper millers were not permitted to set up shop in a street where a scholar lived. In early modern Europe a similar approach was taken: Noisy handworkers were required to perform their activities either in a special quarter of town or outside its borders (Bijsterveld 2008, 56).

The Prussian General Trade Code (*Preußisches Allgemeines Landrecht*), which became binding for the North German League in 1871, required workshops and factories to obtain a license. From 1895 onward, at the height of modern-day industrialization, conditions became stricter: Forges had to put in double walls, double doors, and double windows. Generally, industries could be prohibited or restricted when they severely disturbed the peace of churches, schools, hospitals, or other public institutions (Bijsterveld 2008, 61).

In Germany it was common legal practice to legitimize new noise on the basis of existing noise, and this made it difficult to change matters substantially (Bijsterveld 2008, 67–69). In the early twentieth century, there were antinoise campaigns in the United States, Germany, and other countries that were mainly directed against street noise rather than industrial and occupational noise. In most cases, however, these campaigns had little success (Thompson 2002, 115–30, 157–68).

Industrial Noise and Noise Abatement in Germany in the Late 1920s and the Third Reich

Immediately after World War One, the issue of noise was still neglected. Even during the mid-1920s, the "golden years" of the Weimar Republic, industrial firms concentrated on rationalization and increasing efficiency instead of on noise abatement (Braun 1990, 46–52). In November 1927, however, after pressure from trade unions, physicians, and others, the German Society for Industrial Hygiene began to investigate the issue of health damage by industrial noise. It established a "Committee for Industrial Noise Abatement," made up of employers and representatives from the fields of science, technology, medicine, and public administration. This committee put together a memo with various recommendations regarding the implementation of noise-reducing devices or work processes, but not much was done with them. Next, in August 1928, the Technical Commission of the "Committee for the Fight against Hearing Damage Caused by Industry" was founded, but it soon became clear that business interests, such as the high cost of industrial noise-abatement measures, prevailed (Braun 1998, 255–56).

In the context of Germany's Second Occupational Diseases Regulations of 1928, hearing damage was recognized as an occupational disease, which implied that social security organizations had to compensate workers. This resulted in several industrial noise-abatement measures, but they were implemented in the metal industry only. Furthermore, the professional organization of German engineers, the Verein Deutscher Ingenieure (VDI), also tackled the problem of industrial noise, dealing mainly with issues of physics and technology. In November 1930 the VDI set up a commission for abating industrial noise that cooperated closely with the Berlin Heinrich Hertz Institute for Vibration Research and its president, Karl Willy Wagner, a well-known specialist on acoustics who had close contacts with the Bell Laboratories in the United States. In cooperation with the German firm Siemens, he designed a new noise meter based on a reproduction of the human ear, which was superior to the Barkhausen noise meter widely used in Germany. Although critical of National Socialism, Wagner did not voice his reservations in public. He became Germany's most prominent advocate of the fight against industrial noise and was appointed chairman of the relevant Association of German Engineers' Commission (Braun 1998, 256–58).

Noise abatement was a big topic in National Socialism, at least in its propaganda. The Führer himself claimed that for victory in the struggle for existence he needed people with strong nerves (Wagner 1935; Bijsterveld 2008, 130–31). Noise weakens nerves. This view, which was readily taken up by some leading

Nazi propagandists, seemed to underscore the issue that noise abatement was perfectly suited for propaganda purposes. The Nazi office for "Beauty of Labor" (Schönheit der Arbeit)—part of the Nazi organization "Strength through Joy," which in turn belonged to the German Labor Front—put in a major effort to bring home the topic of noise to German workers (Welch 2002). One of the reasons was that many German workers who had formally supported the now illegal Communist party or the Social Democrats were still skeptical of National Socialism. It was believed that success in foreign policy or on the "home front" would lure them away from left-wing ideas. Much like the communists, the National Socialists promised a "classless society," and in the German Labor Front employees and employers joined hands to work for the common good—for the "people's community."

From 1934 onward, different campaigns aimed at satisfying workers' demands for a better and safer workplace were launched. The 1935 "Fight against Industrial Noise" campaign was followed in 1936 by the "Good Light, Good Work" campaign and in 1938 by one on "Fresh Air in the Workplace." In the 1930s several journals reported on noise-abatement measures in industry. These included not only so-called primary measures, geared to the design of machinery in a "noise-friendly" way or to the replacement of hammering by rolling in metal working, but also secondary measures, aimed at noise reduction through encasement or encapsulation (Hasse 1939).

From April 7 to April 13, 1935, a noise-abatement week took place in Germany (Friemert 1980). Although there had been much propaganda in advance, its organizers soon found out that their job was far from easy. The "Beauty of Labor" office had failed to compile a survey of noise-abatement measures in Nazi Germany, and it soon became clear that the legal basis for protecting factory workers from industrial noise was rather weak (Friemert 1980, 133). One could merely appeal to factory owners to improve shop-floor conditions, but the office itself was not authorized to issue regulations. If the noise-abatement week did alert workers to the problem of industrial noise, partly with the help of mass-media coverage, the project's impact was small. One of the underlying reasons was that it proved difficult to *visualize* the issue of noise (Friemert 1980, 135).

A few years later, in 1938, the Association of German Engineers and the chairman of its noise-abatement committee, Karl Willy Wagner, in cooperation with the Ministry of Labor and the German Society for the Protection of Workers, founded the Committee for Noise Abatement in Industry (Hasse 1939). Together with the Museum for Industrial Hygiene they organized a conference on noise abatement in industry that was geared to the interests of engineers, engineering designers, factory managers, and officers of employers' liability-insurance associations. Although the specialists' discussions seemed to have been lively, practical effects were negligible because the German government gave priority to other issues, including increased armament efforts (Friemert 1980).

Masking Industrial Noise and Increasing Productivity: "Music while You Work"

Several recent publications have focused on the use of music to influence consumer behavior in retail stores, but so far the role of music at work has been little studied (but see Bijsterveld 2008, 81–87). As early as 1915, Thomas Alva Edison used a programmed selection of phonographic music for factories to find out to what degree it could mask hazardous drones and boost morale, but early loudspeaker transmission technology was still too weak for this purpose (Lanza 1994, 13; Thompson 2002, 145). After experiments on the use of music in an architectural drafting room in 1921, American psychologist Esther L. Gatewood argued that familiar, instrumental music influenced productivity in a positive way (Gatewood 1921; Uhrbrock 1961). In their widely quoted 1937 study, British researchers S. Wyatt and J. N. Langdon argued that music diverted the mind from monotonous working conditions, made time pass more quickly, and created a more cheerful attitude toward work (Wyatt and Langdon 1937).

In this context the BBC developed its "Music while You Work" program, which was broadcast from June 1940 onward in two daily half-hour editions at 10:30 A.M. and 3 P.M., with a third half-hour program at 10:30 P.M. for night-shift workers. Numerous tests were done to find out more about the right kind of music to play. These tests showed that the music most appropriate for both masking noise and making people working in factories feel happy had to aim for consistency in rhythm, volume, or tempo. The emphasis was on "bright, cheerful music." Vocal music counted as distracting and could even affect discipline negatively. "Music while You Work" producers optimized their broadcast in order to mask factory noise (Korczynski and Jones 2006).

Masking industrial noise with music was also a topic in the Third Reich. The key researcher in this field was Rudolf Bergius, a psychologist whose 1939 doctoral dissertation dealt with experimental studies on "Distraction from Work by Noise and Music and Its Structural-Typological Context" (Wehner 1992). Familiar with Anglo-Saxon research, Bergius reported on the outcome of experiments that showed that work performance improved with rhythmic noise, whereas it deteriorated when noise was arhythmic. He argued that Germanic people were more susceptible to noise than others, particularly Romanic people. Life in southern countries was generally louder than in northern ones. Contrary to the Romanic race, which tended to be extroverted, Germanic people were generally introverted and had more problems coping with noise (Bergius 1939).

In the late 1930s several industrial firms in Germany used music on the shop floor to increase workers' productivity. However, there was still no formal program similar to the British "Music while You Work" broadcast. Bergius claimed to have found out that in cases in which music in factories was about equal in loudness to machine noise, music prevailed and that, in the listeners' perception, industrial noise seemed to adapt to it rhythmically. Although industrial noise may be predominantly rhythmical and not cacophonous, music had the great advantage of being "designed" and therefore stood out. In Bergius's view, however, it would be a cultural scandal if industrial noise were to be suppressed by music. After all, music was a product of culture, preferably of German culture. Although "designed" music prevailed over industrial noise, this did not mean that industrial noise had no harmful effects on the human organism (Bergius 1939, 112).

From 1937 to 1939 music from recordings and broadcasts was played for a few hours every day in German factories, including one in Berlin that employed 130 tailors (108 women and 22 men), most of whom were working on sewing machines to produce military uniforms. The Nazi "Strength through Joy" organization and the "German Labor Front" supported music at work primarily because it fostered a sense of community rather than actually boosting output. The bonds within the factory community, which formed part of the larger "German peoples' community," were to be strengthened by this practice. Obviously, the "factory leader" and the government also welcomed any increases in productivity. Although the exact number of loudspeakers in the factory is not given in Bergius's study, it must have involved a considerable number in order to mask the din of the numerous sewing machines. Contrary to the practice in Britain, music would accompany the start of the morning shift for half an hour and was again played at the end. There seems to have been no research on the most appropriate time for playing music; this particular sequence was chosen because the "factory leader" believed that employees needed special encouragement in the morning and also when the shift was about to end (Bergius 1939, 130–39).

As in Britain, light music, dance music, and popular songs were played, but, in Germany, marching music, reflecting the popularity of the military, was also in demand. Unlike in British factories, songs were favored, and employees were allowed and even encouraged to sing together during work so as to boost group morale. It seems, then, that in this case Nazi ideals were not only preached programmatically but also put into practice, possibly at the cost of increased productivity.

Bergius also studied the acceptance of music in factories and found out that, as in Britain, most employees welcomed music on the shop floor. However, in his empirical research, some of his questions were different from those of the researchers in Britain. This had to do with the particular interests of German academic psychology at the time. Apart from his interest in the allegedly different behavior of northern and southern Europeans, Bergius also made a distinction between different body types, following the German psychologist Ernst Kretschmer: pyknian, leptosome, and athletic. He claimed that the leptosome types were (positively or negatively) affected by music the most, more so than the pyknians, whereas the few athletic types among the employees, who exhibited a "tough temperament," did not seem to care very much whether there was music on the shop floor. Women were, according to Bergius, generally more affected by music than men (Bergius 1939, 138).

To realize the goal of increasing labor productivity and masking industrial noise, the music played in factories should be just marginally louder than the machines. If it was not loud enough, workers had to concentrate too much on listening to it, which distracted them from work. Workers generally welcomed the music program, but one female worker was opposed to it. Why was this? Bergius first assumed that this woman exhibited a stubborn, oppositional attitude toward the community spirit and probably needed some refresher course in Nazi ideology. Simply asking her for her reason, however, yielded a much more favorable result, which doubtlessly pleased Bergius, a man of culture: The woman proved to be a great music lover and often listened to music at home, concentrating solely on the notes. She felt that listening to music in a factory did not do justice to music as a lofty product of culture (Bergius 1939, 139).

"Music while You Work" also played a role in the German Democratic Republic (GDR), although, according to Soviet and GDR authors, its function was different from music in capitalist countries. In the latter the main aim was to increase output, whereas in the socialist states the well-being of the workers came first. In the Soviet Union several industrial companies started introducing music in their factories, but their number remained small. As several authors explain, the music-selection procedure and loudspeaker installations in factories were amateurish, which undermined the benefits (Klughard and Schwabe 1975). In the early 1970s the USSR Labor Research Institute in Moscow carried out a research project on the music most appropriate to enhance job satisfaction among Soviet workers. The factories that adopted industrial music dealt mainly with assembly work and simple, monotonous labor processes. Research results showed that light music and folk music were particularly well received by the workers (Klughard and Schwabe 1975, 610–12).

Based on these findings, GDR researchers conducted a study in a film factory in Wolfen (Saxony Anhalt) to find out what kind of music was most appropriate for enhancing workers' job satisfaction. Their hypothesis, taken from the results of investigations in the Soviet Union, was that music on the job increases the well-being of workers, reduces fatigue, and enhances the ability to concentrate and therefore the quality of the product. Workers should not need to concentrate much on the music, implying that instrumental pieces were to be preferred over vocal ones. Even though the latter sometimes distracted from work, they were particularly popular, especially when sung in German. The music should be rhythmic and have little variation in tempo (Klughard and Schwabe 1975, 617). Since we have little information on the extent to which these findings were applied in other companies, it seems that industrial music did not play a large role in the GDR.

In the Federal Republic of Germany (FRG), music at work was much more widespread. In 1966 it was estimated that fifteen thousand factories used music programs on the shop floor or allowed workers to listen to their own radio sets (Meissner 1976). One of the organizations that supported the aim of increasing productivity through music was the German industry's "rationalization committee." From 1960 onward, it sponsored a research project on the effects of industrial music, in which nine companies from several branches—textiles, machine tools, printing machines, food processing, precision engineering—participated. This resulted in a well-documented dissertation on industrial medicine by Günter Last from the University of the Saarland in Saarbruecken, published in 1966. Last's main aim was to test the hypothesis put forward by supporters of industrial music, notably in Britain and the United States, who claimed that music at work increased job satisfaction and labor productivity significantly (Last 1966).

The most general finding of Last's investigation was that, as in most other companies in which these experiments were carried out, job satisfaction did in fact increase, albeit to different degrees. As a whole, industrial music enhanced the working atmosphere, reduced unnecessary conversations on the shop floor, countered monotony, and reduced fatigue (Last 1966, 89). In fact, in a factory producing machine tools and printing machines, several workers were prepared to work overtime only if industrial music were provided. Although musical preferences varied, there was, as in Great Britain and the GDR, a penchant for pop songs and dance music. Songs that workers could hum or even sing along with were particularly popular, if not with their employers, who worried about distracted employees, lower output, and defects caused by lack of concentration.

A problem arose in factories in which employees were exposed to noise levels that differed significantly in the various areas. Introducing industrial music only in areas with a comparatively low noise level caused resentment in those employees who worked in noisier surroundings, so music was broadcast either in the factory as a whole or not at all. Last found out that after having listened to music at work for a long time, workers got used to it, and its activating effect disappeared.

Contrary to the outcome of earlier research on the effects of industrial music in Britain and the United States, Last came to the conclusion that industrial music did not have a significant effect on labor productivity, absenteeism, job fluctuation, or defect frequency. If there were any such effects, they were but small and statistically hardly relevant. Slight increases in labor productivity in the presence of industrial music could be due to less conversation between workers on the shop floor or to their not leaving their workplace because of the monotony of work. Last felt it to be impossible to make more detailed claims about the effect of music on labor output because numerous other factors were involved. Therefore, the role of industrial music could not be conclusively determined (Last 1966, 79).

It is no surprise that there were differences regarding concept and application of industrial music in Britain, Nazi Germany, the GDR, and the FRG. In Nazi Germany and the GDR the ideology of enhancing the workers' well-being by playing industrial music was to the forefront, but in practice industrial music on the shop floor did not play a large role, most probably because the productivity effects were considered small. In capitalist Great Britain and the FRG, not so much the well-being of labor but job satisfaction, a notion closely related to the former but different from it, was highlighted. The champions of "Music while You Work" in Britain claimed a notable increase in labor productivity through music, which was also due to the masking of noise, whereas Last's research on the FRG in the early 1960s established a negligible productivity effect of music.

Hearing Protection Devices at the Shop Floor

As early as 1907 the "Ohropax" earplug was introduced, but workers were generally not very eager to use it or similar devices. Many of them regarded the use of earplugs as "unmanly" and were not worried about noise-induced hearing loss, which came only slowly and gradually. On a symbolic level, noise in industry also had positive connotations. As Mark Smith also shows in his contribution to this volume, a noisy factory meant that industry was booming and earnings were good. Besides, wearing a hearing-protection device made workers insecure about where the noise came from while also causing communication problems. Listening to machine noise (or sound—because noise/sound was often not perceived as unpleasant) gave workers a feeling of security; they would hear it when something was wrong. Furthermore, wearing ear protection could even be dangerous because one could also fail to hear particular warning signals. Only from the late 1950s onward did protective devices like earplugs and earmuffs grow more common. Employers and liability-insurance associations insisted on their use mainly because they faced rapidly increasing compensation claims for employees' hearing losses (Bijsterveld 2008, 69–76).

One of the reasons this took so long was that it was difficult to establish how occupational noise was linked to hearing loss. Employers and employees often argued that a partial loss of hearing was not critical to performance on the job. Until legislation set noise-exposure limits in the late 1960 and 1970s in many European countries, it was largely left to the industrial workers themselves to use hearing protection. Physicians' advice failed to impress them because, as hinted at earlier, such caution clashed with their cultural values and the symbolism of noise, with a shop-floor culture of listening to machines and with the overall positive meaning of loud sound, as well as for their routines on the shop floor (Bijsterveld 2008, 77–81).

A 2005 study assesses the influence of hearing-protection devices (HPs) on the recognition of warning-horn signals during work on railway tracks (Lazarus 2005).[2] The outcome was that, with normal hearing, a frequency-independent hearing protector, a plug, improved hearing performance, whereas a frequency-dependent HP, a muff, tended to worsen it, particularly with low-frequency noise. Regarding employees with a hearing impairment, hearing performance diminished when plugs were worn. For subjects with normal hearing, low-frequency interference noise led to a major deterioration in hearing performance when earmuffs were worn, whereas hearing performance with high-frequency interference noise remained almost unchanged. If the subjects, whose average hearing loss measured at frequencies of 0.5/1/2 kHz did not exceed 20 dB, wore earplugs, the warning-horn signals could be heard equally well or even slightly better. However, if hearing loss was greater than 20 dB and the subjects wore earplugs, the higher their hearing loss, the more their perception of warning-horn signals deteriorated. From this it follows that when selecting a hearing protector, the frequency distribution of noise, signal attenuation, and hearing loss all have to be taken into account. In normal cases, in which hearing protectors cannot be adapted to the spectrum of signals and noise, hearing-protection devices that offer frequency-independent attenuation (plugs) should be used, but only if the subjects do not suffer severe hearing loss.

Protecting Comrades: Industrial Noise Abatement in the GDR

When considering noise and noise abatement in the German Democratic Republic, the first socialist state on German soil, the question arises whether the GDR cared more about this problem than nonsocialist states. In theory this should have been the case because, after all, a prime aim of the socialist revolution was to ensure the well-being of the working class. Indeed, apart from outperforming capitalist countries by means of a more competitive industry, the government of the GDR considered it a humanistic obligation to improve labor conditions.

Attempts at noise abatement in industry were led not only by governmental bodies but also by institutions such as the "Chamber of Technology," the equivalent of the Federal German Society of Engineers, which were very active. The Dresden institute in charge of improving labor conditions spent much time and effort carrying out relevant research. The "Protection against Noise" bureau offered counseling and made recommendations, worked out standards and guidelines, and coordinated R&D on the acoustics of machinery. In line with international efforts to curb industrial noise at the source (primary noise-prevention measures), the foremost goal was to replace old machinery with new, designed to reduce noise on the shop floor (Scheuren 2002; Parthey 1989).[3]

At the company level, noise-abatement activities in the GDR were part of the organization in charge of occupational safety and protection against fire.[4] In a company, the foreperson or department manager served as leader of a collective, and that person had to make sure that employees heeded the rules and regulations regarding protection against noise. The company director, to whom the foreperson was accountable, had a supervising and controlling function. The crew leader had to acquire the relevant knowledge of occupational hygiene and had to attend courses on that topic to obtain a certificate of qualification, which had to be renewed every two years.

Moreover, the companies established inspections of workers' safety. The safety inspector made sure that the foreperson fulfilled all work-safety duties and reported to the company director. The foreperson was also authorized to issue directives on occupational safety. The safety inspector cooperated with the company's health services: In small companies with the company doctor, in medium-sized companies with an infirmary on the premises, and in large companies with the company's hospital. If deemed necessary, the safety inspector ordered noise measurements or auditory checks.

The GDR bureaucracy concerning industrial hygiene had even more to offer: Another organization dealing with the issue of occupational noise abatement was the trade union. In the GDR there was only one trade union, the "Free German Labor Union," a unified labor union. Virtually all employees, including forepersons and directors, were members of this organization. At the company level, the union appointed a representative in charge of occupational safety. On a more general level the Free German Labor Union appointed occupational safety inspectors who cooperated with their counterparts who were employed by the company but also checked them and gave them instructions. They tried to ensure realization of the labor unions' objectives, which in theory—but not always in practice—were identical to the company's goals. The company and the trade union inspectors had to make sure that the employees knew the relevant regulations, were alerted to new ones, understood them, and received answers to their questions. In case of an accident the first questions normally were these: Who was responsible? Who had to be alerted? Had the workers involved in the accident heeded the work safety regulations, which were regulations of a legal character that governmental bodies had worked out in cooperation with the trade union? In addition, the technical standards had legal status as well. All of this shows that, organizationally and institutionally, there was, at least officially, a tight net of checks and balances in the GDR noise-abatement bureaucracy.

But how well did it work? As we know from GDR economic and political history, being tied to five-year plans and a stifling bureaucracy did not make for industrial efficiency (Bähr and Petzina 1996; Bentley 1984). At a conference in 1987 celebrating the thirtieth anniversary of the GDR's Noise Abatement Commission, one of the main speakers, praising the achievements of noise abatement in the GDR, gave productivity increase a more prominent place than the welfare of the workers. Because the GDR desperately needed currency, machine noise, he argued, had to be reduced not least because of exports partly to oil-producing countries such as the United Arab Emirates. These export countries often had ambitious standards regarding machine noise, and, of course, the GDR had to compete with capitalist countries, which had better financial resources to meet those standards. A low noise level emitted by machinery, then, was a vital factor in selling machines to other countries.[5] Nevertheless, some of the other socialist countries caused problems. Most of the GDR exports went to the COMECON (Council for Mutual Economic Assistance) countries. In the late 1980s the Soviet Union had a standard of 80 dB (A) maximum for new machinery, whereas the GDR worked with a maximum of 85 decibels.[6] Reducing the noise level of the GDR to 80 decibels would have increased the financial difficulties of the GDR firms even more.

However, apart from the obligatory praise of the GDR concept and practice of noise abatement, there was criticism of various sorts. For example, machinery producers gave out unreliable information on noise emitted. If a company was "well connected," it was possible to obtain permits to manufacture machinery that did not meet the noise-emission requirements. Sometimes factories also had a shortage of noise-abatement specialists.[7]

Although at the anniversary meeting in 1987 participants leveled criticism at how GDR authorities dealt with the problem of machine noise, the reality on the shop floor was generally more severe. After attempts at reform in the 1960s to make the five-year plan more flexible and give material incentives to innovators, business as usual prevailed. The recurring problem was lack of funds. Keeping up full employment in the GDR and paying for the social security system, the army, and the huge bureaucracy were expensive indeed (Braun 1996; Wienhold 2006, 2008; Jarausch 2008). Conversely, productivity was low. Whereas in the late 1980s FRG researchers had assumed that productivity in the East German industry was about 65–70 percent of that in West Germany, research after German reunification showed that GDR productivity in the 1980s had reached only about 35 percent of that in West Germany. Therefore, not many resources were available for designing new noise-reduced machinery. Whereas the FRG imported much technological knowledge not only from the United States but also from the many federal German multinational corporations that were instrumental in technology transfer, little of that took place in the GDR (Schweres 1991).

Although several firms in the GDR could compete internationally, and they were duly highlighted by the government, the full picture was much less favorable. For this reason there was hardly any development of new, noise-reduced machinery. The common picture in GDR factories was that machinery was seldom replaced, was repaired again and again, and was used until it fell apart. Funds were seldom available because they had to be used for importing energy and raw materials, a sector in which the GDR was rather weak. For the standards of COMECON countries regarding machine noise, the products the GDR had to offer were generally good enough. After reunification the enormous environmental problems that had accumulated during the GDR times also became visible (Schweres 1991).

As early as 1978, Herbert Hartig, vice director of the Dresden Institute for the Protection of Labor, had reported on important issues of industrial noise and noise abatement in the GDR. He made it clear that protection of workers against industrial noise not only pertained to individual welfare but also had economic ramifications. Which of these aspects involved in industrial noise and noise abatement were decisive depended on the social system; in marked contrast to the capitalist system, in socialist countries the humanistic aspect always came first (Hartig 1978).

However, also in socialist countries like the GDR, the compensation paid to workers who suffered from hearing difficulties caused by industrial noise was significant. In addition, according to Hartig, workers with hearing impairment were less productive and, if they quit, had to be replaced by new workers, who also had to be trained. Moreover, "noise jobs" were not popular and could be filled only by offering higher wages (Hartig 1978, 333–34). Noisy machinery for the export market caused losses in foreign trade because foreign customers did not accept them or were prepared to buy them only at reduced rates. Of course, the efforts by the Socialist Unity Party (SED) party, the government, and the health-protection officers led to some improvement in the industrial noise situation in the GDR. As to primary measures of noise abatement, machines were designed that emitted less noise.

However, the situation left much to be desired, and several goals to improve the situation had not been realized. In particular the claim that industrial progress and efficiency increases made for better working conditions had proved to be a myth. Productivity increases had led to a spatial concentration of machinery and to significantly higher speeds being used in drive belts. Research at the Dresden Institute showed that a doubling of revolutions per minute in machine drive caused a 12-dB increase in sound emission. The institute had also tried to work out a mathematical formula on the relationship between the reduction of sound emission and an increase of labor productivity. Because of the many different factors involved and the large differences in the jobs investigated, this proved impossible. However, as a rule of thumb, the institute stated that a reduction of industrial noise emission by 5–10 dB resulted in a 5-percent increase in individual job performance. This, however, applied only to relatively simple, repetitive work, whereas in work that demanded more concentration, such as that of crane operators, performance loss caused by industrial noise was significantly higher (Hartig 1978, 335).

If it had not been possible to satisfactorily implement primary industrial noise-abatement measures through machine design or, for that matter, to apply secondary measures such as encapsulation, what about the application of individual ear-protection devices like earplugs or mufflers? The legal and regulatory framework for this existed: In the GDR, industrial workers' hearing difficulties were regarded as an occupational illness from 1958 onward, and a 1970 regulation made it obligatory for all factory managers to introduce ear-protection devices whenever a certain threshold value was surpassed. In 1977 there were about 250,000 industrial "noise jobs" in the GDR (i.e., jobs where primary or secondary noise abatement measures had not been successful). Theoretically, workers could still have been protected at least to a significant degree by individual ear-protection devices. As it turns out, however, the industrial health organizations' campaigns were far from successful. Investigations of some 90,000 industrial noise workers in the GDR showed that only 18.2 percent regularly wore individual ear-protection devices. Moreover, 16.5 percent did not wear them at all, and 65.3 percent wore them only intermittently or not according to official requirements. So only about 25 percent of industrial workers in noise jobs wore ear-protection devices properly (Hartig 1978, 337–38).

Aside from simple earplugs, more sophisticated devices were available, but they found only limited application. These were otoplastics that sealed the ear completely. They put higher demands on their manufacture and had to be fitted by an otologist. Earplugs made from synthetic material and fixed on a support bracket seemed to have several advantages, including that they sealed the ear canal tightly, were relatively easy to produce, and had to be worn only when necessary. However, they often exerted an unpleasant pressure on the ear, especially when the wearers moved their head. They were particularly useful in dealing with intermittent noises.

Mufflers have the advantage that they cover the whole ear and offer a high level of sound protection. However, wearing them can be uncomfortable, especially when temperatures are relatively high and when they are worn in combination with a protective helmet. In wearing ear-protection devices one factor is particularly relevant: It takes some time to get used to them, mainly because many of these devices have different degrees of sound abatement at varying frequencies. Moreover, the warning function of ears is severely impaired by wearing an ear-protection device.[8]

The example of the Diesel Engine Works Rostock, which produced marine diesel engines, illustrates the general remarks in this chapter. This company was located in the southern part of the city of Rostock, Mecklenburg, close to the Baltic Sea, and was established in the early 1950s. At that time the buildings and the machinery were state of the art, but the problem was that, due to shortage of funds, little money was left for modernization and replacement. The factory produced marine diesel engines with 4,000–20,000 horsepower, and it repaired smaller diesel engines. The main features of the production process were mechanical production (chipping), metal working (cutting, welding), hot treatment (postweld heat treatment, founding of bearings), assembly, transport within the company, and test runs (Spychala 2006, 2008).

Most of the production units had a medium noise level, which did not require any hearing-protection devices. Metal working was an exception, especially when, for example, pneumatic chisels had to be used or straightening work after welding became necessary. To reduce workers' exposure to noise, an adaptable noise-protection ceiling was installed above each workplace, and workers had to wear earplugs or earmuffs.

The situation in Rostock, as described thus far, differed little from that in other machine-building companies. But one problem was particularly enervating: The test runs could be extremely loud, especially at frequencies that humans find difficult to cope with. The company at Rostock utilized different kinds of test stands, including one for engines manufactured in the Soviet Union with a high noise level. After repair, those engines were installed in ships of the People's Navy. Because noise emission at test runs was quite high, a new, compact building was erected with a special control room for the operators. Even outside the test area the noise level was high and affected not only the employees at the diesel engine works but also nearby residents. Particularly unpleasant were the high frequencies emitted as a result of the engines' high number of revolutions and, especially, of the exhaust gas turbochargers. There do not seem to have been any restrictions on these testing procedures; any complaints were probably countered with reference to national defense concerns.

The company at Rostock also had a test stand hall for medium-sized diesel engines (4,000 hp) to repair engines for civil purposes, as well as a newly developed test engine of 3,000 hp. This hall contained about ten test stands that, apart from the new engine, came without noise protection such as encapsulation. In order to cope with the deafening noise, the personnel in charge of the new engine's test stand built a noise-reduced room by themselves containing the most important measuring and control instruments. However, the other workers in this hall and in the two other areas were subject to the noise in varying degrees.

In the large hall with four test stands for engines up to 20,000 hp only one was working at a time. These stands were located in the northern part of the three linked halls. The middle part was used for subassembly and some parts of mechanical production, while the southern part of the building was used only for mechanical production. Two-thirds of the outer walls of this hall consisted of glass, whereas the interior of the hall had no dividing walls whatsoever. None of the designers at the time of its construction seemed to have thought about noise protection. This meant that noise emissions at test runs affected the whole hall, as well as employees working in the hall's extensions. As in the other halls with test runs, it was not so much the low frequencies of the engine itself but the piercing high frequencies of the exhaust gas turbocharger that created the most problems. In the 1950s, when the hall was built, its design was regarded as particularly innovative and progressive, but later the hall seemed to many workers to sit there for eternity, unalterable and grand. The only way out was to avoid test runs in this hall altogether and to build a new one close to the water, which would also have had the advantage of easier transport. Plans to build this hall started in late GDR times. Lack of funds, however, prevented them from being pursued, and they were not taken up again until the early 1990s, when the Diesel Engine Works Rostock became part of the Vulcan Shipyard of Bremen. Vulcan built the new hall, but the company went bankrupt shortly afterward, and today the hall is owned by the Caterpillar Diesel Engine Company of Rostock.

This brief case study shows that a large gap existed between the ambitious aims of industrial noise abatement in the GDR and actual shop-floor practice. This was not due to a lack of expertise in noise abatement but to lack of funds and to the government's attitude that, at least in the Rostock case, industrial noise abatement played a secondary role to other goals such as maintaining production. In the Rostock case the future of the factory was severely constrained by the way it had been constructed in the early 1950s. Until the late 1980s, it seemed impossible to make substantial changes in matters of noise abatement. Significant noise-reduction designs rarely occurred because they would have required substantial R&D funds; as a result, little was done even regarding measures such as encapsulation of machinery. Consequently, employees resorted to noise-reducing devices like earplugs and earmuffs, which, even when used regularly and properly, could only partly alleviate the problem.

In the Rostock case, employees took noise regulations seriously. With regard to test-run installations they had no other choice because the piercing noise emitted by exhaust gas turbochargers was unbearable. However, also in areas with less noise, ear-protection devices were generally worn: If caught without them, employees faced severe reprimands from the foreperson and were publicly criticized at the next session dealing with work hygiene instructions.

What about the "Working-Class Exploiters"? Industrial Noise Abatement in the FRG

Having been somewhat harsh about the practice of industrial noise abatement in the German Democratic Republic, the question arises as to whether things were much better in its capitalist counterpart, the Federal Republic of Germany. In theory, and arguing from a GDR socialist point of view, this could not have been the case because in capitalist countries workers are exploited, and profit is all that counts. However, if entrepreneurs were to profit from noise-abatement measures, they would, according to this reasoning, surely introduce efficient noise-abatement devices in their factories. What about the state, governmental institutions, and others bodies? According to Marxist-Leninist lore, they would do what big business expected from them.

In the 1950s, industrial noise abatement in the FRG in fact did not play a large role. In this time of "reconstruction" similar to the immediate post–World War One period, other matters took priority. However, along with rising material welfare and a growing "environmental consciousness" from the early 1960s onward, topics such as noise and noise abatement began receiving more attention. In the FRG the Association of German Engineers again became active, and in 1960 it founded the Technical Noise Abatement Commission, which developed guidelines for acceptable noise levels on the shop floor and in public spaces. In 1968 the Technical Directive regarding Noise was established; it stipulated that, as far as noise emission was concerned, machinery used in factories had to be state of the art. This referred to new machinery, whereas in the case of existing plants secondary measures of noise abatement had to be taken. In 1974 the federal government enacted the Federal Emission Law, which articulated general requirements within which the different German states and factories had some freedom to act (Koch 2002, 235–36; Liedtke 2005; Scheuren 2002).

However, it was not so much an amorphous rise of "environmental consciousness" but hard facts, or, put simply, money that prompted industrial plants to do something about noise. As hinted at earlier, in Germany in the late 1920s hearing impairment and hearing loss were recognized as occupational diseases in the metal industries. This was extended in 1961 to other branches of industry. It meant that the employer's liability-insurance associations and pension offices had to pay for the problems these disabilities created, while at the same time the number of workers taking early retirement rose quickly. This is why, from the early 1960s onward, primary and secondary measures to curb industrial noise increased. The German engineering industry, a large exporter in the world market, had sufficient R&D funds to develop new noise-reduced machinery because this was also, as already mentioned in the case of the GDR, a hallmark of quality (Scheuren 2002, 208–10).

But what happened on the shop floor? This is much more difficult to assess than delineating the debate on setting standards, on legislation, or on the acousticians' differences in measuring noise. Research on this issue yields an uneven picture. Not surprisingly, the noise-abatement activities in German industrial firms differed considerably depending on the industrial branch, on firm size, and on managers and owners. Still, one thing is clear: In terms of noise abatement, the West German industry did not simply outperform the GDR. Thyssen in Duisburg, a large metal-working company that seems a representative example to investigate for our purposes took until 1976 to initiate a large campaign on noise abatement on the shop floor. Its main purpose was to convince workers to wear noise-protection devices. Its outcome was a notable improvement, without creating an optimal situation: Before the campaign only about 20 percent of the workers had used ear-protection devices; afterward, this figure had risen to about 60 percent.[9]

Let us look more closely at this campaign and the workers' reactions to it: In late 1976 the Employer's Liability Insurance Association of the federal German metallurgy industry started a campaign to fight occupational disease number 1, occupational hearing loss. It distributed posters in plants, organized talks, and took other measures to increase employees' awareness of this problem. "Not to wear hearing-protection devices shows a primitive attitude; people who enjoy living in noisy environments, who cause noise themselves, and who regard noise as proof of their industriousness should change their attitude as soon as possible. Protection against noise is protection of the self."[10] The shop council and the industrial relations director of the Thyssen steel company in Duisburg pointed out that noise-measurement activities had been conducted at the company since 1969. Also, the insurance association's 1974 regulation concerning occupational noise, which had legal status, determined that an ear-protection device had to be worn in noisy areas in the company. Employees not complying with this committed a regulatory offense and were liable to a fine of up to 20,000 Deutsche Mark, a substantial sum. In cases in which employees' hearing had already been damaged because of occupational noise, pension payments could partly or completely be stopped if employees did not wear ear-protection devices.[11]

The campaign undertaken by the employers' liability insurance association tried hard to counter employees' reasons not to wear ear-protection devices. Although they might be uncomfortable to wear at the beginning, the insurance association argued, it would only be a matter of time before employees would become accustomed to them. In this the devices were similar to a pair of safety glasses that wearers also had to get used to. Moreover, why, the employers' liability insurance association asked, did people not create a similar fuss when required to wear protective glasses? With ear-protection devices, it seemed, the issue of manliness was at stake. The employers also dealt with the argument that, wearing ear protection, employees were no longer able to distinguish between different sounds. This, according to them, was not a valid argument. After all, machine sounds did not vanish completely; they were only abated, and their sound was changed. Getting used to detecting them merely involved a learning process. After a short time, employees would learn to recognize the now fainter, altered sound; it was similar to the situation an apprentice faced in an industrial company: The trainee, too, had to learn to make sense of the machine sounds. Moreover, according to the insurance association, the argument that speech communication was not possible when wearing earplugs was unsound. Physical laws showed, the association argued, that only high-frequency sounds were abated but not sounds relevant to speech communication. Still, it was necessary of course to speak loudly and distinctly in order to make oneself understood in a noisy environment.[12]

Although these arguments sounded plausible, they, being arguments voiced by one side, the employers, were somewhat simplified and could not completely offset the reservations of many in the workforce. This applied even more to the complaint many employees voiced about inflammations of the auditory channel. This happened quite frequently and prompted employees not to wear earplugs. However, according to the employers' argument, these inflammations were caused by badly fitting earplugs or, even more frequently, by earplugs that had not been kept properly clean. In cases of auditory-channel eczemas or chronic middle-ear diseases, wearing earplugs was in fact forbidden. In these cases an otoplastic device was recommended.

The campaign at Thyssen did have some effect. As stated earlier, workers' acceptance of ear-protection devices significantly improved, but the ambitious aim to convince every worker to wear earplugs was not reached. As indicated, before the campaign only about 20 percent of the workers had used ear-protection devices, while after the campaign this figure had risen to about 60 percent.

Conclusion

This chapter has pursued only a few aspects of a large subject. Comparison of noise and noise abatement in industry during the Third Reich, the FRG, and the GDR has revealed that only in the FRG was industrial noise abatement moderately successful and that this came about only from the 1970s onward. The German engineering industry, a large exporter in the world market, had sufficient R&D funds available to develop noise-reduced machinery. Noise reduction was also regarded as a hallmark of quality. On the shop floor the implementation of noise-reducing measures improved mainly because of stricter insurance association regulations.

For insurance organizations the rapidly increasing numbers of insurance claims had made it necessary to strengthen their efforts in reducing occupational hearing loss. Undertaking primary efforts to reduce machine noise through engineering design was a matter of funds. During the Third Reich the incentive to do so was low because other priorities prevailed. In the GDR the relevant know-how was certainly there, but scarce funds and the fact that only a small portion of exports went to capitalist countries did not lead to any serious effort to reduce the problem.

Were the employees responsible for the fact that not more was done earlier? As my argument demonstrates, the reticence of workers vis-à-vis hearing-protection devices made sense, at least to a considerable extent. Their attitude can be accounted for with reference to legitimate safety concerns, while notions of "manliness" also played a role. Moreover, music on the shop floor served to mask industrial noise and, as some believed, increased productivity.

Much more research is needed to enhance our understanding of industrial noise and noise abatement in the twentieth century. A glance at the annual bibliography of the *Journal of the American Society of Acoustical Engineers* makes it clear that in recent decades a great deal of research has been carried out on various aspects of this topic. This work may provide the basis of an interesting comparative study of industrial noise and noise abatement measures in industrialized countries in various parts of the world. Topics such as the history of sound-measurement devices and standardization in an international historical perspective deserve thorough analysis. Moreover, there is hardly any research on what actually took place on the shop floor. Race, class, and gender are crucial categories in today's historical scholarship, but we have little knowledge about them in the context of industrial noise and noise abatement. In other words, much work remains to be done.

NOTES

1 Karin Bijsterveld (2008) has "Turning a Deaf Ear to Industrial Hearing Loss" as one of the subheadings in her chapter on industrial noise and noise abatement.

2 I am grateful to Professor Hans Lazarus for several hints regarding industrial noise and noise abatement (email, Sept. 4, 2009; telephone conversation, Aug. 27, 2009).

3 See also Kammer der Technik: Bezirksverband Gera und AG (Z) Lärmschutz, *7. Konferenz Lärmschutz: Fortschritte der Lärmbekämpfung—30 Jahre Lärmschutz.* 2 vols., Gera, Germany: Kammer der Technik. Kampf dem Lärm, 1987. Also see *Unsere ATH* 22(11/12) (1976).

4 For the following information and for information on conditions in the Diesel Engine Works Rostock, with which I deal later, I am very grateful to Professor Franz Spychala of Rostock. He was employed at the Diesel Engine Works Rostock from 1960 until 1990 in increasingly responsible positions: engineering designer (1960–1961), development engineer (1968–1970), head of the development department (1970–1977), and head of development-controlling shipbuilding (1978–1990) (email, Aug. 26 and 28, 2009).

5 See Kammer der Technik, vol. 2, 83.
6 See ibid., vol. 1, 14.
7 See ibid., vol. 2, 15.
8 "Lärmbekämpfung mehr denn je aktuell." *Sozialversicherung/Arbeitsschutz* 7 (1977): 16.
9 I wish to thank Professor Manfred Rasch and Andreas Zilt, MA, Thyssen Archives Duisburg, for useful hints on the Thyssen case.
10 "Kampf dem Lärm." *Unsere ATH* 22(11/12) (1976): 3.
11 Ibid., note 10: 17.
12 Ibid.

REFERENCES

Bähr, Johannes, and Dietmar Petzina. *Innovationsverhalten und Entscheidungsstrukturen: Vergleichende Studien zur wirtschaftlichen Entwicklung im geteilten Deutschland 1945–1990*. Berlin: Duncker and Humblot, 1996.
Bentley, Raymond. *Technological Change in the German Democratic Republic*. Boulder: Westview, 1984.
Bergius, Rudolf. "Die Ablenkung durch Lärm." *Zeitschrift für Arbeitspsychologie und praktische Psychologie im Allgemeinen* 12, pt.1, no. 4: 90–114; pt. 2, no. 6 (1939): 133–51.
Bijsterveld, Karin. *Mechanical Sounds: Technology, Culture, and Public Problems of Noise in the Twentieth Century*. Cambridge, Mass.: MIT Press, 2008.
Braun, Hans-Joachim. "Einleitung Themenheft: Technik im Systemvergleich. Die Entwicklung der Bundesrepublik und der DDR." *Technikgeschichte* 63 (1996): 279–84.
———. *The German Economy in the Twentieth Century: The German Reich and the Federal Republic*. New York: Routledge, 1990, Routledge Revivals, 2011.
———. "Lärmbelastung und Lärmbekämpfung in der Zwischenkriegszeit." In *Sozialgeschichte der Technik: Ulrich Troitzsch zum 60. Geburtstag*, ed. Günther Bayerl and Wolfhard Weber. 251–58. Münster: Waxmann, 1998.
Friemert, Chup. *Produktionsästhetik im Faschismus: Das Amt "Schönheit der Arbeit" von 1933 bis 1939*. Munich: Damnitz, 1980.
Gatewood, Esther L. "An Experiment on the Use of Music in an Architectural Drafting Room." *Journal of Applied Psychology* 5 (1921): 350–58.
Hartig, Herbert. "Lärmschutz: Eine humanistische, ökonomische, und technische Aufgabe." *Sozialistische Arbeitswissenschaft: Theoretische Zeitschrift für arbeitswissenschaftliche Disziplinen* 22 (1978): 330.
Hasse, Albrecht. "Lärmbekämpfung: Eine hygienische und wirtschaftliche Forderung." *Gesundheits-Ingenieur* 62 (1939): 165–69.
Jarausch, Konrad, ed. *Das Ende der Zuversicht? Die siebziger Jahre als Geschichte*. Göttingen: Vandenhoeck and Ruprecht, 2008.
Klughard, Heiderose, and Christoph Schwabe. "Der Einsatz von Musik zur Beeinflussung von Ermüdungserscheinungen unter spezifischen Arbeitsbedingungen." *Sozialistische Arbeitswissenschaft: Theoretische Zeitschrift für arbeitswissenschaftliche Disziplinen* 19 (1975): 610.
Koch, Hans-Joachim. "Fünfzig Jahre Lärmschutzrecht." *Zeitschrift für Lärmschutz* 49(6) (2002): 235–44.
Korczynski, Marek, and Keith Jones. "Instrumental Music? The Social Origins of Broadcast Music in British Factories." *Popular Music* 25(2) (2006): 145–64.

Lanza, Joseph. *Elevator Music: A Surreal History of Muzak, Easy Listening, and Other Moodsong*. New York: Picador, 1994.

Last, Günter. *Musik in der Fertigung: Untersuchungen zur Problematik der Musik am Arbeitsplatz in Industriebetrieben*. Berlin: Beuth, 1966.

Lazarus, Hans. "Signal Recognition and Hearing Protectors with Normal and Impaired Hearing." *International Journal for Occupational Safety and Ergonomics* 11(3) (2005): 233–50.

Liedtke, Martin. "Was bringt die neue EU-Lärm-Richtlinie?" *Sicher ist sicher: Arbeitsschutz aktuell* 6 (2005): 248–51.

Meissner, Roland. "Funktioniert die funktionelle Musik? Zur Musik am Arbeitsplatz." *Musik und Bildung: Zeitschrift für Musik in den Klassen 5–13* 8 (1976): 573.

Parthey, Wolfgang. "Bedeutung der Lärmbekämpfung und Lärmschutzvorschriften für Maschinen und Produktionsstätten." In *Lärmbekämpfung: Maßnahmen an Maschinen und in Produktionsstätten zum Schutz des Menschen vor Lärm und Schwingungen*, ed. Werner Schirmer, 8–20. Berlin: Tribüne, 1989.

Scheuren, Joachim. "100 Jahre technische Lärmminderung in Deutschland." *Zeitschrift für Lärmbekämpfung* 49(6) (November 2002): 201–18.

Schweres, Manfred. "Zur DDR-Forschung im Felde der Arbeitswissenschaft." *Zeitschrift für Arbeitswissenschaft* 45(4) (1991): 239–43.

Smith, Mark M. *Listening to Nineteenth-Century America*. Chapel Hill: University of North Carolina Press, 2001.

Spychala, Franz. "Die Entwicklung des Schiffbaus in der SBZ/DDR 1945–1990." In *Die zweite industrielle Revolution: Schiffbau seit dem Ende des 19. Jahrhunderts*, ed. Hans-Joachim Braun, 65–78. Freiberg: Georg-Agricola-Gesellschaft, 2008.

———. *Kolben, Pleuel, und Losungen: Das Dieselmotorenwerk Rostock*. 2 vols. Rostock: MV Wissenschaft, 2006.

Thompson, Emily. *The Soundscape of Modernity: Architectural Acoustics and the Culture of Listening in America, 1900–1933*. Cambridge, Mass.: MIT Press, 2002.

Uhrbrock, Richard S. "Music on the Job: Its Influence on Worker Morale and Production." *Personnel Psychology: A Journal of Applied Research* 14 (1961): 9–28.

Wagner, Karl Willy. "Fortschritte in der Geräuschforschung und Lärmabwehr." *Zeitschrift des Vereins deutscher Ingenieure* 79(18) (1935): 531–40.

Wehner, Ernst G., ed. "Rudolf Bergius." *Psychologie in Selbstdarstellungen*, vol. 3, 33–36. Göttingen: Huber, 1992.

Welch, David. *The Third Reich: Politics and Propaganda*. London: Routledge, 2002.

Wienhold, Lutz. "Arbeitsschutz." In *Deutsche Demokratische Republik 1961–1971: Politische Stabilisierung und wirtschaftliche Mobilisierung. Geschichte der Sozialpolitik in Deutschland seit 1945*, vol. 9, ed. Christoph Kleßmann, 187–224. Baden-Baden: Nomos, 2006.

———. "Arbeitsschutz." In *Deutsche Demokratische Republik 1971–1989: Bewegung in der Sozialpolitik, Erstarrung, und Niedergang. Geschichte der Sozialpolitik in Deutschland seit 1945*, vol. 10, ed. Christoph Boyer, Klaus-Dieter Henke, and Peter Skyba, 201–42. Baden-Baden: Nomos, 2008.

Wiethaup, Hans. *Lärmbekämpfung in der Bundesrepublik Deutschland*. Cologne: Heymanns, 1967.

Wyatt, S., and J. N. Langdon. *Fatigue and Boredom in Repetitive Work*. London: HMSO, 1937.

CHAPTER 3

"SOBBING, WHINING, RUMBLING": LISTENING TO AUTOMOBILES AS SOCIAL PRACTICE

STEFAN KREBS

Introduction

On a hot summer day in 1950, a Bavarian commercial traveler decides to bring his car to a nearby repair shop. When driving, he occasionally hears an inexplicable slapping noise, and he worries that this might be an audible indication of a technical problem. The mechanics at the garage search for the source of the noise. In particular they check the carburetor and the ignition, which often cause trouble, but they are unable to find the slightest malfunction. When the motorist returns that same afternoon to pick up his car, the mechanics ask him to describe the slapping sound in detail, as well as when and where it is most noticeable. He explains: "Every day I drive down a long, straight street lined with poplars, and there I always hear the slapping noise" (Anonymous 1950b, 353).[1] The mechanics burst out laughing because they immediately understood that the driver was merely hearing the regular echo of the car as the sound was being reflected by the trees.

This anecdote, together with three similar stories, appears in the August 1950 issue of the German trade journal *Krafthand*—a major periodical for car mechanics and dealers. All four accounts emphasize the superior knowledge car mechanics possess. If the opening anecdote suggests that even experienced drivers, such as commercial drivers, lack the necessary *techniques* to diagnose audible malfunctions, it also underscores the general importance of listening as a means of noticing technical problems and the general difficulties of locating a sound source and attributing meaning to it.

In sharp contrast to the "tree echo" episode from the 1950s, German motorists in the 1920s in fact appeared to possess the necessary listening skills to diagnose malfunctions. As I show in this chapter, they even questioned the sonic expertise of their car mechanics. So when did listening first become an exclusive domain of German car mechanics? Furthermore, why did drivers of the 1920s and their counterparts of the 1950s listen to their cars differently? To tackle these questions, this chapter describes the relevant listening practices and explains how and why they differ. I focus on what Kevin Borg calls technology's middle ground, the "ambiguous space between production and consumption" (Borg 2007, 2–3) and explore the complex relationship between German auto mechanics and motorists during the interwar period and the first years after World War Two.

My first hypothesis is that the two actor groups (motorists and auto mechanics) gradually developed two different "modes" of listening:[2] "listening while driving" and "diagnostic listening." To investigate the genealogy of these modes I conceptualize them as social practices in accordance with Pierre Bourdieu's theory of practice. As a *technique du corps*, or technique of the body, listening is a practical sense (Bourdieu 1990, 66–79). Marcel Mauss (1936) describes techniques of the body as part of the habitus; they vary with educational method, decency, fashion, and prestige. Following Mauss, Bourdieu conceptualizes techniques of the body as "the *socially informed body*, with its tastes and distastes, its compulsions and repulsions, with, in a word, all its *senses*, that is to say, not only the traditional five senses . . . but also the sense of necessity and the sense of duty, the sense of direction and the sense of reality" (Bourdieu 1977, 124).[3] For Bourdieu, agents incorporate into their habitus the social logic that constitutes the field. He posits a homologous relation between habitus and field (Bourdieu 1999, 138–46). Accordingly, I assume that listening to automobiles as a social practice is structured by the habitus of the actors and the field they act in.

In addition to Bourdieu's theory of practice, I suggest using the notion of the dispositive to further investigate the genealogy of the listening practices. According to Michel Foucault, the *dispositive* is a heterogeneous ensemble "consisting of discourses, institutions, architectural forms, regulatory decisions, laws, administrative measures, scientific statements, philosophical, moral and philanthropic propositions—in short, the said as much as the unsaid" (Foucault 1978, 119–20). Within the dispositive discourses and institutionalized power structures affect nondiscursive practices (Jäger 2001). In my case study, I assume that the nondiscursive practice of listening to automobiles is shaped by the power structure between mechanics and automobilists.

This brings me to my second hypothesis: The formation of a particular German auto mechanic's dispositive in the 1930s altered the power structure in the field, giving rise to the differentiation of the two listening practices by excluding motorists from the realm of "diagnostic listening." In the first two sections I explore the listening *techniques* of motorists and mechanics in the 1920s. The next section describes the repair crisis at the end of that decade and the new legislation concerning the auto mechanics' trade in 1934. Subsequently, I focus on the automotive technology of that time. The final section discusses the differentiation of the two listening practices. I have analyzed different trade journals for automotive engineers, auto mechanics, and garage owners, as well as special-interest journals for automobilists. Particularly the periodicals of automobile clubs turned out to be rich historical sources as they contained hundreds of letters from car drivers, who described technical problems—often mentioning their listening practices. In addition, I considered contemporary handbooks for car owners, chauffeurs, and auto mechanics.

Listening to Automobiles: The Driver's Experience

My inquiry starts at the end of World War One, during which the private use of automobiles had virtually ground to a halt (Ruppel 1927, 10). At that time, members of the German automobile lobby, such as the Imperial Automobile Club (Kaiserlicher Automobil Club), anticipated a growing demand for private motorization in peacetime. They argued that the military use of cars and trucks had proven the usefulness and reliability of automobiles as a means of mass transportation (König 1919). The postwar car was imagined to be small, light, economical, and effortlessly drivable without the help of a mechanic (Ledertheil 1919).

Experts projected that the self-driving tradesman and other middle-class people would be the most likely future car owners. Rudolf Hessler, author of *Der Selbstfahrer* (The Self-Driver) (Hessler 1926, 7), presupposed a reasonable level of technological knowledge as common among these potential motorists. Handbooks and driving manuals were intellectually demanding, and the division between professional and popular automobile journals was not as sharp as today.[4] Accordingly, there was a sense that the new categories of car owners were going to replace the chauffeur in his two roles as driver and mechanic (Borg 2007, 13–30). This continuum between chauffeurs and postwar motorists is visible in the specific concepts used in the literature of the time. A great number of instruction books equally addressed professional chauffeurs and self-driving automobilists (Parzer-Mühlbacher 1926; Hacker 1932; Martini 1938). In concepts and technical level, even specialized almanacs for chauffeurs did not differ considerably from more general instruction books (Martini 1922).

It seems worthwhile, then, first to examine the chauffeurs' practice in terms of their concerns with listening and automobile sound. Besides driving, their main duties included maintenance and repair work. Only major engine jobs that needed special equipment were to be left to repair shops (Martini 1922, 13). Chauffeurs needed a high level of technical knowledge and driving know-how (the latter was acquired exclusively through experience). In addition, chauffeurs had to have a good sense of hearing: "The ear should rigorously register the finest deviance of the engine sound" (König 1919, 12). While driving, the chauffeur had to listen carefully to the machine (Küster 1907, 10), and if he detected any dissonance, he had to look for its source and decide whether it required immediate repair (Martini 1922, 207). One could develop a fine sense of the engine's rhythm and timing only through experience, but theoretical understanding of automotive technology counted as a prerequisite for proper diagnosis (Küster 1907, 10). In summary, professional drivers had to learn to listen to automobiles through hands-on experience, and they had to acquire the appropriate technological knowledge. This *technique* of listening and the embodied cultural capital were requisite for success as a good chauffeur.

As already indicated, self-driving car owners would find similar advice in the contemporary sources of the day. In *Ohne Chauffeur* (Without Chauffeur), one of the earliest handbooks that appeared in as many as thirteen editions between 1904 and 1930, the motorist was urged to "listen to the desires of his engine" (Schmal 1912, 10). Other manuals stressed the importance of regularly listening to the sound of the engine (Hacker 1932, 62). Once the driver noticed any discord, he was to drive carefully and "open up his ears" (Hacker 1932, 83). Furthermore, drivers were instructed to avoid any unnecessary noise because only when the engine "runs as quietly as possible, any malfunctions that might arise can be noticed plainly and early" (Hessler 1926, 217). One even suspected that too much noise would harm the drivers' sense of hearing, thereby affecting their sensitivity to the "desires" of the machine (Hessler 1926, 203). In general, all motorists could train their ears:

> With growing experience and habit even the beginner learns to focus his attention on other things, especially his own car, without being distracted from the road. It is primarily the rhythmic and silent run of the engine that requires his attention. The regular humming of the gearbox or chain drive indicates that everything is in best order. He will soon notice that every engine and every car has its own pace and that even the slightest technical problem alters this lovely rhythm. He will involuntarily listen to this pace very closely, thereby avoiding any greater malfunctions. A knock or rattle of the engine, a crunch of the chain, a rattle of a bolt will indicate the spot where the car needs maintenance, and he will do well to follow the slightest hint to repair malfunctions in time before they grow worse. (Küster 1919, 304)

As this quotation from the handbook *Das Automobil und seine Behandlung* (The automobile and its maintenance) suggests, motorists needed time to get to know their car and to develop the necessary listening skill to understand what the engine "said" to them. They had to learn to distinguish the familiar sounds of a properly running vehicle from the deviant sounds that indicated problems, a practice Karin Bijsterveld calls "monitory listening" (Bijsterveld 2008, 77–78; 2009).

Such monitory listening techniques, which constitute the social practice I call "listening while driving," are just one side of the coin. As mentioned earlier, automobilists were to undertake supplementary diagnoses of the malfunctions they noticed while driving. Here, too, listening played an essential role: "[N]oise is most important for the detection of technical fault" (Hessler 1926, 216). In comparison to the *technique* of monitory listening, the practice of "diagnostic listening" was much harder to achieve: "This skill, to make the correct diagnosis out of a knocking sound, requires tremendous experience and exact knowledge of the type of engine construction" (Hacker 1932, 83). Maintenance almanacs for motorists provided help, however; they offered a systematic overview of possible malfunctions, symptoms, and ways of repairing. Oskar Hacker's manual (1932), for example, categorized malfunctions by the senses: seeing, hearing, smelling, and feeling. He needed no less than twenty-nine pages to list a wide array of audible failures, a systematic approach that was also adopted in journal articles. Other manuals ordered the maintenance section by components or combined the explanation of the engine's function with potential malfunctions (Schmal 1912; Küster 1919). Implicitly, handbook authors began codifying car sounds by transforming their auditory experience into communicable signs and meanings, if only with limited success, as we will see later. Jens Lachmund has described similar problems that physicians encounter when listening to a patient's body (Lachmund 1999, 420).

All of the authors emphasized the importance of technical knowledge and clear thinking: Motorists had to carry out a systematic inquiry to achieve their objective. As guidelines, the manuals contained tables, lists, and fault trees. They could be used to perform a differential diagnosis to narrow down the list to a single condition and provide a basis for a hypothesis of what was ailing the "patient" (Hacker 1932, 18). It is no coincidence that this resembles the physician's routine: "As in the relation between a physician and his patient, the diagnosis itself is the most important thing to a driver" (Hessler 1926, 216).

Further, the motorist was advised to ask specific questions about the different symptoms, the components that might be "infected," or the specific driving conditions under which the malfunction was most noticeable—just like a physician taking a patient's medical history. "If the physician cannot make his diagnosis by the appearance of the patient, he will take his stethoscope and listen to the patient's body. This is how you ought to proceed with your car engine as well" (Hessler 1926, 216). To listen at definite spots, the driver could use a screwdriver or a long metal pipe as a simple ear trumpet (Hacker 1932, 81; chapter 12). A driving manual of the German Association of Motorists (Allgemeiner Deutscher Automobil Club) stated that, with a stethoscope, a trained ear could indeed locate a single dry-running bearing (Dietl 1931, 324). Corresponding hearing devices, such as the Auto-Doktor, were advertised in motoring journals (Anonymous 1929a).

As mentioned earlier, handbook authors recognized automobilists as having the ability to make a range of significant distinctions between acoustic phenomena.

Figure 3.1 Cartoon of a car mechanic as a physician, from the *Briefkasten* section of the *Allgemeine Automobil-Zeitung* 33(42) (1932): 19 (Jonny).

Furthermore, they were seen as eager to acquire the expertise to repair their cars themselves. Nonetheless, why should they want to acquire the needed knowledge and experience, including the hard-to-learn technique of diagnostic listening? The magazines and handbooks of the time agreed on two intersecting aspects: First, self-repairing was simply cheaper and, if done in time, useful in avoiding more serious technical problems. Second, car owners were obliged to monitor their cars to avoid any accidents that might occur due to technical problems. Even if they employed a chauffeur, they were still legally obliged to monitor their car's condition, as well as the chauffeur's driving habits. Thus, employers needed the appropriate technical knowledge and experience to supervise their chauffeurs, in particular if they doubted their employees' trustworthiness (Anonymous 1926).

From 1928 onward, the *Allgemeine Automobil-Zeitung* carried a section called "letter box" (Briefkasten), which regularly addressed legal and technical problems. Until the outbreak of the Second World War, this journal received some fifteen thousand letters, a selection of which was published in each issue. Some reveal concerns about motorists' listening practices. In general, car owners followed the advice given in the driving manuals; they carefully monitored the running of their machines—they "listened while driving." Their descriptions were detailed accounts of the specific driving conditions under which a suspicious noise occurred. To illustrate their cars' audible idiosyncrasies, drivers used a wide range of adjectives: Their cars were sobbing, whining, rumbling, as well as stuttering, hammering, knocking, singing, howling, growling, ticking, hissing, and droning (Briefkasten 1928c). The letter writers additionally referred to other common sounds to depict a specific noise, such as the chirps of a cricket (Briefkasten 1930). Other letters described the frequency and pitch of the noise (Briefkasten 1929a). In addition, they listed tests and repairs they had already carried out and claimed their own expertise by referring to their personal experience: "I have been a self-driver for twenty-two years now, and I know a lot about engine designs, but this time I am helpless," one driver grumbled (Briefkasten 1928d).

Often the editors had difficulty making sense of the written accounts because motorists, despite the attempts to codify car sounds in handbooks and journals, shared no standardized vocabulary to describe their auditory experiences. A knocking, for example, could indicate different malfunctions: spontaneous ignitions as well as worn-out piston bearings. In difficult cases the editors gave general suggestions or described diagnostic strategies to narrow down the range of possible faults (Briefkasten 1929b). Sometimes the communication failed completely: A "hot noise," as one reader wrote, made no sense (Briefkasten 1938). To stress their own expertise, the editors stated that experienced mechanics needed more than ten years to develop a "trained ear," allowing them to trace any noise to its exact source (Briefkasten 1933). Despite this emphasis, the Briefkasten advisers always treated the readers as technically competent and believed them capable of major repairs such as cleaning the oil-carbon deposit in the combustion chamber (Briefkasten 1928a).

To summarize, motorists in the 1920s developed two intertwined listening practices. First, they monitored their cars and "listened while driving" to detect technical problems in time. Second, they acquired the technique of diagnostic listening. A 1926 manual suggested that the experienced driver and the efficient chauffeur could both be recognized by their diagnostic abilities (Parzer-Mühlbacher 1926, 353). Clearly, motorists were oriented toward the chauffeur's techniques of listening, which, in the case of middle-class automobilists, was mediated and reflected in written instructions (Bourdieu 1990, 74). This explains some of the problems they encountered while making sense of what they had heard. However, it was a technological necessity for self-drivers to learn to recognize audible signs of malfunctions. Motorists embodied both listening practices as *techniques du corps*; they engraved these techniques into their habitus. Displaying these techniques became a means of social orientation: They considered car repair as matching their upper-middle-class social standing. Car repair became a bourgeois cultural technique: While car driving in Germany was still an exclusive and expensive occupation, the associated technical expertise was a distinctive sign and a sign of distinction (Wetterauer 2007, 155–66). Furthermore, motorists who acquired diagnostic capabilities could compensate for their distrust of chauffeurs and auto mechanics.

Listening to Automobiles: The Mechanic's Experience

Despite great expectations after the war, mass motorization did not take off in Germany in the interwar period. The number of passenger cars rose from 60,876 in 1914 to 216,300 in 1925. At that same time, there were 17,726,507 passenger cars in the United States; 778,211 in the United Kingdom; and 573,397 in France (Ruppel 1927, 10, 15).[5] Still, the car-repair business saw strong growth in Germany, with some 20,000 repair shops in business in 1929 (Reparatur-Werkstatt 1929, 1–2). This figure included a huge number of workshops run by blacksmiths, tinsmiths, and fitters, who repaired cars on the side, which is why Kevin Borg appropriately refers to this group of auto mechanics as ad-hoc mechanics (2007, 31–52).

In contrast to other trades in Germany, the auto mechanics failed to be legally regulated, and it was regarded as socially legitimate for chauffeurs, retired army drivers, or craftsmen from a wide range of trades to repair cars. The blurred boundaries of this field are reflected in journals such as *Auto-Technik*, which, in the first postwar decade, catered to car dealers, garage owners, mechanics, and automotive engineers. Renamed *Automobiltechnische Zeitschrift* in 1929, it became the landmark of the German automotive engineering profession. In 1928 a monothematic journal, *Die Reparatur-Werkstatt*, renamed *Krafthand* in 1930, was initiated to support the ad-hoc mechanics, as well as specialized auto mechanics, with news from the trade, articles on automotive technology, and repair tips.

Journals for German auto mechanics frequently ran articles on car-related sounds, noises, and listening. A special section in *Auto-Technik* titled "For the Repair Shop," for example, gave advice on how to get rid of minor noises, such as the rattling of the brake linkage. More in-depth articles explained the technological background of new noiseless chain drives and engine knock (Ostwald 1921, 1922). In this context, authors discussed the knock resistance of antiknock fuels. They described engine knock as starting with occasional plinking and increasing up to loud and frequent detonations and claimed that, for the time being, in the absence of measuring devices, the best instrument to determine the antiknock quality of fuels remained a well-trained ear, mainly because of its great sensitivity to differences in frequency (Enoch 1928). Interestingly, this ties in with Marcel Mauss's notion of the body as the first and most natural technical means (1936) and underscores the crucial significance of listening in the field of auto mechanics.

As with motorists, mechanics relied heavily on their own listening skills to diagnose technical malfunctions. The article "How to Diagnose Malfunctions of Passenger Cars" explained different noises, their technical sources, and ways to repair them (S. 1928). Another contribution distinguished a multitude of abnormal sounds. The author categorized noises emanating from the engine, the drive train, the body, and the chassis. He gave details on how to diagnose every sound, he described particular driving conditions typical for this sound, and he advised the disassembly of certain components to rule out other possible sound sources. As one anonymous source put it: "For diagnosing engine sounds a very fine sense of hearing is required indeed, especially when the sounds are very faint and if several sounds from different sources have to be distinguished simultaneously, which is often the case" (Anonymous 1932, 81). The author elaborated that experience was essential for "diagnostic listening" and stated that a lack of practice could not be compensated for by using a hearing device: "Those who are unaccustomed to the use of a wooden rod or a stethoscope, which are both put against the engine from the outside, will easily be misled by the effect of resonance" (Anonymous 1932, 81).

Other authors were much more optimistic that hearing devices could assist the auto mechanic—especially when it was important to focus on definite spots. *Auto-Anzeiger*, a trade journal not distributed to ordinary motorists, presented a special stethoscope with two sensors: the Tektoskop and the Tektophon. This type of construction should enable the mechanic to examine two engine spots at once, thus allowing him to compare two sounds in great detail (Anonymous 1929b). Labeled "the ideal troubleshooter," the Meccano-Stethoskop was described as a brilliant tool: With the help of the stethoscope one could "clearly observe the processes inside the engine" and save the time by not having to disassemble the engine when diagnosing problems (Anonymous 1930). The Meccano-Stethoskop provides another indication of the general importance of listening in the auto mechanics trade. The claim that it could help to speed up the diagnosis appealed

Figure 3.2 Photograph illustrating the use of a Tektoskop: Anon. "Wo entsteht das Geräusch?" *Auto-Anzeiger* 4(40) (1929b): 2–3.

to workshops, as customers expected a prompt and exact estimate. However, because of the difficulty of making a precise sound diagnosis, *Auto-Technik* warned the auto mechanics among its readership to give only nonbinding estimates (Walkenhorst 1926).

Auto-Technik's warning about making estimates indicates two crucial intersections of the practice of diagnostic listening: mapping and verbalization. The notion of *sound mapping* comes from the field of psychoacoustics. It refers to the linking of sounds to the information they represent (Fricke 2009, 55–56), and it is a useful metaphor for grasping the practice of diagnostic listening as the association of particular car sounds with specific malfunctions. The concept of mapping helps to explain why stethoscopes and other hearing devices alone were of little help, as it is not sufficient simply to amplify sounds; one must assign meaning to them as well. But how would auto mechanics be able to learn to map sounds? Was it largely a matter of theoretical knowledge? Or was bodily knowledge the key to diagnostic listening?

As in the case of motorists' listening practice, the literature they had available agreed on the precondition of theoretical knowledge and the status of practical experience (Anonymous 1928; O. Winkler 1928). Tables, lists, and fault trees were published to foster methodological investigation of audible defects (Fischer 1927). Regular sections in auto mechanics trade journals, such as "From the Workshop Practice" (Auto-Technik) and "Do You Already Know?" (Die Reparatur-Werkstatt), emphasized the significance of practical knowledge. Regardless of whether the articles had a more theoretical or practical orientation, however, both journals struggled to put the audible indications of different malfunctions into words even though such codification of car sounds was a prerequisite for the written transmission of the diagnostic listening skill (chapter 4). Engine knock is a good example to illustrate this. An article specifically devoted to the hammering and knocking of the engine described engine knock as a hammering sound, whereas only the audible malfunction of a piston bearing or rod was called knocking (Anonymous 1919). Another 1932 article distinguished no fewer than seven types of knocking, including "metallic knocking," "high knocking," "damped knocking," and a "muffled clang" (Anonymous 1932). Apparently, a standardized set of subtly nuanced terms to verbalize the audible characteristics of malfunctions was not available. Nonetheless, how should the mechanic in his everyday practice know exactly what this author's variously described knocking sounds referred to? Similarly, it was crucial to distinguish between these sounds, as the "metallic" one was just a nuisance, but the "high" one indicated a serious problem.

This raised the question of whether written advice on how to listen to malfunctions was of any use to auto mechanics. Despite his own work, Eugen Mayer-Sidd (1931b), a regular contributor to *Krafthand*, questioned the use of written repair accounts: "[I]t is exceptionally difficult to give someone else a detailed and graspable description of a technical work or method that enables him to do it himself later on." Especially with regard to malfunctions he was very pessimistic: "[I]t is even more difficult to describe a malfunction so comprehensibly that someone else, without physical inspection, may give the appropriate advice with certainty on how to repair it" (Mayer-Sidd 1931b). The unachieved codification of car sounds and the general difficulties of giving written advice underscore the relevance of the tacit dimension of car repair knowledge (Polanyi 1958, 69–245). If written accounts largely failed to convey "diagnostic listening" skills, tacit knowledge could be passed on only through apprenticeship.

It is safe to argue that the motorists' techniques of listening and the auto mechanics' practice of diagnostic listening were in fact quite similar. This is perhaps hardly surprising because specialized auto mechanics shared their knowledge of their field with many inexperienced craftsmen from other trades, former chauffeurs, and other ad-hoc mechanics. Until the mid-1920s, they competed as more or less competent and legitimate rivals, and all of them might gain advanced expertise in diagnosing and repairing passenger cars. If theoretical knowledge generally served as a necessary basis, only through practical experience could one really develop diagnostic skills. All of these individuals recognized car sounds as an indispensable source of information on the engine's condition and as a means of locating particular malfunctions. The ability to diagnose by listening had to be incorporated as an embodied technique, and the mechanics' literature emphasized the considerable effort this took. Only a practically trained auto mechanic could diagnose a malfunction by listening to it, but even then partial disassembling was still regarded as advisable.

The "Repair Chaos" and the Formation of the Auto Mechanics Dispositive

In December 1926 the German Chambers of Industry and Commerce published an assessment of the situation in the field of auto mechanics. The study summarized a series of complaints about excessive prices for spare parts and repair work. In response, the car dealer association denied these shortcomings, but the discussion about cheap and reliable repairs did not die down (Anonymous 1926). Several letters published in the *Allgemeine Automobil-Zeitung* articulated the rising distrust and dissatisfaction among car owners (Briefkasten 1928a, 1928b).

In 1928 *Auto-Technik* published an editorial titled "The Great Repair Misery." Apparently, its author considered it his duty to raise this delicate issue, and he even felt obliged to criticize parties that paid for ads in this same journal. He leveled criticism mainly at auto mechanics and their expertise and trustworthiness, including those employed by car dealers and manufacturers, as well as independent repair shops. During the warranty period, for example, the significance of audible technical problems was often played down if not altogether denied by the manufacturer's mechanics: "[T]hey try to persuade the customer that the abnormal sound, which indicates an upcoming problem, is of no significance—'this does not mean anything,' 'this is just an imperfection' " (Loewe 1928, 11). In other cases motorists would bring in a car that obviously had an audible problem, but after the repair the difficulty continued. The author of the editorial interpreted this as a sign of the lack of the necessary expertise. In sum, he argued that the repair misery posed a serious threat to the whole automobile system because dissatisfied motorists might simply abandon their cars.

A corresponding editorial in the *Allgemeine Automobil-Zeitung* articulated "the outcry of the automobilist" and asked readers to propose a way out of the "repair chaos." As evidence, the article mentioned the numerous complaints published in the journal's "letter box." Motorists, it claimed, preferred to carry out major repairs themselves because of their distrust of auto mechanics (Anonymous 1931b). Coming from the other end, advocates of the auto repair business saw the origin of the "repair chaos" in the unregulated access to the trade: "The blacksmith, the bicycle or sewing machine mechanic, the fitter, they all have to learn their trade for four years, but anybody who learned to handle a file and followed a six-week course is able to repair the complicated and valuable engine of a car flawlessly?!" (Testor 1931). To put things right, they proposed the establishment of an independent auto mechanics trade together with mandatory membership in a guild (Anonymous 1931a).

In this respect it is relevant to briefly discuss the overall German professional trade system. Importantly, the laws governing trade and industry in Germany had been amended in 1897 and 1908. From then on, the right to enter a trade was restricted. To practice a trade, a three- to four-year formal apprenticeship was required. The passing of a final exam, which earned one a journeyman's certificate (Gesellenbrief), served as one's entry ticket to a particular trade. After accumulating experience for three to five years, depending on regional customs, journeymen achieved the right to take a second exam to obtain a master craftsman's certificate (Meisterbrief). Only with this second certificate did one have the right to train apprentices. Moreover, every workshop had to become a member of a trade guild. Because this was mandatory, the system led to a high degree of organization in the trades (Greinert 1994). The journeyman's certificate was not only a form of institutionalized cultural capital that regulated and restricted access to one's field; it also allowed German craftsmen to take up a position of trust, a kind of symbolic capital, which structured their social relationships with nonmembers in the field regardless of individual skills. German craftsmen thus cultivated a particular preindustrial mentality, a habitus grounded in a long and painstaking apprenticeship, whereby ultimately a sense of "master craftsman's honor" and the ideology of "high-quality workmanship" served as guiding values (Holtwick 1999; Sennett 2008).

The craftsman's position of trust was also part of the discussion in the auto mechanics' community, sparked as it was by complaints about the "repair chaos."

Thereby the "physician" played a major role. It was claimed that during their apprenticeship, future auto mechanics would develop—under the strict guidance of a master craftsman—a distinct sense of responsibility. Just like a physician, the mechanic would embody the necessary expertise, as well as commitment to good practice. The advocates of an independent trade of auto mechanics argued that limited repair courses, as offered in the United States (Borg 2007), were insufficient to instill this particular sense of responsibility (Schiff 1930).

The auto mechanics' fight for an independent profession was further embedded in the political action of representatives of the trade organizations to defend their members' social standing and economic position. They argued that the "high-quality workmanship" guaranteed by the craft trades was a public good that deserved special protection through strict regulation of the field. Andrew Abbott has described how such public claims of jurisdiction are often used to enforce the specific interests of a profession (Abbott 1988, 59–85). He has also emphasized the role of legal regulations, "which can confer formal control of work" (Abbott 1988, 59). Despite several initiatives, the craft trades failed to put through their program during the Weimar period, but the demands were taken up by the National Socialists. After they came to power, they enacted new legislation in 1934 that introduced obligatory guilds for all recognized trades. Furthermore, in 1935 an amendment abolished the right to practice a trade, establishing the master craftsman's certificate (grosser Befähigungsnachweis) as a precondition for starting a workshop (H. Winkler 1972, 184–85; Saldern 1979). The first to profit from the new legislative framework were auto mechanics. This is hardly surprising as they maintained good relations with the National Socialist Motor Corps (NSKK). Furthermore, the discourse about the future of the auto mechanics trade meshed well with the National Socialists' plans for mass motorization in Germany (Zeller 2007). Friedrich Stupp, member of the Nazi Party and NSKK-Obersturmführer, was elected first president of the Berlin auto mechanics guild in September 1933, and only a few years later 80 percent of the auto mechanics master craftsmen were members of the NSKK (Hochstetter 2005, 115–16).

In response to the *urgency* of the repair crisis, different groups of actors engaged in reorganizing the auto mechanics trade.[6] This led to the formation of the car mechanics guild, a new corporate body with the symbolic power to approve car repair experts (Bourdieu 1989, 23). The newly established system of obligatory trade guilds, apprenticeships, and workshops led by master craftsmen can be understood as the formation of an auto mechanics dispositive—as the readjustment of power relations within the field of auto mechanics. The formation was determined by the structure of the German system of trades: its common set of legal restrictions, institutions, cultural values, social relations, and practices. The societal conditioning through the tradition of this system resolved the crisis of confidence between motorists and auto mechanics by guaranteeing the trustworthiness of the latter. Henceforth, auto mechanics received theoretical training in vocational schools, and they embodied the necessary techniques, as well as a distinct habitus, during their four-year apprenticeship. The new cultural capital of the journeyman's certificate not only regulated access to the trade but also equipped auto mechanics with an unquestionable expertise: The truth value of that expertise was independent of the individual skills of a single mechanic as it was guaranteed by the "honor" of the craft professions. Foucault has described how the encounters between individual actors become structured by the power relation inscribed in the dispositive (1976, 120).

In comparison, in the United States the status of auto mechanics as trustworthy technical experts remained contested due to the unregulated access to the trade. As a possible solution, the auto mechanics sought to delegate diagnostic authority to instruments and measuring devices—but failed. Also, the introduction of a flat-rate system did not solve the problem of distrust because mechanics started to work hastily and thus did a shoddy job under the pressure of standardized work times (McIntyre 2000, 292). In contrast to their German counterparts, American auto mechanics have suffered an endless crisis of confidence (Borg 2007).

The Ideal of the Silent Car

Before describing the differentiation of listening practices attributable to the formation of the auto mechanics dispositive, I first consider two strands of discourse on the "silent car" that are closely connected to the contemporary automotive technology: noise as an audible sign of technological inefficiency and silence as a cultural expression of modernity and distinction.

After 1900, engineers showed increasing interest in reducing the noise of production equipment. As Karin Bijsterveld has shown, "noisy machinery" became an "indication of mechanical inefficiency" (Bijsterveld 2006, 328–29). This special discourse entered the automotive engineering community, too: Research on noiseless car components, especially noiseless sprocket chains, pertained to theoretical concerns, as well as the production process. Chain drives had the advantage of being quieter than spur wheels. Corresponding articles in trade journals described sprocket chains as efficient and modern (M. 1919).

Advertisements for "silent car components" in consumer magazines clearly show the circulation and significance of the special discourse beyond the field of engineering. Manufacturers informed automobilists that *Fichtel & Sachs* roller bearings "reduced fuel and lubricant consumption" and "guaranteed the silent run of the engine"—or that *ZF* gearwheels were the "wheels of choice for the silent gear box" (Fichtel & Sachs 1928; ZF 1930). Science journalist Walter Ostwald (1921, 11) commented on the "silent car" craze: "Noiselessness is more than a fashion. A noise always points to a waste of energy. Therefore, the longing for noiselessness, if fashionable, is at least a good and useful fashion."

Figure 3.3 Detail from advertisement for silent-running pistons: "Running smoothness matters!" ***Allgemeine Automobil-Zeitung*** **41(49) (1940): 839.**

Furthermore, the advertising of car manufacturers promoted noiselessness as a distinguishing mark of modern and elegant passenger cars. The Citroën-Phaeton stood out with its "noiseless and flexible working engine" (Citroën 1928), Brennabor's Juwel had an "inaudible engine full of adaptable power" (Brennabor 1929), and a positive feature of the Primus was its noiselessness, achieved by the "generous use of rubber for sound insulation" (Adler 1932). Articles offered advice to motorists and auto mechanics on how to get rid of squeaks and squawks (Ostwald 1923). These little noises were not just a nuisance but also an embarrassment to the car owner. An ad for suspension covers came with an illustration of a young driver whose car's suspensions screeched so loudly that his spouse, as the accompanying poem clarified, preferred to take the train back home (Vogelsang 1937).

"Noiselessness was [not only] the mark of a good car" (K. 1927) but also a sign of good manners and distinction as contemporary antinoise advocates stressed (Bijsterveld 2008). Significantly, the propaganda of noise-abatement campaigns, class habitus, and engineering discourse converged in this set of values. Mechanical efficiency and social distinction urged car owners to listen to their cars even more carefully and to complain about each tiny noise. However, as we will see later, automobilists who insisted on the promised "silent run" of their machines were often denounced as "noise fanatics." In the eyes of the mechanics these drivers took the advertisements far too seriously.

Differentiation of Listening Practices

From 1934 onward, the expertise of German auto mechanics was basically guaranteed through their obligatory four-year apprenticeship. During this period the future mechanic worked in very close contact with a master craftsman or a journeyman to learn by observing and imitating. To ensure the quality of their bond, a master craftsman was not allowed to train more than two apprentices, while a journeyman could supervise only one apprentice. During the first six months, apprentices only lent a hand to their mentors, after which they continued with simple tasks like lubricating or checking the tire pressure. In the second year, they learned to diagnose uncomplicated malfunctions, but they were not allowed to carry out maintenance work themselves until the third year of their apprenticeship. In addition, apprentices had to learn a wide range of metalworking skills, such as filing, drilling, milling, turning, and welding. This set of manual skills emphasizes the status of *bodily knowledge* in the auto mechanics trade (Kümmet 1941; Zogbaum 1937).

Training manuals did not explicitly mention specific auto mechanics' skills, such as diagnostic listening. Apparently, such skills were considered part of the practical learning process—based on imitation of the master craftsman. Inherent in the notion of the apprenticeship system was the substitution of codified knowledge through the practical skills apprentices would pick up in their years of practice: They learned by doing. Douglas Harper has shown how car repair knowledge is informally passed on in this manner (1987, 24–31). It involves a practical repertoire of imitated actions rather than models, a process of acquisition Bourdieu calls *mimeticism*: "Bodily hexis speaks directly to motor function, in the form of a pattern of postures that is both individual and systematic, being bound up with a whole system of objects, and charged with special meanings and values" (Bourdieu 1990, 74). In a similar way, Harry Collins speaks of the "unconscious emulation" of uncognizable knowledge (Collins 2001, 72).

During the formative period of the auto mechanics dispositive, individuals questioned the position of car owners as knowledgeable amateur mechanics. Under the programmatic title "Hands Off," in May 1933, the readers of the *Allgemeine Automobil-Zeitung* were urgently requested not to repair their cars themselves because they lacked the necessary abilities. Furthermore, drivers were instructed to ignore little noises that are just a nuisance. Only if they heard a "threatening noise" (Gefahrengeräusch) should they bring their car to a specialized repair shop (Anonymous 1933). Both articles supported the professionalization of the auto mechanics trade. As such they glorified the "high-quality workmanship" and devalued the knowledge of all nonexpert mechanics.

Concomitantly with the stabilization of the dispositive through the 1934 legislation, the new picture of the lay motorist became clearer. As one author put it,

"The rising spread of automobiles goes hand in hand with the diminishing number of knowledgeable motorists" (Rdl. 1936). One cause was seen in the technical level reached in the automotive industry, which made cars more and more reliable, but more important, it was assumed that most people were just "terribly clumsy" (Rdl. 1936). The shifting balance between motorists and auto mechanics can further be observed in the rhetorical shift from *reparieren* (repairing) to *basteln* (tinkering). This semantic turn excluded automobilists from the discourse on automobile repairs. Furthermore, it sustained the new boundary between expert mechanics and lay drivers along the line of technological knowledge (Franz 2005). Automobilists' journals no longer published articles on how to acquire diagnostic skills or provided real repair instructions. Instead, they advised the automobile enthusiasts on how to tinker with their cars (Rdl. 1938).

On the other hand, letters that ended up in the *Briefkasten* section suggest that motorists changed their practices only gradually. In the latter part of the 1930s they still listened closely to their cars, sent in their self-diagnoses, and asked for advice on how to tackle the problems. However, in contrast to the situation several years before, when readers were advised to "listen in" themselves, they were now urged to consult expert mechanics. In addition, the automobile club journal published lists of reliable workshops and informed its readers that most of their trouble with mechanics could be traced back to the readers themselves: They were blamed for having unrealistic ideas about prices and repair times (Dill 1936). Another article argued that motorists should learn not to bother auto mechanics with their self-diagnoses: "If your car is more important to you than your rhetorical exercises, let the master craftsman do the job" (W. 1938).

In the late 1930s an interesting discussion arose about the right of motorists to observe the repair process inside the workshop. The *Allgemeine Automobil-Zeitung* argued that, in particular cases, such as the repair of a rare make, a driver could potentially provide help to the mechanic because of his own special knowledge of the car (Peter 1938). In a direct response, *Krafthand*, the trade's official journal, dismissed this as an offensive intrusion into the mechanics' backstage domain—and as a blunt articulation of distrust (Goffman 1959, 111–21). The journal advised its readers to follow a strict policy of trust: "The motorist who distrusts a workshop should look for another one. One does not go to a physician whom one does not trust, either" (Anonymous 1940). The car mechanics' exclusive access to the actual workshop became part of the dispositive. The new official master craftsman's handbook proposed a ground plan for garages that separated motorists and mechanics (Kümmet 1939, 250): The showroom and a special waiting room for customers served as the front domain, where the master craftsman met his clients, whereas the backstage of the workshop was for mechanics only. In this way the establishment of a trusting relationship between mechanics and motorists simultaneously gave rise to new barriers.

The discourse on lay motorists illustrates, moreover, how the trade used this discussion to demarcate its new boundaries. The cartoon series "Kunibald, the Smart Customer" portrayed the "dull" lay driver. Kunibald is a motorist who tries to diagnose and repair his car on his own, and he often disregards his mechanic's advice, but in the end he always fails and must admit that he should have taken his car to the garage (Jonny 1937). Another fictional story describes a neurotic motorist who always does everything himself, but, like Kunibald, he finally ruins his car and needs the help of a professional mechanic (Windecker 1937). I suggest a reading of these narratives as stereotypes in which the emerging auto mechanic's habitus gradually became visible. They exemplified the "social viewpoint" of an average auto mechanic, and at the same time the "shop talk" structured future encounters between mechanics and motorists by suggesting the proper front-stage behavior (Goffman 1959, 175–76).

As mentioned earlier, the struggle for listening expertise culminated in complaints from mechanics about "noise fanatics": "As you know, there are so-called noise fanatics who can drive a busy master craftsman crazy with their accounts of, sometimes real and sometimes imagined, noises they heard" (Anonymous 1938). Another article elaborated: "He is often bothered by noises that exist only in his imagination. The fact that he is always sure where the noise comes from does not make him any more likeable because mostly he is wrong, thus leading the craftsman down the wrong road" (Anonymous 1939). With the topos of the "noise fanatic," auto mechanics reclaimed the practice of diagnostic listening as their exclusive domain: "When looking for a noise source, never ever let yourself be influenced by the customer" (Anonymous 1939). By denouncing them as overanxious and unknowing, mechanics deprived motorists of their listening expertise. This was not a matter of whether or not they actually lacked such expertise; rather, it followed from their newly gained status as nonexperts. As a result, the motorists' practice of diagnostic listening was reconfigured as illegitimate.

At the same time, the narrations about "noise fanatics" demonstrate why ordinary drivers, though expelled from the realm of diagnostic listening, should nevertheless listen to their cars. One article noted: "The expert's trained ear knows the sound of the engine; he distinguishes between healthy sounds and noises that indicate an upcoming problem. In contrast, the layman is often anxious about harmless noises. But it is still better for him to consult an expert in vain than to disregard noises until the engine has a serious problem" (Anonymous 1936). The latter type of driver was also called the "noise phlegmatic" (chapter 2). Another author explained: "He is not worried when his car rattles and squeaks at all ends and when the chassis together with the engine plays a free concert. . . . Because the noise phlegmatic has a tin ear, this symphony does not disturb him; he will not do anything until his heap breaks down. We find these people just as disagreeable" (Anonymous 1939). Following this rhetorical confrontation, motorists should on no account "listen to diagnose" malfunctions, but they should always "listen while driving" to recognize technical problems in time. In other words, the "sonic contradiction" urged automobilists to listen carefully but not to listen thoroughly.

The ban of private car use in Germany in February 1942 temporarily suspended the struggle between motorists and car mechanics—at a point where the latter had gained the upper hand. After the war, journals for automobilists, such as

ADAC Motorwelt, predecessor of the *Allgemeine Automobil-Zeitung*, and *Auto Revue*, revived some of their old features and columns. For example, the "travel box" section advised readers on their travel plans (Anonymous 1950a). Nevertheless, the technology-oriented *Briefkasten* section was not taken up again in the 1950s. Nor did these journals continue to publish detailed instructions on how to diagnose and repair malfunctions as they had done in the interwar period. Apparently they no longer recognized readers as interested and competent in these matters: Regardless of individual listening techniques, the motorists' practice of diagnostic listening was apparently dismissed as irrelevant or inappropriate. On the other hand, trade journals for auto mechanics continued to publish articles on how to locate audible malfunctions. This underscored the role of listening as a diagnostic technique (Anonymous 1955). As indicated in the introduction, trade advocates also continued their polemics against ordinary motorists, but they were no longer the subject of editorials. Instead, these narrations were taken up in cartoons and small columns on "real" and instructive incidents. Thus, they became part of a folkloristic reassurance of the auto mechanics' own exclusive expertise. The craft professions habitus barred motorists from car diagnostics and contested even their monitory listening skills—without denying them the latter practice.

Conclusion

As I argue in this chapter, listening to automobiles was a common technique among motorists and auto mechanics during the 1920s. At the end of the decade, however, the repair crisis posed a threat to the growth of the automobile system, and in response to this urgency different groups argued in favor of a rigid organization of the trade. This strategic movement led to the formation of an auto mechanics dispositive that altered the balance of power between motorists and auto mechanics. The new social logic devalued the knowledge of ordinary automobilists, which can be read as an adjustment to the new dispositive (Foucault 1978, 121). In the new situation, motorists listened while driving to recognize technical problems in time, and mechanics listened to thoroughly diagnose these malfunctions. The genealogy of listening practices shows that the claim for expertise in diagnostic listening was merely certified by the auto mechanics' position of power (Foucault 1976, 120). So, rather than personal skills, their habitus decided on the truth value of their knowledge, and concomitantly the motorists' habitus was altered by the submission to the *doxa* of the field, which limited their scope of action accordingly.

NOTES

1 My translation. I have also translated all of the sources cited in this chapter that were not available in English.

2 For the distinction between different modes of listening (monitory listening, diagnostic listening, exploratory and synthetic listening), see Bijsterveld (2009).

3 Bourdieu uses the term *hexis* instead of *technique du corps*.

4 In the early postwar years, for example, the same articles appeared in periodicals for mechanical engineers and ordinary motorists (Ostwald 1921).

5 This amounted to 290 persons per car in Germany, 6 per car in the United States, 60 in the UK, and 71 in France.

6 For Foucault a societal crisis, *urgency*, is the trigger for the formation of a dispositive, see Foucault 1978, 120; Bührmann and Schneider 2008, 53, 61.

REFERENCES

Abbott, Andrew. *The System of Professions: An Essay on the Division of Expert Labor*. Chicago: University of Chicago Press, 1988.

Adler. "Advertisement: Primus and Trumpf." *Motor-Kritik* 12(8) (1932).

Anonymous. "Das Hämmern und Klopfen des Motors." *Allgemeine Automobil-Zeitung* 20(20) (1919): 17–19.

Anonymous. "Automobilhandel, Automobilreparaturwerkstätten, Garagengewerbe, und Publikum." *Allgemeine Automobil-Zeitung* 27(51) (1926): 17–21.

Anonymous. "Braucht man für Automobil-Reparaturen theoretische Kenntnisse?" *Die Reparaturwerkstatt* 1(3) (1928): 26–27; 1(4) (1928): 41–42.

Anonymous. "Anzeige: Abhorch-Apparat Auto-Doktor." *Allgemeine Automobil-Zeitung* 30(23) (1929a): 7.

Anonymous. "Wo entsteht das Geräusch?" *Auto-Anzeiger* 4(40) (1929b): 2–3.

Anonymous. "Ein idealer Störungssucher." *Das Kraftfahrzeug-Handwerk* 3(4) (1930): 88.

Anonymous. "Die andere Seite." *Allgemeine Automobil-Zeitung* 32(10) (1931a): 17–18.

Anonymous. "Reparaturen-Chaos, und wer hilft heraus?" *Allgemeine Automobil-Zeitung* 32(2) (1931b): 7.

Anonymous. "Abnorme Fahrgeräusche und ihre Ursachen." *Das Kraftfahrzeug-Handwerk* 5(6) (1932): 81–82.

Anonymous. "Hände weg." *Allgemeine Automobil-Zeitung* 34(14) (1933): 5.

Anonymous. "Der Kunde soll selbst aus- und einbauen." *Das Kraftfahrzeug-Handwerk* 8(19) (1935): 315.

Anonymous. "Der Motor hat einen 'Ton.' " *Das Kraftfahrzeug-Handwerk* 9(11) (1936): 330.

Anonymous. "Zu Ende denken!" *Krafthand* 11(15) (1938): 532.

Anonymous. "Es rattert und quietscht." *Krafthand* 12(26) (1939): 779–80.

Anonymous. "Aufpasser." *Krafthand* 13(18) (1940): 342.

Anonymous. "Der Reiseonkel ist wieder da!" *ADAC-Motorwelt* 3(2) (1950a): 14.

Anonymous. "Ja, gibt's denn dös a." *Krafthand* 23(15) (1950b): 353.

Anonymous. "Auf der Suche nach Geräuschen." *Krafthand* 28(12) (1955): 405.

Bijsterveld, Karin. "Listening to Machines: Industrial Noise, Hearing Loss, and the Cultural Meanings of Sound." *Interdisciplinary Science Reviews* 31(4) (2006): 323–37.

———. *Mechanical Sound: Technology, Culture, and Public Problems of Noise in the Twentieth Century*. Cambridge, Mass.: MIT Press, 2008.

———. "*Sonic Skills: Sound and Listening in the Development of Science, Engineering, and Medicine, 1920s–Now*." Unpublished proposal for the NWO-VICI competition in the Netherlands, 2009.

Borg, Kevin L. *Auto Mechanics: Technology and Expertise in Twentieth-Century America.* Baltimore: Johns Hopkins University Press, 2007.
Bourdieu, Pierre. *The Logic of Practice.* Cambridge: Polity, 1990.
———. *Outline of a Theory of Practice.* New York: Cambridge University Press, 1977.
———. *Pascalian Meditations.* Stanford: Stanford University Press, 1999.
———. "Social Space and Symbolic Power." *Sociological Theory* 7(1) (1989): 14–25.
Brennabor. "Advertisement: Juwel." *Allgemeine Automobil-Zeitung* 30(43) (1929): 31.
Briefkasten. "Nr. 36. Tackende Geräusche im Motor." *Allgemeine Automobil-Zeitung* 29(7) (1928a): 31.
———. "Nr. 45. Der Motor klopft." *Allgemeine Automobil-Zeitung* 29(19) (1928b): 38.
———. "Nr. 93. Schluchzende Geräusche." *Allgemeine Automobil-Zeitung* 29(19) (1928c): 38.
———. "Nr. 155. Das laute Getriebe." *Allgemeine Automobil-Zeitung* 29(8) (1928d): 29.
———. "Nr. 269. Heulender Wagen." *Allgemeine Automobil-Zeitung* 30(18) (1929a): 30.
———. "Nr. 385. Rätselraten." *Allgemeine Automobil-Zeitung* 30(43) (1929b): 28.
———. "Nr. 808. Es zirpt." *Allgemeine Automobil-Zeitung* 31(39) (1930): 22.
———. "Nr. 2673. Der Neuling als Sachverständiger." *Allgemeine Automobil-Zeitung* 34(39) (1933): 18.
———. "Nr. 12318. Ein merklich heißes Geräusch auf dem Gaspedal." *Allgemeine Automobil-Zeitung* 39 (1938): 999.
Bührmann, Andrea D., and Werner Schneider. *Vom Diskurs zum Dispositiv.* Bielefeld, Germany: Transcript, 2008.
Citroën. "Advertisement: Citroën-Phaeton." *Allgemeine Automobil-Zeitung* 29(19) (1928): 5.
Collins, Harry. "Tacit Knowledge, Trust, and the Q of Sapphire." *Social Studies of Science* 31(1) (2001): 71–85.
Dietl, Stephan. *Die Fahrtechnik. Ein Hilfs- und Nachschlagewerk mit dem Kraftwagen richtig umgehen und ihn sicher lenken zu können.* Berlin: Schmidt, 1931.
Dill. "Der Aerger mit dem Abholtermin." *Allgemeine Automobil-Zeitung* 37(35) (1936): 12.
Enoch, O. "Die Klopfgrenze." *Auto-Technik* 17(14) (1928): 10–11.
Fichtel & Sachs. "Advertisement: Roller Bearings." *Motor* 16(11) (1928): 5.
Fischer, Joachim. "Wie erkennt man die Ursachen der Motorpannen?" *Der Dienst am Kunden* 1(13) (1927): 99–101.
Foucault, Michel. *Dispositive der Macht.* Berlin: Merve, 1978.
———. *Mikrophysik der Macht.* Berlin: Merve, 1976.
Franz, Kathleen. *Tinkering: Consumers Reinvent the Early Automobile.* Philadelphia: University of Pennsylvania Press, 2005.
Fricke, Nicola. "Warn- und Alarmsounds im Automobil." In *Funktionale Klänge,* ed. Georg Spehr, 47–64. Bielefeld, Germany: Transcript, 2009.
Goffman, Erving. *The Presentation of the Self in Everyday Life.* Garden City, N.Y.: Doubleday Anchor, 1959.
Greinert, Wolf-Dietrich. *The "German System" of Vocational Education: History, Organization, Prospects.* Baden-Baden: Nomos, 1994.
Hacker, Oskar Hans. *Panne unterwegs: Ein Hilfsbuch für Kraftfahrer.* Vienna: Steyrermühl, 1932.
Harper, Douglas. *Working Knowledge: Skill and Community in a Small Shop.* Chicago: University of Chicago Press, 1987.
Hessler, Rudolf. *Der Selbstfahrer: Ein Handbuch zur Führung und Wartung des Kraftwagens.* Leipzig: Hesse and Becker, 1926.

Hochstetter, Dorothee. *Motorisierung und "Volksgemeinschaft."* Munich: Oldenbourg, 2005.

Holtwick, Bernd. " 'Handwerk,' 'Artisanat,' 'Small Business.' Zur Formierung des selbständigen Kleinbürgertums im internationalen Vergleich." *Jahrbuch für Wirtschaftsgeschichte* 1 (1999), 163–81.

Jäger, Siegfried. "Discourse and Knowledge: Theoretical and Methodological Aspects of Critical Discourse and Dispositive Analysis." In *Methods of Critical Discourse Analysis*, ed. Ruth Wodak and Michael Meyer, 32–62. London: Sage, 2001.

Jonny. "Kunibert, der kluge Kunde: Die Achseinstellung." *Das Kraftfahrzeug-Handwerk* 10(1) (1937): 24.

K., O. "Weniger Lärm beim Autofahren." *Allgemeine Automobil-Zeitung* 28(42) (1927): 29.

König, Ad. "Die experimentelle Psychologie im Dienste des Kraftfahrwesens." *Allgemeine Automobil-Zeitung* 20(10) (1919): 11–13; 20(12): 15–17.

Kümmet, Hermann. *Lehrling im Kraftfahrzeughandwerk*. Berlin: Krafthand, 1941.

———. *Meister im Kraftfahrzeughandwerk*. Berlin: Krafthand, 1939.

Küster, Julius. *Chauffeur-Schule: Theoretische Einführung in die Praxis des berufsmäßigen Wagenführens*. Berlin: Schmidt, 1907.

———. *Das Automobil und seine Behandlung*, 7th ed. Berlin: Schmidt, 1919.

Lachmund, Jens. "Making Sense of Sound: Auscultation and Lung Sound Codification in Nineteenth-Century French and German Medicine." *Science, Technology, and Human Values* 24(4) (1999): 419–50.

Ledertheil, Hans. "Vom kleinen Wagen und seinen Zukunfts-Aussichten." *Allgemeine Automobil-Zeitung* 20(24) (1919): 15–17.

Loewe, Adolf Gustav von. "Die große Reparaturmisere." *Auto-Technik* 17(26) (1928): 11–14.

M., B. "Die moderne Gelenkkette und ihr Entwicklungsgang." *Auto-Technik* 9(5) (1919): 9–10.

Martini, Bernhard. *Praktische Chauffeur-Schule*. Berlin: Schmidt, 1922.

———. *Praktische Kraftfahrkunde: Eine Kraftfahrfibel für jedermann über Störungen, Stegreifreparaturen, Werkzeugkunde, Prüfungsfragen und –antworten*. Berlin: Schmidt, 1938.

Mauss, Marcel. "Les techniques du corps." *Journal de Psychologie Normale et Pathologique* 32(3–4) (1936): 271–93.

Mayer-Sidd, Eugen. "Die Auto-Reparaturwerkstätten der kommenden Jahre." *Das Kraftfahrzeug-Handwerk* 4(4) (1931a): 79.

———. "Reparaturen nach schriftlicher Anleitung." *Das Kraftfahrzeug-Handwerk* 4(3) (1931b): 57.

McIntyre, Stephen L. "The Failure of Fordism: Reform of the Automobile Repair Industry, 1913–1940." *Technology and Culture* 41 (2000): 269–99.

Ostwald, Walter. "Autotechnisches Notizbuch: Isolierung der Karosserie vom Rahmen durch Gummibuffer." *Allgemeine Automobil-Zeitung* 24(10/11) (1923): 36.

———. "Geräuschlose Zahnräder." *Auto-Technik* 11(15) (1921): 11; appeared also in *Allgemeine Automobil-Zeitung* 22(31) (1921): 29.

———. "Selbstzündungs-Klopfen." *Auto-Technik* 11(3) (1922): 5–6.

Parzer-Mühlbacher, Alfred. *Das moderne Automobil: Seine Konstruktion und Behandlung*. Berlin: Schmidt, 1926.

Peter. "Beim Reparieren dabeibleiben." *Allgemeine Automobil-Zeitung* 39(7) (1938): 105–106.

Polanyi, Michael. *Personal Knowledge*. London: Routledge and Kegan Paul, 1958.

Rdl. "Am Sonntag wurde gebastelt." *Allgemeine Automobil-Zeitung* 39(38) (1938): 1147–49.

———. "Mehr Freude am Basteln." *Allgemeine Automobil-Zeitung* 37(23) (1936): 18–19.
Reparatur-Werkstatt. "Es geht weiter vorwärts!" *Die Reparatur-Werkstatt* 2(1) (1929): 1–2.
Ruppel, Ernst. *Die Entwicklung der deutschen Personen-Automobil-Industrie und ihre derzeitige Lage*. Berlin: Wittenberge PDM, 1927.
S., M. "Wie findet man Störungsfehler bei Kraftwagen?" *Die Reparatur-Werkstatt* 1(3) (1928): 25–26.
Saldern, Adelheid von. *Mittelstand im "Dritten Reich."* Frankfurt: Campus, 1979.
Schiff, H. "?" *Die Reparatur-Werkstatt* 3(1) (1930): 2–3.
Schmal, Adolf. *Ohne Chauffeur: Ein Handbuch für Besitzer von Automobilen und Motorradfahrer*. Vienna: Beck, 1912.
Sennett, Richard. *The Craftsman*. London: Lane, 2008.
Testor. "Reparaturenchaos—Geburtswehen eines neuen Handwerks." *Allgemeine Automobil-Zeitung* 32(4) (1931): 18.
Vogelsang. "Advertisement: Drevo." *Das Kraftfahrzeug-Handwerk* 10(12) (1937): 399.
W. "Vom Umgang mit Meistern." *Allgemeine Automobil-Zeitung* 39(22) (1938): 695–96.
Walkenhorst, P. "Die Handhabung des Autoreparaturgeschäftes." *Auto-Technik* 15(13) (1926): 17–19; 15(14): 19–21; 15(15): 17–18.
Wetterauer, Andrea. *Lust an der Distanz. Die Kunst der Autoreise in der Frankfurter Zeitung*. Tübingen: Tübinger Vereinigung für Volkskunde, 2007.
Windecker, Carl Otto. "Der Mann, der alles selber machte." *Das Kraftfahrzeug-Handwerk* 10(16) (1937): 496.
Winkler, Heinrich August. *Mittelstand, Demokratie, und Nationalsozialismus*. Cologne: Kiepenheuer and Witsch, 1972.
Winkler, Otto. *Automobil-Reparaturen*. Halle a. S.: Knapp, 1928.
ZF. "Advertisement: Gearwheels." *Allgemeine Automobil-Zeitung* 31(6) (1930): 6.
Zeller, Thomas. *Driving Germany*. New York: Berghahn, 2007.
Zogbaum, Emil. *Unter Motor und Fahrgestell*. Berlin: Krafthand, 1937.

CHAPTER 4

SELLING SOUND: TESTING, DESIGNING, AND MARKETING SOUND IN THE EUROPEAN CAR INDUSTRY

EEFJE CLEOPHAS AND KARIN BIJSTERVELD

Introduction

Since the late 1990s, leading automobile manufacturers have advertised the sonic qualities and interior tranquility of their vehicles with increasing fervor. The new owners of a Chrysler Voyager will find a "whispering stillness" waiting for them. The Jaguar S-type is propelled by a "silent force," while in a Toyota Avensis "even the silence comes standard." Amid the bombardment of noises our ears have to put up with, we will experience "tranquility" in a BMW diesel, Mercedes-Benz, and Volkswagen Passat. Better still in the Ford Focus, everything we "touch" comes with "the sound it is supposed to make."[1]

Clearly, these advertisements express automakers' growing confidence in their ability to enhance the overall quality of car driving with respect to its various auditory dimensions. Indeed, manufacturers have invested considerable time and money in making sure that switches, warning signals, direction indicators, windshield wipers, the opening of car windows, the locking of car doors, or the crackle of the leather upholstery come with the *right* sound. Such sounds, or "target sounds," should express their make's identity and attract particular groups of consumers. Two simple figures demonstrate just how much today's automotive industry invests in its products' sonic characteristics: For instance, BMW employs more than 150 acoustical engineers, and Ford has an acoustical department of 200 employees (Jackson 2003, 106; Van de Weijer 2007, 9; RH, 20).

It appears to be far from easy, however, to define the kind of interior sounds specific groups of consumers prefer. It turns out that many customers, when acting as test subjects, find it hard to articulate their priorities in the area of sound, while experts diverge significantly over the best methods for measuring and evaluating sound. No wonder, then, that a distinct field of research has emerged, one that has close ties with the car industry and aims to find "the holy grail of psycho-acoustics" (Vetter 2004). Vehicle simulators and specific techniques designed to make test subjects express what they find difficult to articulate—how they think about sound—are center stage in the search for a high sonic quality and the most attractive target sounds.

Our chapter focuses on the rise of a new tradition of testing car sound in the European automotive industry in the 1990s. We explore three issues: How do ways of defining the "reality" of sound perception and the differences between expert and lay listeners affect the dynamics of testing in car manufacturing? Why does extensive testing of car sound not automatically result in the design of new target sounds—sounds, such as "sportive" ones, for specific groups of consumers? And where does this increasing significance of sound design in the consumer industry come from in the first place?

We address the first two questions by drawing on sound studies and science and technology studies (STS). While the field of sound studies helps us to unravel the early history of car sound and its cultural meanings, as well as to understand the complex phenomenon of how sounds are talked about, STS provides theories on testing and simulation. To answer our third question, we capitalize on the notion of *Erlebnisgesellschaft*, or an experience-driven society: a society that constantly flags the significance of the immediate and subjective inner experience of its members (Schulze 1992). Together, these approaches clarify how new sense-oriented ways of marketing and designing cars have prompted new ways of testing. We go further, however, and also show that a mismatch can occur between testing and design of consumer products especially at moments of transition in marketing priorities. In other words, our case does not just exemplify the recently increased sonic sensitivity in design settings; it also provides insight into the consequences of this sonic sensitivity for traditions of testability.

Empirically, our chapter is based upon semistructured qualitative interviews with twenty-two participants in the European automotive industry, European research companies, projects devoted to the study of car sound perception, and organizations that focus on either the standardization of vehicle sound measurement or sensory branding (see table 4.1).[2] We have interviewed staff members from the acoustical engineering departments of Ford, Opel, Renault, BMW, and Porsche. Most of our interviewees were formerly involved in OBELICS, an EU-funded project on car sound perception and design, coordinated by HEAD acoustics GmbH (hereafter: HEAD acoustics), a firm in acoustic consulting. Those not affiliated with OBELICS were all experts working on car sound testing, design, and marketing. In addition, our analysis relies on field notes from the observation of testing sites, documents produced in the context of OBELICS, and publications by acousticians and the automotive industry on car sound design.

A Short History of Sounds and the Automobile Industry

In the early days of motoring, the loud noise of gasoline cars largely had a positive connotation. The controlled explosions of the late nineteenth-century internal combustion engine generated far more noise than the steam-driven and electric cars still around at that time (Volti 2004, 7–8). However, for the men who drove a gasoline car in those days—such as wealthy engineers, physicians, and businessmen—the noise was hardly a problem. On the contrary, noise stood for power, and it allowed one to impress bystanders (McShane 1994, 169; Mom 1997, 475). Only in specific contexts did the noise of the gasoline car prove to be a concern. In the United States, for instance, it was not deemed acceptable to use cars in funeral ceremonies before the 1930s (Berger 1979, 183).

From the 1920s onward, manufacturers of cars and car parts started advertising the "silence" of their cars' engines, exhausts, steel bodies, chains, brakes, and gears. This was tied in part to the fact that in the preceding two decades, engineers had begun stressing that mechanical friction and noise were two sides of the same coin and that noise reduction implied greater motor efficiency and longer engine lifespan (Dembe 1996, 195). "Inside of the smoothly gliding mechanism of the motor car," a Hoover Steel Balls advertisement said in 1920, "is the well-poised ball bearing insuring locomotion luxury that is free from sounds which offend the hearing and shorten the life of the unit."[3] "As friction and wear vanish," manufacturer Hyatt claimed about its roller bearings, "so does noise. Cars, trucks, and buses which operate quietly roll more easily, handle more lightly, perform more economically."[4]

Early advertisements strived to show in addition that nothing should distract a customer from buying a particular car. The implication was that users should not be annoyed by extraneous rattles, creaks, and bleeps. This argument gained increasing credibility after the economic crisis of the late 1920s hit the automotive industry. Manufacturers were desperate to sell vehicles, and any marginal advantage was welcome. This gave rise to a new form of sound craft (Krebs, this volume). Western Felt Works secured "composite body silence" by eliminating "squeaks at metal to metal and wood to metal contacts."[5] Budd's "one-piece body" promised "permanent quiet." As Budd explained: "Every man who sells or services automobiles knows the grief that can be caused by body noises. 'Squeaks and squawks' run free service costs way, way up. They're deadly to the new owner's enthusiasm for his car. They're responsible for many a switch to another make of car"[6] (figure 4.1).

45

To the man who is tired of listening to squawks about squeaks

Every man who sells or services automobiles knows the grief that can be caused by body noises.

"Squeaks and squawks" run free service costs way, way up. They're deadly to the new owner's enthusiasm for his car. They're responsible for many a switch to another make of car.

• • • • •

The automobile body that has no joints just can't squeak or rattle. By eliminating parts and by the extensive use of flash welding, Budd eliminates joints. The Budd All-Steel one-piece body is made from only eight major parts—flash-welded into one single jointless unit.

This construction was originated and perfected by Budd engineers. No other type of construction can equal it for quiet—permanent quiet.

The silence of the Budd one-piece body enables dealers to eliminate their free service costs on body noises. More than that, its quietness is an important factor in selling cars—and in helping to keep them sold.

★BODIES BY BUDD★

Originators of the All-Steel Body. Supplied to Manufacturers in the United States, Great Britain, France and Germany

Automotive Industries *January 2, 1932*

Figure 4.1. Budd's advertisement: *Automotive Industries* (Jan. 2, 1932): 45.

In the latter half of the 1930s, increasing protests against urban traffic noise were another major factor in the growing concern with car sounds. They reached such intensity that producing quieter cars became a goal in its own right. In many parts of Europe, it became mandatory to equip exhaust pipes with mufflers, and even the maximum number of decibels produced by car sounds and horns was regulated (Bijsterveld 2008). Car manufacturers responded to these antinoise campaigns by presenting their cars as silent. In a booklet accompanying the British Noise Abatement Exhibition of 1934, Ford Motor Company advertised its new Ford V-8 with a text that perfectly expresses the three themes mentioned earlier: the efficiency of silence, the attraction of luxurious sound, and the need for less noise:

> Oft, in the Stilly Night, You see the lights of an obviously large, luxurious car. *You hear nothing* but the whisper of its generously dimensioned tyres, rolling over the road. Notable for its Engine and Transmission Silence, Smoothness of Progress, Freedom from Vibration, Altogether and in Every Way the Embodiment of Efficiency, as Well as of Surprising Economy, the Car of Tomorrow for the Lover of Noiselessness is THE NEW FORD V-8. (Noise Abatement Exhibition 1935, 69)

The same booklet carried advertisements for "silent" vacuum cleaners, typewriters, and water toilets, but cars were among the first objects that were subjected to noise-abatement regulation.

This early noise-abatement regulation gave rise to the first research on the perception of car sound. In 1929, the National Physical Laboratory (NPL) in London tried to find criteria for the "stridency" of motor horns by order of the British government. By measuring the loudness, acoustical pressures, and average waveform of the sounds produced by a wide variety of motor horns, as well as by having these sounds evaluated by listeners, it became clear that it was largely loudness that accounted for stridency.

In the subsequent years, the NPL established a tradition of jury testing for the evaluation of sound. In these tests, members of listening juries—ranging from dozens of participants to many hundred—would judge the noise of horns, motor vehicles, or aircraft outdoors (Bijsterveld 2008, 204–207). In the early 1960s, D. W. Robinson, one of the leading acoustical experts at NPL, claimed that such jury tests offered a perfect "middle way" between laboratory studies and social surveys, between preciseness and validity, and "some compromise between scientific rigour and realism." Outside, untrained listeners had no trouble rating the sounds of actual motor vehicles in terms of loudness, but when the sounds were recorded and played back in a sound-absorbent room, "our judges did not seem to be able to make up their minds at all." Moreover, Robinson was not sure whether "unpleasantness" could be tested in the same way as "loudness." One problem was to choose the right words. The "inelegant" words *disturbingness* and *bothersomeness* might be synonymous to one person yet dissimilar to another. And would "*lästigkeit*" in Switzerland mean the same thing as "noisiness" in England?[7] Robinson's remarks went not just to the heart of the verbalization problem in evaluating sound; they also went to the heart of the very phenomenon of testing.

Testing in Science and Technology Studies

Since the early 1980s, scholars such as Edward Constant (1980, 1983), Donald MacKenzie (1989), Walter Vincenti (1990), and Trevor Pinch (1993) have identified testing as a highly significant research area for inquiry in science and technology studies. Because the engineers doing the testing consider test data as "a final check on whether the expert's conception of reality conforms to the physical world at hand," there is really something "at stake" in testing (Pinch 1993, 26).

All kinds of testing follow similar patterns. For instance, a highly significant property of testing is projection. Pinch clarifies this concept as follows: "If a scale model of a Boeing 747 airfoil performs satisfactorily in a wind tunnel, we can project that the wing of a Boeing 747 will perform satisfactorily in actual flight." Such an "act of projection" from the small to the large and from the present to the future "depends crucially upon the establishment of a similarity relationship. It is assumed that the state of affairs pertaining to the test case is *similar* in crucial respects to the state of affairs pertaining to the actual operation of the technology." Disputes about tests often resolve around the issue of whether a particular test can be judged as sufficiently similar to the actual situation or is deemed to be significantly different from it. Moreover, Pinch emphasizes that such judgments of similarity and difference "always rest within a broader framework of commitments and assumptions about how a technology will operate," adding that valid tests results "depend upon the acceptance of a similarity relationship, and such a relationship can only be constructed within a body of conventions or within a form of life" (Pinch 1993, 28–29).

In this sense, debates about the quality of tests concerning technologies are analogous to those about the character of replications of experiments in science, as Donald MacKenzie has stressed (MacKenzie 1989). Whether a replication is considered sufficiently similar to the original experiment in order to count as valid proof often depends on the judgment of whether the experimental procedures have been executed properly and competently. Yet since many of such procedures involve tacit knowledge, the discussion about replication often leads to the problem of what Harry Collins (1985) calls "the experimenters' regress" in science: a seemingly endless and detailed discussion of how the replication has been similar or dissimilar to the original experiment without resolving the academic issue at hand. Such similarity discussions also form the crux of disputes over testing.

MacKenzie reminds us, however, that most testing is "routinely accepted as fact" (1989, 415). He does so by referring to the work by Edward Constant (1980), who has shown that "traditions of testability" are at the heart of the routine practices of communities of technological practitioners. With the origins of *and* resistance to the turbojet revolution in the aircraft industry in mind, Constant claimed the following:

> Particular techniques, technologies, and practices relating to testability ordinarily will be commensurate with, and to some degree will define, a specific community's normal technology. The dominant system and conventional modes of testability usually are mutually reinforcing. To overthrow a conventional system frequently requires, in addition to the formulation of an alternative system and the definition of new performance parameters, the creation of new or much refined testing techniques. (Constant 1980, 22)

Accordingly, methods of testing—or traditions of testability—and the dominant design characteristics of a particular technology are closely interconnected.

How this works in practice has also been illustrated by Walter Vincenti in his book *What Engineers Know and How They Know It* (1990). In one chapter Vincenti examines the establishment of design requirements surrounding the flying qualities of American aircraft in the interwar years. Vincenti unravels the way in which a new tradition of testability had to be established before the aircraft industry was able to identify flying-quality specifications that both reflected subjective pilot opinion and were sufficiently precise to be useful in the process of design.

The idea behind the specification of flying qualities was to produce aircraft "possessing adequate stability, responsiveness to the controls, lacking in eccentricities or sudden changes of behavior, and generally satisfactory to the pilot" (consulting engineer Edward Warner, quoted in Vincenti 1990, 81). The initial problem, however, was that test pilots expressed their opinions of flying qualities in highly subjective and qualitative language. In their terms, the controls would respond "very readily," for instance, while the pressure exerted on the controls was "normal," and the longitudinal, lateral, and directional stability were considered "good" (Vincenti 1990, 64). This bore no relation, however, to the formal engineering criteria for describing particular types of aircraft stability or to what the engineers could measure in flight.

Only after prolonged close cooperation between engineers, instrument makers, and pilots at a particular laboratory did the aircraft industry manage to develop a set of precise instruments and piloting techniques to make measurements in flight, as well as increased refinement of pilot opinion, so that the pilot "*knew* what he liked" and could *say* so" (Vincenti 1990, 103; our emphasis). In this process, the use of "standardized terminology and definitions for research pilots and engineers and standard rating scales for quantification of pilot opinion" was paramount (Vincenti 1990, 98). This shared vocabulary would not have been possible, however, without the iterative testing in which small groups of experienced pilots, instrument makers, and engineers—expert testers—collaborated for long periods of time.

As we argue in the following sections, the new experience-oriented marketing trend in the automotive industry has elicited a new strand of "testability" in the area of car sound evaluation that resembles this tradition of the testability of aircraft flying qualities in several, yet not all, respects. Just like the aircraft engineers, the acoustical engineers who aimed at concrete sound-design specifications for cars had to construct a "realistic" setting for *listening* to car sounds in which the input of sound could be measured precisely. Similar to the aircraft engineers, moreover, the acoustics people needed to find ways to have the test subjects *verbalize* their evaluation of sound in a maximally sophisticated manner. As one of the researchers involved put it graphically, they had to find tools to "squeeze out" the test subject's experience of sound as if this subject was an "orange" (NC, 4). To this end, the engineers collaborated with psychoacousticians. The test subjects themselves, however, were less expert testers than the pilots. This, we will see, proved to be a crucial difference.

Artificial Heads and SoundCars: Constructing Reality in Test Settings

The OBELICS research project, launched in 1997 and funded by the European Union, lasted for three years. Its playful acronym stands for Objective Evaluation of Interior Car Sound. The project's main goals were to understand the subjective evaluation of car sound ("the basis of sound language and sound perception"), to establish "methods and tools for an objective evaluation" of automotive sound, and to define "target sounds for different driving situations" (OBELICS 1999, 3, 6). Or, in the words of one of its participants: "What is a sportive car? How should a luxurious limousine sound?" (WK, 4). The institutes and companies collaborating in the project represented the automotive industry (Renault and Fiat), academic acoustics (Oldenburg University and Bochum University in Germany), and consultancy companies that specialized in acoustic measurement and analysis (HEAD acoustics) or in testing and simulation.

Looking back on the project, several of its members were highly positive about what it had delivered in terms of the subjective and objective evaluation of sound and the tools for such evaluation. They were much less enthusiastic, however, about what it had meant for the definition of target sounds (NC, 7–10; KG1, 2–3; WK, 4; RS1, 9). One reason mentioned by the interviewees involved the difficulties of collaborating with competing car-manufacturing companies—who were not fully forthcoming about their marketing data—in one project (NC, 10; KG1, 3; WK, 5; RS1, 15). One interviewee added that a European project like this simply could not do without the "rhetoric" of heading for a "concrete goal" that could convince "Brussels," even though not everyone considered this aim "realistic" (NC, 7–8).

Other comments, however, hinted at the complexities of linking the results of car sound evaluation to the specifications for preferred car sound and the actual target sound design. We know from the story about the flying qualities of aircraft that connecting subjective evaluations to concrete design requirements is difficult, yet not impossible. What, then, happened in the case of automotive sound?

Let us first focus on how the OBELICS members and the research projects preceding OBELICS dealt with projection and similarity issues concerning the car sounds to which the test subjects were asked to listen. Which testing sites and instruments did they consider necessary in order to make listening in the testing situation sufficiently similar to listening to car sounds in a real-life situation? And which tools, in their view, could measure in great detail what the test subjects listened to?

In 1986 HEAD acoustics, the coordinating company behind OBELICS, was established. In the early 1980s, Klaus Genuit, the company's founder, collaborated with Daimler-Benz on car sound analysis. Daimler had been confronted with clients' complaints about noise, but the company—relying on a conventional setup of a microphone, a sound-pressure-level meter, and a frequency analyzer—failed to find the origin of the problems (KG1, 1, KG2, 2). Genuit therefore developed an artificial head, or *Kunstkopf*, that could record and play back sound. His claim was that the recordings resembled human hearing much more closely. In contrast to traditional techniques, the artificial head took directional patterns and dynamic range into account. In addition, since the artificial head could both record and play sounds, it could be used for listening tests. Moreover, its software enabled modifications and analysis of the sounds. One could thus "manipulate something and recognize [it] directly by listening : 'oh, this pattern, that is disturbing me. Not this, not this, but this' " (KG2, 2). On the basis of this idea, Genuit established HEAD acoustics, which now works for the automotive industry worldwide, as well as for the information technology and domestic appliances industry (KG2, 6).

Soon, however, the HEAD acoustics personnel decided that just sitting in a seat and listening to car sounds over loudspeakers or with binaural headphones[8] was not realistic enough. First, they developed a SoundCar that mimicked a car in which the test subjects would not only hear car sounds—over headphones—but would also feel their seat and wheel *vibrating* at the very same time. For quite some time it had been known that noise, vibration, and harshness were interrelated in the evaluation of vehicle sound (which in fact had led this field of research to be labeled noise vibration harshness [NVH]). For this reason, the test site should be constructed accordingly so as to offer the test subjects an experience of the full range of effects (KG1, 4). Yet the disadvantage of the original SoundCar was that when sitting in the test car and opening the throttle, one "would indeed hear the right sounds, but . . . not feel the acceleration, yet stupidly stand still in the hall" (KG1, 4). As one interviewee added, "It was even a bit comical: we drove about 120 and looked out the window, and there were our colleagues standing next to us" (RH, 3).

This is why HEAD acoustics positioned the SoundCar on a *fahrbar* (moving belt) and also transformed it into a tool in which the test listeners could actually *interact* with the car and hear the car sounds accompanying their actions. For this simulation, HEAD acoustics developed its H3S software (HEAD 3-dimensional Simulator), or Sound Simulation System. This system, as the firm claims in its advertising, "accurately and interactively simulates different driving situations. In a driving simulator [like a SoundCar] it creates not only engine, tire, wind, and other vehicle or background sounds but also structure-borne sound in the form of vibration" (*HEADlines* 2003, 5). Instead of being passive recipients of car sounds, the test subjects were now expected to interact with the setting. The headphones or speakers would produce new sounds only when the test driver acted, such as by shifting gears or putting on a turn signal.

The third and most recent step in the process of creating more realistic test conditions has been the construction of a *moving* car sound simulator, the Sound Simulation Vehicle, which HEAD acoustics developed in collaboration with Ford's Acoustics Center in Cologne. While actually driving a real car along the road, different sounds could be channeled to the driver's ears via headphones. The increasing similarity between the test site and actually driving a car created new problems for the evaluation of sound, however. First, a classic trade-off between the need for a "realistic environment" and the need for control popped up—between "realism" and "rigor" in Robinson's terms or between external and internal validity in today's quantitative research vocabulary. In the Sound Simulation Vehicle, the researchers, unlike in the original SoundCar, are unable to control the car's vibrations, and thus they do not have the option of modifying vibrations and experimenting with the relationships between the perception of vibration and sound (KG1, 5; RS1, 20). Moreover, when driving the Sound Simulation Vehicle, unlike in the lab car simulation, one cannot control test-driving conditions, such as specific weather or road conditions (SP, 9; RH, 14).

In addition, the automotive engineers actually working with the mobile sound car, such as Ford, stumbled upon new "realisms" when doing the tests. In one such test the sound of a glamorous 500 PS Aston Martin was added to a relatively ordinary make, a Ford Focus. Test drivers opening the throttle of the Ford Focus would hear the sonic feedback of an Aston Martin: "And acoustically you get such feedback that you think, 'Wow, now we're really driving,' but then not much really happens because the car altogether lacks performance." This could be quite dangerous in fact when trying to pass another car (RH, 7). In this way, the testing situation itself created shifting conceptions of what listening while driving was really about.

If it was a challenge to create testing sites in which the listening situation resembled the *real* situation of listening to car sounds as closely as possible, it was quite another challenge for both automakers and testing consultancy companies to acquire knowledge about how test subjects thought and felt about the sounds to which they were exposed. The problem they faced was how to squeeze out listeners' inner experiences and emotions, especially when they were untrained listeners rather than experts in acoustics.

Finding Words for Sound: The Expression of Sonic Perception

For a long time, as Robinson's remarks from the 1960s already suggest, it was not self-evident to use lay listeners when evaluating sound. In acoustics research in general, many early laboratory tests were executed by the acousticians themselves, and it was only in the course of time that the researchers started utilizing lay listeners. Moreover, the car industry continued to use expert listeners (and still does today)—often the acoustical engineers themselves—for the testing of particular sounds, such as car door sounds (BL, 9; SP, 9; KG2, 19). Using lay listeners became increasingly important, however, as the car manufacturers wanted to know how potential customers evaluated their car sounds.

However, how did the automotive industry consider it possible to measure what these lay listeners experienced in terms of car sound? Initially, both the car industry itself and HEAD acoustics used so-called pair comparison tests and scale assessments. In pair comparison, or A/B tests, test subjects compare two sounds (sound A and sound B) that immediately follow each other, for instance in terms of agreeableness, and subsequently compare other pairs: A and C, B and C, and so on. The advantage of this method is that the test subjects do not need to remember the sounds for a prolonged time and can focus on very small differences. Yet, as a HEAD acoustics employee claimed, it is "artificial" since in everyday situations "one does not [constantly] jump from one car into another" (NC, 4). Because the test subjects are "forced" to make a choice, they often report feeling insecure about whether they "have done the assignment well" (RH, 17), or they "just think of something else" and simply decide to not to participate in the test (RS, 8).

The conventional alternative for the pair comparison test is the scale assessment test. Scale assessments allow test subjects to evaluate a particular phenomenon with the help of a scale. Whereas pair comparison is seen as *relative* testing, scale assessment is considered *absolute* testing. Again, test subjects find it rather hard to do such scale ratings in the case of sound (SP, 6). As one of us experienced herself at the lab of HEAD acoustics, it is difficult to rate the quality of a particular sound without the option of comparing it to other sounds. Furthermore, how does one choose between the third, fourth, or fifth box on a scale from one to seven?

In psychoacoustics research, the semantic differential, a specific form of scale assessment, was introduced for sound perception in 1954. The original test focused on the meaning of underwater sounds to navy sonar operators. It displayed fifty bipolar attributes, such as "powerful–weak," for the evaluation of complex auditory stimuli. Many of these attributes came from descriptions of sonar sounds by journalism students who had received specialized training in verbal expression and from "sonar recognition cues" used in teaching methods for sonar experts (Solomon 1954, 17–18). Since the 1970s, these lists have been fine-tuned several times, for instance for vehicle acoustics. The members of OBELICS used a set of thirty-four bipolar attributes, such as "boring–stimulating," "cheap–expensive," "loud–soft," and "tiring–relaxing" (Hempel 2001, 170) (figure 4.2).

One particular problem in tests based on the semantic differential, however, is this use of language. Since the project involved several European countries, the test subjects spoke different languages. However, as Robinson suggested as early as the 1960s, words for sonic attributes that seemed to be translatable from one language to another might not have the same connotation in the culture using that language. To make the project truly European, one member of OBELICS developed a thesaurus in which each set of expressions in German had its counterpart in English,

Beziehen Sie die folgenden Beschreibungen bitte nur auf das **Fahrzeug** bei 130km/h Autobahnfahrt.

("Das Fahrzeug klingt ...")

mächtig	machtlos
häßlich	schön
dynamisch	undynamisch
langweilig	anregend
lästig	angenehm
defensiv	offensiv
lebhaft	müde
alt	neu
komfortabel	unkomfortabel
abstoßend	anziehend
angenehm	unangenehm
gefährlich	harmlos
leidenschaftlich	leidenschaftslos
bedrohlich	unbedrohlich
luxuriös	einfach
langsam	schnell
beweglich	unbeweglich
empfindlich	unempfindlich
billig	teuer
sanft	wild
schwungvoll	schwunglos
vertraut	fremd
aufdringlich	unaufdringlich
störend	erwünscht
behaglich	unbehaglich
freundlich	unfreundlich
sportlich	unsportlich
laut	leise
zuverlässig	unzuverlässig
sicher	unsicher
beruhigend	aufregend
werthaltig	einfach
anstrengend	erholsam
stark	schwach

Figure 4.2 The semantic differential used in the OBELICS project: Thomas Hempel. *Untersuchungen zur Korrelation auditiver und instrumenteller Messergebnisse für die Bewertung von Fahrzeuginnenraumgeräuschen als Grundlage eines Beitrags zur Klassifikation von Hörereignissen* (Munich: Herbert Utz Verlag, 2001, 170). Courtesy Herbert Utz Verlag.

French, and Italian (WK, 4). Indeed, the thesaurus and cross-country comparison enabled the researchers to show that some car sound attributes had quite different meanings in different countries. The sonic understanding of *sportiveness* in Germany differed from that Italy, and *loudness* did not have the same significance in France that it did in Germany (NC, 3). This, Klaus Genuit suggested, was dependent on the traditions of car engines and gear systems—engine power, automatic shifting or not—in use in different countries (KG2, 16–18). Differences in technological development itself thus seemed to undermine the cross-country equivalence of rating words for sound in different languages.

The OBELICS members developed a similar sense of variety when working with the semantic differential test in Germany. The words used for the bipolar attributes were unknown to many of the lay test subjects or were interpreted in a highly arbitrary manner:

> For example, what is *rau* [rough]? Do you know what sound *rau* is? What is a rough sound? *Rau* consists of the letters *R* and *A*, rrrrrrrrr, is this rough? Is aaaaaa rough? [I (KV) reply: "Rather R than A."] Psychoacoustically, though, exactly the opposite is the case; A is very rough psychoacoustically because it comes with a high intonation. R comes with a deep intonation. It is not at all as rough, from a psychoacoustical angle, but in practice you are right, for most people think of *Rauhigkeit* as rrrrrrrrr. (KG1, 11)

Moreover, during the semantic differential tests, the researchers noticed that the test subjects were inclined to tell the test leader much more than was actually asked for. One subject insisted on telling about her vacation experiences while traveling by airplane since the car sounds were "similar" and just as "pleasant." Another left the test site crying "because a specific sound reminded her of an accident" (NC, 4). In order to record these additional stories, the researchers introduced associated imagination of sound perception (AISP), which was later reworked into explorative vehicle evaluation (EVE). In the construction of these tests, Brigitte Schulte-Fortkamp, a psychoacoustics and sociology professor at Oldenburg University (later Berlin) played a significant role.

In the AISP tests, the subjects, while positioned in a SoundCar, were asked to comment on the car sounds freely and mention every association that might pop up in their minds. Subsequently, they were interviewed about their comments, and all of them "found sentences to express themselves" (NC, 19; SP, 5). Through such tests the researchers claimed to reach "another dimension in the description of noises," one much closer to the test subjects (NC, 5–6; AF, 3; RS2, 4). Instead of mere evaluations of sound, one now had a way to understand their origin—the *why* behind the *what*. Each person's assessment of a particular sound, they said, was based on *gesammelte Erfahrungen*, a series of earlier experiences and memories and expectations built upon them. Two persons, for instance, might both consider a car sound loud. Yet, while the first one would love it, claiming to be a "fan" of the car tested, the second considered it "too loud . . . for I imagine myself having to drive this car for five hours, which I do regularly, Paris–Aachen, for instance" (NC, 6). The explorative vehicle evaluation tests, created after the completion of the OBELICS project, were tailored to elicit such stories in situations in which test subjects would actually drive a car while listening.

The usefulness of the AISP and EVE tests for sound design was highly contested within the automotive engineering and acoustics community, however, and not only because the tests were time consuming and expensive (WK, 13, RS2, 3). Scientists and engineers from Oldenburg University considered the tests too vague and doubted their usefulness (NC, 12–13; KG1, 14). Many acousticians and acoustical engineers had hoped to set up a sort of universal sound quality index, in which a series of numbers would indicate which design specifications would lead to sounds of high quality (RS, 4–5). They did not see how the outcomes of the AISP tests could bring the sound quality index any closer (KG1, 14–15). An Opel engineer explained that he actually preferred a sound quality system that would get rid of differences in taste (BL, 2).

Many of the OBELICS personnel involved in subjective testing considered such a sound quality index a bridge too far, at least for automotive sound. In their view, car sound quality depends simply too much on context (TH, 3; WK, 10; RS1, 6; SP, 14), and ideas about the best sound quality will probably evolve over time (NC, 10). They were similarly skeptical of the possibility of designing a few widely loved car sounds, the "right" sounds for many. One of them even considered it "completely utopian," especially for interior sounds. A target sound will work only under strictly confined conditions, say, if "I drive a 4-cylinder, midsize, 30 km/h, I (being under twenty-five years of age) would prefer something like a sporty car, without exhaust or anything" (NC, 8). In fact, he and others claimed, one can define preferred sounds only for very specific groups of consumers. This implies that one should also invite narrowly defined target groups for the testing procedures (NC, 8; RS1, 10; SP, 11). Moreover, particular expectations will affect the evaluation of sound: "When I drive a Mercedes-S, I expect something different from what a Porsche driver expects from his Porsche. Were I to interbreed both cars, both drivers would feel unhappy. So I cannot simply say 'this sound is okay' " (KG1, 7). HEAD acoustics' president Klaus Genuit therefore felt that AISP and similar tests can be commercially exploited only for quite specific versions of target sounds or "benchmarking" (KG1, 3).

Although Genuit expressed disappointment over the engineers' inclination to consider everything measurable (KG1, 14), he is an engineer himself—as are many OBELICS members, even those involved in subjective testing. This is why the line between proponents and opponents of AISP was never simply one between engineers and nonengineers. Rather, there was, by and large, a distinction between nonengineers and engineers involved in *consulting* on sound design on the one hand and those involved in *implementing* sound design on the other (WK, 10–11). Given their reliance upon numbers responsible for implementing new designs, the "implementation" engineers considered AISP not easily "management-compatible" (TH, 5).

Some makes, such as the Harley Davidson motorcycle and the Porsche sports car (RS, 12, 15; PE, 3) already had exterior target sounds connected to their image.

Yet in these cases, several interviewees claimed, engine sounds originated first, while the idea to treasure these as target sounds came only much later. Such "acoustical fingerprints" might work (RH, 14). Porsche, for instance, had worked hard to keep its engine sound stable even when it shifted from its boxer air-cooled engine to the 6-cylinder, water-cooled system (KG1, 9). Moreover, representatives from the luxury car manufacturers on the ISO committee for the standardization of vehicle noise measurement (Technical Committee 43, Working Group 42) preferred taking tire noise into account when measuring car sound because this would give them more leeway in dealing with the engine sound (US, 4–5; FR, 31–40). If the measurement procedure had focused solely on engine noise, manufacturers like Porsche would have been forced to change the characteristic sound of their car engines.

Yet designing a target sound from scratch was widely considered to be a completely different challenge. What's more, who should have the final word in the actual design? A Ford employee explained that one simply could not leave the conception of target sounds in the hands of the test subjects only:

> When you ask customers how a sound should be, you will get as many different answers as the number of customers you ask. This means that in part you cannot have the customer decide; after all, you need to hold on to the characteristic Ford sound. As a distinctive sound it also has to be embedded in a whole range of different things. The sound should fit in with everything we want to convey through our car philosophy. (RH, 8)

In the ideal world of this Ford sound engineer, future Ford customers would be able to upload a series of car sounds (for the turn signals, seat-belt warning, windshield wiper, and so on) of their choice, just like uploading a ringtone for a cell phone. Yet, and this was a crucial twist, this option should be available to the customers only after the Ford sound engineers had created a full "sound composition" in which all of the sounds would both be typically Ford and go together extremely well (RH, 8; 12, PE, 4). Moreover, design needed to be accompanied by marketing techniques. When Ford had designed a new fancy turn-signal sound ("pock-pock-pock") for its Ford Focus, many test subjects and automotive journalists initially rejected it because of their nostalgia for the venerable "click-clack" sound originally linked to the relay. Only after the Ford Focus had been advertised as a first-rate modern car and a story had been linked to its futuristic turn-signal sound did the new sound come to be accepted (RH, 9–10).

Yet even aside from the sonic cultural conventions the automakers had to deal with, it is clear that in the debate over sound quality, acoustical engineers desiring an index and researchers heralding the significance of subjective testing by lay listeners were pitted against each other. One way to understand the tensions between the two is suggested by the work of linguistic anthropologist Tom Porcello (2004). Porcello analyzed a conversation between an experienced, professional producer and a student, a novice, in a recording studio by focusing on their talk about sound during a recording session. While the professional was used to talking in terms of musical instruments and technologies such as microphones to explain the kind of sound he wanted, the student referred to particular bands and songs to express the sounds he preferred. This led to profound misunderstandings between the two despite the producer's efforts to bridge the gap by using metaphors and mimicry.

These misunderstandings seem to be similar to those between the index-loving acoustical engineers and the protagonists of subjective lay testing, with the test subjects in the role of novices and the acoustical engineers in the role of the producers. The discursive clash that Porcello analyzed also reminds us of the misunderstandings that originally kept apart the aircraft engineers and the pilots in the flying qualities case described by Walter Vincenti. In Vincenti's example, this was solved by having the pilots and the engineers work closely together—even in the aircraft itself—to find connections between the pilots' ways of expressing themselves and the engineers' technology-related thinking. In the car sound evaluations, however, there was little sustained collaborative effort between the engineers and the lay listeners. We explain this by referring to the marked differences in status between test pilots and test car drivers. While the first are members of a profession with, in the terms of sociologist Andrew Abbott (1988), an "exclusive jurisdiction" concerning the flying of aircraft, everyday car drivers cannot claim such a special position. The officially acknowledged exclusiveness of the pilots' skills and certainly *test* pilots' skills fueled the interest of the engineers in a sustained collaboration with these pilots (Vincenti 1990, 70–71).

We have already explained why the acoustic consultants at OBELICS considered it important to work with lay listeners. Preferably, these lay listeners should be highly specific groups of car consumers tailored to the type of car tested. As one engineer said, it was quite absurd to have the sound of a hundred thousand Euro cars assessed by students—not exactly the usual candidates for driving such cars (BL, 3). At the same time, however, many of the interviewees stressed that expert listeners noticed far more than lay listeners (RS2, 5; WK, 14; BL, 5; KG2, 13). In contrast to lay listeners, experts were able to determine within the laboratory which sounds would sound *natural* outside the lab's walls (BL, 6–7). Yet outside the lab, the interviewees preferred lay listeners. Why had this come to have such crucial significance?

Marketing Sound in the Experience Society

In the 1960s and 1970s, national governments and European bodies imposed increasingly stringent limits on the maximum noise emission of cars (Sandberg 2001; DC, 1–2). Between 1970 and 2000 the European maximum for passenger cars decreased from 86 to 74 dB (EL and VM, 6). As a result, automobile makers were forced to further limit car noise. Engineers had their hands full when it came to reducing the low-frequency sound in cars (Stockfelt 1994). There were many such sounds: the rustle of the wind, the engine vibrations, the sound of tires touching the road surface, the hydraulic system of power steering, and the whining of the fuel pump. Acoustical engineers tried to tackle these by, for instance, applying acoustic glass or introducing an innovative design for car tires (Dittrich 2001; Steketee 2006; Kouwenhoven 2002). Time and again, however, they would find that removing one noise rendered another one audible (Schick 1994; Freimann 1993; NC, 20). So, even though car manufacturers had been interested in interior car sound since the early 1920s, it was the obligatory reduction of exterior noise that brought formerly masked interior sounds into the limelight again after the 1970s (WK 8; BL, 11; RS, 11; SP, 13).

Both the car industry and cultural sociologists, however, tend to explain the increasing focus on sound design by referring to the rise of experience society–related issues. German sociologist Gerhard Schulze (1992) argues that sensory experience plays a large role in the selling of consumer goods today. This emphasis has emerged because many products have been perfected to such a degree that differences in technical specifications between brands have decreased over time. Since the 1980s most products now simply *work*. Moreover, most consumers today choose from an enormous array of consumer goods on account of the booming postwar economy. The many options available, however, also give rise to uncertainty. To compensate for the absence of differences in technical quality and make selection easier for consumers, products are increasingly sold by cashing in on the emotional meaning and inner experience products evoke in buyers.

In the marketing of new products, therefore, sensory experience has become crucial. How appliances feel, smell, or sound and how that fits the buyer's identity have thus become as relevant as how they look (Marks 2002, 114; Wenzel 2004). All this has led to a strong "aestheticization" of everyday life, described by Schulze as the emergence of the *Erlebnisgesellschaft*, or experience society, as explained earlier. In such a society, experience value trumps functional value both in selling strategies and in the motivation of consumers (Schulze 1992, 59).

Many of our interviewees within or related to the automotive industry presented a similar analysis of today's society (KG1, 9; RH, 20–21; BL, 10–11; AF, 4; RS, 8). Now that there are many wealthy people, one interviewee claimed, their concern for quality of life has come into play. In addition, since they no longer have to worry about whether their cars actually run, they may start complaining about its sound (NC, 11). What's more, the automotive industry considers sound more easily observable by consumers than other aspects of automotive quality, such as safety (Zeitler and Zeller 2006, 1). Dependent on customers' specific preferences, cars should have "decent," "luxurious," "dynamic," or "sporty" sounds (Bernhard 2002). Renault, for one, made its Clio 3 "silent" and gave its Megan Sport a "dynamical sound" (EL and VM, 2).

Although there may not be sufficient evidence to warrant an analysis of contemporary culture in terms of the experience society, what is relevant to our understanding of the tensions between subjective car sound evaluation and the search for objective car sound specifications is that the world of marketing *believes* in such a society and in sound as a "marketable vehicle attribute" (Repik 2003, 5). Designers are now obliged to "ascertain what emotional values they want the consumer to attach to the product. They then develop forms which instigate the associations to, hopefully, inculcate those feelings" (Guy Julier, quoted in Lury 2004, 87). Or, as brand experts Hajo Riesenbeck and Jesko Perry explain, fully in line with Schulze's analysis, "The price of a new car is so high that buyers will have the basics covered anyway.... This makes it imperative for automotive advertising to appeal to the heart and soul rather than reason (Riesenbeck and Perry 2009, 29). This has grown even more important after an increasing number of different car makes and types began to be put together on the basis of the exact same set of car parts (Gottfredson et al. 2001).

This shift in marketing strategies needs qualification to prevent misunderstanding. We do not claim that until recently marketing people have had no eye for things such as the *image* of a product or the *lifestyle* to be associated with the product. On the contrary, Wolfgang König has convincingly traced attempts to create a positive product image by means of symbolic messages in advertising to the years prior to World War I (König 2000, 405). Nor do we claim that companies did not have their products tested by consumers: In Western Europe, firms started establishing testing labs and design departments in the 1950s (Oldenziel, de la Bruhèze, and de Wit 2005, 118). Yet, while manufacturers had long been interested in what *the consumer did with the product*, since the 1970s they have become obsessed by what *the product does to the consumer*. What manufacturers now want to have tested is which emotions the product stimulates in test subjects, the future consumers.

It is in this context that an extensive research industry has evolved that concentrates exclusively on developing testing methods for tracing these feelings. One component of this new research industry is HEAD acoustics, providing the automotive industry (and increasingly the domestic appliances and personal computer industry) with testing methods and actual testing of sound. In turn, companies such as HEAD acoustics collaborate with universities, whose staff members equally address the changes associated with today's experience society (Schulte-Fortkamp, Genuit, and Fiebig 2007, 12). The notion that sound "is well known to enhance or detract from our pleasure in possessing or using a product" (Boulandet et al. 2008, 1) has thus been reinforced by an emerging and growing network of manufacturers, designers, testing companies, marketers, and academics who reciprocally *spread the word* of sensorial branding and design.

In light of the trends in marketing just mentioned, it is understandable that HEAD acoustics endeavored to develop qualitative tests that would enable the full expression of the rich variety of test subjects' emotions with regard to car sounds. Because these tests had to reflect the evaluation of untrained listeners, that is, the everyday consumer of cars, HEAD acoustics increasingly selected lay listeners. It is ironic that while the automotive industry considered these consumers highly sensitive to sound, it took new ways of testing to have these consumers properly express their assessments of particular sounds. The experts' idiom of sounds did not speak to these lay listeners. The new tests in the OBELICS project did not straightforwardly lead to the specification and design of target sounds. Nor did it generate a sound quality index, however. Why this happened becomes clear when we compare it with the project on flying qualities specifications as described by Vincenti (1990). Unlike this project, OBELICS encompassed no close and prolonged collaboration between engineers and testers. Unlike the pilots, the car drivers were not members of an acknowledged profession with whom the engineers had to build a long-term relationship.

Conclusions

In the 1980s and 1990s the automotive industry displayed a widespread belief in the rise of the new experience society. This resulted in new marketing strategies that highlighted the significance of car sounds. The driver's emotional experience of quality car sound, so it was believed, would help to sell a car. Moreover, specific target sounds would enable new connections between a car make and particular groups of consumers.

This belief led to an emerging *tradition of testability* in the European automotive industry. The crux of this testability, in which the German company HEAD acoustics played the lead role, is subjective testing, which enables free, associative verbalization of car sound perception. Moreover, testing car sounds while driving (in a mobile SoundCar) became the heart of *projection*: the idea that the testing was "similar in crucial respects" to actual driving.

The OBELICS project, in which representatives from the automotive industry, acoustics departments at universities, and testing consultancy companies collectively aimed at linking subjective testing to objective target sounds, did not immediately lead to the definition of such target sounds, however. In contrast to a similar research project on the specification of flying qualities for airplanes, HEAD acoustics used *lay* rather than *expert* testers. Although this was fully in line with the marketing strategies of the automotive industry, it did not easily facilitate fruitful collaboration between automotive engineers and sound designers. Whereas test pilots and aircraft engineers had found a common language for the design specification of aircraft flying qualities through long-term, close, and interprofessional collaboration, the lay testers expressed their sonic preferences in the absence of the designers. In the designers' perspective, the lay testers behaved like the novice in Tom Porcello's study of studio engineering: The words they employed for sound were incomprehensible to the sound designer.

If the marketers are right in suggesting that the testing of consumer goods will increasingly concentrate on the feelings that products elicit in their users, scholars in STS and sound studies should take the specific difficulties of testing the sensory qualities of products into account when studying the coproduction of technology by consumers and producers. Michel Callon, when reflecting on the new "economy of qualities," seems to be optimistic about this coproduction. In his view, the increasing focus on the construction of qualities in design and the organization of "real-life experiments on preferences" related to it "*facilitate* the intensification of collaboration between supply and demand in a way that enables consumers to participate actively in the qualification of products." This process implies consumers who are "capable of perceiving differences and grading them" (Callon, Méadel, and Rabeharisoa 2002, 212; our emphasis). Callon presents particular forms of Information and Communication Technology, which constantly tracks consumers' actions as preferences, as his ideal example and also notes that consumers are supported in their evaluations by suppliers and their intermediaries. He assumes, however, that consumers in fact are capable of articulating what they want.

As our argument underscores, a similar interpretation of the new sensory-oriented testing traditions is unwarranted because, in the settings involved, consumers cannot directly vote for their preferences. Speech about sound (Krebs and Rice, this volume) is so invested with cultural habits, training, and differences in cultural capital that a good match between testing and designing car sound seems to still be in the distant future. We need to acquire more knowledge about the new traditions of testability related to the new economy of qualities and sensory-oriented marketing. In this respect sound sounds like the perfect case to start with indeed.

NOTES

1 "Let your car do the talking" (magazine ad, Chrysler Voyager, published between 2002 and 2007); "U rijdt geen diesel, u rijdt Jaguar" (brochure, Jaguar, Netherlands, 2006); *Zelfs de stilte is standaard*, commercial, Toyota Avensis, videotape (Talmon: AV Communicatie, 1999); "Een mens krijgt 469.082 geluidsprikkels per minuut," *Elsevier* (Oct. 23, 1999): 12–14; *Business Travel*, commercial, Mercedes-Benz, videotape (Stuttgart: DaimlerChrysler AG, 1990); *Blind Brothers Three*, commercial, Volkswagen Passat, videotape (Almere: TeamPlayers, 1999); "De nieuwe Ford Focus: Voel de verfijning," *Dagblad de Limburger* (Jan. 24, 2005): B11. All translations are by the authors or language editor, Ton Brouwers. We thank him, Klaus Genuit, and Trevor Pinch for their comments.

2 We thank Kristin Vetter and Fleur Fragola for the interviews they undertook.

3 "Hoover Steel Balls" advertisement, *Automotive Industries* (Feb. 12, 1920): 61.

4 "Hyatt Quiet Roller Bearings" advertisement, *Automotive Industries* (Jan. 30, 1932): 5.

5 "Western Felt Works" advertisement, *Automotive Industries* (Jan. 2, 1932): 64.

6 "Bodies by Budd" advertisement, *Automotive Industries* (Jan. 2, 1932): 45.

7 Archives National Physical Laboratory, Papers of D. W. Robinson, "Recent Advances in the Subjective Measurement of Noise," [1963]: 157–77, at 164, 163, and 161, respectively.

8 Binaural headphones replay recordings made with the help of two microphones, one for each ear, thus mimicking human hearing.

REFERENCES

Abbott, Andrew. *The System of Professions: An Essay on the Divison of Expert Labor.* Chicago: University of Chicago Press, 1988.

Berger, Michael L. *The Devil Wagon in God's Country: The Automobile and Social Change in Rural America, 1893–1929.* Hamden, Conn.: Archon, 1979.

Bernhard, Ulrich. "Specific Development of a Brand Sound." *AVL Engine and Environment* (2002): 103–15.

Bijsterveld, Karin. *Mechanical Sound: Technology, Culture, and Public Problems of Noise in the Twentieth Century.* Cambridge, Mass.: MIT Press, 2008.

Boulandet, Romain, Hervé Lissek, Patrick Monney, Jacques Robert, and Sylvain Sauvage. "How to Move from Perception to Design: Application to Keystroke Sound." Paper presented at Noise-Con 2008 and the Sound Quality Symposium, Dearborn, Mich., July 28–30, 2008.

Callon, Michel, Cécile Méadel, and Vololona Rabeharisoa. "The Economy of Qualities." *Economy and Society* 31(2) (2002): 194–217.

Collins, Harry M. *Changing Order: Replication and Induction in Scientific Practice.* London: Sage, 1985.

Constant, Edward W. *The Origins of the Turbojet Revolution.* Baltimore: Johns Hopkins University Press, 1980.

———. "Scientific Theory and Technological Testability: Science, Dynamometers, and Water Turbines in the 19th Century." *Technology and Culture* 24 (1983): 183–98.

Dembe, Allard E. *Occupation and Disease: How Social Factors Affect the Conception of Work-Related Disorders.* New Haven, Conn.: Yale University Press, 1996.

Dittrich, Michael. "Sound of Silence." In *TPD in 2000—Projecten*, 18. Delft: TPD, 2001.

Freimann, Raymond. "Das Auto—Klang statt Lärm." In *Der Klang der Dinge*, ed. Arnice-Verena Langenmaier, 45–57. Munich: Silke Schreiber, 1993.

Gottfredson, Mark, Elias Farhat, Paul Rogers, and John Smith. "The Ultimate Testing Laboratory: Carmakers Mutate from Heavy Manufacturers to Consumer Goods Companies." *European Business Journal* 13(2) (2001): 66–73.

HEADlines. Newsletter of HEAD acoustics. Herzogenrath, Germany: HEAD acoustics, April 2003.

Hempel, Thomas. *Untersuchungen zur Korrelation auditiver und instrumenteller Messergebnisse für die Bewertung von Fahrzeuginnenraumgeräuschen als Grundlage eines Beitrags zur Klassifikation von Hörereignissen.* Munich: Utz, 2001.

Jackson, Daniel M. *Sonic Branding: An Introduction.* New York: Palgrave Macmillan, 2003.

König, Wolfgang. *Geschichte der Konsumgesellschaft.* Stuttgart: Steiner, 2000.

Kouwenhoven, Erik. "Autoband moet fluisteren." *Algemeen Dagblad* (May 4, 2002): 43.

Lury, Celia. *Branding: The Logos of the Cultural Economy.* London: Routledge, 2004.

MacKenzie, Donald. "From Kwajalein to Armageddon? Testing and the Social Construction of Missile Accuracy." In *The Uses of Experiment: Studies in the National Sciences*, ed. David Gooding, Trevor J. Pinch, and Simon Schaffer, 409–36. New York: Cambridge University Press, 1989.

Marks, Laura U. *Touch: Sensuous Theory and Multisensory Media.* Minneapolis: University of Minnesota Press, 2002.

McShane, Clay. *Down the Asphalt Path: The Automobile and the American City.* New York: Columbia University Press, 1994.

Mom, Gijs. *Geschiedenis van de auto van morgen: Cultuur en techniek van de Elektrische Auto.* Deventer, the Netherlands: Kluwer, 1997.

———. *Noise Abatement Exhibition*. London: Anti-Noise League, 1935.
OBELICS (Objective Evaluation of Interior Car Sound). Synthesis Report, 1999.
Oldenziel, Ruth, Adri Albert de la Bruhèze, and Onno de Wit. "Europe's Mediation Junction: Technology and Consumer Society in the 20th Century." *History and Technology* 21(1) (2005): 107–39.
Pinch, Trevor. "Testing: One, Two, Three . . . Testing! Towards a Sociology of Testing." *Science, Technology, and Human Values* 18(1) (1993): 25–41.
Porcello, Thomas. "Speaking of Sound: Language and the Professionalization of Sound-Recording Engineers." *Social Studies of Science* 34 (2004): 733–58.
Repik, Edward P. "Historical Perspective on Vehicle Interior Noise Development." Paper presented at Noise and Vibration Conference and Exhibition (SAE Technical Paper Series) at Traverse City, Mich., May 5–8, 2003.
Riesenbeck, Hajo, and Jesko Perry. *Power Brands: Measuring, Making, and Managing Brand Success*. Weinheim, Germany: Wiley-VCH, 2009.
Sandberg, Ulf. "Abatement of Traffic, Vehicle, and Tire/Road Noise—the Global Perspective." *Noise Control Engineering Journal* 49(4) (2001): 170–81.
Schick, August. "Zur Geschichte der Bewertung von Innengeräuschen in Personenwagen." *Zeitschrift für Lärmbekämpfung* 41(3) (1994): 61–68.
Schulte-Fortkamp, Brigitte, Klaus Genuit, and André Fiebig. "A New Approach for Developing Vehicle Target Sounds." *Sound and Vibration* (October 2007): 2–5.
Schulze, Gerhard. *Die Erlebnisgesellschaft*. Frankfurt: Campus, 1992.
Solomon, Lawrence Norval. *A Factorial Study of the Meaning of Complex Auditory Stimuli (Passive Sonar Sounds)*. PhD diss., University of Illinois, 1954.
Steketee, Menno. "Hightech autoruit." *Elsevier Thema Auto* (April 2006): 68–69.
Stockfelt, Ola. "Cars, Buildings, and Soundscapes." In *Soundscapes: Essays on Vroom and Moo*, ed. Helmi Järviluoma, 19–38. Tampere, Finland: Tampere University, 1994.
Van de Weijer, Bard. "Kalm tuffen of bronstig cruisen." *De Volkskrant* (February 3, 2007): 9.
Vetter, Kristin. Humming, Hissing, and Human Values—Sound Evaluation Tests in Car Acoustics as Mediators between Technological Artifacts, Scientific Methods, and Culture. Master's thesis, European Inter-University Association on Society, Science, and Technology, 2004.
Vincenti, Walter. *What Engineers Know and How They Know It*. Baltimore: Johns Hopkins University Press, 1990.
Volti, Rudi. *Cars and Culture: The Life Story of a Technology*. Baltimore: Johns Hopkins University Press, 2004.
Wenzel, Silke. "Vom Klang zum Lärm." *Neue Zeitschrift für Musik* 165(2) (2004): 34–37.
Zeitler, Alfred, and Peter Zeller."Psychoacoustic Modeling of Sound Attributes." *Proceedings SAE World Congress*, April 3–6, 2006, Cobo Center, Detroit.

Table 4.1 Overview of Interviews

Name of Interviewee	*Background, Affiliation at Time of Interview, Former Affiliations*	*Date and Place*	*Interviewer (Original Language)*
Nicolas Chouard (NC)	acoustic engineer, former Renault employee, formerly involved in OBELICS	June 7, 2004 Aachen, Germany	Kristin Vetter (KV) (German)
David Delcampe (DD)	European Commission, DG Environment	October 23, 2007 Brussels, Belgium	Fleur Fragola (FF) (French)

Table 4.1 Continued

Name of Interviewee	*Background, Affiliation at Time of Interview, Former Affiliations*	*Date and Place*	*Interviewer (Original Language)*
Peter Ehinger (PE)	head of acoustics department, Porsche, former member of ISO TC 43, WG 42	December 10, 2008 Weissach, Germany	Eefje Cleophas (EC) (English)
André Fiebig (AF)	employee, HEAD acoustics	July 31, 2009 Kohlscheid, Germany	EC (English)
Klaus Genuit (KG)	engineer, president, HEAD acoustics, formerly involved in OBELICS	June 3, 2004 (KV) Aachen, Germany November 18 (EC) Kohlscheid, Germany	KV & EC (German & English)
Ralf Heinrichs (RH)	engineer, Ford employee	June 8, 2004 Cologne, Germany	KV (German)
Thomas Hempel (TH)	engineer, formerly involved in OBELICS	May 26, 2004 Munich, Germany	KV (German)
Ian Knowles (IK)	European Commission, DG Enterprise	October 9, 2007 Brussels, Belgium	FF (English)
Winfried Krebber (WK)	engineer, formerly involved in OBELICS	May 6, 2004 Aachen, Germany	KV (German)
Bernhard Lange (BL)	engineer, Opel employee	June 23, 2004 Rüsselsheim, Germany	KV (German)
Eric Landel (EL)Virginie Maillard (VM)	Renault employees	October 16, 2007 Guyancourt, France	FF (French)
Stephan Paul (SP)	PhD student at HEAD acoustics	June 17, 2004 Bonn, Germany	KV (German)
Foort de Roo (FR)	acoustic engineer, former member ISO TC 43, WG 42	June 26, 2008 Delft, the Netherlands	EC (Dutch)
Gerhard Thoma (GT), Manuel Reichle (MR), Alfred Zeitler (AZ),	members of acoustics department BMW	November 20, 2008 Munich, Germany	EC (English)
Ulf Sandberg (US)	engineer, member of Swedish National Road and Transport Research Institute (VTI), chair of several committees of ISO and CEN	October 1, 2007 Brussels, Belgium	FF (English)
Wolfgang Schneider (WS)	European Commission, DG Enterprise	October 18, 2007 Brussels, Belgium	FF (English)
Roland Sottek (RS)	engineer formerly involved in OBELICS	May 14, 2004 Aachen, Germany	KV (German)
Rob Vermeulen (RV)	director of Total Identity (marketing company)	August 7, 2008 Landgraaf, the Netherlands	EC (Dutch)
Charles Zhang (CZ)	acoustic engineer, Renault employee	November 13, 2008 Guyancourt, France	EC (English)

SECTION II

STAGING SOUND FOR SCIENCE AND ART: THE FIELD

CHAPTER 5

SOUND STERILE: MAKING SCIENTIFIC FIELD RECORDINGS IN ORNITHOLOGY

JOERI BRUYNINCKX[1]

INTRODUCTION

> It is a mad, reckless song-fantasia, an outbreak of pent-up, irrepressible glee. He begins bravely enough with a number of well-sustained tones, but presently he accelerates his time, loses track of his motive, and goes to pieces in a burst of musical scintillations. (Mathews 1904, 49)

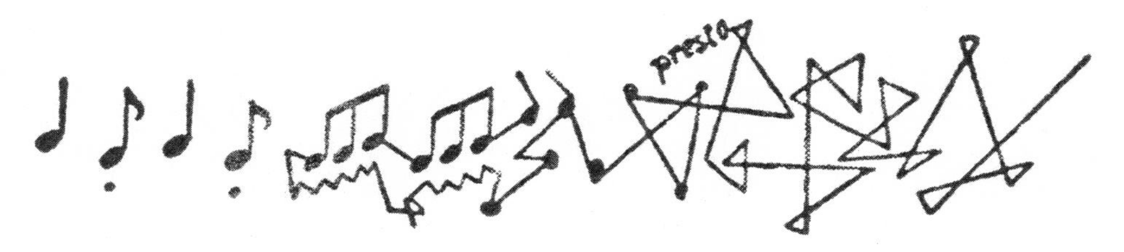

Figure 5.1. Transcript of bobolink song. From Ferdinand S. Mathews, *Field Book of Wild Birds and Their Music: A Description of the Character and Music of Birds* (New York: Putnam, 1904, 51).

When writing this in 1904, American naturalist and composer Ferdinand S. Mathews was not reviewing one of his fellow artists' latest compositions. In fact, he was describing a motif of the "musical fireworks" produced by a bobolink, an American songbird. In his elaborate *Field Book of Wild Birds and Their Music* (1904), Mathews had made impressively detailed transcripts of the songs of almost 130 North American wild bird species. He had transcribed them all by ear and in musical notation because "only the musical staff expresses [them] accurately" (Mathews 1904, xxi). With that claim, however, Mathews found himself caught up in an emerging controversy.

Animal sounds had been recorded in musical notation for centuries—the very terminology of *song* and *singing* testifies to their original appreciation in musical terms. In the first decades of the twentieth century, however, a range of other techniques was developed to record such sounds "after nature" (Ranft 2001). Some naturalists found bird sound better represented by nonsense syllables or phonetic vowels; others devised their own graphic notation. In unison, these recordists rejected musical notes as inaccurate and unscientific descriptions. They argued that musical scores represented the pitch and harmony of a song but that other features were more important. Such features, like timbre and sound quality, were rendered more clearly by their respective systems. Mathews in turn sneered at his skeptics and claimed that they were just not knowledgeable about musical theory (Mathews 1904, i).

However, in a twist of irony, the song that inspired Mathews to write such powerful prose appeared unsusceptible to his favored approach. Acknowledging that "the difficulty in either describing or putting upon paper such music is insurmountable," he quoted a fellow naturalist who had failed to write down this song as early as 1892: "We must wait for some interpreter with the sound-catching skill of a 'Blind Tom' and the phonograph combined, before we may hope to fasten the kinks and twists of this live music-box" (Mathews 1904, 49).[2] Recording and representing the sounds of nature clearly came with its own set of problems—problems that mechanical recording, or so these naturalists hoped, could help to solve.[3] Earlier, in 1898, the audience of the American Ornithologists' Union congress had been played a sound recording of a captive bird, indeed suggesting "great possibilities to be looked for in the future" (Sage 1899, 53). It was not until three decades later, however, that university ornithologists and the recording industries in Britain and the United States began organizing frequent bird-sound recording expeditions. From the late 1920s onward, electrical recording devices began to feature prominently as tools for the study of birds' singing behavior in the field.

With this shift from recording by ear to a mechanical recording with microphone, the culture of scientific recording also changed. The emergence of electrical recording not only presented ornithologists with different material and technical conditions of their work. It also forced them to rearticulate the role of recording in their scientific studies, as well as the practical conventions guiding it: How was one to deal with sound recording as a scientific practice? In this chapter I connect these material conditions and scientific conventions by asking how the mechanization of sound recording has altered the ways ornithologists dealt with natural sound as an object of scientific study.

I show that electrical recording confronted the ornithologists with a different way of listening in the field—and later in the lab. Recording through mechanical "ears" rather than human ears gave ornithologists certain advantages, such as replay or "faithful reproduction," but they also came to experience the world as noisier, more contingent, and more uncontrollable than they had before. Electrical amplification now forced ornithologists to decide what sounds to record, what to discard as noise, and how to eventually represent those sounds. I show that although these recordings were made outdoors, some scientists tweaked them such that ultimately they transformed a "messy" field sound into "cleaned-up" imagery. Through specific recording equipment and techniques such as the parabolic microphone and the spectrographic visualization of sound as data, ornithologists gained a particular form of control over field sound. As a result, the distinction between field and laboratory practices becomes blurred. This, I suggest, results in a consistent sonic sterilization of field sound, the intentional erasure of noises.

Thus, in investigating sound recording as a scientific practice, this chapter traces dynamics between field and laboratory, between science and music, between wanted sound and unwanted noise, and between the audible and the visual. I do so by looking into the actual recording work of a small number of recordists and ornithologists in the United States and the United Kingdom between the 1910s and the early 1960s. Empirically, this study draws on a longitudinal reading of five leading British and American ornithological journals, *The Condor*, *The Auk*, *The Wilson Bulletin*, *Journal of Field Ornithology*, and *British Birds*, combined with archival work at the British Library and a selection of memoirs by the ornithologists and recordists under study. The goal is decidedly not to survey a long history of wildlife sound recording. Instead, I single out two periods, in the 1930s and the 1950s, in which the entanglement of the production of sound recordings and scientific culture and conventions becomes especially clear. Additionally, the focus of this chapter on the techniques of nature recording and the involvement of the recording industry and academia presents a relevant contribution to the recording histories collected in this handbook (Helmreich, Kursell, Sterne, Taylor, this volume).

THE SOUND OF FIELD SCIENCE

This chapter signals a history of sound recording *in the field*, outside the typical concert hall or recording studio. This outdoor focus is not uncommon in science studies, where the field site is considered not just a random place. Crucially, like the laboratory, it is also found to be a site of scientific labor, as well as a distinct cultural category (Kuklick and Kohler 1996; Burkhardt Jr. 1999). In scientific work, the field site and the laboratory are recognized as two ways of legitimizing knowledge, each associated with different occupants, vocabularies, materials, and virtues (Gieryn 2006). Both sites have been considered opposites that mutually defined each other even though in practice they have sometimes been hard to distinguish between (Kohler 2002).

For natural scientists, the field site has remained a place where they encounter their object of study in its natural and unspoiled state. The ornithologists quoted in this chapter all drew heavily on their recordings in the field. Being *there* granted them the unique authority of personal witnessing. Yet phenomena in the field are often also considered particular, unrepeatable, and uncontrollable (Collins and Pinch 1993). This makes scientific work in the field a complex task. Moreover, public field sites are almost never exclusively scientific domains: Other actors, processes, and materials are after all difficult to fence out (Kuklick and Kohler 1996). Consequently, the field is associated not just with immediate experience but also with distraction, lack of control, and imprecision.

Modern laboratories, on the other hand, are claimed to grant scientists control over their objects of study (Latour and Woolgar 1986). Laboratories create their objects of investigation by detaching natural phenomena from their original context and environment. It is this isolation of the object against the standardized and sterile background of the laboratory that is said to yield experimental results that are "objective" and "universally true" (Kohler 2002). At the same time, the lab enables scientists to measure, manipulate, and control natural objects (Knorr-Cetina 1999).

Recent studies have shown this issue of control and isolation to be in place also for laboratory *sound*. Cyrus Mody (2005), for instance, describes how sound can be perceived as a sonic contaminant that potentially disturbs scientists' experimental setups. Droning infrastructure or outside traffic noises challenge the conviction that laboratories not only look and perform alike but may also *sound* alike. Moreover, this perception of acoustic pollution is by no means exclusive to high-tech, modern-day laboratories. Henning Schmidgen (2003) shows that even in the nineteenth century, psychological experiments were found to be heavily disturbed by noise from outside and even inside the isolated rooms of subject and experimenter. A lack of what could be termed *sonic sterility* in this laboratory context apparently challenges what scientists regard as faithful, reliable, and objective practice.

Laboratories, in contrast to field sites, are places where, also sonically, control is desired. This study expands this work on sound in scientific practices by focusing on a "nonlab" environment. It investigates the ways in which ornithologists and recordists have dealt with a working environment where isolation of and control over natural acoustic phenomena are particularly hard to achieve. Work by Thomas Gieryn (2006) can suitably serve to conceptualize the relation between laboratory and field site. As Gieryn demonstrates, this relation is a dynamic one. Urban sociologists would in their papers continuously oscillate between such categories of lab and field. By rhetorically constructing their work as an experimental laboratory practice, as well as an observational field practice, these sociologists have managed to relate to the epistemic authority of both places. Here, laboratory and field feature not so much as physical workplaces but as textual constructions or frames of mind that "legitimize" these scientists' results.

Gieryn identifies three dimensions to schematize how researchers shuttle back and forth between their constructions of the city as a lab or field. First, at the field site, objects are *found* naturally, while in the laboratory they are *made*, crafted first into suitable objects of research. Furthermore, at the field site, objects of research connect intimately with the *place* of research, while the laboratory practice intentionally constructs environments and objects that are both *generic* and *universal*. Finally, field and lab position analysts in different ways vis-à-vis their objects of study. At the field site, researchers get *immersed*—their research object is everywhere around—while the laboratory creates a distance between researcher and object: "[A]long with the white coat comes a *detached* objective view from nowhere. Elements . . . are manipulated in a passionless, mechanical, and antiseptic way" (Gieryn 2006, 11).

Drawing upon these dimensions (ready-to-use, specific context, and immersion versus craft-to-use, generic context, and detachment), I argue that although nature recordists necessarily recorded their specimens in the field, their recordings were also made to fit Gieryn's category of the lab. Using mechanical recording techniques, these recordists crafted the sounds they collected in the field in ways that ultimately problematizes the distinction between lab and field.

Naturalists Listening: Graphic Scores of Birdsong

Since the end of the nineteenth century, nonprofessional naturalists[4] and university biologists have increasingly turned from museum collections to field sites and zoological laboratories to systematically observe living animals' behavior (Burkhardt Jr. 2005).[5] Along with the rise of ethology and behavioral biology in the first decades of the twentieth century, the field gained special importance for ornithologists interested in the function of birdsong. Which setting, however, granted the best scientific insights into the animals' natural behavior remained an issue of contention (Burkhardt Jr. 2005). In the 1920s, for instance, the young recordist Ludwig Koch, whom I discuss in more detail later, was told by his mentor, Professor Oscar Heinroth—one of ethology's founding fathers—that his attempts at recording birds in the open were "merely a circus performance" (Koch 1955). After all, Heinroth argued, birds behave the same in captivity as in natural surroundings. However, even then, naturalists appealed to the field site to make their observations of songbirds.

In doing so, field ornithologists in the first decades of the twentieth century devised a whole array of recording techniques to replace the musical notation technique that was common then. Simpler graphic and phonetic techniques were thought to be more practical for the demanding job of field recording and identification. In addition, American naturalist Saunders (1916) claimed, existing methods of recording birdsong gave no satisfactory impression of the actual song. Musical notation only made a *natural* song fit the rules of an *artificial* system: "We cannot fit wild bird songs to our standards of music" (Saunders 1916). Thus, a practical method was needed that discarded the "mechanical rules of human music, without losing any of its scientific accuracy" (Saunders 1915). Saunders proposed to represent the pitch and duration of a song on a schematic graph along two axes, which, he claimed, did not obscure variations in sound as a musical script would.

His proposal met with controversy, however. A musician, naturalist, and avid advocate of musical notation in the American Ornithologists' Union, Robert Moore objected that Saunders's graphical system was actually less comprehensive because it obscured those features of interest to the student of birdsong that musical notation had represented far better (1915). Moore's skepticism triggered an exchange of letters that led Saunders eventually to discard his opponent's arguments as nonscientific: "[W]e must realize that it is our intention to study bird songs, not from the standpoint of a musician but from that of a scientist." Moreover, "the science necessary to the student of bird songs consists almost entirely of the physics of sound, not the use of technical musical terms" (Saunders 1916, 103). Saunders may have exaggerated here, for his graphic method still plotted sound against a musical scale rather than in acoustic frequencies. Still, this remark does suggest a desire to systematize the analysis of bird sound by recording it in a "more scientific" way—different from the musical techniques naturalists usually applied.

This was a quarrel over what constituted the right scientific methodology and how to listen to sound. Moore seemed to recognize in the bird utterances harmonic relations and a distinct meter. Saunders, on the other hand, rendered pitch in microtones—*without* key or meter (figure 5.2). His graphic technique, in other words, was not just easier for ornithologists without any musical background to apply; it also provided a different way of analyzing the sound. Many such claims emerged around the time: some believed the quality of a bird's sound should be compared to musical instruments, and Hunt (1923, 202), for instance, proposed to represent bird sounds into English phonetics "since bird sounds are essentially human speech-like." Although all of these recordists stressed the need for a uniform method of recording, they did not seem to agree on what it was they were actually listening to. At the same time, however, even when modeling bird vocalizations on speech, these critics of a "musical" approach to birdsong did not abandon the common term *birdsong* to denominate their unit of analysis.

This confusion over the preferred way of recording and analyzing birdsong persisted well into the twentieth century. At the same time, however, some ornithologists had begun to expect much of mechanical recording in their field. Writing in 1929, Aretas Saunders commented: "Perhaps some day we can devise a phonograph that can amplify bird songs sufficiently to record those of wild birds. Then we

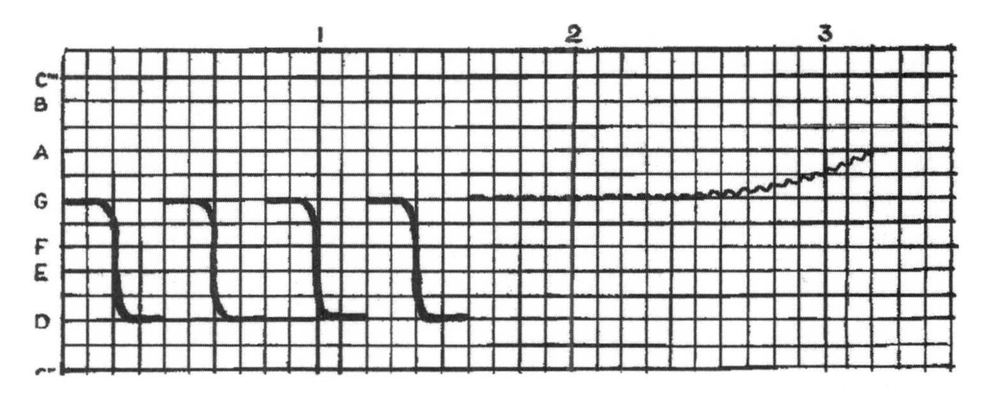

Figure 5.2 A graphic recording of a field sparrow song by Aretas A. Saunders. From Aretas A. Saunders, "Some Suggestions for Better Methods of Recording and Studying Bird Song," *Auk* 32(2) (1915): 173–83, 175.

shall be able to play the records over as much as we like and analyze the song in detail" (Saunders 1929, 7). Likewise, ornithologist Lucy Coffin believed that "it would be worth while . . . to thoroughly investigate the latest developments in phonograph recording in relation to the recording of bird songs" (1928, 99).[6] These quotes demonstrate a shared but not very specific optimism in the potential of mechanical recording for field practices.

Mechanical Recording: Out in the Wild

As early as the spring of 1929, for some the advantages of mechanical recording over other techniques took a more concrete form. A movie production company appealed to the expertise of Dr. Arthur A. Allen, an eminent scholar of bird behavior at Cornell University. The Fox-Case Movietone Corporation sought to record birds in the wild to accompany one of their films and asked the Cornell ornithologists to help them approach their subjects. Eager to collaborate, the ornithologists found themselves confronted with a truck equipped with microphones, electrical amplifiers, and, most important, a "sound camera" that recorded sound waves into a black-and-white inscription on movie film.[7] Allen and his colleague Peter Kellogg saw some exciting implications of these devices for their own field studies. In hindsight, as their associate at the American Museum of Natural History Albert Brand put it, "previous methods . . . were, at best, only makeshifts, awaiting the time science should have advanced sufficiently so that *faithful* reproductions . . . could be made" (Brand 1932, 436).[8]

In the same period, the German and British entertainment industries also sought the cooperation of ornithologists to start mechanically taking birdsong out of doors. Together with ornithologist Oscar Heinroth and under the supervision of producer Ludwig Koch, the German record company Carl Lindström published a first bird sound book in 1935. Koch soon moved to Great Britain, where he set up long-term collaborations with well-known British ornithologists Julian Huxley and Max Nicholson and also with the Parlophone record company (Koch 1955). These producers relied on ornithologists' field expertise to guide them in the field, while the professional ornithologists were able to familiarize themselves with the latest sound-recording technologies: sound film photography at Cornell and Parlophone's disc-cutting phonographs in the UK. Mechanical recording practices now required expertise in engineering and physics rather than musical theory.

However, while Koch and his team of ornithologists and engineers continued recording for Parlophone and the British Broadcasting Company until the 1950s, the contacts between the Cornell University Ornithological Laboratory and the movie industry were rather short lived. Although the Cornell ornithologists thus were no longer able to draw on the expertise of specialized recording engineers and technicians, the department decided to purchase its own equipment for scientific research.[9] After all, Brand noted, "many of the secrets of avian life are hidden in an understanding of the meaning of the song." With phonograph recordings, "the serious student of ornithology could study song in a way that has been impossible heretofore." So, he added, "if the motion picture industry can take sound out-of-doors, the naturalist should also be able to do so" (Brand 1932, 436).

Still, despite the availability of electrical amplification and microphones, recording birdsongs outdoors remained a daunting task. So much equipment was required that it all had to be transported in a van that was so big that "the police [would] not allow it to be left on the public highway" (Nicholson and Koch 1937, 30). The sound truck carried the amplifier and multiple microphones, as well as a great number of storage and dry cell batteries for the power supply. Moreover, a phonograph would cut the recording in wax discs. To make the needle run smoothly on the wax, the recordists had to soften it in a large electric oven. Finally, to make the needle run true on the wax, the phonograph and recording van had to be precisely leveled so the equipment could stand upright. Indeed, the recordists themselves would refer to the recording truck as a studio, similar to those at Abbey Road (Nicholson and Koch 1937).

Transporting this studio into the wild was a logistic tour de force not only because of the diverse staff that had to operate the van and its equipment but also to get the studio in place. Whereas the recordist could depart from the tracks to follow the bird to its perch, the heavy recording equipment was less flexible in use. Roads, paths, and trails formed a network of mobility that became vital to the logistics of nature recording but also enforced some restrictions. As Brand warned his readers, "It is not as simple as it would seem to get a location where there is absolute quiet . . . Too great proximity to a traffic road, for instance, makes recording impossible" (Brand 1932, 438). Hence, the recordists' reliance on roads to reach distant areas for recording had an ironic effect: It produced noisy surroundings.

Both Brand and Koch complained that their recordings of bird sound were often interrupted by a passing airplane or turned out to be unfeasible because of the hum of a distant highway (Brand 1932; Koch 1955). The ornithologist Max Nicholson had accompanied Koch on some of his recording excursions into the British countryside and recalled how such infrastructure often drastically complicated their attempts at recording:

> Aeroplanes, motor-cars, lorries, trains, and motor-bicycles combined to shatter the tranquility which had been so perfect a few hours before. Just as smoke pollution helps to swamp a town under fog, so the natural peace of the country was drowned under the indefinable hum of distant engines and wheels . . . . Until one has listened *objectively* to all these sounds coming through the loud-speaker, in what counts still as a peaceful retreat from the bustle of London, it is hard to realize what a noise-ridden world we have managed to make ourselves live in. (Nicholson and Koch 1937, 38)

The comparison of noise to a fog that "swamps" and "drowns out" other sounds evokes a sense of its ungraspable omnipresence. These recordists' aversion to traffic noise was part of a more general resistance to all kinds of mechanically generated noise emerging in the first decades of the twentieth century. Historians of sound and noise Emily Thompson (2002) and Karin Bijsterveld (2001, 2008) have described significant antinoise campaigns in American and European cities between 1910 and 1940. After 1925, the decibel standard and technical instruments such as acoustic meters were called upon for a more systematic and "objective" measurement and comparison of city noise levels.

It is then no coincidence that those who would abate noise increasingly considered noise a problem in this "mechanical age" (Bijsterveld 2001). Not only did mechanical artifacts and "machinery" generate more noise, but mechanical measuring devices or recording technology also made noise increasingly audible.[10] As such, the recordists' audible experience changed into one of a noisy and uncontrollable field site. As Koch explained, "the sensitive microphone takes up all noises, often within several miles' radius, and exaggerates them" (Nicholson and Koch 1937, 20). At Cornell, Brand concluded that "a mechanical ear . . . transmits all the sounds it hears, without discrimination. Though man has been able to invent a mechanical ear—the microphone, and it is an extremely sensitive one, too—he has not been able to equip it with a brain" (Brand 1934, 19).

A mechanical ear recorded sound not just as it was heard by humans: It also assumed an indiscriminate, unselective registration. "Heretofore," Brand stated, "the bird song student has had to rely entirely on auditory impressions," but with mechanical devices such as the sound camera, "an objective medium of study is now available" (Brand 1935, 192). This straightforward equation of objectivity with a mechanical apparatus Lorraine Daston and Peter Galison (2007) have termed *mechanical objectivity*. Until the mid-nineteenth century, they found, scientists called upon their experience and personal judgment to decide what to represent and how to do so. Yet increasingly, scientists considered such practices of selection and idealization to be "merely subjective" projections, disturbed by an artistic fancy. Instead, scientists valued data that were "unmarked by prejudice, skill, judgment, or interpretation" (Daston and Galison 2007, 17). To comply with this emerging morale of nonintervention and abstinence, they delegated as many of their tasks as possible to mechanical procedures or automatic registration techniques. Because mechanical reproductions were believed to be guided by protocol, scientists considered them impartial and objective. Similarly, mechanical recording assumed a superior means of sound reproduction than artisanal notation techniques.

In sum, the introduction of mechanical recording changed the field of ornithology in multiple ways. It did so first in a practical sense: Instead of a naturalist with a good musical ear and a notebook, fieldwork now involved specific, costly equipment and a number of skilled professionals to operate it. However, it also transformed the audible experience of the recordist: Bird sounds were suddenly drowned out by the hum of traffic, industry, and vacationers, qualifying the field as a particularly noisy and somewhat uncontrollable environment. Finally, mechanical devices redefined the understanding of a sound recording into mechanically mediated content. What made a good recording was no longer what Saunders and Moore had called the "accuracy" of a human observer but what Brand and Nicholson perceived as the "objectivity" produced by the recording device. Both aimed to reduce interpretation from the act of recording, but while Saunders relied on a disinterested observer and a nonmusical system, Brand disqualified the observer in favor of a "neutral recording machine." Mechanical devices thus rendered artisanal techniques of recording not just *inaccurate* but also *subjective*.

Mechanical recording partially solved a problem of faithful reproduction with which ornithologists had earlier struggled. At the same time, however, it also created a new one: the perception of noise. As such, mechanical listening generated a new experience of field sound. How, then, did the sound recording teams in London and Cornell come to deal with the presence of such unpleasant yet "objective" noises?

Dealing with the Field: Studio Recordists and Ornithologists

Koch and his British recording team dealt with the realities of field recording by employing a strategy common in the studio recording practices of the day. While in the field, they would adjust sound levels by a strategic placement of microphones and the singing bird (which sometimes had to be persuaded to perch near the microphone). Historian of sound technology Susan Horning (2004) has argued that, since the early days of sound recording, one of the most important skills acoustical recordists had to acquire was to know where to position the sound source with respect to the recording element. Without amplification of the source, this tinkering in real time was the only way to adjust the balance between singer and the various instruments on the recording. However, even when recording turned electric in the 1920s and the sound quality drastically improved, the greater sensitivity of the microphone also demanded greater skill in placing it in relation to the sound source, and this expertise could be acquired only by trial and error. It was this particular skill that the recording engineers and technicians in Koch's unit had learned in the studio. Unlike Koch, these men had little or no experience of working outdoors.

Koch and his team outlined an approach similar to that used in studio recording when he advanced that a real birdsong hunter must "with skill and much patience go about the job of bringing the songster as near as possible to his microphone, or rather the microphone as near as possible to the songster" (Koch 1955, 3). At all times, Koch had to arrange for the microphone to be near the bird "as if he were a performer in a studio" (Koch 1955, 73). Despite these arrangements, however, he also realized that the conditions of the field were considerably different from those in a studio. In fact, he preferred the former. "If one wishes to avoid such natural disturbances, then only recording in a closed room can be considered, and, as I have found in my own experience, in such altered acoustical conditions the bird-notes are distorted and no longer make a natural record" (Nicholson and Koch 1937, 21). The natural environment was an important asset for this kind of recording. As Julian Huxley repeatedly stated in his introduction to Koch's first British sound book, Koch's records had "the quality for evoking the bird's environment" and "a true picture of the birds' voices." This "natural" sound was what Koch was after. "In my nature recording I invariably observe the principle of getting the bird-notes in question in the foreground by all means, but always with their natural background" (Nicholson and Koch 1937, 21).

However, in the given circumstances, making such a record required a few crucial interventions. The team had to distinguish between the bird's voice in the foreground and the noisy environment in the background. To do so, the recordists would arrange up to six different microphones to enclose the bird's song perch, thus improvising a large-scale recording room outdoors (figure 5.3). Each of these microphones was linked to a control panel in the van several hundred yards away, where an incoming sound to one of them would be cut right into the wax disc. Occasionally the bird would fly away, and then the whole disposition of the microphones would have to be changed, including tests to make sure the microphones "cut out as much as possible of interfering noises" (Nicholson and Koch 1937, 37). These quotes make it clear that while the recordists considered noise part of the reality of field recording, the intended purpose of the recordings also dictated that interference was to be reduced as much as possible. Furthermore, as Sterne (2003) shows, the act of recording always implies mediation. To make a "natural-sounding" recording required at least some engineering and staging of the recorded sound, guided by an implicit preference for how nature should actually sound (also see Helmreich, this volume).

Figure 5.3 Parlophone recording van and setting up several microphones. From Max Nicholson and Ludwig Koch, *More Songs of Wild Birds* (London: Witherby, 1937, 5).

On the other side of the Atlantic, the recordists at the Cornell Laboratory of Ornithology were facing similar problems but eventually took a rather different approach. After experimenting for some months with the recording equipment, the ornithologists experienced some success but also had to acknowledge that noise drowned out most of the recorded sounds. "If we were to achieve any kind of perfection," Kellogg recounted, "we would have to develop our own specialized equipment and techniques for fieldwork" (Kellogg 1962, 38). The group allied with Cornell's electrical engineering department to rebuild its equipment for this purpose. In addition, after some experimentation, the group manufactured a parabolic reflector that became an important instrument for the ornithologists at Cornell. According to Kellogg, the "parabola for picking up and concentrating bird songs on the microphone was probably our greatest piece of good fortune. Recording would have been possible without a reflector, but the results with it were so superior as to make the instrument a universal tool in this field" (Kellogg 1962, 39).

Although Kellogg and Allen would use the reflector for all of their expeditions, the instrument was not quite yet universally used, but parabolic reflectors did become more common in the 1950s. By then, the microphone could be combined with a tape-recording setup—Kellogg helped design the Magnemite model, which would become the standard recorder for the next decade. The tape recorder provided an acceptable semiportable alternative to the heavy sound camera or disc recorder. Since the 1920s, this costly and complex equipment had made sound recording the almost exclusive preserve of professionals, but the tape recorder and parabolic reflector made field recording more affordable and easier to do, attracting also amateur recordists with an interest in natural sound (Kellogg 1962).

Especially in the first years of the laboratory's Library of Natural Sounds, Kellogg would instruct amateur recordists on the proper use of the equipment made available to them.[11] In this way, parabolic recording could indeed be successfully implemented as a standard in field recording.

But what then, made the parabolic reflector so superior to the ears of the Cornell ornithologists? The principle was such that the surface of the parabola reflected sound waves to a dynamic microphone at its focal point (figure 5.4). Focusing the sound waves like this drastically increased the input to the recording equipment. As a result, the reflector would help to record sound at a much greater distance than microphones in an ordinary setup. Consequently, bird sounds could be recorded at distance, in flight or in inaccessible places—an added advantage to students of animal behavior in the wild. Equally important was that directional recording amplified only those sounds at which the recordist directed the parabola, which permitted a greater selectivity in recording. Albert Brand explained that "the outside noises are very nearly shut out, while the sounds wanted are greatly increased" (Brand 1934, 23). As an article in *Science News Letter* read, it was a way to "overcome

Figure 5.4 Focusing a parabolic recorder. From Albert R. Brand, *Songs of Wild Birds* (New York: Nelson, 1934, 2).

much of the handicap imposed by lack of soundproof studios where the wild birds sing."[12]

Recordists could more easily eliminate interference from other ambient sounds. By these means both the Cornell ornithologists and Koch's team attempted to minimize noise interference on their recordings. The parabolic reflector, however, did enable a differ kind of control. While the natural environment featured prominently in the background of Koch's recordings, the parabolic reflector technique excluded most other sounds. It produced a close-up recording: a sound that was sonorous and had a high signal-to-noise ratio but provided little information about its context. The analogy of binoculars or a gunsight mounted on the reflector is especially useful here. Just as the binoculars at the same time frame and magnify a distant image, so the parabolic microphone intensifies only those distant sounds that occur within its limited and directional scope. The technique had an interesting impact on the recording: depending on the level and proximity of ambient noise, it created a "sterile" background—indeed, a sonic sterility that seems typical of the soundproof studio that came into being in the 1930s, a space that constituted clear and controlled, direct and nonreverberant sound (Thompson 2002).

The parabolic reflector produced a kind of sound studio that was different from Koch's approach. More specifically, it also seemed to "laboratorize" the recording in a way that evokes Gieryn's schematic distinction between the laboratory and the field site. For the parabolic reflector was not only used to reduce some of the ambient noise in the recording. By reinforcing only those sounds at which the reflector was aimed, it also isolated and detached them from their surroundings, enforcing an aesthetic of the individual. Although in practice it may have been hard to eliminate *all* superfluous sounds from the recording, the technique was different from Koch's in that it did not aim to record the bird against a natural background but to catch the sound of the bird and nothing but that. These recordings revealed little about the local context. They did not evoke the bird's environment in the way ornithologist Julian Huxley appreciated Koch's recordings; they were not created to have properties such as the "fullness, immediacy, and emotional completeness" of an aural immersion (Nicholson and Koch 1936, xiv).

This is of course not to suggest that Koch's recordings could not sound sonorous or close up. The point is that the parabolic recorder was designed to create close-up recordings within a panoramic, immersive, and noisy natural soundscape. In effect, bird sound was mechanically *detached* from a specific context. Sound recording was physically detached by enabling a greater distance in the field between the recordist and the singing bird. However, by suppressing contextual sound, it also detached the sound by making it resound "without perspective" or sound as though it were "coming from nowhere." It is this particular way of crafting a sound by suppressing its context and surroundings that enhances sonic sterility and frames recorded natural sound in a "laboratory" way.

To further conceptualize this "laboratorization" of the field, an essay by art historian Svetlana Alpers (1998) is revealing. In exploring an analogy between the artist's studio and the scientific laboratory, she regards both not only as physical workplaces of artists and scientists but more fundamentally also as an instrument and "a frame of mind." Her discussion of the relation between the landscape and the artistic studio is especially noteworthy here. In an accepted historical account of nineteenth-century landscape painting, painters freed themselves from studio conventions and began painting "real landscapes" instead. However, Alpers claims, whether painted outside or inside, their work was not necessarily *studio free*. Even in landscape painting, she finds, a landscape *found* outdoors could still be transformed into what seemed to be an object *composed* in the studio: Even outside at the field site, the landscape could be represented as a still-life motif, reducing the expanse of the landscape to the static and focused presence of an object.

I do not intend to evaluate this observation on its art historical merit but use it heuristically to highlight two things. First, following Alpers's account, the "studio frame" can be operated as effectively by the artist in the field as it can by the studio painter. Likewise, a "laboratory frame" need not be restricted to experimental cultures within laboratory walls. Even a contingent, often distorted and hard-to-control field site may be materialized, conceptualized, or presented as a space that incorporates laboratory-like features (Bont 2009; Gieryn 2006; Kohler 2002). However, as the case of parabolic recording demonstrates, such features may also remain very much implicit and hidden in the material conditions of recording. Second, Alpers's discussion of landscape painting calls for an analogy to wildlife sound recording. The landscape painters, whether or not adopting a studio frame in their work, translated a three-dimensional surrounding onto a flat canvas. Similarly, the directional way of parabolic sound recording seems to position the listener *in front of* a framed auditory canvas rather than *in* it. Like the painting and the observer, the design of the sound recording does not immerse the listener.

Intriguingly, it is difficult to articulate the effect of directional recording without resorting to visual analogies such as "close up," "perspective," "focus," or "frame." Of course, this is not to suggest that ornithologists themselves consciously employed such analogies to develop this recording technique. Rather, this example puts into a critical perspective what media historian Jonathan Sterne once dubbed an "audiovisual litany" (2003, 18). Many theorists and historians of sound, Sterne claims, have taken refuge in a transhistorical idealization of the nature of sound. In this customary conception, only hearing puts the individual into close contact with the world; it is *essentially* immersive, subjective, and affective. This has been contrasted to vision—generally considered directional, enhancing distance and objectivity. The history of parabolic recording, in which hearing is not immersive but directional, distancing, and objectifying, warns against a conception of sound and hearing that emphasizes its putative transcultural and transhistorical features (see also Smith, this volume).

Analyzing Sound: Listening with Eyes and Hands?

Mechanical recording changed not only the way ornithologists in the field experienced bird sound but also how they treated it. Yet, although mechanical devices provided a way to experience and record sound more "objectively" and to deal with the consequences of such "faithful reproduction," it did not do away with the initial problem of how to *describe* or even *analyze* bird sound when out of the field (Daston and Galison 2007). In a paper titled "Why Bird Song Can Not Be Described Adequately" (1937), Brand relates how listening to his own phonograph records would stir discussion with Dr. Allen. Although both were experienced students of bird sound, they each heard different resemblances between the vocalizations they compared. However "objective" the recording might be when played in the lab, he concluded that "the subjective reaction of the listener is much the same as in the open" (Brand 1937, 14).

Brand had personally witnessed a series of experiments at Cornell that demonstrated that considerable "fading points" existed in the hearing spectra of most people. Moreover, listening was shown to have a significant psychological factor: "[W]e hear what we are listening for and what we expect to hear," for "it is impossible to separate the hearing apparatus from the thinking mechanism" (Brand 1937, 12). If earlier Brand regretted not being able to equip the mechanical ear with a brain, he now suspected that the brain permeated human hearing with the flaw of individual perception. To Brand these experiments proved that it was not just recording but also *hearing* that was to be mistrusted: Listening after all was a subjective and individual experience (Brand 1937, 12).

In contrast, he advanced recording on sound camera film as "a medium from which sounds can be studied objectively" (Brand 1938, 63). The camera would reproduce an image of recorded sound that, once developed, could be analyzed under a microscope. This way, the "extremely short notes and those of very high pitch, often inaudible to the human ear, can be clearly seen and studied" (Brand 1935, 40). In an experimental test in 1938, Brand and one of his students, Harold Axtell, used a sound film recording to determine that it took an accomplished musician about four hearings to correctly estimate the notes of an ordinary utterance (1938). Their finding suggested that some of the observations Saunders (1929) had collected in what still counted as an authoritative study might actually be up for revision. Moreover, the implication of this juxtaposition of the human listener and the visualization by mechanical means was that listening to sound was more difficult than *looking* at it.

Although this did not keep ornithologists, including Axtell (1938) himself, from recording bird sounds by ear, in the following years a series of mechanical devices for the visualization of sound, such as the "Vibralyzer," the oscillograph, and the audiospectrograph, were introduced into the analysis of bird sound.[13] From the early 1950s onward, these technologies would provide what was (again) considered to be "objective data that are more detailed and accurate than those obtained by most methods heretofore used" (Borror and Reese 1953, 276). Most of these technologies had been developed during wartime experiments to detect the sounds emitted by enemy submarines. Yet after the war, more peaceful applications, such as the analysis of animal sounds, were quickly recognized (Potter, Kopp, and Green 1947). Eventually, one of these technologies, the audiospectrograph, even inspired some to compare its value for the science of bird sound to that of the microscope for (cell) biology (Marler and Slabbekoorn 2004; Baker 2001).

Why did the audiospectrograph become such a success? One way to account for this would of course be to understand the device as another stage in the development of mechanical recording—technology now also contributed to sound *analysis*. Additionally, the audiospectrograph connects to other recording technologies and techniques in that it inscribes sound. Such inscriptions—nonmechanical such as musical scores or mechanical such audiospectrograms—we could call with Bruno Latour (1986) "immutable mobiles." Like the printing press, the recording device makes mobilization and immutability possible at the same time by casting many identical copies of a unique sound event and enabling those to be circulated widely. By conserving, multiplying, and displacing sound, a recording challenges what may be called the "ephemerality" of sound, that is, its uniqueness in time and place. Unrecorded sound "evaporates," while recorded sound is detached from the field and made available for others to be repeated at will. Such mechanical inscriptions are not necessarily limited to *visual* technologies such as the sound camera or the audiospectrograph. Yet, if with Brand one finds that *listening* to sound may be deceptive and more difficult than *looking* at sound, what advantage did the audiospectrograph then provide?

According to Latour (1986), inscriptions are not just immutable and mobile. They may also be *flat* and are thus more easily cut up, recombined, compared, or superimposed—in short, dominated. An audiospectrographic comparative analysis of Western meadowlark songs by Lanyon and Fish (1958) demonstrates how flat and printed inscriptions work differently from sonic inscriptions. The article presents a plate of four spectrograms of a Western meadowlark call note (figure 5.5). The calls were recorded in different field locations from California to Wisconsin. Referring to the combined spectrograms, the authors concluded that all notes were identical. They then stated that "*collectively*, [the spectrograms] suggest that Western meadowlarks render the same characteristic call note regardless of their geographical location" (Lanyon and Fish 1958, 340).

This quadruple comparison of evidence *is* Lanyon and Fish's central argument (Lynch 1988), made possible only because the original recordings were processed into inscriptions, which can be juxtaposed and combined especially because they are flat. To understand its value, one has only to imagine the same fourfold presentation of the original recordings but played out loud simultaneously. In comparison, two-dimensional, printed inscriptions can be made to "tune" only when the observer looks at them—one at a time or all at the same time.

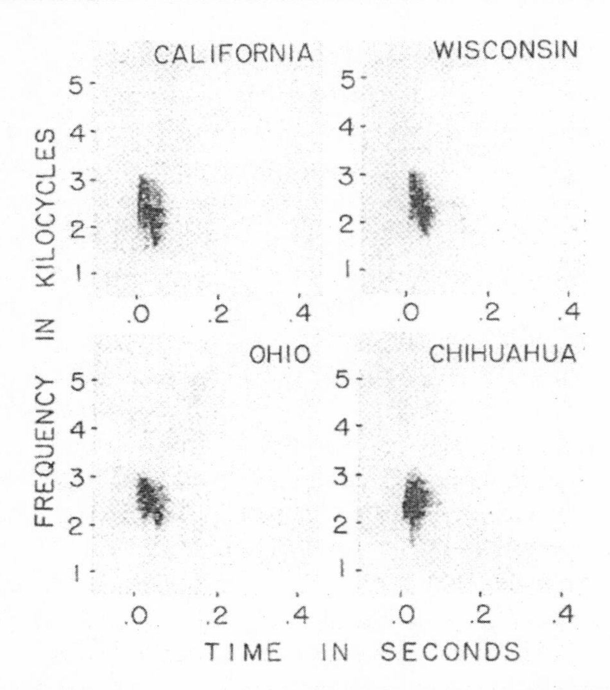

Figure 5.5. Audiospectrographic analysis of geographic variation of Western meadowlark calls. From Wesley Lanyon and William Fish, "Geographical Variation in the Vocalization of the Western Meadowlark," ***Condor*** **60 (1958): 339–41, 340. Courtesy Cooper Ornithological Society.**

Hence, the Meadowlark inscriptions do something besides representing sound, in Latour's words (Latour 1986), "synoptically." They enabled Lanyon and Fish to deal with the sequentiality of sound. After all, sound is not only tied to a specific time and place (it is ephemeral) but also happens "in time" (it is sequential). Even when a recorded bird sound is played again, every note is evidently replaced by the next and the next. Thus, where auditory analysis necessarily takes place *in time* and thus *takes* time, the audiospectrograph stabilizes time through combination and scaling. Spectrographic analysis, in other words, allows one to go back and forth and cut across sound sections. Recorded tape, on the other hand, may be easy to circulate, but it is cumbersome and time consuming to select, cut, paste, listen, rewind, and listen.

The analogy to the microscope may lead one to mistake the new "objectivity" of the audiospectrograph for a purely visual matter—just as the phonograph was not more "objective" simply *because* it was mechanical. Not just a mechanical inscription machine, then, the audiospectrograph seems to represent another phase in a process of gaining control over sound—a process that started with musical notation scripts. An explanation for the authority of inscriptions should not be sought in the single *visual* inscription but in a continuous process, a cascading of inscriptions. The audiospectrograph does not make other recording technologies obsolete. Instead, it processes recorded sound further into even flatter and more controllable, comparable units of sound.

Crucially, visualization enabled ornithologists to deal with the dimension of time in a more felicitous way. However, it also permitted them to deal with the problem of noise, by erasing it visually. An article by Marler, Kreith, and Tamura (1962) signals a specific approach to visual noise. After their parabolic recordings had been processed by the audiospectrograph, the originals had again been traced by hand in pen and ink (figure 5.6). In another article that same year, the authors indicate that their images are in fact "photographs of sonograms which were retouched with white paint to mask inscriptions and some background noise" (Marler and Tamura 1962, 369).

Both the parabolic microphone and the spectrogram were thus geared toward the erasure of noise from the recording. Ornithologists processed recorded sound into clean and sterile representations by manually eliminating the mechanically recorded noise. What Michael Lynch (1988) observed to be sequential transformations of an original image into a presentable image then concerns the whole process of sound recording. It is not just visual displays that are made more coherent with each other: Sound recordings—audible and visual—are schematized, simplified, and marked more clearly. In this process of schematization, tracings of the field conditions disappear; place-specific sounds and noises make place for a generic—white and mute—background.

Mechanical recording categorized noise as a concrete problem—rustle on a recording or a few printed spots. Of course, noise did not come into existence with mechanical recording, but it certainly did turn problematic. As naturalists heard "what they were listening for," precluding wind, foliage, and traffic, the "indiscriminate and objective" mechanical ear of the microphone had begun to guide a different perception of the environment. Yet in whatever way noise was experienced,

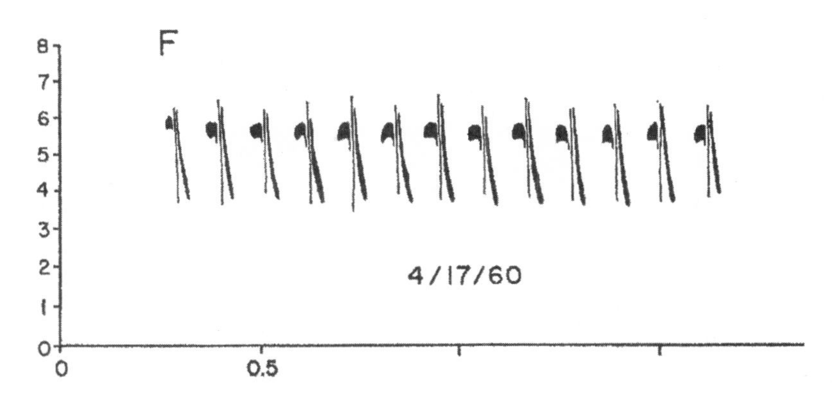

Figure 5.6. Ink tracing by hand of an audiospectrogram. From Peter Marler, Marcia Kreith, and Miwako Tamura, "Song Development in Hand-Raised Oregon Juncos," *Condor* 79(1) (1962): 12–30, 15. Courtesy Cooper Ornithological Society.

both mechanical and nonmechanical inscriptions remained silent about it. This does not mean that the field site itself became meaningless. Paradoxically even, the detailed and time-economic analysis of sonograms at the same time also enabled the geographical variation in birdsong to become a topic of growing interest (Baker 2001). However, while the field since the early decades of the twentieth century gained relevance as a place where ornithologists could observe the natural behavior of birds, it also emerged as a place that could be actively controlled and silenced. The impact of this conception of the field site on bird sound recording is best illustrated by the fact that the influence of a noisy habitat on bird behavior has only recently become an important topic of research. In the last two decades, researchers have begun paying systematic attention to changes in birds' singing behavior because of interference from city or traffic noises (Slabbekoorn 2004).

Conclusion

In this chapter I have explored ways in which ornithologists appropriated and redefined listening and recording as meaningful modes of acquiring scientific knowledge in the twentieth century. By analyzing two periods of time, I investigated how ornithologists made conscious decisions on *what* to record and *how* to record it. These decisions were structured by recording techniques and technologies. As listening became entangled with recording, editing, measuring, and reading, the parabolic microphone and the spectrograph represented cascading technologies of increasing control over sound, exemplified by a silenced and white background. It is through this sterile environment that the field site connects to the laboratory, as unwanted sound is intentionally eliminated.

At the same time, the point at which the field ends and the lab begins became blurred. In 1950s, for instance, ethologists Bill Thorpe and Robert Hinde (1956) at Cambridge University fashioned the construction plan of their ornithological field station in Madingley after a standard soundproof room. Sound insulation enabled them to do their experimental work on the learning of birdsong, they said, without too much intrusion from outside noises. Insulating the field station against unwanted sound, Thorpe and Hinde seem to merge the strategy of a modern laboratory with that of the modern studio (Thompson 2002). As Kohler (2002) has argued, a field station blurs the categories of laboratory and field. It becomes a way of drawing on the unique characteristics of the field site, while simultaneously enforcing a laboratory-like control over its contingencies. Indeed, also at Madingley a lablike control of sound was focused on particular frequencies of sound for specific experiments, while purposely disregarding others.

Like the Madingley soundproof room, the recording techniques analyzed in this chapter blur the boundaries between field, studio, and laboratory. They merged these categories not only by introducing a mobile studio *in* the field but even more intricately through the parabolic microphone and audiospectrogram. While dealing explicitly with sound located and recorded in the field—whether in New York or California—these instruments fabricated a crisp, controlled, and context-ridden sound. These features may be associated with the sonic control of a studio, but they can also be integrated in the "passionless" crafting, generic context, and detached approach of the scientific laboratory. These mechanical instruments illustrate a scientific culture in which places of science are demarcated by sterility and silence and underscore the need to understand how this scheme is enforced by and reinforces wider cultural expressions of modern sound control.

NOTES

1 I would like to thank Karin Bijsterveld, Tessa Fox, Fabian de Kloe, Julia Kursell, Mara Mills, Trevor Pinch, and Alexandra Supper for their feedback on earlier draft.

2 "Blind Tom" refers to an African American musical savant who lived in the latter half of the nineteenth century. In this context, he exemplifies the exceptionally skilled listener.

3 At that time, in other scientific disciplines the potential of phonograph recording for fieldwork was also acknowledged: Radick (2000) describes the recording of captive and wild primate sounds on wax cylinder and their use in evolutionary philology and biology since 1892. In addition, Stangl (2000), Ames (2003), Brady (1999), and Sterne (2003) discuss the use of phonography for ethnology, comparative musicology, and anthropology since the late nineteenth century.

4 In the United States and especially in Britain, tenured positions in behavioral biology were limited. Most work in ornithology was performed by amateur naturalists (Burkhardt Jr. 2005).

5 Before the late nineteenth century, so-called armchair natural history was most dominant. Ornithological study consisted mainly of taxonomic classification based on museum collections of specimens.

6 Although recordings of captive birds' songs had been available as early as the late nineteenth century, they were of little interest to Saunders and other field ornithologists. These recordings, often of talented songsters and mimics, were intended for entertainment, not for naturalistic research (Copeland, Boswall, and Petts 1988; Brand 1934).

7 The photographic sound recorder transforms sound waves into electrical energy, which makes a small glow tube flicker. The light, which passes through a small air slit, triggers a chemical change in a passing film. Thus being exposed at a constant speed, the film produces a black-and-white inscription that can later be played back as sound again. For a technical history of photographic sound recording, see Kellogg (1955).

8 Emphasis mine.

9 Even though the Cornell ornithologists intended their recordings primarily for scientific use, between 1932 and 1936 they also published a series of gramophone records of wild bird songs that were taken from their original film recordings. The royalties of these popular recordings helped to finance further study and recording of birdsong (Kellogg 1962).

10 Natural noise, such as wind, water, rain showers, foliage, and other birds calling, provided an equally challenging soundscape to work in.

11 See, for instance, Stillwell (1964).

12 "Reflectors like Airplane Detectors Catch Bird Songs," *Science News Letter* (Nov. 10, 1934): 299–300.

13 However, not all of these technologies visualized sound in the same way. They represented recorded sound on a time scale, indicating amplitude but not frequency or pitch. The sonagraph, on the other hand, represented recorded sound also on a time scale, indicating frequency and giving some impression of amplitude. See also Rieger (2009).

REFERENCES

Alpers, Svetlana. "The Studio, the Laboratory, and the Vexations of Art." In *Picturing Science, Producing Art*, ed. Caroline A. Jones and Peter Galison, 402–17. New York: Routledge, 1998.

Ames, Eric. "The Sound of Evolution." *Modernism/Modernity* 10(2) (2003): 297–325.

Axtell, Harold. H. "The Song of the Kirtland's Warbler." *Auk* 55(3) (1938): 481–91.

Baker, Myron C. "Bird Song Research: The Past 100 Years." *Bird Behavior* 14(1) (2001): 3–50.

Bijsterveld, Karin. "The Diabolical Symphony of the Mechanical Age: Technology and Symbolism of Sound in European and North American Noise Abatement Campaigns, 1900–40." *Social Studies of Sciences* 31(1) (2001): 37–70.

Bijsterveld, Karin. *Mechanical Sound: Technology, Culture, and Public Problems of Noise in the Twentieth Century*. Cambridge, Mass.: MIT Press, 2008.

Bont, Raf de. "Poetry and Precision: Johannes Thienemann, the Bird Observatory in Rossitten, and Civic Ornithology, 1900–1930." *Journal of the History of Biology* (November 2009).

Borror, Donald J., and Carl R. Reese. "The Analysis of Bird Songs by Means of a Vibralyzer." *Wilson Bulletin* 65(4) (1953): 271–76.

Brady, Erika. *A Spiral Way: How the phonograph Changed Ethnography*. Jackson: University Press of Mississippi, 1999.

Brand, Albert R. "Recording Sounds of Wild Birds." *Auk* 49(4) (1932): 436–39.

———. *Songs of Wild Birds*. New York: Nelson, 1934.

———. "The 1935 Cornell-American Museum Ornithological Expedition." *Scientific Monthly* (August 1935): 187–90.

———. "Vibration Frequencies of Passerine Bird Song." *Auk* 55(2) (1938): 263–68.

———. "Why Bird Song Can Not Be Described Adequately." *Wilson Bulletin* 49(1) (1937): 11–14.

Burkhardt, Richard W., Jr. "Ethology, Natural History, the Life Sciences, and the Problem of Place." *Journal of the History of Biology* 32(3) (1999): 489–508.

———. *Patterns of Behavior: Konrad Lorenz, Niko Tinbergen, and the Founding of Ethology*. Chicago: University of Chicago Press, 2005.

Coffin, Lucy V. B. Individuality in Bird Songs. *Wilson Bulletin* 40(2) (1928): 95–99.

Collins, Harry M., and Trevor Pinch. *The Golem: what You Should Know about Science*. New York: Cambridge University Press, 1993.

Copeland, Peter, Jeffery Boswall, and Leonard Petts. *Birdsongs on Old Records: A Coarsegroove Discography of Palearctic Region Bird Sound 1910–1958*. London: British Library, 1988.

Daston, Lorraine, and Peter Galison. *Objectivity*. New York: Zone, 2007.

Gieryn, Thomas F. "City as Truth-Spot: Laboratories and Field-Sites in Urban Studies." *Social Studies of Sciences* 36(1) (2006): 5–38.
Horning, Susan Schmidt. "Engineering the Performance: Recording Engineers, Tacit Knowledge, and the Art of Controlling Sound." *Social Studies of Science* 34(5) (2004): 703–31.
Hunt, Richard. "The Phonetics of Bird Sound." *Condor* 25(6) (1923): 202–208.
Kellogg, Edward W. "History of Sound Motion Pictures, First Installment." *Journal of the Society for Motion Picture and Television Engineers* 64 (June 1955): 291–302.
Kellogg, Peter P. "Bird-Sound Studies at Cornell." *The Living Bird Annual* 1 (1962): 37–48.
Knorr-Cetina, Karin. *Epistemic Cultures: How the Sciences Make Knowledge.* Cambridge, Mass.: Harvard University Press, 1999.
Koch, Ludwig. *Memoirs of a Birdman.* Boston: Branford, 1955.
Kohler, Robert E. "Labscapes: Naturalizing the Lab." *History of Science* 40(4) (2002): 473–501.
———. *Landscapes and Labscapes: Exploring the Lab-Field Border in Biology.* Chicago: University of Chicago Press, 2002.
Kuklick, Henrika, and Robert E. Kohler, eds. *Science in the Field. Osiris, vol. 11. n.s.* Chicago: University of Chicago Press, 1996.
Lanyon, Wesley, and William Fish. "Geographical Variation in the Vocalization of the Western Meadowlark." *Condor* 60(5) (1958): 339–41.
Latour, Bruno. "Visualisation and Cognition: Thinking with Eyes and Hands." In *Knowledge and Society Studies in the Sociology of Culture Past and Present*, ed. Henrika Kuklick, Jai Press: Greenwich, Conn. 6 (1986): 1–40.
Latour, Bruno, and Steve Woolgar. *Laboratory Life: The Construction of Scientific Facts.* Princeton, N.J.: Princeton University Press, 1986.
Lynch, Michael. "The Externalized Retina: Selection and Mathematization in the Visual Documentation of Objects in the Life Sciences." *Human Studies* 11(2/3) (1988): 201–34.
Marler, Peter, Marcia Kreith, and Miwako Tamura. "Song Development in Hand-Raised Oregon Juncos." *Condor* 79(1) (1962): 12–30.
Marler, Peter, and Hans Slabbekoorn, eds. *Nature's Music: The Science of Birdsong.* Boston: Elsevier Academic Press, 2004.
Marler, Peter, and Miwako Tamura. "Song 'Dialects' in Three Populations of White-Crowned Sparrows." *Condor* 64(5) (1962): 368–77.
Mathews, Ferdinand S. *Field Book of Wild Birds and Their Music: A Description of the Character and Music of Birds.* New York: Putnam, 1904.
Mody, Cyrus C. M. "The Sounds of Science: Listening to Laboratory Practice." *Science, Technology, and Human Values* 30(2) (2005): 175–98.
Moore, Robert Thomas. "Methods of Recording Bird Songs." *Auk* 32(4) (1915): 535–38.
Nicholson, Max, and Ludwig Koch. *More Songs of Wild Birds.* London: Witherby, 1937.
———. *Songs of Wild Birds.* London: Witherby, 1936.
Potter, Ralph K., George A. Kopp, and Harriet C. Green. *Visible Speech.* London: Macmillan, 1947.
Radick, Gregory. "Morgan's Canon, Garner's Phonograph, and the Evolutionary Origins of Language and Reason." *British Journal for the History of Science* 33(1) (2000): 3–23.
Ranft, Richard. "Capturing and Preserving the Sounds of Nature." In *Aural History: Essays on Recorded Sound*, ed. Andy Linehan, 65–78. London: British Library, 2001.
Rieger, Stefan. *Schall und Rauch: Eine Mediengeschichte der Kurve.* Frankfurt: Suhrkamp, 2009.
Sage, John H. "Sixteenth Congress of the A O.U." *Auk* 16(1) (1899): 51–55.

Saunders, Aretas A. *Bird Song*. Albany: University of the State of New York, 1929.
———. "Correspondence." *Auk* 32(1) (1916): 103–108.
———. "Some Suggestions for Better Methods of Recording and Studying Bird Song." *Auk* 32(2) (1915): 173–83.
Schmidgen, Henning. "Time and Noise: The Stable Surroundings of Reaction Experiments, 1860–1890." *Studies in History and Philosophy of Biological and Biomedical Sciences* 34(2) (2003): 237–75.
Slabbekoorn, Hans. "Singing in the Wild: The Ecology of Birdsong." In *Nature's Music: The Science of Birdsong*, ed. Peter Marler and Hans Slabbekoorn, 178–205. Amsterdam: Elsevier Academic Press, 2004.
Stangl, Burkhard. *Ethnologie im Ohr: Die Wirkungsgeschichte des Phonographen*. Vienna: WUV Universitätsverlag, 2000.
Sterne, Jonathan. *The Audible Past: Cultural Origins of Sound Reproduction*. Durham, N.C.: Duke University Press, 2003.
Stillwell, Norma. *Bird Song: Adventures and Techniques in Recording the Songs of American Birds*. New York: Doubleday, 1964.
Thompson, Emily A. *The Soundscape of Modernity: Architectural Acoustics and the Culture of Listening in America 1900–1933*. Cambridge, Mass.: MIT Press, 2002.
Thorpe, William H., and Robert A. Hinde. "An Inexpensive Type of Sound-Proof Room Suitable for Zoological Research." *Experimental Biology* 33(4) (1956): 750–55.

CHAPTER 6

UNDERWATER MUSIC: TUNING COMPOSITION TO THE SOUNDS OF SCIENCE

STEFAN HELMREICH

INTRODUCTION

How should we apprehend sounds subaqueous and submarine? As humans, our access to underwater sonic realms is modulated by means fleshy and technological. Bones, endolymph fluid, cilia, hydrophones, and sonar equipment are just a few apparatuses that bring watery sounds into human audio worlds. As this list suggests, the media through which humans hear sound under water can reach from the scale of the singular biological body up through the socially distributed and technologically tuned-in community. For the social scale, which is peopled by submariners, physical oceanographers, marine biologists, and others, the underwater world—and the undersea world in particular—often emerges as a "field" (a wildish, distributed space for investigation) and occasionally as a "lab" (a contained place for controlled experiments).

In this chapter I investigate the ways the underwater realm manifests as such a scientifically, technologically, and epistemologically apprehensible zone. I do so by auditing underwater music, a genre of twentieth- and twenty-first-century composition performed or recorded under water in settings ranging from swimming pools to the ocean, with playback unfolding above water or beneath. Composers of underwater music are especially curious about scientific accounts of how sound behaves in water and eager to acquire technologies of subaqueous sound production. We can learn much about how the underwater domain has been made sonically perceptible by attending to how composers adapt their practice to scientific language and technique in ways both rigorous and fanciful. We can learn how sound has been abstracted from the water medium to reveal and produce resources imagined as musical. We can track how technologies of underwater audition are often adjusted to deliver aesthetic experiences in line with the way composers imagine submerged sound *should* sound; how, to take one example, the notion of water as sublimely immersive can be reinforced in compositions that make use of hydrophonic listening and playback. We can also sometimes discern a querying of dominant thinking about the symbolism of underwater sound.[1]

One tradition in the history of sound tells us the ocean was once taken to be a place of silence (thus, in 1953, Jacques Cousteau's book *The Silent World*). Auguste and Jacques Piccard the same year described travel two miles down in their bathyscaphe *Trieste* as surrounded by the "quiet of death" (Long 1953, quoted in Schwartz, forthcoming). That tone had been set in 1896, when Kipling wrote in his poem "The Deep-Sea Cables," "There is no sound, no echo of sound, in the deserts of the deep" (resonating with early nineteenth-century theories of the deep as a lifeless "azoic zone"). However, there has existed a more sonorous imagination of the sea—think of singing mermaids and sea monsters. In Charles Kingsley's 1863 novel, *The Water-Babies*, the boy protagonist, approaching a submarine volcano, comes "to the white lap of the great Sea-mother, ten thousand fathoms deep . . . aware of a hissing and a roaring, and thumping, and a pumping, as of all the steam engines of the world at once" (quoted in Kaharl 1989, xiii). As we will hear, the underwater world was, even in its first scientific manifestations, full of sound—even music—echoed in the poetic descriptions such vibration often called forth. Such soundful seas found expression in Romantic musical efforts to evoke underwater realms, which bequeathed a store of symbolism to later music meant to be realized under water. Notions of the immersive and sublime continue to saturate audio work. However, listening closely to such work, as this chapter does, also reveals how underwater music tracks shifting perceptions of the sea (from a space of cold-war mystery to a commons imperiled by global warming), changing ways of inhabiting swimming pools (primarily implicating gender), and fashions of connecting sound, art,

I thank Karin Bijsterveld and Trevor Pinch for inviting me to the Sound Studies workshop. Participants in Maastricht provided invaluable comments, particularly Eefje Cleophas. For careful readings, I also thank Etienne Benson, Caitlin Berrigan, Beth Coleman, Kieran Downes, Douglas Kahn, Ernst Karel, Heather Paxson, Tara Rodgers, Sophia Roosth, Michael Rossi, Hillel Schwartz, Nick Seaver, Malcolm Shick, Nicole Starosielski, and Peter Whincop.

and science in contemporary practices such as field recording, sampling, and sound art.

Evoking, Invoking, Soaking

I distinguish three modes through which music meets water. In the first, musical composition or performance *evokes* water symbolically, metaphorically, or timbrally—in the arrangement of notes, the organization of rhythms, or the choice of instrument.[2] In the second, music *invokes* water as a material instrument or sonic element. In the third, music *soaks* in water—that is, music is immersed in actual water as an encompassing medium within which it is performed, recorded, played back, or listened to (see the end of the chapter for a musicography organized according to *evoking*, *invoking*, and *soaking*). This chapter concentrates on the third mode.

Before focusing on soaking, however, here are some notes on evoking and invoking:

Evoking: Romantic composers Berlioz, Debussy, and Ravel, along with modernists Satie and Schoenberg, are known for portraying tranquil and tumultuous seas with an orchestral palette.[3] Acoustic ecologist R. Murray Schafer tested the waters of scientific seas in 1978 with String Quartet no. 2, *Waves*, which offered "dynamic, undulating wave patterns, the rhythm and structure of which [were] based on his analysis of wave patterns off both the Pacific and Atlantic coasts of Canada" (Knight 2006, 58).[4] The undersea world came into palpability in soundtracks to science documentary and science fiction film and television. Cousteau's *The Silent World* (1956)—not at all silent (bubble noises and scuba breathing abound)—features an Yves Baudrier score, towed down by dropping cello lines, suspended with gurgly horns, and buoyed up with tinkling harp notes (indebted to Liszt's 1877 "The Fountains of the Villa d'Este," full of rippling piano[5]).[6] In soundtracks to *20,000 Leagues under the Sea* (Paul J. Smith 1954) and *The Deep* (John Barry 1977), composers evoke the undersea with arpeggiating harps and minor-keyed swelling strings. In line with apprehensions of the ocean as a feminized, mysterious other, Western composers often employ orientalist motifs. In Angela Morley's 1969 "Martineau and Organ," from *Captain Nemo and the Underwater City*, the Mellotron organ and the Theremin evoke mellow swirls and half-forgotten siren songs. A lexicon for music evocative of the underwater world comes into being.[7]

In the mid-twentieth century, composers move from iconic to onomatopoetic, seeking to create sounds that *sound* as though they originate in water. Electronic effects become important.[8] Ussachevsky's 1951 "Underwater Waltz" employs reverberated piano. Synthesizers burble onto the scene: For Jean Painlevé and Genevieve Hamon's 1965 film, *The Love Life of the Octopus*, *musique concrète* composer Pierre Henry offers oozy synthesizer noodlings.[9] The electric guitar is adapted for surf music. The "sopping-wet 'surf' sound" (Bogdanov 2001, 105), realized by "[l]iquid guitar drenched in deep-tank reverb" (Priore 2007, 72), results from reverberation: "Surf guitarists are noted for extensive use of the 'wet' spring reverb sound and use of the vibrato arm on their guitar to bend the pitch of notes downward."[10] Jamaican dub drips reverb. David Toop suggests that "sonar transmit pulses, reverberations and echoes of underwater echo ranging and bioacoustics" constitute the "nearest approximation to dub" (1995, 116; cf. Henriques 2003). In the late twentieth century, Detroit techno outfit Drexciya conjured an imagined underwater sonic universe. Their homage to Kraftwerk's "Autobahn," "Aquabon," guides German electronica into a disturbing fantasy of a black Atlantis founded by Africans thrown overboard during the Middle Passage.[11] Important in these works is how the underwater world is imagined, for this imagination remains influential when composers work under "real" water as they coax out of the medium that which they imagine they should hear. Underwater soundworlds are unearthly, evanescent, all encompassing, dreamlike, alien.[12] Aesthetics saturates technique.

Invoking: Douglas Kahn's *Noise Water Meat* is the essential scholarly work about modernist music that calls upon water. "The first notable use of wet percussion was Erik Satie's use of *boutelliphone* (a series of tuned bottles suspended from a rack, 'a poor man's glockenspiel') in *Parade* (1918)" (Kahn 1999, 247).[13] John Cage's 1952 *Water Music* "included among its forty-one events a duck whistle blown into a bowl of water and two receptacles for receiving and pouring water" (Kahn 1999, 242). The genre of "drip music," starting in the 1950s, dipped into nonsense, babbling.[14] However, "Drops of water were [also] conducive to music because they could comfortably assume musical speeds and were amenable to total organization by the composer" (Kahn 1999, 251), as in Hugh Le Caine's 1955 "Dripsody: An Etude for Variable Speed Recorder." Yoko Ono's 1963 "Water Piece" (followed by her 1971 "Toilet Piece") keys us into the context for drip music, the art movement known as "Fluxus," Latin for "flow."[15] In the late 1960s, flows and drips—artificially created in laboratories of modernist aesthetics—were joined by a new genre of field recordings. In 1966–1967 New Zealand composer Annea Lockwood began a project called "The River Archive," recording sounds of rivers around the world.

The aesthetic aim in pieces that *evoke* and *invoke* water is a sense of immersion (cf. Grimshaw, this volume), and one purpose of this chapter—especially as I turn in the remainder of this text to underwater music that *soaks*—is to think critically about how that immersion is achieved in controlled lablike spaces such as swimming pools and in the wilder field of the ocean. It turns out that immersion is accomplished through composers' appropriation of scientific and technical models of underwater soundworlds, as well as their tweaks of those models to align with ideas about how the underwater domain should sound. If the sounds of science (Mody 2005; Pinch and Bijsterveld 2004) saturate underwater music, these sounds are multiply mediated and manipulated.

Tacking between Field and Lab in Underwater Music: Submarine Noises and Whale Songs

Musical language shapes descriptions of early experiments in underwater sound propagation. In a 1708 issue of the Royal Society's *Philosophical Transactions*, Francis Hauksbee published an "Account of an Experiment Touching upon the Propagation of Sound through Water," in which he pronounced that a bell under water sounded "much more mellow, sweet, and grave at least three notes deeper than it was before" (1708, 372).[16] Imagery of music played under water reverberates through early technoscientific inquiries into submerged sound.

By the early and mid-twentieth century, as oceanographers and antisubmarine warfare researchers listened closely to the underwater realm—realizing that it was not a place of silence—music and its metaphors continued to shape sea sound description. Maritime military research history entwines with musical history:

> In World War I the composer and conductor Sir Hamilton Harty was called in by the British Admiralty's Board for Invention and Research to identify the most likely frequency bands of hull and propeller noises. . . . Ernest Rutherford also took a colleague with perfect pitch out in a small boat as part of the war effort. At a prearranged spot one of the great names in atomic physics took a firm grip of his companion's ankles while this man stuck his head into the Firth of Forth and listened to the engine note of a British submarine. Hauled back into the dinghy and toweling his head he announced it was a submersible in A-flat. (Hamilton-Paterson 1992, 114–15)

Submarine pilots used less fanciful discernment. The U.S. Navy created instructional LPs to train submariners to distinguish enemy sub sounds from ambient noise.[17] "Still," writes Hillel Schwartz:

> Sonarmen went "ping-happy." Straining to identify threats within an underwater environment that behaved "very much like a large empty room with bad acoustic properties," they heard pings bouncing off what turned out to be whales and schools of fish, heard pips refracting off what turned out to be temperature gradients, heard roars from what turned out to be waves rushing at rocks on distant shores, and heard much better in mid-morning than in the late afternoon. (forthcoming)

In order to endow submarine space with immersive sonic depth, to carve a soundscape for humans out of the subaqueous milieu, it takes technical and cultural work and translation. Equipment must be constructed that can capture submarine vibrations in the audio register and ready them for humans to listen to—equipment like hydrophones, which can capture underwater vibrations using microphones fashioned of ceramic or some other material sufficiently denser than water to allow propagating waves to be impeded. The earliest hydrophones were manufactured in 1901 by the Submarine Signal Company of Boston, which imagined "a network of underwater bells whose sonorous gongs would carry through the water at great distances" (Schlee 1973, 246). The company, seeking an alternative to foghorns, built receivers to capture underwater bell sounds for listeners on surface ships—though plans to use bells for Morse code were scuttled by the turbulence of the submarine medium. Hydrophones came into their own on submarines with sonar (sound navigation and ranging), which was in operation by the 1930s. Sonar works by bouncing signals off the ocean bottom or other boundaries in water, permitting submariners to time echoes to compute distances. It produces a dimensional portrait—not so much a soundscape as a *soundedscape*. Hydrophonic signals were rendered into stereo by the use of devices that transformed signals arriving at separate underwater receivers into "binaurally centered" impressions in headphones, creating spatial relations meaningful to hearing humans (Höhler 2003). If, as Emily Thompson (2002) has argued, the soundscape of modernity is patterned by sounds fed through technological filters, underwater soundscapes do not exist at all for humans without such filtering all the way through. Sea becomes part lab.

It was incumbent on early submariners to be attentive auditors of sonar, and it was through such listening that the crackling of crustaceans and the snapping of shrimp were disclosed, providing a portrait of soundscapes already in existence for underwater creatures with means to hear them (see Iselin and Ewing 1941). Such human listening participated in "field" science in its canonical form—the investigation of a space of shifting boundaries, of natural and cultural agents (Kohler 2002; see also Bruyninckx, this volume). The care with which submariners listened emerged in part from cold-war anxieties about the possibility that missing the faintest signal might be disastrous.

It was through such field listening—filtering "noise" from sound, tuning it to the human auditory range—that whale "songs" were discovered. Biologist William Schevill of Woods Hole was the first to call these sounds whale "music," though musical metaphors—and comparisons to birds—circulated earlier. Once whale sounds had been separated, they were aestheticized as lonely, majestic, ecologically tuned-in arias to the wounded sea (see Madsen 1999, 33). They were pressed onto LP, notably by bioacoustician Roger Payne on *Songs of the Humpback Whale*, in 1970. They were fused with classical composition in Alan Hovhaness's 1970 *And God Created Great Whales*, a mix of recorded and represented, indexical and iconic sounds.

One could describe the trajectory this way: The underwater realm starts out as silent, becomes soundful (eerie with cold-war echoes), occasionally noisy, and, once so revealed, turns out to be full of music from creatures imagined close to humans in cognitive power, creatures whose songs can then be separated from their medium—field—for contemplation. This trajectory is accompanied by another, in which humans experiment with their own subaqueous music to see whether they can move from the iconic and symbolic mode of Ravel and Debussy, from the octopoid onomatopoeia of Pierre Henry, to a more indexical evocation of "actual" immersion, or soaking. This other trajectory for underwater music is less natural historical than "experimental." Having learned lessons from the oceanographic "field," musicians tinker in lablike settings, spaces where they control boundaries and variables. The primary such space is the swimming pool, and one of the first to dive in is John Cage.

The Pool of Experiment: Cage and Neuhaus

Cage—son of an engineer who worked with hydrophones—first brought water sound/noise into his modernist composition

> in his collaboration with Lou Harrison, *Double Music* (1941), in which Cage specified the use of a "water gong (small—12"–16" diameter—Chinese gong raised or lowered into tub of water during production of tone)." . . . Cage . . . traces his use of the water gong to 1937 at UCLA, where, acting as an accompanist, he sought a solution to the problem of providing musical cues to water ballet swimmers when their heads were under water. (Kahn 1999, 249–50; see also Hines 1994, 90)

Cage's approach later became more experimental, mixing subjective and scientific methods. Water was important to Cage because, as he put it, it "prepared me for the renunciation of intention and the use of chance operations" (quoted in Kahn 1999, 249–50). In such renunciations, the authorial self-dissolved—as in Cage's tale of his time in an anechoic chamber, in which he heard his own blood flow as part of the environment. Water stood as a symbol of gentle noise, of a scientific modernist sublime into which an individual might dissipate, a view literalized in the self-experiments of dolphin researcher John Lilly, who reported auditory hallucinations while floating in isolation tanks.

In some measure, this immersive symbolism is read out of how human hearing works under water. Sound waves travel four times faster in water than air—making it nearly impossible for humans to use underwater acoustic vibration to locate themselves in space. This difficulty is compounded by the fact that human eardrums are too similar in density to water to provide the resistance that can interrupt many underwater vibrations so they might be translated into tympanic movement—sound—in the ears; many vibrations pass right through our bodies. For humans, underwater sound is largely registered by bones in the skull, which allow enough resistance—impedance—for vibrational motion to be rendered into resonances in the body. Moreover, conduction of sound by bone directly to the inner ear confounds differences in signals received by both ears, making it impossible to compose a "stereo image." Unaided human ears perceive underwater sound as omniphonic: coming from all directions at once (and because of sound's seemingly instantaneous arrival, often as emanating from within one's own body).

The underwater world is not immediately a soundscape for humans because it does not have the spatiality of a landscape; one might, rather, think of it as a zone of sonic immanence and intensity: a *soundstate*.[18]

Sound installation artist Max Neuhaus was next in the pool:

> In *Water Whistle* [1971–1974], water was forced through whistles under water to produce pitched sounds that could be heard by the audience only when they submerged themselves. In *Underwater Music* [1976–1978], he modified this technique by using specially designed underwater loudspeakers and electronically generated sounds, which were composed through a combination of scientific experiment and intuitive, creative decisions. (Miller 2002, 26)

Simon Miller writes, "Such compositions literally immersed Neuhaus in the medium—the sites were swimming pools—as he adjusted pitch perimeters and envelopes, in effect 'coloring' the sound. They dramatized the spaces of sound, its limits, because the medium contained them" (2002, 26). In planning for these works, Neuhaus made drawings to think through "configurations of the sound sources in the three-dimensional space of each body of water" (see figure 6.1). Neuhaus's experiment, like the use of hydrophones and sonar, took place within a soundscape of modernity (Thompson 2002), depending crucially on standardized knowledge of how sound behaves in water.

In this work, the pool is a lab, a site with clear boundaries within which variables can be manipulated and in which the "nature" of water can be brought "indoors" to be controlled and cleaned of agents that populate it in "the wild" (see Knorr-Cetina 1999; Kohler 2002). It becomes a placeless, universalized space, though because of its artistic repurposing, it also becomes a space in which subjective experience—objectively modulated—can be realized as immersive. So, rather than a space in which scientists generate inscriptions on paper (Latour and Woolgar 1986), it becomes a space for generating *impressions*. In addition, as with the "house of experiment" described by historian Steven Shapin (1999)—a site where gentleman scientists could gather to agree on matters of fact—this "pool of experiment" admits a properly trained public of modest auditors, ready to be immersed.

Conducting Transduction: Redolfi's *Musique Subaquatique*

The aim of immersion is elaborated in work of French composer Michel Redolfi, who in 1982 performed in Dartmouth College's indoor pool. As Redolfi put it, "[L]isteners of the *Underwater Concerts* 'immerse themselves not only in the 90-degree heated swimming pool, but in the sound itself'" (quoted in Charles 1993, 60). How does this immersion work?

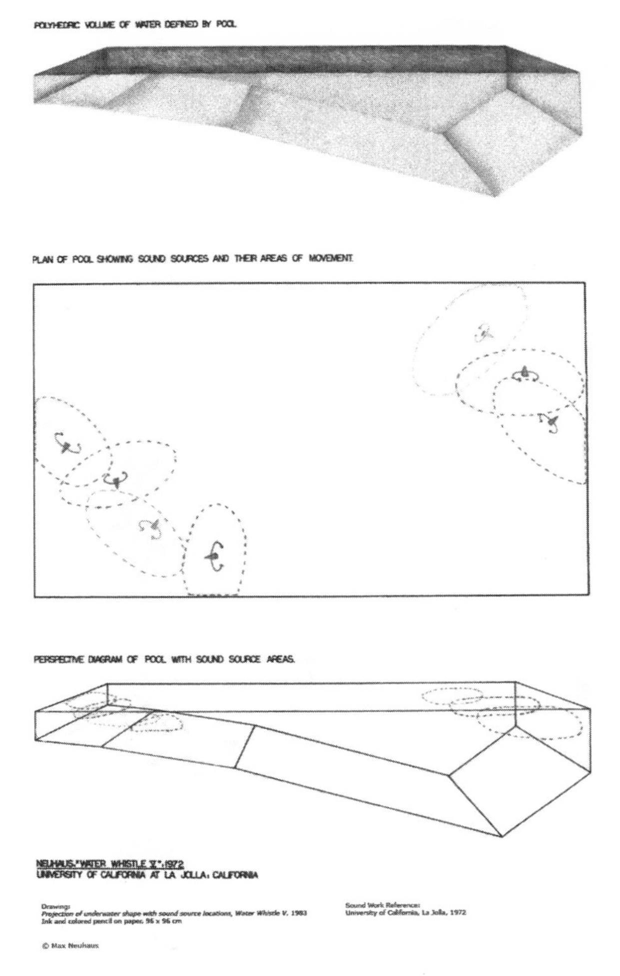

Figure 6.1. Max Neuhaus's Projection of underwater shape with sound source locations, Water Whistle V, 1983, ink and colored pencil on paper, 96 × 96 cm; sound work reference: University of California, La Jolla, 1972 (www.max-neuhaus.info/drawings/waterwhistle/index_drawings.htm).

Redolfi uses modified speakers in his underwater events. Managing the mix of electricity and water is key. Redolfi's sound engineer, Daniel Harris, explains:

> Having people, the audience, and speakers with electrical connections in the water together can be a concern unless proper measures are taken to insure that no harm comes to the audience. The SWSS (Sonic Waters Safety System) was developed in response to our concern about the speaker systems introducing stray electrical currents in the water. The SWSS inserts a 50-kHz pulse wave into all cables entering the water in the same manner as DC power is introduced into microphone cables to power microphones. The return signal is passed through a counter/comparator circuit, which will instantly shut down any line that drops a settable number of pulses, usually 3, indicating a short or other malfunction.[19]

As this list of equipment makes clear, "immersion" is a hard-won pun, a hard-won laboratory effect. It depends on engineering a sequence of *transductions*—translations of signals across various media, acoustic, electronic, watery—so that the transduction itself is inaudible, seamless. The immersive effect would be ruined if Redolfi's listeners were electrocuted.

The symbolism of water as a sublimely immersive medium must be actively realized technologically. Harris writes that "The genius of Michel Redolfi is in how he applies his knowledge of the human acoustics of underwater sound and intimate familiarity with the playback systems to the composing and mixing of his very original and enchanting music." In some cases, however, that knowledge is subordinated to Redolfi's sense of how underwater music *should* sound. Harris writes he writes, of an underwater xylophone:

> Hitting a metal bar under water results in a very unsatisfying "tink," no matter how hard the performer strikes the instrument, nor how heavy the bar. [So] I glued piezo sensors on the bars, which, when struck, triggered samples or other electronic sound sources via MIDI triggers. . . . The resultant audio was mixed with other sources and sent to the SWSS and then to the underwater speakers.[20]

Philosopher of music Daniel Charles hears such modulations as an ideal meeting of water and music: "What is at stake, then, is not the spatial idiosyncrasy of the environment, but the degree of achievement of the blending of a music which can be described as simple, tranquil and transparent, with the physical characteristics of water—its density, temperature, and color" (1993, 63). Redolfi is, of course, alive to how science and symbolism come together, though he doesn't always call attention to the artificiality of that relationship. In connection with a later piece, *Sonic Waters*, he writes:

> [W]ater materializes sound, thickens it, and makes it palpable and penetrable. Water and sound, combined together at the molecular level, create a sonic and fluid substance that can be appreciated not only by looking at its surface reflections, but by sinking oneself into its volume, density, warmth and vibrations (quoted in Charles 1993, 63).

In addition:

> [T]he very concept of underwater sound, Redolfi says, goes all the way back to the songs of sirens, the bells of submerged cathedrals, the voices of lost mariners. "These noisy and eerie myths," he notices, "have been swept away by the XXth Century and replaced by a quiet and sterile belief in the ocean as a silent world and occasionally disturbed by the long song of the Aqua Diva whale. But the fact is that the sea is a cacophony of sound, complete with fish "barking and croaking," shrimp "snapping," dolphins "whistling" and sea urchins "click-click-clicking." (Charles 1993, 63).

Note that Redolfi does not mention sonar surveys, pings, or the like; technological intrusions do not figure in his impression of water. Such intrusions are far from trivial—as the ears of cetaceans damaged by underwater sounds attest (James 2005). Dialing in to deployments of sound deleterious to dolphins and whales might reveal a genre of underwater music no one has yet considered: cetacean death metal.[21]

So, while Redolfi knows that transductions are necessary to immersion, he ontologizes water as dreamy, alive, penetrable. Charles argues that Redolfi has done something radical here, however. Redolfi not only seeks to:

> *reduce all metaphors of presence, but make of presence the metaphor of itself.* In that sense, Redolfi is no more a minimalist: thanks to technology, he transforms the very status of presence until presence becomes, through the acceptance of its reproducibility, an instance which does not need any more to be interpreted or symbolized or displaced, because it entails its own interpretation or symbolization or displacements. (1993, 67)

However, if we thought not of reproducibility but of *transducibility*, we might hear more clearly the material conditions of Redolfi's "presence," which depends, again, on making sure the underwater audience is not flash-fried by subaqueous electronica (cf. Helmreich [2007, 2009] on how transduction bolsters perceptions of "presence" in submersibles; Henriques [2003] on transduction in reggae; and Roosth [2009] on transduction in sonocytology, listening to cellular life).

From 1981 to 1984 Redolfi scaled up his enterprise, moving into the field—the ocean—when *Sonic Waters* was performed off the coast of Southern California. This opening up of the dream lab of the pool into the field was accompanied by campy sea creaturey devices, such as the giant colorful "jellyfish" that kept a low-frequency speaker afloat in La Jolla Cove. Such playfulness is a reminder that Redolfi does not imagine crustaceans, fish, or marine mammals as audiences: "Redolfi is concerned with humans: 'Every dolphin, he says, has a person nowadays to take care of him. . . . I prefer to take care of humans' " (Charles 1993, 63). Redolfi's approach looks similar to that of the Florida Keys underwater music festival. Celebrating its twenty-fifth anniversary in 2009, the festival offers to scuba divers music played over Lubell Laboratory speakers attached to boats floating near the reef.[22] Attendees dress up as fish.

Listening to the Sounds of Science: The Wet Sounds Festival

In 2008, announcements of the United Kingdom's first underwater sound festival, "Wet Sounds," asked people to dip into municipal swimming pools to listen to music specially composed by industrial, electronica, and noise performers, as well as sound artists (Blanning 2008). Sound art—a genre of art that creates sound objects to be experienced in galleries, in public spaces, and via headphones—made a good match with this festival, which treated the pool not just as a chill-out room but also as a gallery space/lab (on sound art, see Labelle 2006;

Licht 2007). Listeners' attention was persistently drawn to technoscientific frames of reference.

Wet Sounds pieces are archived online. A few are hydrophonic recordings—so that the fact that they were played back under water raises the question of whether these are compound or redundant underwater pieces (and what happens when we listen in air?). The attempt to superimpose one underwater space (the ocean) on another (the pool) makes particularly explicit the multiple meanings of "medium" in this practice of *schizophonia*—the splitting of sound from its source.[23] Slavek Kwi's "Sonafon," for example, is "a structural and textural exploration of echolocation sounds made by Pink Dolphins recorded in Rio Jaupeperi in Amazonas, Brazil, using ultrasound range hydrophones." Klaus Osterwaldt's "Donatus Subaqua" offers "Recordings made in a quiet lake in the forest using hydrophones placed two meters below the surface. There are sounds of the underwater environment like gas bubbling up from the bottom, plants producing oxygen, insects and even the calls of waterboatman." Amie Slavin's "Wave Play" is "an abstract interweaving of pulse, saw, and triangular waves designed to invoke the playfulness and the latent power of waves, in both sound and water."[24] In addition, Disinformation's "Ghost Shells" plays back sferics and whistlers, sounds produced by storm disturbances in Earth's electromagnetosphere (and captured on very low-frequency radio bands): "These phenomena are referred to as 'hydrodynamic' because the math used for modelling their behaviour has been extrapolated from observations of how equivalent wave phenomena behave in the variety of fluid media" (see Motschmann, Sauer, and Baumgaertel 1984).[25] By 2008, then, scientific representation—as a warrant for underwater realism, as an aestheticized device for delivering other worlds, as a fetish for formatting serious art—was sharply in presence as a passage point for making underwater music and sound art.

Queering the Mermaid: Snapper, Oleson, Leber, Chesworth

In May 2009 singer Juliana Snapper premiered an underwater opera titled "You Who Will Emerge from the Flood" (written with composer Andrew Infanti) at the Victoria Baths, a historic public swimming pool in Manchester, England. Snapper took on the challenge of singing under water, fusing technique and aesthetic: "Maximizing bone conduction and controlling bubble output as part of a new vocal fabric," her website reports, "Snapper merges extended techniques with Baroque tropes that represent human longing and passion as aspects of weather. Pre-recorded sounds from oceanic bubble fields and birdcalls throb above the water as Snapper's voice (amplified by an underwater microphone) presses through the soundscape."[26]

There is a virtuoso, alien effect to be won from doing something under water that requires breathing.

As with the work of Neuhaus, Redolfi, and Wet Sounds participants, this is music that requires *research*. About the field. About the body. Snapper writes as follows:

> I have researched underwater acoustics by reading, consulting at the Scripps Institute of Oceanography, and spending hours submerged in my bathtub and borrowed pools. My experimentation up to now has allowed me to control my voice for long stretches under water, negotiate changes in depth and pressure, and to invent a new expressive vocal language that I call mouth-to-water singing. . . . Vocalizing in water involves working with air pressure shifts that compress the air in the lungs and effect the sound of the phonation at varying depths. Mouth-to-water singing relies heavily on bone conduction to transmit voice, and that vibration is also the basis for sensing/hearing sounds in the water. The interference of bubble sounds in the breath-born phonation begets lightly percussive rhythmic textures, and it may be possible to control the pitch material and rate of bubble noise through breath pressure and buccal aperture so that as I sing I am also releasing a secondary melody in duet with my vocal cords.

Snapper imagines herself a modern day mermaid—though one schooled in critical feminism. Snapper, who performed at the "Queer Up North" festival in the UK, does not stage herself as a deliquescent delicacy for a heterosexual masculine imagination, a woman merging with a uterine medium, a möbius mother/lover in the way that, as Douglas Kahn has argued, "submerged women" have persistently been posed, particularly in surrealist representation (see Redolfi's website, full of pictures of bikini-clad women floating in pools).[27] Snapper modulates water as feminized other into a critical substance, one that can detour the way water and waves have been symbolized as feminized flux (see Rodgers forthcoming). She is an active rather than a passive part of the medium—the aim of her work is thus distinct from Cage's, which sought to eliminate intentionality. She is interested, too, in the pool as a simultaneously public and intimate space in which sex and sexuality have historically been subject to disturbance—and also strictly ordered, policed, and ranked (she reports that, historically, the water used on "women's days" in the Victoria Baths was recycled from "men's days"!); what better place to play with gender and sexuality?

Snapper's critique of canonical mermaid models becomes crisp in work in private spaces. In "Aquaoperas," Snapper joins Jeanine Oleson to visit home bathrooms and perform minioperas, with Snapper in the tub and Oleson singing into the toilet (see figure 6.2), their voices fusing through a "snorkelabra." Snapper and Oleson's performance of what they call the "hot lez flower duet" from the opera *Lakmé* by Léo Delibes combines a lesbian sublime with a heterodox dabbling with the abject space of the toilet. In this "pool of experiment," this soak opera, the relation of water to public/private, feminine/feminist/queer is up for grabs.

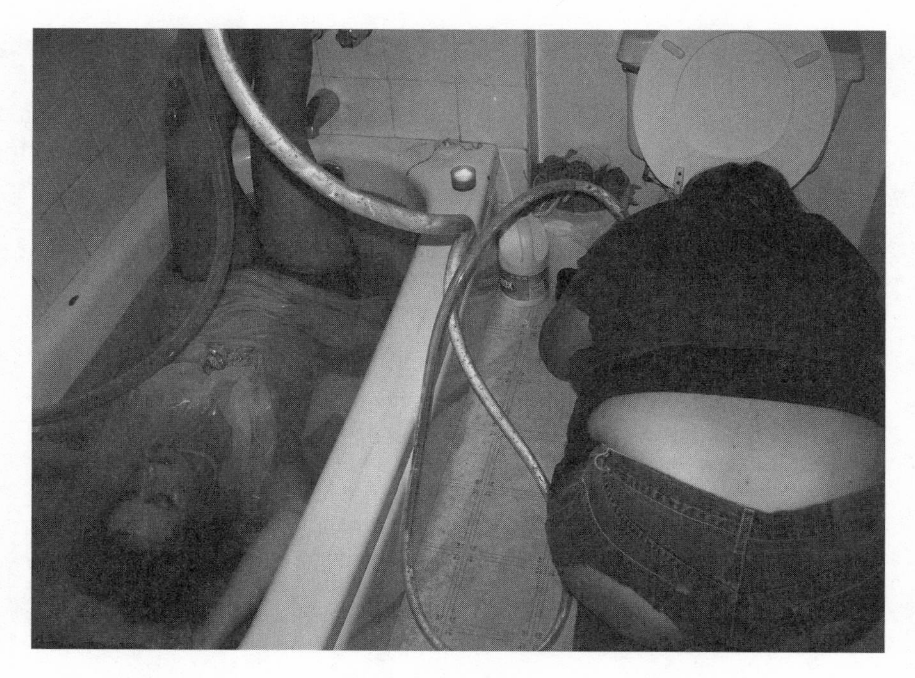

Figure 6.2. Photo of Juliana Snapper and Jeanine Oleson performing Aquaopera #2 (SF/Lakme Redux at Shotwell Shack, San Francisco).

Snapper and Oleson's work is kin to the sinister work of Australian sound artists Sonia Leber and David Chesworth, who in *The Gordon Assumption* (2004) offer a recording installed in the subterranean toilets of Gordon Reserve at Parliament Station, Melbourne:

> An incessant outpouring of female voices lures passers-by down the stairwell to the cave-like subterranean toilets. At the lower gates, they are confronted with an asynchronous chorus of female voices in infinitely rising pitch. The voices gather and thicken without respite, in upwards glissandi, constantly trailing upwards. . . . The voices recall the mythologies and mysteries of voices heard in caves, where the voices of spirits, sibyls and oracles are believed to announce predictions and warnings from the mouth of a cave. (Leber and Chesworth 2004, quoted in Kouvaras 2009, 101)

Linda Kouvaras reads *The Gordon Assumption* as a feminist critique of mythological siren songs; the songs float indeterminately above signification, yes, but are sited in an abject locale. The piece recalls Duchamp's urinal. Pools, baths, and toilets become laboratories for rethinking water, gender, and sexuality.

But this laboratory research longs to go into the field. Snapper writes, "I am now fully prepared to realize opera in the ocean depths." Why? Snapper writes, "I am interested in the rub of cultural metaphors in which water represents a dangerous zone of 'pure' emotion and now-urgent specters of drought and drowned cities." Though she doesn't mention it, Snapper's words summon memories of Hurricane Katrina (recalled in the image of an "underwater jazz funeral" used in a review of a 2008 play staged by Tulane Environmental Law Society students in New Orleans). Snapper suggests that "the material tension between water and air (breath, foam, thirst) . . . speaks to an unknown ecological future." Ecology is of signal interest in new underwater music. This is another arena in which science and sentiment, substance and symbol join.

Return to the Sounded Sea: Winderen

Tracking back from lab to field, we find Norwegian field recordist Jana Winderen. Winderen uses hydrophonic recordings "sourced from beneath the oceans surrounding Norway, Greenland, and Iceland" (White Line [UK]): "The music here is assembled from various auditory documents gathered from research trips, all treated as improvisational material, and morphed into elaborate sound collages" (Boomkat [UK]). *Heated* (2008) is a recording of warming water. Like other underwater composers, Winderen is interested in technologies of underwater audition. *Heated*'s liner notes report the use of the following: "2 × 8011 DPA hydrophones, 2 × DolphinEAR/PRO hydrophones and 2 × 4060 DPA microphones." Winderen is artist and empirical researcher both.

What is all this water music in search of? Again: immersion, though now in the sea—though a sea accessed by treating it as a lab, a recording studio. One reviewer writes as follows of Winderen: "With my headphones on and my eyes closed it sounds as if you are really in the middle of this water" (Earlabs [Netherlands]). However, the possibility of imagining oneself immersed depends, press rewind, on *transduction*. It depends on a cyborg sensibility—one indexed in David Toop's *Ocean of Sound*, which he concludes with this cybernetically inflected contention: "Music—fluid, quick, ethereal, outreaching, time-based, erotic and mathematical, immersive and intangible, rational and unconscious, ambient and solid—has anticipated the aether talk of the information ocean" (1995, 280).

However, if immersion is a submersion of self in water, water is controllable only in swimming pools. The "field" is different, full of other critters. Winderen, in "The Noisiest Guys on the Planet," uses snapping shrimp to shift us to a multispecies soundscape.[28] Unlike Redolfi, who ignores underwater nonhumans, Winderen calls us back to these creatures, sentries now in climate change. The field, Winderen's work suggests, is becoming more like a lab not only for artists like herself but also for humanity, experimenting on its ecosystem.

What Winderen is creating, then, is not just music but—in the idiom of sound art—*documents* as well, one reason she and others find technological access to and scientific models of the sea compelling. If earlier generations of composers sought simply to replicate a submarine sublime, today's sound artists hope not just to soak in sound but also to broker ear-opening accounts of human relations with the water around us.

Dunn's *Chaos and the Emergent Mind of the Pond*

Skip back to David Dunn's 1992 *Chaos and the Emergent Mind of the Pond*, a collage of recordings of aquatic insects in ponds in North America and Africa. This collage—fusing recordings from different ecologies—stages a different construal of underwater worlds, one that hears a percolating intelligence in water. In *Insectopedia*, Hugh Raffles writes of Dunn's practice:

> Listening to the pond with two omnidirectional ceramic hydrophones and a portable DAT recorder, he hears a rhythmic complexity altogether greater than that in most human music . . . The sounds can't be arbitrary, he decides. These animals are not simply following their instincts. "The musician in me cannot help but hear more." . . . He begins to hear the pond as a kind of superorganism, a transcendent social "mind" created from the autonomous interaction of all the life within it, terms not dissimilar to those used by complexity theorists to describe the nest colonies of the eusocial insects . . . [His] soundscape is more than a recording, more even than a composition. It is also a research method, one that flows easily from a principle of wholeness. (2010, 323–24)

The piece is inspired by anthropologist Gregory Bateson's model of mind as a phenomenon present in worldly relations, not locked in people's heads. Dunn's insect recordings posit water not simply as a medium but also as organically enlivened, cogitating: "[W]hen I see a pond, I think of the water's surface as a membrane enclosing something deep in thought" (quoted in Ingram 2006, 129). Unlike Cage, who would have advocated listening to these sounds "in themselves," Dunn wants to preserve sounds' referentiality, their link to empirical ecological processes (Ingram 2006)—as on his recordings of bugs in piñon trees, in which he tracks how their sounds flag global warming. While there exists a risk of romanticizing balanced nature, Dunn's Batesonian approach tunes into how water may not only contain life but also be constituted through living things. There can be no purely lablike pool of water; water is made of vitality—which gets us to field recordings of the vital signs of ocean Earth.

From the Cold War to Global Warming: Under Arctic Seas

Andrea Polli's *Sonic Antarctica* CD consists of "recordings of the Antarctic soundscape made during the author's seven-week National Science Foundation residency in Antarctica during the 2007/2008 season." Polli mixes field recordings of melting glaciers with audio translations of scientific data ("sonifications" [Supper, this volume]) on climate and peppers these documents with snippets from interviews with climate scientists. The sound of ice melting and of data about ice melting are signs of global warming. This aesthetic production has a scientific-political point.[29]

Charles Stankievech's "DEW project" has a scientific-political point as well. As Stankievech explains, "As much ideological deterrent as defense infrastructure, the Distant Early Warning (DEW) Line constructed between 1954–56 [near the Arctic Circle] was a joint venture between the US Air Force and the Royal Canadian Air Force. A long-distance radar and communication system, the DEW Line created an electromagnetic boundary able to detect airborne invasion."[30] Stankievech revisits questions of territorial sovereignty, listening not *up* for enemy others but *down* for global humanity's ecological depredations: "The radio station [at Stankievech's Yukon river site] monitors the sounds of the river's ice and underwater flow on a continual basis, transmitting the signals to Dawson City, where the field-recordings are processed and broadcast via the internet."[31] April 6, 2009, saw Stankievech deliver "a sound performance using live samples from the river installation, electromagnetic microphones, radios and computer." Submarine sounds such as these melt the distinction between music and data collection.

However, even data collection is available for aesthetic contemplation. The PerenniAL Acoustic Observatory in the Antarctic Ocean (PALAOA) transmits sound from the Antarctic Ocean. In addition, PALAOA (Hawaiian for "whale") detects marine mammals and provides a research baseline for relatively quiet underwater environments.[32] The PALAOA MP3 audio stream at the Alfred-Wegener-Institut für Polar- und Meeresforschung website anticipates listeners, warning that its sound is "not optimized for easy listening, but for scientific research. . . . [B]eware of sudden extremely loud events."[33] The sound travels from hydrophones to a research station and then to Germany, where it is put online. In addition, PALAOA's hydrophones, hanging just below the hundred-meter-thick Ekström ice shelf (near Neumayer Station, a German research center), are placed well above the SOFAR (Sonar Fixing and Ranging) channel, a layer of seawater in which the speed of sound reaches its underwater minimum. Low-frequency vibrations can travel long distances through this conduit (about 800 to 1,000 meters deep at midlatitudes and higher toward the surface in temperate zones) before they dissipate.[34] Sounds from this region are essential for ocean acoustic tomography, the study of ocean temperature using sound (Munk, Worcester, and Wunsch 1995). Submarine sounds in art and science now echo concerns not about the Cold War but about global warming. Rather than evoking, invoking, or soaking, they broker connections between the ocean understood as natural field and considered, for better or worse, as a lab for global ecopolitical futures.

Conclusions

Summing up the arc of underwater music over the last half-century: There have been two primary venues for underwater music: field settings of the ocean (with rivers, ponds, and lakes less frequently used) and lab settings of swimming pools. For the field tradition, underwater music emerges from the noise of the Cold War, which reveals the songs of whales. Those songs then become submerged within the worked-over subaqueous soundscapes of modern human enterprise—soundscapes that harbor evidence of global warming, of sea creatures under stress, evidence that becomes source material for composers who mine scientific idioms for artistic and political statements. For the lab tradition, the pool begins as a stage to realize the ascetic aesthetic of Cagean modernism. It then becomes a space to play with meanings of water—either, as with Redolfi, to reinforce canonical symbolisms of dreamy, meditative, womblike space or, as with Snapper, to queer such imagery, to experiment with gender, sexuality, and public and private. However, while the field setting is "wild" and entangled with nonhuman sounds and the lab setting is more social, cultural, or anthropocentric, artists working in both settings seek to *evoke* an "immersive" experience. Moreover, in both settings the transductive properties of water must be managed in order to *invoke* water as a material accomplice in this enterprise, this aim of *soaking* listeners in the sublime surround of sound submerged.

NOTES

1 This is not a history of underwater sound, which historians of oceanography and acoustics have already delivered. Höhler (2002) describes a shift from sounding the sea with metal ropes to sounding with reflected sound, a practice that turned the ocean into a three-dimensional volume (*sound* as fathoming has moorings in the Old English *sund*, "sea," whereas *sound* as vibration reaches back to Old English *swinn*, "melody.").

2 Kahn calls such evocation "programmatic, depicted, or discursive water" (1999, 245).

3 Wagner linked water and music: "If rhythm and melody are the shores on which music touches and fertilizes the two continents of the arts that share its origin, then sound is its liquid, innate element; but the immeasurable extent of this liquid is the sea of harmony" (quoted in Kahn 1999, 246). This chapter leaves aside evocations of sea creatures, such as Camille Saint-Saëns's Aquarium movement in his 1886 *Le Carnaval des Animaux*, or Edward MacDowell's 1898 Nautilus in *Sea Pieces* (see Shick 2007).

4 Toru Takemitsu's *I Hear the Water Dreaming* (1987) evokes water in its Debussy-styled sound and also formally, through an E-flat - E - A motif that—spelled in German as Es-E-A—transliterates the word *sea*.

5 See Chen (2008, 6) on "tremolos to describe the shimmering effects of light on a still water surface; ascending and descending arpeggiated figures and glissandi to depict the undulating of waves."

6 Cousteau's show "The Undersea World of Jacques Cousteau" was soundtracked by Walter Scharf, who sought evocative music for animal subjects. For a special on sea elephants, "The music had a wonderful reedy quality, the same as the sea elephants, but with added feeling" (quoted in Shaheen 1987, 96). Cousteau's editors made modifications: "The sounds of the manatees in *The Forgotten Mermaids* were too abrasive so the music editor rearranged several notes to help soften the manatee munches" (Shaheen 1987, 99).

7 See www.filmscoremonthly.com/board/posts.cfm?threadID=59157&forumID=1&archive=0. Sound recordist Darren Blondin (2007) writes: "Films have taught us that

underwater sounds [are] muffled, echoing, and bubbly. In actuality water is alive with high frequencies, but a bright sounding recording tends to come off as less realistic."

8 Though an earlier—mechanical—precedent is Italian futurist Luigi Russolo's "gurgler" ("gorgogliatore"), an *intonarumori* (noisemaker) described in his 1913 *Art of Noises* manifesto, which called for nonrepresentational sounds—splashing, sirens—in composition. More satirically, Spike Jones's 1948 "William Tell Overture" features gargling. In experimental composition, Francisco Lopez's 1993 *Azoic Zone*, with pieces such as "A Vibrational Trip From Bathyal to Hadal Zones," sounds sourced from hydrophonic recordings but is merely timbrally evocative of them.

9 The indie band *Yo La Tengo*, resoundtracking Painlevé, retains burbly emphases.

10 Http://en.wikipedia.org/wiki/Surf_music. "Wetness" is now taken as such an obvious term that even a sober book on architectural acoustics contains index entries such as "reverberation, excess creates aural soup" and "reverberation, soup and mud" (Blesser and Salter 2007, 428). See Doyle 2005.

11 Listen to Parliament's 1978 "Aquaboogie" for more drowning sounds from the black Atlantic. Compare Sun Ra's vision of a black utopia in outer space (Williams 2001).

12 On FreeSound, a database for sound effects, an MP3 titled "under_alien_ocean" is described as "beneath alien waves, its [*sic*] liquid, but it ain't water . . . oh and its [*sic*] from 1950." www.freesound.org/samplesViewSingle.php?id=14260.

13 Henry Cowell "used '8 Rice Bowls' tuned to no definite pitch using water for *Ostinato Pianissimo (For Percussion Band)* (1934)" (Kahn 1999, 248). South India's Jalatarangam ("water waves") is a carnatic instrument that uses bowls filled with water. Steven Feld (1991) describes flowing water in songs about the Kaluli of Papua New Guinea. Satie wrote satirically on the move from evocative to invocative: "The hydrographic engineers tell us that all the waterfalls of the earth, whatever their social standing might be, yield a low F, clearly audible, upon which it so happens is built a perfect chord in C Major. . . . The Water Company is elated: it is going to install carefully calibrated conduits in all the concert halls to offer musicians an entire chromatic scale of little cascades" (quoted in Kahn 1999, 247). Satie would not have been surprised when "Cage composed . . . *Water Walk: For Solo Television Performer* (1959). . . . The water-related instructions and properties . . . include a bathtub of water, an operating pressure cooker, a supply of ice cubes, a garden sprinkling can, a soda siphon" (Kahn 1999, 250).

14 "Aldous Huxley . . . in 'Water Music' (1920), anticipated the importance of dripping water in Fluxus and, later, in chaos theory: 'Drip drop, drip drap drep drop. So it goes on, this water melody forever without an end. Inconclusive, inconsequent, formless, it is always on the point of deviating into sense and form' " (Kahn 1999, 252).

15 See Dunn and Young (2008), an anthology of drip music. Christian Marclay's 1990 *Bottled Water* is an installation consisting of bottles filled with tape containing the sound of dripping water. The "hydraulophone" uses streams of water in a flutelike apparatus: http://hackedgadgets.com/2007/02/17/ontario-science-centre-hydraulophone-musical-keyboard-water-fountain/.

16 See also Colladon (1893) on 1826 experiments at Lake Geneva.

17 Consult www.hnsa.org/sound/, an archive of navy training sounds.

18 Differences between fresh water and seawater are consequential, too.

19 Http://danielharrismusic.com/Underwater_Music_I.html.

20 Compare Alan Silvestri's soundtrack for *The Abyss* (1989), which uses "electronic pinging and underwater clanging effects" (www.filmtracks.com/titles/abyss.html).

21 Compare the use of whale songs to frighten seals away from Lincolnshire river fish stocks: "To drive the seal family back into the North Sea, the National Rivers Authority

[*sic*] have been playing recordings of killer-whale songs under the surface of the Glen." *The Times*, Oct. 31, 1994, quoted in Toop 1995, 3.

22 Listen also to Erik DeLuca's "The Deep Seascape: The Sonic Sea," which "explores the underwater soundscape of South Florida" (www.erikdeluca.com/).

23 So dimensional has the underwater world become that artist Bill Fontana imagines the transplant of a watery soundscape—as in his 1994 *Ile Sonore*, transposing sound from the beaches of Normandy to the Arc de Triomphe in Paris, making the "white noise of the sea" (Madsen 1999, 29) surround a traffic island.

24 "There has been a long-standing association of water and sound in observational acoustics from antiquity through Chaucer to Helmholtz and beyond, with the sound of a stone hitting water producing a visual counterpart, which was then mapped back onto the invisible movements of sound waves" (Kahn 1999, 246; Rodgers, 2009). The correspondence between water and sound waves is at the heart of "cymatics" (Jenny 1967, 1974; Lauterwasser 2006), which employs sound vibration to generate patterns in watery substrates.

25 The 2009 Wet Sounds festival promised "deep listening" (a term coined by Pauline Oliveros in 1991, which referred less to water than to a mode of attention) (see Allan 2009). Sound artists continue to organize festivals of underwater audio. "Hydrophonia," a festival of hydrophone sound art dedicated to raising public awareness of ocean noise" has been held in 2009 and 2010 in Italy and Spain. Refer to http://hydrophonia.com/

26 Quoted on http://maryrachel.wordpress.com/2009/06/19/underwater-microphones-and-juliana-snapper/

27 And in tourist sites such as Florida's Weeki Wachee mermaid attraction, though women there only pretend to sing in the giant fishtank in which they perform. One wonders whether in future they will use Stachowski's (1999) apparatus for talking under water.

28 See Stocker (2002/2003). Sheila Patek studies lobster and shrimp sound and hearing in Biology @ Berkeley: http://www.bio.umass.edu/biology/pateklab/home.

29 Compare DJ Spooky/Paul D. Miller's *Sinfonia Antarctica*, which uses field recordings to "capture the acoustic qualities of Antarctic ice forms, [which] reflect a changing and even vanishing environment under duress" (http://djspooky.com/art/terra_nova.php). Miller's piece is an homage to Ralph Vaughn Williams's 1952 *Sinfonia Antarctica*, which evoked the austral landscape with orchestral arrangements. Miller suggests that his *Sinfonia* will bypass "metaphor" and "go to Antarctica and record the sound of the continent." Compare Peter Cusack's *Baikal Ice (Spring 2003)*, which contains hydrophonic recordings of the springtime sound of ice thawing at Siberia's Lake Baikal.

30 Www.stankievech.net/projects/DEW (accessed 9 September 2009)

31 "A remote transmission station housed in a geodesic dome on the Yukon and Klondike rivers continually records and transmits the sounds of the rivers flowing and the ice shifting using hydrophones embedded in and under the ice."

32 Www.awi.de/en/research/new_technologies/marine_observing_systems/ocean_acoustics/palaoa/. Christine James's 2005 "Sonar Technology and Shifts in Environmental Ethics" charts the rise of noninvasive sonar in concert with environmental movements.

33 w.awi.de/en/research/new_technologies/marine_observing_systems/ocean_acoustics/palaoa/palaoa_livestream/.

34 Garry Kilworth's science fiction story "White Noise" tells of a haunted undersea cable station. Characters find that the deep retains noises from ages past because (in a fabulous piece of SF logic), " 'Cold, dense water is less likely to disperse or be infiltrated by warm currents. The circular currents weave their way intact around the ocean floor like blind worms.' 'And they retain sound patterns . . .' 'Like magnetic tape' "(1990, 513). The main characters, listening in on a microphone attached to a deep-sea cable beneath the Red Sea, hear Moses leading his people out of Egypt.

MUSICOGRAPHY

Evoking

Barry, John. 1977. *The Deep*. Motion picture soundtrack.

Berlioz, Hector. 1844. *Le Corsaire*, op. 21.

Bridge, Frank. 1911. *The Sea*, H. 100.

Carpenter, John Alden. 1914. "The Lake" (*Adventures in a Perambulator*, no. 4).

Crackle. 2008. *Heavy Water*. Slowfoot.

Debussy, Claude. 1903–1905. "Jeux de vagues" (*La Mer*, L. 109, no. 2).

———. 1910. "La cathédrale engloutie" (*Préludes* [Book 1], L. 117, no. 10).

Drexciya. 1997. "Aquabon." On *The Quest*. Submerge.

Dunn, Alan, and Jess Young, eds. 2008. *Music for the Williamson Tunnels: A Collection of the Sound of Dripping Water*. Arts Council, England, edition of 1000.

Elgar, Edward. 1899. *Sea Pictures*, op. 37.

Giant Squid. 2006. *Metridium Fields*. The End Records.

Golijov, Osvaldo. 2004. *Oceana*. Deutsche Grammophon.

Hovhaness, Alan. 1970. *And God Created Great Whales*, op. 229.

Ligeti, György. 1961. *Atmosphères*.

Liszt, Franz. 1877. "Les jeux d'eau à la Villa d'Este" (*Années de Pèlerinage: Troisième Année*, S. 163, no. 4).

Lopez, Francisco. 1993. *Azoic Zone*. Geometrik.

MacDowell, Edward. 1898. *Sea Pieces*, op. 55.

Montgomery, Will. 2008. "Submarine." www.touchradio.org.uk/touch_radio_36.html.

Morley, Angela. 1969. *Captain Nemo and the Underwater City*. Motion picture soundtrack

Parliament. 1978. "Aqua Boogie (A Psychoalphadiscobetabioaquadoloop)," single. Casablanca.

Ravel, Maurice. 1901. "Jeux d'eau," M. 30.

———. 1908. "Ondine" (*Gaspard de la Nuit*, M. 55, no.1).

Ritchie, Anthony. 1993. *Underwater Music*.

Russolo, Luigi. 1913. "Gorgogliatore." On *Musica Futurista: The Art of Noises 1909-1935*. LTM, 2004.

Saint-Saëns, Camille. 1886. "Aquarium" (*Le Carnaval des Animaux*, no. 7).

Satie, Erik. 1913. *Embryons desséchés*.

Schafer, R. Murray. 1971. *Miniwanka—Moments of Water*.

———. 1978. String Quartet no. 2, *Waves*.

Schoenburg, Arnold. 1909. "Farben" (Five Pieces for Orchestra, op. 16, no. 3).

Sibelius, Jean. 1914. *The Oceanides*, op. 73.

Slotek. 1999. *Hydrophonic*. WordSound.

Smith, Paul J. 1954. *20,000 Leagues under the Sea*. Motion picture soundtrack.

Takemitsu, Toru. 1987. *I Hear the Water Dreaming*.

Ussachevsky, Vladimir. 1951. "Underwater Waltz."

Vaughan Williams, Ralph. 1903–1908. "A Sea Symphony" (Symphony no.1).

Waterjuice. 2001. *Hydrophonic*. Vaporvent.

Xela. 2009. *The Dead Sea*. Type Records.

Yo La Tengo. 2002. *The Sounds of the Sounds of Science*. Egon Records.

Ziporyn, Evan, with I Wayan Wija. 2003. "Ocean." On *Shadowbang*. Cantaloupe Records.

Invoking

Brecht, George. 1959. "Drip Music."
Cage, John. 1952. *Water Music.*
———, and Lou Harrison. 1941. "Double Music."
Cowell, Henry. 1934. *Ostinato Pianissimo.*
DJ Spooky. 2009. *Sinfonia Antarctica.* www.djspooky.com/art/terra_nova.php.
Feld, Steven. 2006. *Suikinkutsu: A Japanese Underground Water Zither.* VoxLox, Earth Ear.
Feld, Steven, ed. 1991. "Relaxing by the Creek." On *Voices of the Rainforest: A Day in the Life of the Kaluli People.* Rykodisc.
Le Caine, Hugh. 1955. "Dripsody: An Etude for Variable Speed Recorder."
Lockwood, Annea. 1989. *A Sound Map of the Hudson River.* Lovely Music, Ltd.
Ono, Yoko. 1963. "Water Piece."
———. 1971. "Toilet Piece."
Satie, Erik. 1917. "Prestidigitateur Chinois" (*Parade* [*Ballet Réaliste sur un Thème de Jean Cocteau*], no. 2).
Spandau Ballet. 1982. "Innocence and Science." On *Diamond.* Chrysalis.

Soaking

Blackburn, Philip. 2007. Symphony in Sea. http://www.philipblackburn.com/Compositions.html.
Dunn, David. 1992. "Chaos and the Emergent Mind of the Pond." On *Angels and Insects.* Nonsequitur/What Next Recordings.
Harris, Yolande. 2009. *Now Stripe Time.* DNK Amsterdam: A Concert Series for New Live Electronic and Acoustic Music, September 14. http://www.yolandeharris.net/.
Humid, Bob. 2006. "It's Warm Besides [sic] the Submarine Cables." On *Second Wind Phenomenon.* Suburban Trash.
Polli, Andrea. 2009. *Sonic Antarctica.* Gruenrekorder.
Redolfi, Michel. 1989. *Sonic Waters #2 (Underwater Music) 1983–1989.* Hat Hut Records.
Sauvage, Tomoko. 2009. *Ombrophilia.* Either/OAR.
Snapper, Juliana. 2008–2010. *Five Fathoms Opera Project.* www.julianasnapper.org.
Stankievech, Charles. 2009. *The DEW Project,* Dawson City, Yukon Territory: www.stankievech.net/projects/DEW/stream/index.html.
Winderen, Jana. 2009a. *Heated: Live in Japan.* Touch Music.
———. 2009b. *The Noisiest Guys on the Planet.* Ash International. Cassette only.
———. 2009c. *Submerged.* Touch Music.

REFERENCES

Allan, Jennifer. "Wet Sounds." *Wire* 307 (2009): 77.
Blanning, Lisa. "Wet Sounds." *Wire* 296 (2008): 81.
Blesser, Barry, and Linda-Ruth Salter. *Spaces Speak, Are You Listening?: Experiencing Aural Architecture.* Cambridge, Mass.: MIT Press, 2007.
Blondin, Darren. "Recording Underwater Ambiences." www.dblondin.com/101507.html, 2007.
Bogdanov, Vladimir. *All Music Guide: The Definitive Guide to Popular Music,* 4th ed. Ann Arbor, Mich.: Backbeat, 2001.

Charles, Daniel. "Singing Waves: Notes on Michel Redolfi's Underwater Music." *Contemporary Music Review* 8(1) (1993): 57–69.

Chen, Karen. Water Reflection upon Four Piano Works—Liszt, Debussy, Ravel, and Griffes. Diss. in Musical Arts, Claremont Graduate University, 2008.

Colladon, Jean-Daniel. "Experiments on the Velocity of Sound in Water." In *Acoustics: Historical and Philosophical Development*, ed. and trans. R. Bruce Lindsay, 194–201. Stroudsburg, Penn.: Dowden, Hutchinson, and Ross, 1973. Orig. pub. in *Souvenirs et Mémoires: Autobiographie de Jean-Daniel Colladon* (Geneva: Aubert-Schuchardt, 1893).

Cousteau, Jacques, with Frédéric Dumas. *The Silent World.* New York: Harper and Brothers, 1953.

Crone, Timothy J., William S. D. Wilcock, Andrew H. Barclay, and Jeffrey D. Parsons. The Sound Generated by Mid-Ocean Ridge Black Smoker Hydrothermal Vents. *PLoS* ONE 1(1) (2006). www.plosone.org/article/fetchArticle.action?articleURI=info:doi/10.1371/journal.pone.0000133 (accessed May 20, 2007).

Doyle, Peter. *Echo and Reverb: Fabricating Space in Popular Music Recording, 1900–1960.* Middletown, Conn.: Wesleyan University Press, 2005.

Hamilton-Paterson, James. *The Great Deep: The Sea and Its Thresholds.* New York: Random, 1992.

Harris, Daniel. "Underwater Music Engineering." www.danielharrismusic.com/Underwater_Music_I.html.

Hauksbee, Francis. "An Account of an Experiment Touching the Propagation of Sound through Water." *Philosophical Transactions* 26 (1708): 371–72.

Helmreich, Stefan. "An Anthropologist underwater: Immersive Soundscapes, Submarine Cyborgs, and Transductive Ethnography." *American Ethnologist* 34(4) (2007): 621–41.

———"Submarine Sound." *Wire* 302 (2009): 30–31.

Henriques, Julian. "Sonic Dominance and the Reggae Sound System Session." In *The Auditory Culture Reader*, ed. Michael Bull and Les Back, 451–80. Oxford: Berg, 2003.

Hines, Thomas S. "Then Not Yet 'Cage': The Los Angeles Years, 1912–1938." In *John Cage: Composed in America*, ed. Marjorie Perloff and Charles Junkerman, 65–99. University of Chicago Press, 1994.

Höhler, Sabine. "Depth Records and Ocean Volumes: Ocean Profiling by Sounding Technology, 1850–1930." *History and Technology* 18(2) (2002):119–54.

Ingram, David. " 'A Balance That You Can Hear': Deep Ecology, 'Serious Listening,' and the Soundscape Recordings of David Dunn." *European Journal of American Culture* 25(2) (2006): 123–38.

Iselin, Columbus O'D., and Maurice Ewing. *Sound Transmission in Sea Water: A Preliminary Report.* Woods Hole, Mass.: Woods Hole Oceanographic Institution for the National Defense Research Committee, 1941.

James, Christine. "Sonar Technology and Shifts in Environmental Ethics." *Essays in Philosophy: A Biannual Journal* 6(1) (2005). commons.pacificu.edu/cgi/viewcontent.cgi?article=1172&context=eip.

Jenny, Hans. *Cymatics: A Study of Wave Phenomena and Vibration.* Newmarket, N.H.: MACROmedia, 2001. Orig. pub. 1967 and 1974.

Kaharl, Victoria A. *Water Baby: The Story of Alvin.* New York: Oxford University Press, 1989.

Kahn, Douglas. *Noise Water Meat: A History of Sound in the Arts.* Cambridge, Mass.: MIT Press, 1999.

Kilworth, Garry. "White Noise." *Year's Best Fantasy and Horror,* vol. 3, ed. Ellen Datlow, 508–16. New York: St Martin's, 1990.

Kindermann, Lars, Olaf Boebel, Horst Bornemann, Elke Burkhardt, Holger Klinck, Ilse van Opzeeland, Joachim Plötz, and Anna-Maria Seibert. A Perennial Acoustic Observatory in the Antarctic Ocean. Paper presented at the International Expert Meeting on IT-Based Detection of Bioacoustical Patterns. December 7–10, 2007, International Academy for Nature Conservation (INA), Isle off Vilm, Germany, 2007, http://hdl.handle.net/10013/epic.28799.

Knight, David B. *Landscapes in Music: Space, Place, and Time in the World's Great Music.* Lanham, Md.: Rowman and Littlefield, 2006.

Knorr-Cetina, Karin. *Epistemic Cultures: How the Sciences Make Knowledge.* Harvard: Harvard University Press, 1999.

Kohler, Robert E. *Landscapes and Labscapes: Exploring the Lab-Field Border in Biology.* Chicago: University of Chicago Press, 2002.

Kouvaras, Linda. "Toilets, Tears, and Transcendence: The Postmodern (Dis-)Placement of, and in,Two Water-Based Examples of Australian Sound Art." *Transforming Cultures eJournal* 4(1) (2009): 94–107. epress.lib.uts.edu.au/ojs/index.php/TfC/article/download/1062/1201

Labelle, Brandon. *Background Noise: Perspectives on Sound Art.* New York: Continuum, 2006.

Latour, Bruno, and Steve Woolgar. *Laboratory Life: The Construction of Scientific Facts,* 2nd ed. Princeton, N.J.: Princeton University Press, 1986.

Lauterwasser, Alexander. *Water Sound Images.* Newmarket, N.H.: MACROmedia, 2006.

Licht, Alan. *Sound Art: Beyond Music, between Categories.* New York: Rizzoli, 2007.

Long, James M. " 'Absolute' Calm Two Miles Down." *San Diego Evening Tribune* (October 1, 1953).

Madsen, Virginia. "The Call of the Wild." In *Uncertain Ground: Essays between Art and Nature,* ed. Martin Thomas, 29–43. Art Gallery of NSW, Sydney, 1999.

Miller, Simon. *Visible Deeds of Music: Art and Music from Wagner to Cage.* New Haven, Conn.: Yale University Press, 2002.

Mody, Cyrus C. M. "The Sounds of Science: Listening to Laboratory Practice." *Science, Technology, and Human Values* 30 (2005): 175–98.

Motschmann, U., K. Sauer, and K. Baumgaertel. "Whistler Wave Amplitude Oscillation and Frequency Modulation in the Magnetospheric Cavity." *Astrophysics and Space Science* 105(2) (1984): 373–77.

Munk, Walter, Peter Worcester, and Carl Wunsch. *Ocean Acoustic Tomography.* New York: Cambridge University Press, 1995.

Pinch, Trevor, and Karin Bijsterveld. "Sound Studies: New Technologies and Music." *Social Studies of Science* 34(5) (2004): 635–48.

Priore, Dominic. *Smile: The Story of Brian Wilson's Lost Masterpiece.* Bobcat Books, 2007.

Raffles, Hugh. *Insectopedia.* New York: Pantheon, 2010.

Rodgers, Tara. "Toward a Feminist Epistemology of Sound: Refiguring Waves in Audio-Technological Discourses." Invited plenary lecture. Luce Irigaray Circle, State University of New York at Stony Brook, Manhattan, September 12, 2009.

Roosth, Sophia. "Screaming Yeast: Sonocytology, Cytoplasmic Milieus, and Cellular Subjectivities." *Critical Inquiry* 35(2) (2009): 332–50.

Schlee, Susan. 1973. *The Edge of an Unfamiliar World: A History of Oceanography.* New York: Dutton.

Schwartz, Hillel. *Making Noise: From Babel to the Big Bang & Beyond.* Zone, forthcoming.

Shaheen, Jack G. "The Documentary of Art: 'The Undersea World of Jacques Cousteau.' " *Journal of Popular Culture* 21(1) (1987): 93–101.

Shapin, Steven. "The House of Experiment in Seventeenth-Century England." In *The Science Studies Reader*, ed. Mario Biagioli, 479–504. New York: Routledge, 1999.
Shick, Malcolm. "Siren Song." *Chamber Musings: Newsletter of the Chamber Music Society of the Maine Center for the Arts* 4(2) (2007): 1, 3–4.
Stachowski, Richie. "Device for Talking Under Water." U.S. Patent #5877460, March 2, 1999.
Stocker, Michael. "Ocean Bio-Acoustics and Noise Pollution: Fish, Mollusks, and Other Sea Animals' Use of Sound, and the Impact of Anthropogenic Noise on the Marine Acoustic Environment." *Soundscape: Journal of Acoustic Ecology* 3(2)/4(1) (2002/2003): 16–29.
Toop, David. *Ocean of Sound: Aether Talk, Ambient Sound, and Imaginary Worlds.* London: Serpent's Tail, 1995.
Urick, Robert J. *Principles of Underwater Sound,* 3rd ed. New York: McGraw-Hill, 1983.
Williams, Ben. "Black Secret Technology: Detroit Techno and the Information Age." In *Technicolor: Race, Technology, and Everyday Life*, ed. Alondra Nelson and Thuy Linh N. Tu with Alicia Headlam Hines, 154–76. New York: NYU Press, 2001.

CHAPTER 7

A GRAY BOX: THE PHONOGRAPH IN LABORATORY EXPERIMENTS AND FIELDWORK, 1900–1920

JULIA KURSELL

Introduction

In 1878, the year after Thomas Alva Edison submitted the first patent for his phonograph, two British physicists, Fleeming Jenking and J. Alfred Ewing, published a report in the scientific journal *Nature* of a recent experiment they had conducted. This experiment used Edison's device to address the nature of spoken language:

> Let a set of vowel sounds, as A E I O U (pronounced in Italian fashion), be spoken to the phonograph in any pitch and with the barrel of the instrument

Acknowledgements: This chapter was written with the support of the VolkswagenStiftung and the Max Planck Institute for the History of Science, Berlin. I wish to thank both institutions, as well as the Phonogram Archive, Berlin, and, for comments and support, Susanne Ziegler, Britta Lange, Kelaine Vargas, Armin Schäfer, and the editors of this volume.

> turned at a definite rate. Then let the phonograph be made to speak them, first at the same rate, and then at a much higher or lower speed. The pitch is, of course, altered, but the vowel sounds retain their quality. (Jenkin and Ewing 1878, 384)

As Jenkin and Ewing noticed, the faster or slower speed of the phonograph altered the pitch of the vowels, making them appear to be spoken by a higher or lower voice. Yet, they found the quality of the voice remained the same: They could still recognize the recorded vowels. This experiment was soon contested. Charles Cross, working in Boston, challenged these findings with his own experiments using one of the first prototypes of the phonograph built by Edison. Cross (1878), in a reply that was also published in *Nature*, asserted that a change in speed actually distorted the vowels to the extent of making them unintelligible. For Jenkin and Ewing, however, their experiment had a second purpose, one that is easily overlooked. Most of their report was devoted to the reconstruction of "the instrument." Unlike Cross, they did not have access to the instruments Edison had built himself. For Jenkin and Ewing it was therefore a major concern whether they would manage to construct the phonograph at all based on the inventor's description. They struggled with every aspect of the complicated apparatus, such as finding a foil that would allow the recorded sound to be replayed more than once and making a spring that would not only hold the speed of the barrel constant but also allow for controlled deviation. Such concerns are part and parcel of laboratory work and the well-known difficulty in replicating pioneering experiments (Collins 1985). Their article also demonstrates that they understood the phonograph to be a scientific instrument whose technical functions had yet to be defined. The mass production of phonographs, either for the laboratory or for commercial purposes, had not yet occurred.

Some twenty years later, a group of experimental psychologists, based at the Berlin Institute of Psychology, started to integrate the phonograph into their scientific work on a systematic basis. In 1900 Carl Stumpf, founder of the institute and its director until 1921, recorded a Siamese court theater group on twenty wax cylinders. These recordings laid the foundation for what was to become the Berlin Phonogram Archive. In 1908, when the existence of this archive was officially announced, the collection contained more than one thousand items—the number of phonographic rolls would eventually grow to thirty thousand.

In 1901 Stumpf extensively discussed his first recordings in the institute's journal, *Beiträge zur Akustik und Musikwissenschaft*. Summing up, he touched upon the issue of methodology. In order to preserve samples of music that were as yet unknown to researchers and were in danger of being lost forever, a double strategy was recommended. Traditional notation, though necessary for discussion within the scientific community, was not sufficient but had to be backed up by phonographic recordings (Bruynincks, this volume). However, Stumpf was well aware of the difficulties posed by phonographic recordings. Among the major problems he listed was the recording speed: "Most importantly, one has to provide for constant rotation speed . . . The difficulty lies in . . . avoiding the troubles caused by arbitrary diminishing of the rotation speed, as this will produce sounds that are too low and tend to blur" (Stumpf 1901, 135–6).

Twenty-five years later, in 1926, Stumpf summed up his research of the past two decades in a treatise on the sounds of language, titled *Die Sprachlaute* (1926). One chapter was devoted to phonographic experiments with vowels. The initial question that had driven Jenkin and Ewing to reconstruct the phonograph had since been solved, Stumpf declared. It was now agreed that reproducing recorded vowels at a different speeds would alter their sound. This was because the characteristic sound of the vowels resulted from resonances that depended on the shape of the mouth cavity. The pitch of these so-called formants was fixed. A change in the rotation speed would thus alter the pitch of the recorded resonance frequencies and therefore distort the vowels' characteristics.

Stumpf's very detailed experiments, however, did more than just provide yet another verification of this already accepted theory. As his report demonstrates, even by 1926, when wire recording had started to replace it as a recording device, many aspects of the use of the phonograph were still in need of careful study. The phonograph thus appears as a site of intersection for technology, experimental practices, and ways of hearing and listening, none of which were stable. Rather, they continually changed in the course of the experimental work, and the means of recording and the means of listening were constantly adjusted in relation to each other. Practices of fieldwork involving phonographs, tests in the laboratory, and the organization of storage and access in the phonogram archive were finely tuned to each other. For their experiments, psychologists developed technologies for the production of multiple copies and standards for identifying the rotation speed. They tested the degree of variation in recording quality, and they eventually discovered the interrelatedness of ways of recording and ways of listening.

Among the thirty thousand wax cylinders preserved in the Berlin Phonogram Archive today, with musical recordings from all over the world, there are also several items made for experimental use. These "experimental cylinders" (*Experimentalwalzen*), as they are classified in the lists of the archive (Ziegler 2006), are the focus of this chapter.[1] They show that the phonograph was not just a new apparatus ready at hand. Rather, experimental work was needed to create uses of the phonograph, to gauge its products (the wax cylinders), and ultimately to enable recurrence and comparison as the basic operations of the phonogram archive. Tracing the history of the phonograph means tracing the "becoming" of a medium.

Eight of the experimental wax cylinders were used to study the tone color of vowels. Taking this series of experiments as an example, my chapter addresses the following: First, I show that a medium is not a given entity but rather an unstable and heterogeneous object. In the 1980s, German media studies borrowed Michel Foucault's notion of the "historical *a priori*" (Foucault 1969) to analyze discourses of media culture. Most prominently, Friedrich Kittler transformed "historical *a priori*" into "technological *a priori*" and discussed how "discourse networks" grow from new technologies (Kittler 1990, 1999). By showing how media produce inscriptions, such analyses investigated the way in which technologies selected and shaped the utterances possible in a network by forcing communication into specific formats.

To be sure, sound recording, as made possible by the phonograph, can be said to have created such a discursive network. On the one hand, the phonograph immediately suggested new ways of asking questions even though none of its material aspects were yet fixed. On the other hand, some level of technical reliability and established modes of use made it possible to study the phonograph itself, turning it from a technical object into an object of investigation (Rheinberger 1997). There is thus an inherent ambiguity in the use of a medium, seen in the process of its emergence. While in the beginning, the phonograph was immediately accepted as a scientific instrument, the commercial use soon dominated, turning the phonograph into something completely different. Scientists, however, did not stop using the phonograph but rather integrated its new functions into their own work. This chapter traces this ambiguity, looking at the emergence of the Phonogram Archive in Berlin. More specifically, I show that the researchers had to adjust, calibrate, and standardize the new functions again before integrating them into the scientific context.

I thus propose a view of the complex history of sound as it moves through the laboratory. Science and technology studies (STS) offer the framing for a history of media that can accommodate the idea of change and allow for the analysis of the phonograph as a site of intersections between its various uses.

Instruments

In 1878 Edison's phonograph entered the scientific community. Experimenters such as Jenkin, Ewing, and Cross were by no means unprepared to accept this new instrument. "Self-recording instruments" had had a role in the laboratories of the life sciences since the 1840s, when the "kymograph" was invented; sound recording, though without a rendition of the graphically recorded pressure wave, soon became one of the kymograph's specific uses (Sterne & Akiyama, this volume). In his first report on this new device, Carl Ludwig included a sample of the resulting curves, showing traces of two bodily functions. The registration of blood pressure was paralleled by the registration of lung pressure for the same individual. Each of these movements had been traced onto the blackened surface of a turning cylinder at the same time. This simultaneous registration of two curves opened up the possibility of correlating body processes more generally, an effort numerous laboratory researchers engaged in during the second half of the nineteenth century.

It is easy to overlook the fact that what made this instrument so important was the correlation rather than the simple fact of recording. The success of the new recording method lay in its promise of mathematically formalizing the body's processes instead of merely visualizing them (de Chadarevian 1993). In later developments of the kymograph, this correlation became standard, although it was featured less and less on the visible surface of the graphic tracings. Soon after the kymograph was introduced, tuning forks were integrated into the apparatus (Jackson 2006 and this volume). Their regular vibration, registered in parallel with the observed body functions, served as an indicator for the speed of the turning cylinder. The shape of the tuning fork's curves enabled the observer to detect any irregularities occurring in the rotating movement. This new arrangement enabled the experimenters to register only one bodily function and then to correlate it with the time indicated by the fork. As the fork vibrated at, for instance, one hundred times per second, its curve related the observed body process indirectly to chronometric time. At this stage, the two elements necessary for correlation were still present in the shape of two registered curves. The function of reference was, in later kymographs, displaced by the supposed regularity of the cylinder's movement. The function of correlating eventually thus became part of the technical device. The tuning fork, which at first seemed to replace the registration of a second bodily function, turned out, with hindsight, to have been an externalized control device. After the tuning fork, its curve, and its soft humming disappeared from the assemblage, the function of control persisted but was internalized in the apparatus. The cylinder of the kymograph was now understood to represent a system of coordinates that allowed the curves to be read directly (i.e., as a visualization of the movement), which did not necessarily require mathematical calculation.

The phonograph posed a new challenge to laboratory work: Experimenters had to use their ears. Jenkin and Ewing carefully embedded this new task in the more common procedures of gauging and comparison. Their experiment on the recording speed was actually part of a larger project that consisted of two series of experiments. Their short announcement about vowel quality in 1878 was followed by several longer reports on extensive studies of vowel traces, in which the phonograph was seemingly understood as yet another device for producing visible traces. However, the traces were more difficult to obtain, as the authors explain:

> The experiments were made as follows:—The vowel under consideration was spoken or sung at a given pitch, determined by a piano, while the barrel of the phonograph was turned at a definite speed, regulated by means of a metronome. The indentations made in the tin-foil were then mechanically transcribed, so as to give curves representing a magnified section of the impressions.... All transcripts were rejected if the tin-foil did not continue to give the sound clearly after being used to produce these curves. (Jenkin and Ewing 1878, 340)

On first impression, hearing seems to be an addendum to a technique of visualization in these experiments, turning the visible recordings into authentic traces of sound. Like many researchers before them, Jenkin and Ewing produced curves that they hoped would explain the nature of sound. It seems here that the audio recordings functioned mainly to provide further verification of the data recorded in the curves. Indeed, the bulk of the experiments were dedicated to the analysis of curves. In this respect the two researchers continued what had been done with devices such as the phonautograph (Sterne and Akiyama, this volume). Similar to the many varieties of the kymograph—including the labiograph, laryngograph, or logograph, which recorded movements in vocal sound production and in the air—the phonautograph turned vibrations into a curve on blackened paper (Panconcelli-Calzia 1994; Rieger 2009).

The phonograph, however, shifted the focus to hearing. This was due, first, to technical constraints, as the tin foil was designed to carry only the traces of the sounds. No other (i.e., visual) inscription could be made in that same medium. Second, this forced the experimenters to carry out various functions of adjusting, calibrating, and controlling the correct operation of the phonograph with their ears. Hearing was required both for comparing the pitch of the recorded sound to some given instrument and for controlling the speed of the barrel by means of a metronome. Listening to the recorded sounds after they had been mechanically transformed into a shape was supposed to guarantee that the indentations still referred to the same sounds. Instead of correlating two simultaneously recorded processes, another mode of correlating was now developed that involved the sense of hearing. It successively addressed different—visual and audible—recordings of the same process rather than correlating simultaneous recordings of different processes.

The new way of producing data—data that in fact served as a token for listening—remained almost imperceptible as long as the main method of correlation still involved the familiar analysis of visible curves. In the experiments on the distortion of vowels, however, the correlation of what was actually heard during the experiment came to the fore. Perhaps the most important shift between these experiments and Jenkin and Ewing's later work was the following: A change of speed was applied to recordings. Thus, correlation now applied to audible data alone. The issue at stake here was not comparison between nature and recording but the transformation of recorded sound itself.

Vowels

The phonograph enabled experimenters to answer a still-unresolved question about the nature of sound. Scientific investigations in sound color did not begin in earnest until the nineteenth century, although already in 1761 Leonard Euler critically remarked in one of his *Letters to a Young German Princess* that the timbre of tones had escaped the attention of philosophers. As he explained to his addressee, listeners easily discriminated the loudness of tones. Also, music had taught them that tones vary in pitch, and it is on this differentiation that musical harmony was based. There was, however, another property of sounds that people often experienced when listening to music: "Two sounds may be of equal force, and in accord with the same note of the harpsichord, and yet very different to the ear. The sound of a flute is totally different from that of the French horn, though both may be in tune with the same note of the harpsichord, and equally strong" (Euler 1823, vol. 2, 68). The human voice, "that astonishing master-piece of the Creator" (69), could produce this variety of sounds simply by modifying the shape of the mouth. Although the consonants involved more "organs" than just the mouth cavity, such as lips, tongue, and palate, Euler claimed it should be possible to construct a machine that could articulate the sounds of language: "The thing does not seem to me impossible" (70).

Subsequently, various attempts were made to construct such a device.[2] In 1780 Danish physician and physicist Christian Gottlieb Kratzenstein won a prize awarded by the Imperial Academy of Saint Petersburg for the successful construction of organ pipes that could imitate the vowels of language (Ungeheuer 1983, 157). Wolfgang von Kempelen constructed a talking machine around 1780, an apparatus that emitted entire phrases in various languages. In this machine, bellows sent air through a variety of devices that would perform different aspects of articulation, such as phonation and the formation of vowels and sonorous and noisy consonants. In the final stage, the air passed through a malleable leather bell whose shape could be changed by hand, thus altering the quality of the "vowel" sounds (Kempelen 1791). The phonation device in the machine could produce only one pitch—Kempelen did not intend his machine to sing. Nevertheless, he noticed that a sequence of vowels would sound like a melody. This observation, however, was a by-product of his work for which he did not provide an explanation. By the 1830s the earlier attempts to imitate speech had attracted the attention of experimental researchers. Physicist Robert Willis conducted a series of experiments on vowel sounds and published his results in 1830. He criticized his forerunners for considering vowel sounds only with regard to articulation:

> Kempelen's mistake, like that of every other writer on this subject, appears to lie in the tacit assumption, that every illustration is to be sought for in the form and action of the organs of speech themselves, which, however paradoxical the assertion may appear, can never, I contend, lead to any accurate knowledge of the subject. (Willis 1830, 233)

In a long article, "On Vowel Sounds, and on Reed-Organ Pipes," Willis emphasized that the means to produce the sounds did not have to resemble the organs of speech. One of the experiments he described broke completely with the idea of such similarity. While the article is mostly about air columns in reed pipes, this one experiment did not involve any wind instrument. Holding a piece of watch spring against a revolving toothed wheel, an alternation of sound qualities was produced that depended on the length of the vibrating portion of the spring. As Willis observed:

> In effect the sound produced retains the same pitch as long as the wheel revolves uniformly, but puts on in succession all the vowel qualities, as the effective length of the spring is altered, and that with considerable distinctness, when due allowance is made for the harsh and disagreeable quality of the sound itself. (249–50)

Some thirty years later, Helmholtz reported on this experiment to strengthen his own argument that sound color is independent of a particular sound source. Singling out Willis's experiment as the one in which the similarity of sound sources was most clearly abandoned, he remained critical: "Willis's description of the motion of sound for vowels," he commented, "is certainly not a great way from the truth; but it only assigns the mode in which the motion of the air ensues, and not the corresponding reaction which this produces in the ear" (Helmholtz 1885, 118).

By 1863, when his comprehensive study on hearing, *On the Sensations of Tone as a Physiological Basis for the Theory of Music*, appeared, Helmholtz had investigated the sounds and functions of musical and acoustical instruments, including the reed instruments (Jackson, this volume). In reed pipes, a stream of air that is regularly interrupted produces the tone. An opening that opens and closes cuts the airstream into parts with the help of a reed. The resulting air "puffs" will be heard as a tone whose pitch depends on the velocity with which the puffs follow each other. This family of instruments included not only some types of organ pipes and the pipes of the harmonium (i.e., the reed organ) but also the human vocal tract, the "singing voice."

In the voice, Helmholtz considered the vocal cords or the "membranous tongs" to perform the part of the reed. The sound of the voice and more specifically of vowels depends both on these "membranous tongs" (i.e., the vocal cords, which can change the velocity of their movement freely) and on the air chamber (i.e., the mouth cavity, which can change its shape). Both parts of the sound production can vary independently. The voice can sing a melody on the vowel *a* alone, or stay on the same pitch level and pronounce *a*, *i*, and *o*. The two varying parameters of the voice thus had their corollary in the mechanism of articulation. The specificity of the vowel sounds lay, it was assumed, in the relationship between the vocal cords and the air chamber. Helmholtz stated that the pitch range of the vocal cords is, in most cases, lower than the resonance tone of the air chamber. Therefore, the air chamber reinforces, he assumed, one of the partials of the sound produced by the vocal cords.

Helmholtz's further investigation of vowels as sound colors divided acousticians into two groups. In the late 1850s he had started publishing his first experiments on the sound color of vowels (Helmholtz 1859). The main impulse for taking up this question was a mathematical theorem originally proposed by Jean-Baptiste Joseph Fourier at the beginning of the nineteenth century. It claimed that any periodic wave could be formally described by its sinusoidal components. Building upon prior assumptions by Georg Simon Ohm, Helmholtz applied this theorem to the periodic waves of musical tones. He claimed that if Fourier's theorem applied to these components of a tone, it predetermined the relationship of the components as forming integer ratios.

The sound color of vowels served as Helmholtz's most prominent object of investigation. He was even able to synthesize a number of vowels from sinusoidal components that he obtained from a set of amplified tuning forks. The sounds of vowels fitted the experiment all too well. They built up a set of sounds that were easy to discriminate even when pronounced under the most varied conditions, such as speaking, singing, or whispering. As it turned out, they were also comparatively easy to synthesize. At first glance, their sound color seemed to be determined by the relative strengths of the different sinusoidal components. This was in keeping with the analysis of musical instruments that Helmholtz carried out. Specific patterns in the components, such as the relative strength of every other component in the series of overtones in the sounds made by a clarinet, made it likely that these patterns would reappear in every note the instrument was able to produce. The vowels Helmholtz synthesized revealed a similar pattern: for an *u* (pronounced in the German manner), only one component was enough; an *o* could be produced with two strong components, while reinforcing the third component in the row of tuning forks produced a sound more similar to an *a*. Vowels thus seemed an exemplary case for a study of sound color in general.

However, the analysis of vowels revealed something different from Helmholtz's findings on the sound of musical instruments. The characteristics of vowels depended not just on a relative pattern of strong components but also on the absolute pitch of one or more of the stronger components in the sound spectrum. This had gone unnoticed as long as the production of sounds was tied to the fixed pitch of the tuning forks. With tuning forks a modification of pitch was almost impossible. Yet, the two aspects of articulation rather easily explained this phenomenon, as the shape of the mouth did not change when a vowel was sung at different pitches. Therefore, the vocal cords alone produced the difference in pitch while the mouth alone produced the differences in sound color. Hence, Helmholtz inferred that vowels differed from musical instruments in that their characteristics involved an element of absolute pitch. For *e* and *i*, he even discovered two such elements that had been too high for his tuning fork synthesizer to produce.

Jenkin and Ewing immediately responded to the invention of the phonograph by seeing it as the device that would shed light on the question of the color of sounds. They perhaps understood the new possibilities for manipulating recordings that the phonograph offered before grasping the meaning of the phonograph as producing audible and manipulable tokens for acoustic research. They saw the phonograph as a laboratory instrument that would enable them to carry out the controlled transposition of frequencies. A change in the rotation speed would alter all recorded frequencies in the same way. In this respect, the phonograph actually resembled an existing acoustic apparatus: the siren (Jackson, this volume; Welsh 2008). This instrument consisted of a rotating perforated disc through which a stream of air was passed. When one of the openings in the disc crossed the airstream, the air could pass through the disc; thus, a regular series of air pulses emerged, which, starting at a certain frequency, were heard as tones. The typical howl of the siren was due to the fact that the rotation had to be started at zero velocity and reach the required frequency only gradually. Assuming that the sound of the siren was composed of partials that were in accordance with Fourier's theorem, researchers such as Helmholtz or Jenkin and Ewing could infer that the transposition of spectra was feasible. In the howl of the siren, one could not make out any apparent change in color.

With its feature of transposing entire sound spectra, the phonograph opened up a new possibility of testing the two explanations for sound color Helmholtz had offered. His explanation for the tone color of musical instruments implied that their sound should not be affected by transposition, while his explanation for the tone color of vowels implied that transposition would cause distortions. The phonograph thus suggested the *experimentum crucis* to decide between these two explanations. Jenkin and Ewing heard no change in tone color. Again, the immediate response, this time by Charles Cross, showed that this experiment involved more than just the observation of physical phenomena. The vowels form a system of differentiations that allows for great deviations, while maintaining the ability to communicate the differences.

As it turned out, the first experimenters had to sort out what the term *quality* actually referred to when applied to the distorted vowels. Soon after they had published the first notice in a letter to the editor of *Nature* Jenkin and Ewing wrote the following:

> We venture, however, to remind any one trying the experiment that a low note followed by a high one suggests a change from *u* (Italian) to *i*. Thus if we whistle a low note and then the octave to it or a note near this, the ear is easily persuaded that the whistle resembles *u i*, but if now, beginning again on the note we just thought was *i*, we go up another octave, the new sequence again suggests *u i*, although the very note which was last taken to represent *i* now stands for *u*. If, therefore, we wish to judge what a sound really is we should not trust much to contrast, especially when a change of pitch is involved in the comparison. (Jenkin and Ewing 1878b, 167)

Howls

In 1914 and 1916, when Stumpf carried out his vowel experiments, his situation was very different. For one thing, he noted the awkwardness of his situation: In the laboratory, he and his assistants listened to almost imperceptible sounds while outside, beyond the city limits, the war roared. This silence is documented on one of the cylinders, which contains a soft noise, showing the attempt to use the phonograph for the investigation of whispered vowels. The analysis of the whispering voice had by then become a standard item in the investigation of the resonances of the mouth cavity. Dutch physiologist Franciscus Cornelis Donders had inferred from Helmholtz's investigation that the mouth cavity would render its tone even without being stimulated by the vocal cords. This gave rise to the question of intelligibility, which was to be investigated in terms of communication technology (Schmidgen 2007).

In a very different sense, however, Stumpf and his team listened to the howl of the siren. All of a sudden it appeared as a constant change in sound quality. A new way of listening posed the question of the vowel sound anew, but the object of listening was now the apparatus and its properties. The phonograph turned from a technical device into an object of investigation. The Berlin laboratory for experimental psychology, after more than ten years of work building the phonogram archive, was well equipped with a number of different phonographs. The laboratory protocols reveal that some experiments were carried out with a device called a "parlograph," designed for office use where it was supposed to substitute for face-to-face dictating.[3] The parlograph thus exemplifies a specialization in the development of the phonograph by focusing on one of the functions its inventors had been busy promoting.[4] Stumpf's interest was more in comparing the recording devices that were available with regard to their potential use in musicology and experimentation. The instrument used in most of the experiments, however, was a phonograph that was more appropriate for the purpose of changing the rotation speed. Its range of possible rotation speeds was large, allowing for a triple augmentation of the lowest speed. Together with his assistants, Stumpf carried out two series of tests—eight wax cylinder rolls in all were recorded.[5] The first series consisted of three cylinders, one for each rotation speed: slow, normal, and fast. These cylinders clearly revealed the experimental nature of the enterprise. On one cylinder, a voice is clearly heard asking for the tuning forks as the next item to be recorded: "und jetzt die Gabeln" ("and now the forks"); a howling sound follows this speech. It appears that the fork had been brought too close to the recording funnel and had thereby distorted the sound.

A list of the recorded sounds is given in the protocols, as well as in Stumpf's report in *Die Sprachlaute*. In addition to sung vowels, the cylinders recorded the sounds of tuning forks, organ pipes, and artificial vowels produced with a vowel tube (a device developed by Robert Willis in the 1830s). Thus, the recorded items combined sounds of the voice with sounds of laboratory instruments. All of these sounds were carefully chosen to enable maximal control of distortion. For instance, on the first series of cylinders not all vowels were recorded, only the *u* and *o* were juxtaposed. With only two vowels the systemic effect of recognizing the whole sequence of vowels could be avoided—a problem that had led former researchers to overhear the actual distortion of the sounds. The choice of these two vowels was sufficient to verify Helmholtz's assumption. Stumpf agreed with Helmholtz in that he took *u* to come closest to a sound with only one component of no fixed pitch. Thus, *u* should remain unchanged when the rate of rotation was altered. An alteration of *u* should in turn indicate that something was wrong in the experiment. In contrast, *o* should easily turn into an *a* when the speed was increased if Helmholtz's hypothesis about absolute pitch characterizing vowels was correct. The change in speed should not only alter the vowels but also create a difference between them on account of their transposition.

The next series of five cylinders focused on sung vowels, juxtaposing them with two spoken words and a series of tuning fork sounds. For these cylinders, the complete series of German vowels, including *ü*, *ö*, and *ä*, was recorded, followed by the words *Kuckuck* (cuckoo) and *Uhu* (owl). Five recording speeds were tested by replaying the recording at the same speed, as well as a number of different speeds. The distortions varied and pointed to different problems in recording. The tuning forks indicated the frequency range in which a recording could be made. Under certain conditions, their sounds would simply disappear. For the other sounds, squeaking and bleating tones were heard, as were resemblances to various musical instruments. In most cases, these changes could be explained by the so-called formant theory, which assumes stable amplitude peaks in the frequency spectrum. When using very low or very high rotation rates, the distortions were less easy to understand. Additional overtones appeared: *Kuckuck* turned into the nonsense syllables *Tretre*. Stumpf concluded that such phenomena must be explained by physical distortions that occurred when the phonograph was used in extreme ways. In other words, as any other medium, the phonograph had a linear characteristic for only a small region. Beyond the limits of this region, nonlinear distortions occurred. Using vowel sounds as objects for their experiments in recording quality, Stumpf and his team found two characteristics of recorded sound that were of the utmost significance for the use of the phonograph in ethnomusicology. The first was that sound color changed when the reproduction speed was altered. As long as the musicological investigation of the recorded sounds had not yet determined which of its characteristics were significant, such distortion could endanger the significance of the whole enterprise. Second, with the use of sung vowels and spoken words as the object under investigation, they also showed that an incorrect reproduction speed could destroy the meaning of a recorded item.

Cylinders

The group of researchers who took part in these experiments was remarkable in itself. Stumpf was assisted by, among others, Max Wertheimer and Erich Moritz von Hornbostel (i.e., by one of the founders of Gestalt psychology and one of the founders of ethnomusicology).[6] The latter had been director of the Phonogram Archive for more than a decade. Although this was still not an official position, Hornbostel had invested a considerable amount of time and money in the enterprise. In his growing compendium of published work, methodology was a central issue. Most notably, two articles he coauthored with Otto Abraham laid the foundation for comparative musicology, both giving center stage to the phonograph: *Über die Bedeutung des Phonographen für die vergleichende Musikwissenschaft* (On the Meaning of the Phonograph for Comparative Musicology) (1904) and *Vorschläge für die Transkription exotischer Melodien* (Suggestions for the Transcription of Exotic Melodies) (1909), where transcription meant mostly transcription from recordings. Most of Hornbostel's subsequent publications on specific musical examples were based on the collection of the archive, often pointing to this origin with titles such as *Phonographierte türkische Melodien* (Phonographed Turkish Melodies) and *Phonographierte Indianermelodien aus British Columbia* (Phonographed American Indian Melodies from British Columbia).

In cooperation with Felix Luschan, the first chair of ethnology at Friedrich-Wilhelms University in Berlin, Hornbostel prepared instructions for travelers on how to use the phonograph in the field. It was integrated into a guide for ethnological travels published in the journal *Zeitschrift für Ethnologie* (Journal for Ethnology). The Phonogram Archive handed out an offprint of this guide to travelers who were willing to make recordings and bring them back to Berlin. In many cases, the phonographs themselves were also provided by the archive, as were tuned whistles to be used to indicate the rotation speed. In the article, significant consideration was given to how the recordings in the archive might be used. In order to safely identify them, some preventive measures had to be taken. As any written information about a roll could get lost or be attached to the wrong item, the travelers were instructed to record the identifying data directly onto the wax cylinder. They were instructed to first blow the whistle and then briefly classify the provenance of the recording. The rotation speed could later be easily identified from the pitch of the whistle. This instruction added an acoustic tag to the recording, which became the basis for comparison. Without this tag, the recording of unknown sounds would be useless for the musicologists using the archive. With the help of the whistle, however, the traveler could safely identify the speed of the recording and thus guarantee that the sound color was correctly reproduced. As a consequence, the traveler could deliberately choose whether greater length or higher quality were more important in a given situation. The texts of epic poetry, for instance, would often require more time, while for music better quality seemed indispensable. This choice was backed up by the pitch tag, which connected the recording to the collection of other items in the archive.

In 1908 the *Phonographic Journal* reported on a recording that had been made in the name of preserving the culture of a Slavic people living in the Lausitz region of Germany. The trip that led to the recording was organized in March 1907 by the author of the report, Baron von Hagen. Hagen had invited the members of the Phonogram Archive to join him in recording "Wendish" music. This was the German name for the Slavic minorities living in German territory.

> On behalf of Geheimrat Stumpf (the Psychological Institute had been invited), Dr. Hornbostel appeared with a small phonograph. However, Dr. von Hornbostel recorded only the spinning songs of an old Wendish woman because Geheimrat Stumpf has not yet shown interest in [recording] speech or in things already set to music (von Hagen 1909, 460).

Baron von Hagen had other interests, as he declared in this and other articles that he and like-minded people published in the journal. He believed it should be the responsibility of a phonogram archive to preserve and foster German culture. This required archiving the voices of prominent people and making a survey of the German dialects and the non-German languages spoken within the territory of Germany. For teaching purposes, a third section of the archive should contain samples of languages other than German. A fourth section should provide materials for physiology and voice therapy. The last section, he suggested, should be a music collection, divided in two parts, one part containing vocal samples for the teaching of singers and the other—corresponding to what the Berlin Phonogram Archive already did—with samples of music from all over the world (von Hagen 1909, 382). In the late nineteenth and early twentieth century, the obsession to inscribe one's own culture into history also turned into an obsession with preserving one's present. The phonograph offered a twist on this desire for preservation, a twist that projected it into the future because the phonograph created as it preserved (Sterne 2003, 332–3; Hoffmann 2004). All over Europe similar activities took place, such as the Musée phonographique in Paris (Brain 1998, 277ff.) or the later activities of German linguist Wilhelm Doegen for the German Sound Archive (Lange 2007).

Von Hagen's somewhat eclectic program was held together by commercial interests. Hagen urgently advised the Phonogram Archive to get in touch with the emerging recording industry, which, he hoped, would support them in collecting samples of languages and music abroad. This link between the professional recording industry and scientific enterprises has been noted elsewhere in this handbook (e.g., Bruynincks, Whittington, this volume). Most important, Hagen tried to convince the Phonogram Archive to abandon the phonograph and to use gramophone technology. The main advantage of the gramophone was that its discs were easier to copy. Copying initiated the double success of the gramophone on the market, and soon it surpassed the phonograph. In contrast to the phonograph and its cylinders, the gramophone and its records thus split into two industrial branches. Selling sounds on gramophone discs became a business in its own right, and the sound quality became subject to competition and secrecy. Duplication technology, of course, was the technological basis for the commercial success of the discs. Phonographic recordings posed greater difficulties in copying. Hornbostel, a trained chemist, eventually succeeded in solving this problem. Electroplating allowed a matrix to be made of the cylindrical shape of phonograph recordings. For a while, the Berlin Phonogram Archive even led the technology of duplicating wax cylinders. Although this method was risky (it destroyed the original in the process of the galvanoplastic duplication), other collectors trusted Hornbostel's method to such an extent that they sent their originals to Berlin in order to obtain "galvanos" (i.e., the matrices) and copies in exchange. This allowed the collection in Berlin to grow even faster as a copy of each electroplated cylinder was kept. For commercial purposes, however, the method was too costly. The material used to make the copies was found only after a long series of trials.[7] The phonogram archive repeatedly appealed for funds, mostly asking the government to support the work.[8]

Baron von Hagen and an anonymous supporter ridiculed the phonogram approach in numerous articles published in the *Phonographische Zeitschrift*.[9] They pleaded for a strong commitment to the gramophone's recording and duplicating methods as this would open up the potential for a self-sustaining basis for the archival work. Gramophone discs could be sold to schools or to individuals interested in learning foreign languages and thus would return some income to the scientific enterprise.

Abraham and Hornbostel, however, opted for the technology that enabled users to record themselves. What was even more important to them was the possibility of manipulating recordings for research purposes. They discussed the pros and cons of the "phonographic technology" at hand in their 1904 article on the significance of the phonograph for comparative musicology:

> In recent times, phonographic technology has made great progress. The phonograph was followed by the gramophone, and both apparatuses contest each other, each has its advantages and shortcomings. Both use membrane vibrations, which are transmitted to a pivot by a lever, which in turn writes its movements into the waxen surface. The phonograph does this vertically, punctuating its movements into the laterally moving wax cylinder. In contrast, the gramophone uses discs instead of cylinders; the pivot draws the undulating shape of the tonal vibrations onto the discs' surface. (Abraham and Hornbostel 1904, 231)

The exact knowledge about the gramophone's recording technology, Hornbostel and Abraham added, remained an industrial secret. For this reason the gramophone could not be used for private recordings.[10] A great advantage of the gramophone was the convenient storing of discs. To this end, the Viennese Phonogram Archive, which was established shortly before the Berlin Archive, had invented its own method of transcribing wax cylinders onto discs, using vertical cutting similar to Edison's phonograph. This technology, however, tied the apparatus to its location. It was so heavy that it had to remain in the rooms of the Austrian Academy of Sciences, which hosted the archive. The phonograph, in contrast, was easy for one person to carry. In his travels to the Lausitz, Hornbostel brought only a small phonograph. However, the recordings he made were futile, not so much because of their allegedly inferior quality but because he was not allowed to play them. These recordings are preserved today in the archive, but in every container there is a note stating "playing forbidden." The right to market and to listen to them belonged to von Hagen only.[11]

In the Berlin archive, the phonograph remained in place as *the* recording device. Although, back in 1900, Stumpf had been given an opportunity to try wire recording—then called "telephonography"—the archive stayed with the wax cylinders. Over the years, methods of identifying the cylinders were developed, the safest being inscription on round labels that were glued to the cylinders' boxes. These labels were printed for the Psychological Institute, allowing for the inscription of categories such as inventory and catalogue number, as well as the tribe, object, person reciting and person recording, and the place and date of the recording. Most often, these inscriptions were repeated on the bottom of the box to minimize the danger of errors. These boxes were also used for the copies that the archive obtained from collectors who had asked for galvanos.

For the experimental recordings, these identifying measures were only rarely used. A collection of song performances initiated by Otto Abraham as early as 1907 to measure how closely performances matched given intervals is still stored in boxes with archive numbers 1401 to 1416 and catalogue numbers Exp. 18 to 34. But the identification of many others still remains questionable. The recordings were not copied, so the boxes contain the original wax cylinders. In addition to damage from mold, the recordings are spoiled in many places by the distortions produced by repeated playback of specific parts, thus clearly pointing to the methods that were used when measuring the recorded songs. When listening to the recordings, the musicologists used the cylindrical shape to create loops. As Hornbostel and Abraham wrote in their methodological instruction, "In order to get to hear a single note in isolation, one has to make the membrane [i.e., the needle] touch the cylinder while hindering its lateral movement by switching off the guide conduct. This method, however, deteriorates the recording" (Abraham and Hornbostel 1909/1910, 17).

Abraham's recordings make this deterioration palpable. Abraham published his results in 1921 in an article titled "Tonometrische Untersuchungen an einem deutschen Volkslied" (Tonometric Study of a German Folksong) (Abraham 1923). In order to get people to sing in an automated manner rather than reflect on the pitches, Abraham chose to have the national anthem serve as the "folksong" in question. Hornbostel, who had perfect pitch, recorded the anthem several times. He whistled it and sung it without words, first in one key, then transposed by one half-tone, and then very low and very high. His performance as it is preserved today is interrupted many times by the traces of measuring isolated notes. The recording eventually gets stuck when Hornbostel sings in the highest register—the cylinder loops his crackling voice when he bursts into laughter at his own singing.

Conclusion

The historiography of recording has done great service in enumerating technical devices: First, the phonograph was invented, shortly after that the gramophone, then around the turn of the century, wire recording comes in, and so forth. Not surprisingly, the straightforward accounts, indispensable and rich in historical and archival detail as they may be, have long been challenged. It has remained difficult, however, to focus on technology at the same time as integrating it with cultural history.

An important achievement in this respect has been Friedrich Kittler's quest for the technical conditions of cultural strata, which brought about a new perspective on technology. Given that at the time of Kittler's first pathbreaking writings (1990, 1999), the computer was opening up a new reality of media use, this search for the technical *a priori* was also pertinent to a new theory of media—one that could describe how symbolic codes were dissolved by self-registering apparatuses only to be turned into symbolic code again, if on a completely different, binary level. Sound, for example, was transformed into curves to be then digitally encoded. For the problem ethnologists encountered when they transcribed the music engraved on wax cylinders, computer technology developed new solutions. A century after Hornbostel and Abraham wrote down their ideas, musicians and ethnologists are now able to use special functionalities for transposition in sound software. These tools use digital encoding to avoid the distortion of pitch and sound quality that occurs when a recording is replayed at a slower speed in the analogous mode.

Media analysis in the wake of Kittler's search for technological *a priori* thus set for itself two stable entities: a contemporary point of departure, such as the computer and the discursive network enabled by the computer on the one hand, and a history of symbolic encoding, binary calculus, the history of investigating temporal resolution on the other hand among the conditions that made computer technology and its uses possible. Moving backward from a given state, the analysis of the technical *a priori* eventually discovered some primary conditions for this state. Around the turn of the twenty-first century, media studies started to question the historical framework of this analysis. Since then, media analysis has been understood to help focus on the instability of the notion of a medium. Focusing, for example, on a historiography of media, such studies show that most accounts of media technology could not work without a notion of medium that took its definition for granted rather than explaining it. While this had been excusable with regard to the simple account of technological development, in later histories of recording technology, such a stable notion of medium was concealed as well. The history of recording and its meaning for an auditory culture often presupposed that, once the phonograph was invented, any of its later uses could have been anticipated.

When viewing technology as the condition of culture one can easily get trapped in this presupposition. Other accounts of recording history have therefore strived to keep the definition of a device open to redefinition. For instance, Emily Thompson has urged researchers in her account of the history of the phonograph to accept that just after its invention "there was no single role or purpose for the invention to fulfill. The phonograph appeared before a need for its function had been identified" (Thompson 1995, 137). With this methodological caveat in mind she did indeed keep her approach to the history of the phonograph open, particularly with respect to an important element in the historiography of media—feedback, in her case feedback between consumers and technology. Obviously, however, the functions of the phonograph were not exclusively defined by consumers. One should not underestimate the importance of finding the modernity of the phonograph in consumption any more than one should also stress that consumption is always bound to some consumable object.

What is important in this story of the phonograph also within the context of science and technology studies is that it is not simply a story of black boxing. When dealing with the phonograph as the object of study, one must keep in mind that a process of standardization covers only some of the important aspects of the phonograph as a medium. A narrative of standardization has an inherent directionality: It tells how a function becomes fixed or, in other words, black boxed. While both the functions of recording and playback and the concrete instrument of the phonograph became standardized to some extent during the time the phonograph existed, their interplay often counteracted standardization. The black box had to be reopened, and some light had to be shed on its functioning, thus turning it into a gray box.

The story of the vowel experiments proposes a different approach to the history of the phonograph because it foregrounds the interaction between scientists and technological development.[12] This story traces the multiple shifts from exploring sounds by using a specific instrument to exploring a technology by using specific sounds. Introducing the phonograph as a scientific instrument allows the shift from an object to a means of investigation to be described and vice versa. As shown, the phonograph was immediately accepted by acousticians as another recording device. Technology thus takes center stage, but it also puts the emphasis on manipulation rather than standardization. Aside from the history of technical development and the history of its uses, there is a history of the phonograph as an unstable object—a gray box—in the context of research.

The story of the vowel experiments mirrors the various histories of the phonograph in a peculiar way. While every experimenter who repeated this experiment usually pointed to the improving quality of the device, the word *quality* describes this change insufficiently. A detail in the functioning, such as the rotation speed, must be understood differently in different historical contexts. Varying the speed meant different things in each case. If for Fleming and Jenkin speed was just one technical element among many that they had to control, rotation speed turned out to be the crucial characteristic of this new device because it allowed for the manipulation of recorded sound. After 1900, the situation changed profoundly. Ethnology brought new uses for the phonograph. Ethnological recordings, however, were worthless without information about the rotating speed. Still, ethnologists, did not want see the speed of recording and replaying be shut away in a black box. They preferred to stay in charge of this technical detail. Asking travelers to measure the speed by recording a known note, the ethnologists similarly avoided losing control of the issue of standardization. With such details, the phonograph points to features of sound recording that would become central in further research on sound and hearing: recurrence and manipulation. As a technological object, the phonograph quickly stabilized. Its basic elements—the needle, the barrel and spring, the membrane and funnel—point to functions such as engraving, transport, and amplification. Because they can be found at any stage of the technical development of sound recording, they are not specific to the phonograph but rather pertain to sound recording in general. Also, particular uses of the device soon crystallized, which proved to be of considerable social impact.[13] On closer inspection, however, none of these functions and uses were as stable as they seemed. Rather, in order to understand what could be done with a phonograph, one had to consider what could *not* be done with it. The specificity of the phonograph did not reside in its stable functions and uses but rather in a specific instability. A historiography of media has to look at the "becoming-media" (Vogl 2007), that is, the gradual coming into being of media, as Joseph Vogl has suggested.[14]

Media involve perception. However, at the same time they replace it with technology. First, the phonograph created sounds that turned perception upon itself: In the context of experimental psychology, the phonograph enabled the experimenter to retain the identity of the investigated recording, while at the same time producing variation in the perceived phenomena. Second, the phonograph thereby created points of reference: In the context of the phonogram archive, recurrence was essential, guaranteeing the repeated accessibility and the identity of the recorded phenomenon in each act of investigation. Finally, the phonographic recording provided a field of possible perception—the archival collection of data that would be open for questions that the researchers could not yet foresee. This collection brought into awareness that it would remain incomplete not so much because there was always more to be collected but more so because there would always be new ways of dealing with the preserved material. For comparative musicology, which would later be renamed ethnomusicology,[15] the phonograph provided the elementary technology for recursion and comparison. Its use in these emerging disciplines was supported, however, by a separate strand of research that intersected with ethnomusicology in the Berlin laboratory. Here, functions of the phonograph were tested (Bijsterveld, this volume), simultaneously revealing functions of hearing. These experiments involved more than science. Consumer uses, political interests, and cultural settings shaped the ways in which the experiments employed the phonograph, as well as the ways in which their results were fed back into culture. These experiments took place at the intersection of music psychology, ethnomusicology, experimental phonetics, and psychology—the phonograph serving as the point at which these disciplines converged.

NOTES

1 Ziegler (2006, 83) mentions 106 cylinders in the collection of "Experimentalaufnahmen" (experimental recordings). A number of digitized recordings from this collection are available at http://vlp.mpiwg-berlin.mpg.de/library/audio.html (accessed 30.09.2009).

2 Accounts of the history of speaking machines are given in Ungeheuer (1962, 1983); Hankins and Silverman (1995, 179–220); and Felderer (2002). Further attempts to construct speaking devices are mentioned by Hankins and Silverman (e.g., by Pope Sylvester III, Robert Hooke, and, in the late eighteenth century, the Abbé Mical and Erasmus Darwin).

3 Cf. Ethnologisches Museum Berlin, Preußischer Kulturbesitz. Berliner Phonogramm-Archiv: Carl Stumpf Papers on Acoustics, Envelope 17, Phonographische Versuche.

4 For the use of the phonograph, see Thompson (1995); more generally on early recording, see Reed and Welch (1994).

5 Cf. http://vlp.mpiwg-berlin.mpg.de/library/data/lit38711, http://vlp.mpiwg-berlin.mpg.de/library/data/lit38712, http://vlp.mpiwg-berlin.mpg.de/library/data/lit38713, http://vlp.mpiwg-berlin.mpg.de/library/data/lit38894, http://vlp.mpiwg-berlin.mpg.de/library/data/lit38895, http://vlp.mpiwg-berlin.mpg.de/library/data/lit38896, http://vlp.mpiwg-berlin.mpg.de/library/data/lit38897, and http://vlp.mpiwg-berlin.mpg.de/library/data/lit38898 (accessed Sept. 30, 2009;, cf. footnote 2).

6 On Wertheimer as a pupil of Stumpf see Ash (1995); on Hornbostel see Klotz (1998).

7 The recipe for this was lost until 1989, when the Berlin collection was reassembled. See Ziegler (2006).

8 Cf. the appendix in Kaiser-El-Safti (2003).

9 Anonymous (1907, 1908); von Hagen (1908); cf. also Stumpf (1908).

10 Like the phonograph, the gramophone was originally a recording *and* a replaying apparatus. Cf. Moore (1999).

11 Cf. Susanne Ziegler's comment in the booklet to *Music! 100 Recordings: 100 Years of the Berlin Phonogramm-Archiv 1900–2000* (Wergo LC 06356 2000), 125–27.

12 In this respect, my account of the phonograph comes close to the accounts of interaction between instruments and their users given in Joerges and Shinn (2001).

13 The use of the phonograph has been critically reviewed by Jonathan Sterne (2003) under the heading of "A Resonant Tomb." Here, Sterne shows how preserving sound for eternity became a leitmotiv in the discourse on the phonograph.

14 Vogl (2001, 2008) and more generally the corresponding issue of the *Archiv für Mediengeschichte* with a focus on media historiography, ed. Lorenz Jäger, Bernhard Siegert, and Joseph Vogl, as well as the issue on "New German Media Studies" in *Grey Room* (29) (Winter 2008), ed. Eva Horn.

15 The discussion about the naming continued until the 1950s; cf., for example, Kolinski (1957).

REFERENCES

Abraham, Otto. "Tonometrische Untersuchungen an einem deutschen Volkslied." *Psychologische Forschung: Zeitschrift für Psychologie und ihre* Grenzwissenschaften 4 (1923): 1–22.

———, and Erich Moritz von Hornbostel. "Über die Bedeutung des Phonographen für vergleichende Musikwissenschaft." *Zeitschrift für Ethnologie* 36 (1904): 222–36.

———. "Vorschläge für die Transkription exotischer Melodien." *Sammelbände der Internationalen Musikgesellschaft* 11(1) (1909/1910): 1–25.

Anonymous. "Das Berliner Phonogrammarchiv." Phonographische Zeitschrift 9 (1908): 382–83.

———. "Wissenschaftliche Phonogramme für die Allgemeinheit." *Phonographische Zeitschrift* 8 (1907): 381.

Ash, Mitchell G. *Gestalt Psychology in German Culture, 1890–1967: Holism and the Quest for Objectivity*. New York: Cambridge University Press, 1995.

Brain, Robert. "Standards and Semiotics." In *Inscribing Science: Scientific Texts and the Materiality of Communication*, ed. Timothy Lenoir, 249–84. Stanford, Calif.: Stanford University Press, 1998.

Collins, Harry M. *Changing Order: Replication and Induction in Scientific Practice*, Beverley Hills: Sage, 1985.

Cross, Charles. R. "Helmholtz's Vowel Theory and the Phonograph." *Nature* 18 (May 23, 1878): 93–94.

de Chadarevian, Soraya. "Graphical Method and Discipline: Self-Recording Instruments in Nineteenth-Century Physiology." *Studies in the History and Philosophy of Science* 24 (1993): 267–91.

Euler, Leonhard. *Letters of Euler on Different Subjects in Natural Philosophy: Addressed to a German Princess*, ed. David Brewster, 2 vols. Edinburgh: Tait, 1823.

Felderer, Brigitte. "Stimm-Maschinen: Zur Konstruktion und Sichtbarmachung menschlicher Sprache im 18. Jahrhundert." In *Zwischen Rauschen und Offenbarung: Zur Kultur- und Mediengeschichte der Stimme*, ed. Friedrich Kittler, Thomas Macho, and Sigrid Weigel, 257–78. Berlin: Akademie, 2002.

Foucault, Michel. *L'archéologie du Savoir*. Paris: Gallimard, 1969.

Hankins, Thomas L., and Robert J. Silverman. *Instruments and the Imagination*. Princeton, N.J.: Princeton University Press, 1995.

Helmholtz, Hermann von. *On the Sensations of Tone as a Physiological Basis for the Theory of Music*. New York: Dover, 1954 (reprint of the London 1885 ed., trans. Alexander Ellis, after the 4th German ed. of 1877).

———. "Ueber die Klangfarbe der Vocale." *Annalen der Physik und Chemie* 108 (1859): 280–90.

Hoffmann, Christoph. "Vor dem Apparat." In *Bürokratische Leidenschaften: Kultur- und Mediengeschichte im Archiv*, ed. Sven Spieker, 281–94. Berlin: Kadmos, 2004.

Jackson, Myles. *Harmonious Triads: Physicists, Musicians, and Instrument Makers in Nineteenth-Century Germany*. Cambridge, Mass.: MIT Press, 2006.

Jenkin, Fleeming, and J. Alfred Ewing. "Helmholtz's Vowel Theory and the Phonograph." *Nature* 17 (March 14, 1878a): 384.

———. "On the Harmonic Analysis of Certain Vowel Sounds." *Proceedings of the Royal Society of Edinburgh* 28 (1876–1878): 745–77.

The Phonograph and Vowel Sounds," *Nature* 18 (June 13, July 25, and August 22, 1878b): 167–69, 340–43, 454–56.

Joerges, Bernward, and Terry Shinn, eds. *Instrumentation between Science, State, and Industry*. Boston: Kluwer, 2001.

Kaiser-El-Safti, Margret, ed. *Musik und Sprache: Zur Phänomenologie von Carl Stumpf*. Würzburg: Königshausen und Neumann, 2003.

Kempelen, Wolfgang von. *Mechanismus der menschlichen Sprache nebst der Beschreibung seiner Sprechenden Maschine*. Stuttgart: Fromman (Holzboog), 1970 (reprint of the Vienna ed., Degen, 1791).

Kittler, Friedrich. *Discourse Networks 1800/1900*. Stanford, Calif.: Stanford University Press, 1990 (German ed., Berlin: Brinkmann and Bose, 1985).

———. *Gramophone, Film, Typewriter*. Stanford, Calif.: Stanford University Press, 1999 (German ed., Berlin: Brinkmann and Bose, 1986).

Klotz, Sebastian, ed. *"Vom tönenden Wirbel menschlichen Tuns": Erich M. von Hornbostel als Gestaltpsychologe, Archivar, und Musikwissenschaftler: Studien und Dokumente*. Berlin: Schibri, 1998.

Kolinski, Mieczyslaw. "Ethnomusicology, Its Problems and Methods." Ethnomusicology 1(10) (May 1957): 1–7.

Lange, Britta. "Ein Archiv von Stimmen: Kriegsgefangene unter ethnografischer Beobachtung." In *Original/Ton: Zur Mediengeschichte des O-Tons*, ed. Nikolaus Wegman, Harun Maye, and Cornelius Reiber, 317–41. Konstanz: Universitätsverlag Konstanz, 2007.

Moore, Jerrold Northrop. *Sound Revolutions: A Biography of Fred Gaisberg, Founding Father of Commercial Sound Recording*. London: Sanctuary, 1999.

Panconcelli-Calzia, Giulio. *Geschichtszahlen der Phonetik: Quellenatlas der Phonetik*, ed. Konrad Koerner. Philadelphia: Benjamins, 1994.

Reed, Oliver, and Walter L. Welch. *From Tinfoil to Stereo: The Acoustic Years of the Recording Industry: 1877–1929*. Gainesville: University Press of Florida, 1994.

Rheinberger, Hans-Jörg. *Toward a History of Epistemic Things: Synthesizing Proteins in the Test Tube*. Stanford, Calif.: Stanford University Press, 1997.
Rieger, Stefan. *Schall und Rauch: Eine Mediengeschichte der Kurve*. Frankfurt: Suhrkamp, 2009.
Schmidgen, Henning. "The Donders Machine: Matter, Signs, and Time in a Physiological Experiment, c. 1865." Configurations 13(2) (2007): 211–56.
Sterne, Jonathan. *The Audible Past: Cultural Origins of Sound Production*. Durham, N.C.: Duke University Press, 2003.
Stumpf, Carl. "Das Berliner Phonogrammarchiv." *Internationale Wochenschrift für Wissenschaft, Kunst, und Technik* 2 (1908): 225–46.
———. *Die Sprachlaute: Experimentell-phonetische Untersuchungen nebst einem Anhang über Instrumentalklänge*. Berlin: Springer, 1926.
———. "Tonsystem und Musik der Siamesen." *Beiträge zur Akustik und Musikwissenschaft* 3 (1901): 69–138.
Thompson, Emily. "Machines, Music, and the Quest for Fidelity: Marketing the Edison Phonograph in America, 1877–1925." *Musical Quarterly* 79 (1995): 131–71.
Ungeheuer, Gerold. *Elemente einer akustischen Theorie der Vokalartikulation*. Berlin: Springer, 1962.
———. "Über die Akustik des Vokalschalls im 18. Jahrhundert: Der Euler-Lambert-Briefwechsel und Kratzenstein." *Phonetica* 40 (1983): 145–71.
Vogl, Joseph. "Becoming-Media: Galileo's Telescope." *Grey Room* 29 (2007): 14–25.
———. "Medien Werden: Galileis Fernrohr." *Archiv für Mediengeschichte* 1 (2001): 115–23.
von Hagen, Baron. "Das Phonogrammarchiv." *Phonographische Zeitschrift* 9 (1908): 404–406, 460–61, 488–89.
———. "Das Phonogrammarchiv." *Phonographische Zeitschrift* 9(15) (1909): 382, 460.
Welsh, Caroline. "Die Sirene und das Klavier: Vom Mythos der Sphärenharmonie zur experimentellen Sinnesphysiologie." In *Parasiten und Sirenen: Zwischenräume als Orte der materiellen Wissensproduktion*, ed. Bernhard J. Dotzler and Henning Schmidgen, 143–77. Bielefeld, Germany: Transcript Verlag, 2008.
Willis, Robert. "On Vowel Sounds, and on Reed-Organ Pipes." *Transactions of the Cambridge Philosophical Society* 3 (1830): 231–68.
Ziegler, Susanne, ed. *Die Wachszylinder des Berliner Phonogramm-Archivs*. Berlin: Staatliche Museen zu Berlin, Preußischer Kulturbesitz, 2006.

SECTION III

STAGING SOUND FOR SCIENCE AND ART: THE LAB

CHAPTER 8

FROM SCIENTIFIC INSTRUMENTS TO MUSICAL INSTRUMENTS: THE TUNING FORK, THE METRONOME, AND THE SIREN

MYLES W. JACKSON

Introduction

This chapter analyzes the ways in which nineteenth-century acoustical instruments that were meant to standardize musical performance and measure various dimensions of sound, such as pitch and beat, were a century later put to use as musical instruments themselves. Metronomes (and their predecessor, the chronometer)

The author is Dibner Family Professor of the History and Philosophy of Science and Technology, Director of Science and Technology Studies at the Polytechnic Institute of New York University and Professor of the History of Science at the Gallatin School of Individualized Study of New York University.

and tuning forks migrated from mechanicians' workshops to bourgeois households, rehearsal halls, and physics and physiology laboratories and then back to concert halls, where they were the primary instruments of a number of twentieth-century compositions. Similarly, sirens, another instrument employed by nineteenth-century acousticians for accurately determining musical pitch, were heard with increasing frequency in the twentieth-century music halls of New York, Berlin, and Paris. Drawing upon a material cultural history of science and technology, this chapter traces the ways in which these objects were redefined by their new roles as the generators rather than the quantifiers of musical qualities by exploring both the use of mechanical apparatus to standardize critical aspects of early nineteenth-century music and the resulting debates surrounding what such standardization meant to the art. Did these machines hinder or enhance expression and creative genius? Could they thwart the attempts of virtuosi to take liberties with the composer's original intentions? Twentieth-century composers such as Györgi Ligeti, Edgard Varèse, and Warren Burt used these same acoustical instruments to subvert the very notions they were created to define and reinforce.

Standardization, the theme that ties together the various portions of this chapter, has been a critical component of science and technology studies (STS) and recent works in the history of science and technology (Frängsmyr, Heilbron, and Rider 1990; Olesko 1996; Schaffer 1992, 23–56; 1995, 135–72; Alder 1995; Gooday 2004; Galison 2003, 84–155). This chapter focuses on a different type of standardization, one of aesthetic qualities relevant to music, such as pitch and beat. Paradoxically, the physical instruments used to standardize music resulted in a plethora of pitches and tempi, thereby vastly increasing the richness of musical expression. Finally, this chapter addresses a topic often ignored by historians and sociologists of science and technology, namely the importance of sound in eighteenth- and nineteenth-century workshops. Most historians of science and technology, as well as STS scholars, have been guilty of fetishizing the visual at the expense of the aural. In so doing, they have missed an opportunity to investigate the importance of sound in early acoustical laboratories.

Nineteenth-Century Physical Instruments

The experimental history of all three instruments described in this chapter originated in nineteenth-century workshops. The tuning fork, invented in 1711 by John Shore, sergeant trumpeter to England's royal court, did not become a scientific instrument until the early nineteenth century. In 1802 German acoustician E. F. F. Chladni explained the principles of a tuning fork based on the longitudinal vibrations of a straight rod, which is bent in the middle. The tuning fork fits squarely into Chladni's research on vibrations and nodal points with a view to applying this knowledge to his invention of two musical instruments, the euphone and the clavicylinder. He also noted that the volume of a sounding tuning fork alternatively increased and decreased as one rotated it 360° (Ullmann 1996, 102).

In 1825 Ernst Heinrich Weber, Leipzig professor of anatomy, and his younger brother Wilhelm Eduard Weber, Leipzig professor of physics, turned to the tuning fork to explore the property of transverse vibrations, or ones that vibrate perpendicular to the direction of the propagating sound wave. They, like Chladni, were interested in the relationship of the volume of a tuning fork and the direction of the propagation of sound waves. They also employed the tuning fork for their investigation of condensed and rarified sound waves (Weber and Weber 1825, 507, 513). Their preliminary research on transverse vibrations led to Wilhelm Weber's subsequent research during the 1820s and early 1830s on the construction of compensated organ reed pipes, whereby the transverse vibrations of a vibrating reed compensated for the longitudinal vibrations of the air column, resulting in a constant pitch regardless of the air pressure blowing through the organ pipe (Jackson 2006, 114–34). In short, the Weber brothers and Chladni used tuning forks as physical instruments to study sound-wave propagation, interference, and musical-instrument design.

Shortly after Wilhelm Weber completed his work on acoustics, J. Heinrich Scheibler, Krefeld's leading silk manufacturer and opera aficionado, was working on his invention of the tonometer, which would revolutionize the tuning of pianos in equal temperament (Scheibler 1834, 53). His reputation began to spread throughout the German scientific community for the precision of his tonometer, an apparatus that comprised fifty-four tuning forks, each of which pulsated at 4 vibrations per second (vps; later Hz) sharper than the previous one. It encompassed the range of one octave, from 220 vps to 440 vps and served as the model for future nineteenth-century tonometers, including those of Georg A. I. Appun and Rudolph Koenig. He was also renowned among scientific (and, to a much lesser extent, musical) circles for his attempt to standardize performance pitch.

From 1840 onward, vibrations were often depicted graphically. Jean-Marie Constant Duhamel was one of the first physicists to realize that a stylus or pin, with ink at its tip, when attached to a tuning fork, can record vibrations, which the pin sketches onto a piece of paper wrapped around a resolving drum. The drum makes one complete revolution in a precise, known period. The tuning fork is struck, and the pin makes sinusoidal curves on the paper. One counts the number of curves and determines the frequency of the tuning fork (Morton 2004, 2). This is the basic idea behind the kymograph, originally invented by Carl Ludwig in 1847 to monitor blood pressure (Kursell, this volume).

In 1855 Jules Antoine Lissajous, French mathematician, experimental physicist, and professor of physics at the Lycée Saint-Louis, invented a method to visualize acoustic vibrations with the aid of tuning forks. He reflected a light beam from a small mirror attached to a vibrating tuning fork and then from a larger, rapidly rotating mirror onto a screen. After further study, he was able to generate his "Lissajous figures" by reflecting a beam of light from mirrors perched on top of two vibrating tuning forks positioned perpendicular to each other. These curves are determined by the relative frequency, phase, and amplitude of the tuning forks' vibrations as depicted on a screen (Jackson 2006, 210).

Throughout the 1860s French physicist Victor Regnault collaborated with the Prussian-born, Parisian instrument maker Rudolph Koenig on measuring the velocity of sound in the Parisian sewers. Regnault invented his own version of Koenig's tuning-fork chronograph, which registered the arrival of a sound pulse with an electric signal at the beginning and end of the pulse. The tuning-fork chronograph could graphically depict with unprecedented accuracy the travel time of the sound wave (Pantalony 2009, 49).

Tuning forks were at the center of a controversy between Koenig and the doyen of German physics, Hermann von Helmholtz (Pantalony 2005, 57–82). Throughout the 1870s and 1880s Helmholtz and Koenig engaged in a lively exchange about a new class of combination tones, or tones that are produced when two pitches are sounded simultaneously. Helmholtz called these new class of tones *summation tones* since they sounded at a frequency equivalent to the sum of the two primary pitches (Pantalony 2005, 62). These summation tones were much more difficult to hear than the other types of combination tones. Indeed, their existence was extremely controversial as they could not be heard by the unaided, untrained ear. Nonetheless, Helmholtz argued for their existence based on measurements obtained by resonators and membranes (Pantalony 2005, 62). Koenig, using his tonometer, came up with a different rule for the generation of combination tones. Drawing upon the older "beat theory" originally proposed by Thomas Young in the early nineteenth century, Koenig argued that these tones were a rapid succession of beats that blended as a tone. Beat tones were his answer to Helmholtz's combination tones. Koenig predicted and detected beat tones that were neither mathematically predicted nor observed by Helmholtz (see also Kursell 2009). In short, the tuning forks of the late nineteenth century became scientific instruments used to measure pitch precisely and describe the results of acoustic interference.

Tuning forks, however, were not limited to the experiments of physicists; they also found their way into the work of nineteenth-century anatomists and physiologists dedicated to elucidating the processes involved in aural perception. In 1827 Charles Wheatstone, a London experimental philosopher and later professor of experimental physics at King's College, discussed the effects of a tuning fork on the ear. The sound of a tuning fork whose base is placed on the forehead when the ears are blocked is louder than when the ears are open. If one ear is open and the other closed, the sound will be louder in the closed ear. If a tuning fork is placed above the temporal bone adjacent to a closed ear, the sound seems to be diverted to the opposite ear. Wheatstone concluded that since the volume of sounds communicated to the ear canal is increased, so too should the volume of external sounds acting on a cavity in a similar fashion (Wheatstone 1879, 30–32). "The great intensity with which sound is transmitted by solid rods, at the same time that its diffusion is prevented, affords a ready means of effecting this purpose, and of constructing an instrument which, from its rendering audible the weakest sounds, may with propriety be named a microphone" (Wheatstone 1879, 32). In short, he discovered the effects of occlusion and lateralization.

In 1834 Ernst Heinrich Weber described the same phenomenon based on the very same technique without referencing Wheatstone (E. H. Weber 1834, 25–44; Feldmann 1997, 321–22). A vibrating tuning fork is placed in the middle of the forehead equidistant from a patient's ears. The patient is asked to report in which ear the sound is louder. The sound should be heard equally loudly in both ears. If not, then the patient is either suffering from one-sided conductive hearing loss or unilateral sensorineural hearing loss due to an improperly functioning cranial nerve VIII.[1] This technique is now referred to as the Weber test. Eduard Schmalz, a Dresden physician and one of Weber's students, was responsible for introducing the Weber test to the field of otology for diagnosing various aural disorders (Feldmann 1997, 322–23). In 1855 Heinrich Adolf Rinne, a Göttingen physician, developed a similar hearing test, which now bears his name. It is often used in conjunction with the Weber test for detecting sensorineural hearing loss. The stem of a vibrating tuning fork is placed on the mastoid bone. Patients then inform the physician when they can no longer hear the tone. The physician notes the time and then immediately places the vibrating prongs of the fork next to the ear. A patient with a normally functioning cranial nerve VIII will hear the tuning fork for a longer period of time when placed next to the ear than when it is resting on the mastoid bone. The procedure is then repeated with the other ear[2] (Feldmann 1997, 324–325).

A second instrument used by nineteenth-century acousticians to measure pitch and to understand the phenomenon of acoustical interference was the siren. In the 1820s it was the preferred scientific instrument for precise pitch measurement in France. Whereas tuning forks sound due to the vibrations of the metal prongs in the air, the siren sounds as a result of rapid pulses of air. Since sirens produce sounds in ways distinct from rods and strings, they provided physicists and physiologists an opportunity to reconceptualize and understand the notion of tone. By the 1840s natural scientists were convinced that the sensation of tone was predicated upon periodic pulses transmitted to the auditory nerve. Hence, tone was now associated with periodicity rather than with the form of the vibration (Vogel 1994, 263).

Originally invented by Edinburgh professor of physics John Robinson during the late eighteenth century and subsequently improved in design by French engineer Charles Cagniard de la Tour in 1819 (Robinson 1822, 403–405; de la Tour 1819, 167–71; Robel 1891, 7–14), the siren is a thin disc, usually made of tin, with a series of concentric holes drilled around its circumference with a stationary pipe covering one of the holes. As the disc revolves around its axis, air travels through the pipe and also through the holes when they are directly over the mouth of the pipe, producing puffs of air. If one knows the number of revolutions of the wheel and the number of holes, the product of the two will be the number of puffs per second. De la Tour's device of 1819 was the most accurate and commonly used siren of the early nineteenth century.[3] As Helmholtz was to argue in his seminal *Die Lehre von den Tonempfindungen* of 1863, "this number is consequently far easier to determine exactly than in any other musical instrument, and sirens are accordingly extremely well adapted for studying all changes in musical tones resulting from the alterations and ratios of the pitch numbers" (Helmholtz 1954, 12).

In the late 1820s and early 1830s sirens were used in numerous physical and physiological investigations. The Parisian acoustician Félix Savart modified de la Tour's siren. His new device, known as the Savart wheel, comprised toothed metal discs. The discs spin and rub against a card at the rim of the wheel. The frequency of the pitch is directly proportional to the rate of the rotation of the wheel. He used this device to test the lower range of human hearing (Savart 1830, 337–52; 1831, 69–74). During that same period, de la Tour turned to his siren to test the entire range of human hearing. He also employed the instrument to calculate wind speed, as the pitch of the siren was proportional to the velocity of the wind entering through the holes (Robel 1891, 16–17). In 1829 Pierre Louis Dulong used a de la Tour siren to confirm Pierre Simon Laplace's equation for the speed of sound through air.

In 1841 August Seebeck's polyphonic siren, a siren that has several circles containing a different number of holes producing different pitches, was the instrument of choice to investigate phase relationship and interference generated by two or three tones. Historically, this is one of the most important and best-known applications of the siren to test physical theories. Seebeck was interested in determining how the ear perceives nonisochronic pulses. In situations where pulses occur in successive intervals of t, t', t, t', he was able to distinguish two tones, one corresponding to t + t', the other representing the octave (t + t')/2. The relative intensity of the two tones depended upon the difference between the two intervals. He concluded from his siren experiments that a tone was "a nearly isochronic series of single pulses of arbitrary form" (Vogel 1994, 263; Seebeck 1841, 417–36). Two years later Seebeck's conclusions were challenged by Georg Simon Ohm, who asserted that any sound could be expressed by a simple harmonic term in a Fourier series (Ohm 1843, 513–65; Jungnickel and McCormmach 1986, vol. 1, 268; Vogel 1994, 264). Seebeck responded by arguing for the importance of empirically researched sound perception, maintaining (contra Ohm) that the definition of a tone should not be limited to those that can be expressed as simple harmonic terms. Seebeck employed the polyphonic siren to produce tones for which the intensity of the vibrations corresponding to the upper partials could either be calculated mathematically or be distinctly heard[4] (Helmholtz 1895, 59).

Metronomes are the third example of physical instruments used to standardize musical qualities, in this instance, beats. The chronometer was invented in 1656 by the Dutch natural philosopher Christiaan Huygens and patented the ensuing year. He drew upon the earlier work of Galileo Galilei, who had realized during the early seventeenth century that pendulums could be used as effective timekeepers. Galileo had determined experimentally that the period of any given pendulum is independent of the weight of the plumb bob and the amplitude and that the square of the period varies directly with the length of the pendulum. Clearly, seconds pendulums, or those with a period of two seconds (one second per direction) were critical to eighteenth- and nineteenth-century astronomy and navigation. The metronome, which in essence is a modified chronometer, employed a compound pendulum pivoting around a central point with a fixed weight at the bottom and an adjustable weight at the top (Levin 1993, 83). It was introduced as a scientific instrument in the 1830s, when Scheibler used it while constructing his tonometer. As mentioned earlier, Scheibler's tonometer, which comprised fifty-four tuning forks made by the Krefeld instrument maker Hermann Kämmerling, encompassed the octave from 220 to 440 vps. With a modified metronome made by Kämmerling, Scheibler now possessed a precise measurement of beats produced by adjacent forks. Kämmerling also created a tuning fork for the tonic, its octave, and each semitone in between (Jackson 2006, 162–63).

Much like tuning forks, metronomes were also used in physiological investigations of the mid-nineteenth century. The Dutch physiologist Franciscus Cornelius Donders employed metronomes, tuning forks, and a kymograph in his research on physiological time. The subject sat at a table with an induction coil and several electric switches. Electrodes were placed on the subject's foot at the anklebone and on the skin of the groin. The subject operated the switches in response to light electric shocks generated by the investigator. A metronome was used to determine the reaction time (Schmidgen 2005, 225). Donders also employed a metronome to measure the rhythm of heartbeats (Schmidgen 2005, 229). In short, much like the tuning fork, the metronome was an instrument central to nineteenth-century physical and physiological research. The borders between music, physics, and physiology were rather porous.

The Standardization of Nineteenth-Century Musical Practice

Tuning forks and metronomes were clearly not restricted to nineteenth-century acoustical workshops and laboratories; they were also critical to musical practice. During the late eighteenth century, chaos apparently reigned in performance pitches. In 1791 flautist J. G. Tromlitz had remarked that "the pitch of all places is not the same but sometimes varies up to a semitone higher or lower" (translated in Powell 1791 and cited in Haynes 2002, 313). The lowest Viennese tuning fork had still been a semitone sharper than the Leipzig pitch (Haspels 1987, 122; Haynes 2002, 338). Not only did different countries possess different pitches for *a'*, but it was a rarity to find orchestras even in the same city sharing the same pitch. For example, in Paris, the pitch used by the Grand Opera for *a'* ranged between 427 in 1811 to 434 in 1829, while the Italian Opera preferred 424 in 1823 (Ellis in Helmholtz 1954, 495–500). The pitch used by the Dresden opera began to rise, albeit rather slowly, in 1821 and reached 435 some time between 1825 and 1830, while the organ of the Roman Catholic Church in Dresden sounded at 415 in 1824 (Ellis 1880, 310, 318).

In addition to the myriad of concert pitches throughout Europe, a gradual sharpening of the various concert pitches was well under way by the 1820s. As early as 1802 Heinrich Koch had noted in his seminal *Musikalisches Lexikon* that *Kammertöne* were gradually rising (Koch 1802, 822; Haynes 2002, 312). Generally, with the formation of large orchestras with a greater variety of instruments and musical scores written to accommodate such orchestras, tonal color catered to the higher instrumental range rather than the human voice, or the *vox humana*, as had been the case during the Middle Ages (Ellis 1880, 309–10). The rapid rise in pitch was perhaps greatest in the German territories. Whereas *Kammertöne* throughout Prussia and Saxony had been consistently lower than those in Vienna and Paris during the early years of the nineteenth century, by the 1830s German pitches were slightly higher than in most French cities and Vienna (Haynes 2002, 349). According to the Romantic music critic G. L. P. Sievers, by 1817 "the pitch of the three great Parisian orchestras is more than a semitone higher than the highest in Germany and Italy. The purely instrumental groups, where no singing is involved . . . [use a] tone [that] is even higher" (*Allgemeine musikalische Zeitung* [henceforth *AmZ*] 19 (1817), 302; Haynes 2002, 330). In the early 1830s, however, J. Heinrich Scheibler took several tuning forks from Berlin, Dresden, and Paris, and demonstrated that the tuning fork from the French capital possessed the lowest pitch. The pitches of these forks varied as much as three-quarters of a full tone. In the *Allgemeine musikalische Zeitung* a concerned reader lamented the "excessively high pitch" of a number of German orchestras (*AmZ* 37 (1835), 206; Haynes 2002, 349). In September of 1834 Heinrich Scheibler presented his work on the determination of the pitch for *a*' at 440 vps (or Hz). After conducting several experiments with various pitches used in Paris, Berlin, and Vienna, Scheibler decided to choose his *a*' at 440 as the middle of the extremes between which the pitch of Viennese pianos rises and falls due to change in temperature. The pitches of these pianos were determined by a monochord, and the pitch 440 vps was checked by his tonometer. This precise value was approved as the official, national German pitch by the physics section of the *Versammlung deutscher Naturforscher und Aerzte* in Stuttgart in September 1834.

Variations in pitch did not merely alter the aesthetic effect of pieces but also had an anatomical consequence. Vocalists in particular were distressed by the lack of pitch standardization. Throughout the first four decades of the nineteenth century, the *Allgemeine musikalische Zeitung* often reported the various diatribes against the rising pitch (*AmZ* 4 [1801], 76; 5 [1803], 529–35; 16 [1814], 772–76; 31 [1829], 285–94; 37 [1835], 205–207). Sopranos in particular protested vehemently: The prima donnas carried along tuning forks to performances and insisted (with limited success) that host orchestras tune to them (*AmZ* 3 [1801], 76; Haynes 2002, 333). In 1814 the philosopher C. F. Michaelis, professor of the philosophy of aesthetics at Leipzig, and J. G. Schicht, cantor of the Thomasschule in Leipzig, had pleaded for a common pitch for singers and orchestras throughout Europe. They argued that the necessity of singers to adjust to orchestral pitches, which differed as much as a full tone, was both exhausting and deleterious to their health (Michaelis and Schicht 1814, 772–76).

The Parisian prima donna Alexandrine Caroline Branchu forced the opera to lower its pitch in the early 1820s as she feared the high pitch would result in a premature loss of her voice (Haynes 2002, 330). Arias that required the mastery of very high pitches, as well as pieces that contained falsettos, were particularly difficult as the singers touring Europe were forced to readjust their ears and vocal cords to the particular *a*' that the orchestra chose. The problem became particularly acute in the mid-nineteenth century, when vocalists and instrumental virtuosi toured Europe at a feverish rate. Standardization was both an aesthetic and a physiological necessity.

Tuning forks were the instrument of choice when Europeans decided to standardize concert pitch. In 1859 a French commission that comprised composers, including Hector Berlioz, Gioacchino Rossini, and Giacomo Meyerbeer, and physicists, including Jules Lissajous, professor of physics at the Lycée Saint-Louis, and César-Mansuéte Despretz, sought a French concert pitch. During his research, Lissajous investigated the pitches of numerous tuning forks that musicians throughout Europe had sent him. He employed his technique of using Lissajous figures (discussed earlier) to determine the pitch of each fork precisely and accurately. When the French commission decided on a standard pitch of 435 Hz, it had the *diapason normal* constructed that sounded at that pitch (Jackson 2006, 207–14). Similarly, when representatives from six European nations traveled to Vienna in November of 1885 to determine a European pitch, two physicists, Vienna's professor of physics Josef Stefan and Rome's professor of experimental physics Pietro Blaserna, informed the international gathering of the necessity of tuning forks for keeping a steady pitch. Their pitches, unlike those of organ pipes, were not greatly influenced by changes in temperature. Hence, the tuning fork remained the physical instrument of choice to convey the proper pitch to European musicians. National scientific and technological bureaus, such as the Physikalisch-technische Reichsanstalt, were required to produce tuning forks that sounded at 435 Hz (Jackson 2006, 218–29).

What in the seventeenth century had started as isolated calls for mechanical instruments to more accurately count the beats of a composition culminated in a deafening crescendo by the end of the eighteenth century. In response, musicians and mechanicians attempted to construct portable chronometers. Magdeburg organist and cantor G. E. Stöckel argued that chronometers were necessary for beginners interested in learning to play music, as well as for those Liebhaber (amateurs) who do not play professionally but simply for their enjoyment (Stöckel 1800, 658). In addition, Stöckel underscored the importance of a correct, standard measure of time for compositions: "[T]he advantage of an as yet unknown, absent, generally accepted standard for musical time, or a chronometer, according to which this time is precisely determined . . . will be clearly recognizable" (Stöckel 1800, 658). The chronometer was, for Stöckel, the ally of the composer, whose "spirit" would be preserved (Stöckel 1800, 660). He also hoped that chronometrical markings might ensure a beat based on the universal standard of time, replacing the constant referral to vague terms such as *andante*, *allegro*, and *presto* (Stöckel 1800, 677).

He argued that the chronometer was just the device to assist pupils, "who by nature do not at first possess a sensitive and sure feeling of the beat (Taktgefühl) to sustain the most exact, rhythmical practice of the pieces" (Stöckel 1800, 679). The chronometer could provide mechanically what the performer lacked organically. In 1806 the piano instructor F. Guthmann penned an article describing his idea for a Taktmesser. He expressed his concerns about using a chronometer as a teaching tool, which students followed throughout the practicing of the entire piece, as this was anathema to "the spirit of true music" (Guthmann 1806, 118). He added, "[The] greatest artistes have also proven that they could not play following such a mechanical instrument. It repulsed their sensitivity, and they deviate from it involuntarily" (Guthmann 1806, 118).

During the early nineteenth century, not coincidentally the period that witnessed an increase in individual interpretation, improvisation by performers, particularly the virtuosi, gave rise to what some would see as the mechanical tyranny of precision. In the second decade of the nineteenth century, debates began to rage within German musical circles as to whether chronometers or timepieces should be used. Critics noted that the mathematical tempo corresponding to chronometer markings might simply be too mechanical. Precision just might not be all that beneficial. By not divulging the exact tempo markings for the piece, the composer allowed for individual interpretation. Mechanical control was diametrically opposed to the art of music—a theme echoed by a number of performers.

Perhaps the most informed piece on the subject was written by Gottfried Weber. Born in 1779, Weber had studied law at Göttingen, subsequently taking up a position in the imperial court of law in Wetzlar. He later became a tribunal judge in Mainz and remained in that position until 1818, when he was elected court legal advisor at Darmstadt, where he was named to a legal commission in 1825 that drafted the civil laws of the region. His civil and legal accomplishments, however, belie his undying passion for music. An accomplished cellist and flautist, he learned much from his friend, the acclaimed composer Carl Maria von Weber. Gottfried Weber, who founded *Caecilia*, an important journal for musicians and connoisseurs of music, authored various books and essays on a myriad of topics, including critiques of Mozart's compositions, a theory of tonality, the acoustics of wind instruments, the improvement of musical-instrument design, and the physiological properties of the human larynx (Schilling, vol. 6 [1838], 832–33).

Weber set up his essay on the chronometer in the form of a dialogue between a composer and a music director (G. Weber 1813, 441–47). The composer wishes to convince the music director that chronometers and timepieces are necessary for the proper performance of a composition. The music director, sensing that he is losing his authority to interpret the piece, protests staunchly. After the music director argues speciously against using new instruments to determine the beat by drawing upon classic authorities such as Sébastien de Brossard's *Dictionnaire de musique*, the crux of the argument ensues. The music director first condemns the use of mechanical instruments that make noise while keeping the beat. The composer, however, wholeheartedly agrees. The music director openly disputes the use of a chronometer to replace his conducting duties (G. Weber 1813, 443). In a sense, the machine would be too perfect for the imperfect art of music and for those who perform it. Once the composer assuages the music director's fear of replacement, he continues the conversation: "I was never of the opinion that a chronometer would be suitable for directing music. No lifeless or insensitive machine would ever be suitable for that" (G. Weber 1813, 443). For the composer (who represents Weber's own views here), the chronometer has the unique function of serving as an interpreter between the composer and the performers. The composer continues: "It is merely a standard of measure for the composer to be able to show the performer or the director of large musical pieces at exactly which tempo he wishes to have his work performed" (G. Weber 1813, 444). The composer simply needs to provide the music director and performers with the length of the pendulum, which produces the required time interval. The music director, by the conclusion of the dialogue, is convinced of the chronometer's importance.

The early chronometers, as ingenious as they might have been, never proved to be nearly as successful as Johann Nepomuk Mälzel's metronome. Mälzel's metronome, which he pirated from the renowned Dutch mechanician Diederich Nicolaus Winkel, possessed three advantages over the earlier chronometers. First, it was much smaller, more portable, and easier to use. Second, it was based on the number of beats per minute, a measure of time (unlike a measure of length), which every country recognized and hence was the best measure possible for standardization. Third, it was relatively inexpensive. Indeed, Mälzel himself made two hundred metronomes and sent them free of charge to the leading musicians and composers of the period: a clever business tactic to ensure that his device was the only one used for counting beats at various tempi (Harding 1983, 27–28).

Composers joined the ever-increasing number of supporters of Mälzel's mechanical contraption. They included Gaspare Spontini, Antonio Salieri, Muzio Clementi, Giovanni Battista Viotti, Luigi Cherubini, J. N. Hummel, Conradin Kreutzer, Ludwig Spohr, and Ignaz Moscheles. More important, we are told that "the aforementioned masters have promised to note all their future compositions according to the scale of Mälzel's metronome in order to confront every argument that can arise over the speed of the movement" (*Wiener Allgemeine musikalische Zeitung*, February 6, 1817).

In 1817 Mälzel's metronome made its grand appearance and received strong approval from an anonymous author in the *Allgemeine musikalische Zeitung* who expressed the hope that Mälzel's machine would free musicians and composers from "such multiheaded anarchy, which rules over these matters" (*AmZ* 19 [1817], 417–18). The metronome was easy to use as it did not presuppose any mathematical or scientific knowledge on the part of the musician, and it had a range of 50 to 160 pendulum swings per minute. The *Quarterly Musical Magazine and Review* of 1821 summed it up perhaps best: The metronome's "divisions are thereby rendered intelligible and applicable in every country: a universal standard measure for musical time is thus obtained, and its correctness may be proved at all times by comparison with a stopwatch" (*Quarterly Musical Magazine and Review* 3 [1821], 303). By 1824 the metronome had become part of life for musicians and composers in Britain: "[A] metronome must be referred to in order to ascertain the movement of the adagio or the allegro" (Cutler and the Editor 1824, 31–33).

Beethoven's first works with Mälzel's metronome markings were his string quartets op. 18, 59, 74, and 95 and the piano sonata op. 106. He then added his markings to all the movements of his symphonies (Wehmeyer 1993, 52; Stadlen 1979, 12–33). Beethoven did his best to force his will on the performers: "Maelzel's metronome gives us an excellent opportunity to do so [i.e., ensure that the composer's intensions were followed]. I give you my word for it here, in my further compositions I shall not use those terms. . . . That one will cry out 'tyrants,' I do not doubt . . . It would still be better than us being accused of feudalism" (Hamburger, 1951, 16; Leonhardt 1990, 146). Beethoven realized the issue was quintessentially one of control. He was particularly angered by the virtuosi, who constantly took liberties with the tempi of his piano pieces (Haupt 1927, 129; Schindler 1840, 212–24). Being infamously a rather whimsical creature, he withdrew his support once he and Mälzel became embroiled in a lawsuit over the ownership of the "Battle of Vittoria," calling the metronome a "dumb thing; one must feel the tempi" (Hamburger, 1951, 161). This view was the major objection to using the metronome. Rhythm should somehow be organic; one needed to feel it from within, not copy it from an external source. In addition, Beethoven famously mocked the metronome in the second movement of his Eighth Symphony. This sentiment was echoed by Sievers, who wrote in September 1819 that the metronome was an "error of the human spirit because music cannot be improved by mechanical inventions" (Sievers 1819, 599). He felt that the performer could use a different tempo; hence, the metronome was merely a triviality (*Spielerei*) (Sievers 1819, 600). He thought it strange that Mozart and other renowned composers were content to compose their pieces without fearing that the tempi of their compositions would be violated (Sievers 1819, 600). According to Sievers, a "philosophical" point was at issue: "A piece of music stops being a product of its composer the moment it is performed by an artiste; then this artiste is its second creator or something much more: an animator of the work, transferring his living spirit into the dead form of the composition" (Sievers 1819, 600).

Sievers's opinion notwithstanding, Mälzel continued to tinker, subsequently improving the design of his metronome in 1818 and 1832 (Anon. 1836, 10). In 1836 an anonymous pamphlet boasted two important attributes of the device. First, it gave the composer control over the performance of the piece. Second, it helped beginners learn the correct "feeling of the beat" (Taktgefühl) (Anon. 1836, 5). The mechanical metronome helped discipline both the youthful beginner and the capricious, unbridled performer.

Twentieth-Century Musical Performance

In the twentieth century, tuning forks continued to be the subject of research in physics and engineering. The use of quartz tuning-fork resonators is still rather popular in a range of scientific instruments, such as optical microscopes, atomic and magnetic-force microscopes, high-sensitivity magnetometers, atomic point contact sensors, and spectroscopic gas sensors (Barbic, Eliason, and Ranshaw 2007, 524). Quartz tuning-fork resonators are quite small: The tines are about 1.5 mm in length and about 100 μm in both width and thickness. Such dimensions permit the tuning fork to be highly sensitive force detectors (Barbic, Eliason, and Ranshaw 2007, 525). In the Accutron watch marketed by Bulova, Max Hetzel employed a 360-Hz steel tuning fork powered by a battery, thereby greatly increasing the accuracy of the watch. The tiny quartz crystal used in modern quartz watches is made in the shape of a tuning fork. The properties of quartz cause the tuning fork to generate a pulsed electric current, which is detected by the computer chip in the watch to keep track of the time.[5] Tuning forks are also still used in physics and engineering courses to teach dampened oscillatory motion (Wolfson and Pasachoff 1999; Serway 1996). They are also employed in primary hearing tests in the same manner discussed earlier in this chapter. In addition, they can be used to test for broken bones in the absence of an X-ray machine. If a sounding fork placed on a bone causes pain, the bone has most likely suffered a fracture (see Rice, this volume).

During the twentieth century, the tuning fork reentered concert halls, this time as a performance instrument rather than a mechanical device to assist one's technique in private. From 1985 to 1987, the American-born Australian musician Warren Burt composed "Music for Tuning Forks." Burt is known for composing in a wide variety of new musical styles, including acoustic music, electro-acoustic music, sound art installations, and text-based music. He constructed the aluminum tuning forks himself in the autumn of 1985 while an artist in residence at the Australian Commonwealth Scientific and Industrial Research Organisation's (CSIRO) National Measurement Laboratory at Monash University under the Australia Council, Artists and New Technology Program.[6] He tuned his forks, which were both treble and bass, to a nineteen-note, just-intonation scale derived from Claudius Ptolemy's *Harmonics* of the second century CE.[7]

As an experimental composer, Burt tinkers with multiple versions of the same pitch. His work is a prime example of microtonal music— music that uses pitches smaller in interval than the twelve Western semitones per octave—which explores the sounds generated from nineteen to fifty-three pitches to the octave. For his CD *The Animation of Lists and the Archytan Transpositions*, Burt wished to expand this scale without making fifty-three tuning forks. His solution was to transpose recorded tuning forks in a computer, creating an additional number of pitches that could not be generated by the tuning forks themselves. His preferred interval for the transposition came out to be "the 28/27 Archytas' large 63 cent version of the quarter-tone."[8] Hence, with a computer he was able to achieve what Scheibler, Koenig, and others had more than a century earlier with a tonometer. In essence, echoing the sentiments of numerous twentieth-century avant-garde composers, Burt wishes to challenge equal-temperament tuning, in which each semitone is the twelfth root of two times sharper than the preceding pitch. Since his work "Music for the Tuning Forks," he has written numerous pieces using tuning forks, computers, and even live-interactive computer pieces. Interestingly, whereas Scheibler used his tonometer of tuning forks back in 1834 to assist piano tuners to tune in equal temperament, Burt uses tuning forks to explore creatively other possible temperaments. One music critic noted, "In Warren Burt's hands, these tuning forks become some strange new instrument, complete with its own exotic tuning system."[9]

Similarly, the siren also became a performance instrument of the twentieth century even though it had previously been solely a physical instrument, unlike the tuning fork and the metronome. Indeed, it is present in many more compositions than the tuning fork and the metronome. The avant-garde French composer Edgard Varèse employed sirens in a number of his compositions, most notably in *Ionisation*, which debuted in 1931. Varèse was fascinated by the intersection among science, technology, and music. He studied Helmholtz's *Die Lehre von den Tonempfindungen* while experimenting with sirens and whistles (Wen-Chung 1966, 165). In 1913 he had met René Bertrand, inventor of the Dynaphone, with whom he would go on to experiment on the possibilities of electronics as a musical medium. Three years after his initial meeting with Bertrand, Varèse argued, "Our musical alphabet must be enriched. We also need new instruments badly . . . In my own works I have always felt the need of new mediums of expression . . . which can lend themselves to every expression of thought and can keep up with thought" (quoted in Wen-Chung 1966, 165; originally appeared in the *New York Morning Telegraph*, 1916). In 1922 he suggested that "The composer and the electrician will have to labor together to get it" (quoted in Wen-Chung 1966, 165; originally appeared in the *Christian Science Monitor*, 1922).

To develop an electronic machine for composing, in 1927 Varèse collaborated with Harvey Fletcher, who was the acoustical research director at Bell Telephone Laboratories at the time. From 1932 to 1936 Varèse applied annually for a Guggenheim Fellowship "to pursue work on an instrument for the producing of new sounds" (quoted in Wen-Chung 1966, 165). Sadly, the Guggenheim Foundation rejected his application each time. As a result, he conducted only a number of low-level experiments with phonograph turntables with motors set at different speeds.

From 1929 to November 1931, while working on *Ionisation*, Varèse had been contemplating the establishment of a music laboratory in which students would be able to compose music. Although the laboratory never came to pass, one can nevertheless see that he was preoccupied with scientific instruments while composing *Ionisation*. He was particularly fascinated by the science of sound and timbre and relentlessly sought out new musical instruments (Ouellette 1968, 103). Much like Beethoven and others of the early nineteenth century who used machines such as the metronome to prevent the virtuosi from taking unrestrained liberties with their compositions, Varèse proclaimed, with much bravado:

> The performer, the virtuoso, ought no longer to exist: he would be better replaced by a machine, and he will be. We shall find new intensities, too, for the realm of sound has still been very incompletely explored. These ideas still shock a great many people, but you will see them become realities in the more or less distant future . . . The composer will have improved, more flexible means at his disposal with which to express himself. His ideas will no longer be distorted by adaptation or performance as all those of the classics were. (quoted in Ouellette 1968, 106)

In 1930 Varèse took part in a discussion on the future of the mechanization of music. His revolutionary views shocked many (Ouellette 1968, 104–105). Echoing the desires of Italian composer Luigi Russolo (Bijsterveld 2002, 121), Varèse went on to express his hope that the innovative mechanical system would allow new timbres, intensities, and frequencies, furthering the work of acousticians on the limits of human hearing.

Ionisation was composed for thirteen musicians playing thirty-seven percussion instruments, including two sirens, one possessing a high pitch, the other a low one, both of which were borrowed from the New York City Fire Department[10] (Ouellette 1968, 105). The fire sirens, which Varèse also used in subsequent compositions, created a glissando effect through a myriad of frequencies without ever resting on any one in particular, thereby using them in a manner quite different from the practices of nineteenth-century acousticians.[11] Rather than implementing a siren to provide a steady pitch as acousticians and physicists had in the nineteenth century, Varèse stressed the instrument's ability to elicit numerous tones with various pitches throughout an octave. Sirens in twentieth-century compositions were not meant to increase precision but to abolish tonal increments in music. As Wen-Chung argues, *Ionisation* is a classic work "because it demonstrates that Varèse's concept is successfully applicable even when no definite pitches are present, the supreme test for his goal of liberating sound" (Weng-Chun 1966, 161, 163). It premiered at New York's Carnegie Hall on March 6, 1933, under the direction of Nicolas Slonimsky, to whom the piece was subsequently dedicated. Varèse's concept was that of the "process of atomic charge as electrons are liberated and molecules are ionized."[12] As the composer later divulged when reflecting on this work, "I was not influenced by composers as much as by natural objects and physical phenomena"[13] (Schuller 1965, 34).

In the twentieth century the metronome also became a performing instrument to test the boundaries of the well-established classical musical qualities of rhythm. Avant-garde composers toyed with the metaphors of machines, clocks, precision, and standardization. The Hungarian Jewish composer György Ligeti, born in Transylvania, Romania, was a pioneer of electronic music who enjoyed challenging well-established musical boundaries. Wishing to remove himself from the classic-Romantic tradition, the young Ligeti embraced the electronic music of Karlheinz Stockhausen and Gottfried Michael Koenig "with regard not only to sonority but also form, unfolding, the 'flow' of the music. It was a liberation for me—a liberation from this thinking in bars, this measured time" (quoted in Griffiths 1997, 22). He developed the musical technique of *micropolyphony*, a form of twentieth-century musical texture involving the use of sustained dissonant chords that shift slowly over time, or "a simultaneity of different lines, rhythms, and timbres."[14] Ligeti himself described micropolyphony as "[t]he complex polyphony of the individual parts [that] is embodied in a harmonic-musical flow, in which the harmonies do not change suddenly, but merge into one another; one clearly discernible interval combination is gradually blurred, and from this cloudiness it is possible to discern a new interval combination taking shape."[15]

Much of his musical compositions dealt with time, its measurement, and its precision. He spoke of "the spatialization of the flow of time, creating a continuous present," adding that "it would be much more worthwhile to try and achieve a compositional design of the process of change" (Griffiths 1997, 32–33). Perhaps the best example of his depiction of precision and the spatialization of the flow of time was his *Poème symphonique for 100 Metronomes* of November 1962. Ligeti himself concedes that the origin of this work dates back to a story with which he had been obsessed as a youth. A widow, whose husband has been dead for years, lives in a house full of clocks constantly ticking away. The walls are bedecked with barometers and hygrometers. "The meccanico-type music really originates from reading that story as a five-year-old, on a hot summer afternoon" (quoted in Griffiths 1997, 40). "Nobody comes, maybe for a hundred years. Nothing happens. So there is a combination of movement, which is machine-like, and absolutely nothing . . . a timelessness . . . no beginning and no end" (quoted in Steinitz 2003, 8).

Ligeti's *Poème symphonique* is seen as his final fluxus composition (the fluxus movement was an international network of artists, composers, and designers who were noted for blending different artistic media and disciplines in the 1960s and were active in neo-Dada noise music and art, literature, urban planning, architecture, and design).[16] The piece debuted at the annual Gaudeamus Music Week in the Hilversum City Hall in Holland in September 1963. Ligeti's piece, which was an experiment in indeterminate rhythmic counterpoint, furthered his exploration of micropolyphonic textures. It offered the interest of individual clockwork mechanisms heard together and gradually thinning out.

Ligeti himself proffered detailed instructions on how he wished to have his piece performed.[17] The metronomes are played by ten players following the conductor's lead. Each player is responsible for operating ten metronomes, which are placed on resonators and brought onto the stage completely rundown. The metronomes are arranged in groups of ten around a microphone connected to a loudspeaker. The conductor gives the signal, ordering the players to wind up the metronomes on stage. The speeds of the pendulums are set: Within each group, each metronome beats out a unique tempo. Two versions of the performance are possible, and the conductor chooses one of the options before the performance. Either all the metronomes are wound equally tightly, or several metronomes in a group are wound unequally. The conductor also determines the interval of silence (between two and six minutes) from the end of the preparation to the commencement of the performance. Another signal from the conductor instructs the performers to set all of their metronomes in motion simultaneously. "To carry out this action as quickly as possible, it is recommended that several fingers of each hand be used at the same time."[18] With practice, a set of four to six metronomes can be set into motion simultaneously. Once the performance starts, the performers and the conductor immediately leave the stage, allowing the audience to observe. The conductor determines the duration of the pause at the end of the performance (i.e., from when the last metronome runs down completely to the moment when the conductor returns to the stage with the players in order to receive their applause).[19] Sadly, such applause was never realized during the premiere in Hilversum, Holland. As Ligeti recalls many years later,

> Following the official speeches, the premiere proceeded according to plan. Since the audience had never heard of Fluxus, and since John Cage, too, was as yet completely unknown to the invited nobility and citizens of Hilversum, the last tick of the last metronome was followed by an oppressive silence. Then there were menacing cries of protest. (quoted in Steinitz 2003, 128)

The concert was scheduled to be televised later that week. "I was still Walter Maas's guest in the curious Gaudeamus Foundation house, which was shaped like an open grand piano—we sat in front of the television awaiting the scheduled broadcast of the filmed event. But, instead, they showed a football game . . . the programme had been prohibited at the urgent request of the Hilversum Senate" (quoted in Steinitz 2003, 128).

The superimposition of pulsation grids produced by the metronomes (which Ligeti referred to as a moiré effect in physics, or an interference pattern created, for example, when two grids are overlaid at an angle or when they have slightly different mesh sizes) would be used later on in numerous of his compositions. At the beginning of the *Poème symphonique*, the grids are so numerous that they coalesce, sounding disorderly and blurred. Ligeti added:

> As soon as some of the metronomes have run down, changing rhythmic patterns emerge, depending on the density of the ticking [i.e., the number of metronomes and their speeds]; until, at the end, there is only one, slowly ticking metronome left, whose rhythm is then regular. The homogeneous disorder of the beginning is called "maximal entropy" in the jargon of information theory (and in thermodynamics). The irregular grid structures gradually emerge, and the entropy is reduced since previously unpredictable ordered patterns grow out of the opening uniformity. When only a single metronome is left ticking in a completely predictable manner, then the entropy is maximal again—or so the theory goes. (quoted in Steinitz 2003, 129)

As years went on, Ligeti altered the performance of the piece. No longer insisting on the fluxus ceremony, he argued that "the metronomes should be started before the audience enters the concert hall, so that the piece truly runs like a machine: metronomes and audience are confronted with each other without any human mediation" (quoted in Steinitz 2003, 129). The performers had completely vanished.

Conclusion

In the twentieth century, the metronome and the tuning fork were transformed from the invisible to the visible. What do I mean? Throughout the eighteenth and nineteenth centuries these physical instruments were seen as mechanical devices used to assist the musician and the composer by standardizing musical pieces. Recall the debate generated by the music director and composer in Gottfried Weber's essay on the chronometer. The virtuoso was center stage. Clearly the metronome and the tuning fork were never to be used during a performance. In addition, debates arose as to whether mechanical devices deprived the art of music of its inherent expression of freedom. During the twentieth century, composers decided to draw upon these instruments to explore the various notions of precision, whether it be pitch or beat. Whereas Scheibler's tonometer was seen as a device to standardize and thwart the individuality of various temperaments, Burt's tuning forks empowered him to explore forms of creative expression using microtonalities. In a very real sense, avant-garde composers such as Ligeti and Burt provoked their listeners to rethink the very musical notions that metronomes and tuning forks had originally been created to reinforce centuries earlier, namely tonality and rhythm. Similarly, the sound generated by sirens in nineteenth-century acoustical workshops and physics laboratories migrated to the concert hall throughout the United States and Europe. No longer committed to generating a constant pitch, they were now performance instruments used to increase the types of pitch available to imaginative composers. They were now employed to thwart the precision of pitch. Varèse wished to stress the range of microtonality that machines could produce, and in the process he rendered the virtuoso superfluous. The mechanical now did not merely threaten to quash the creativity of the musician interpreting the piece; in a number of cases it also became the performer.

NOTES

1 The efficacy of this test is rather controversial, however; http://www.fpnotebook.com/ENT/Exam/TngFrkTsts.htm (accessed Dec. 28, 2009).

2 Franciscus Cornelius Donders and Hermann von Helmholtz used tuning forks in physiological and anatomical experiments. See Schmidgen 2005, 211–56) and Kursell (this volume).

3 It was commonly used by French and British experimental natural philosophers during the 1820s, while German experimentalists began to use it in the mid-1830s (Jackson 2006, 172, 323).

4 Although the importance of the siren as an acoustical instrument began to wane after the Ohm-Seebeck dispute, it was still the instrument of choice for the study of the combination of tones, the ratios of consonances, and interference (Robel 1891, 22; Pantalony 2005, 60; Helmholtz 1954, 162–63, 182).

5 Www.Absoluteastronomy.Com/Topics/Tuning_Fork (accessed Dec. 28, 2009).

6 Warren Burt, "Nonpop New Music Composer"; http://kalvos.org/burtwar.html (accessed Dec. 28, 2009).

7 "Warren Burt: The Animation of Lists and the Archytan Transpositions"; http://home.swipnet.se/sonoloco26/xi/burt.html (accessed Dec. 28, 2009).

8 Http://Home.Swipnet.Se/Sonoloco26/Xi/Burt.Html (accessed Dec. 28, 2009).

9 Http://Www.Forcedexposure.Com/Artists/Burt.Warren.Html (accessed Dec. 28, 2009).

10 Http://dic.academic.ru/dic.nsf/enwiki/668999 (accessed Dec. 28, 2009) and "Varèse's Ionization: The Percussion Revolution," http://www.scena.org/lsm/sm5-10/percu-en.html (accessed Dec. 28, 2009).

11 Http://www.scena.org/lsm/sm5-10/percu-en.html (accessed Dec. 28, 2009).

12 Http://Dic.Academic.Ru/Dic.Nsf/Enwiki/668999 (accessed Dec. 28, 2009) (originally cited in Slonimsky, 1994).

13 Http://dic.academic.ru/dic.nsf/enwiki/668999 (accessed Dec. 28, 2009).

14 David Cope, "Texture-Music" in *New World Encyclopedia*; http://www.newworldencyclopedia.org/entry/Texture_(music) (accessed Dec. 28, 2009).

15 Http://www.newworldencyclopedia.org/entry/Texture_(music) (accessed Dec. 28, 2009).

16 Http://Fluxmuseum.Org/Fluxhibtion3-Intro.Html (accessed Dec. 28, 2009).

17 Http://Www.Artnotart.Com/Fluxus/Gligeti-Poemesymphonique.Html (accessed Dec. 28, 2009).

18 Ibid.

19 For a performance of Ligeti's Poème Symphonique, see http://www.youtube.com/watch?v=X8v-uDhcDyg (accessed Dec. 28, 2009).

REFERENCES

Alder, Ken. "A Revolution in Measure: The Political Economy of the Metric System in France." In *The Values of Precision*, ed. M. Norton Wise, 39–71. Princeton, N.J.: Princeton University Press, 1995.

Anonymous. *Kurze Abhandlung über den Metronomen von Mälzl und dessen Anwendung als Tempobezeichnung sowohl als bei dem Unterricht in der Musik.* Mainz: Schott, 1836.

Barbic, Mladen, Lowell Eliason, and James Ranshaw. "Letter to the Editor: Femto-Newton Force Sensitivity Quartz Tuning Fork Sensor." *Sensors and Actuators A: Physical* 136(2) (May 16, 2007): 564–66.

Bijsterveld, Karin. "A Servile Imitation: Disputes about Machines in Music, 1910–1930." In *Music and Technology in the Twentieth Century*, ed. Hans-Joachim Braun, 121–35. Baltimore: Johns Hopkins University Press, 2002.

Cutler, W. H., and the Editor. "The Metronome." *Quarterly Musical Magazine and Review* 6 (1824): 31–33.

De la Tour, Cagniard. "Sur la Sirène, nouvelle machine d'acoustique destinée à mesurer les vibrations de l'air qui constituent le son." *Annales de chimique et de physique* 12 (1819): 167–71.

Ellis, Alexander J. "On the History of Musical Pitch." *Journal of the Society of Arts* 28 (1880): 293–336, 401–403.
Feldmann, H. "Die Geschichte der Stimmgabel. Teil II: Die Entwicklung der klassischen Versuche nach Weber, Rinne, und Schwabach." *Laryngo-rhino-otologie* 76 (1997): 318–26.
Frängsmyr, Tore, John L. Heilbron, and Robin E. Rider, eds. *The Quantifying Spirit in the Eighteenth Century.* Los Angeles: University of California Press, 1990.
Galison, Peter L. *Einstein's Clocks, Poincaré's Maps: Empires of Time.* London: Norton, 2003.
Gooday, Graeme J. N. *The Morals of Measurement: Accuracy, Irony, and Trust in Late Victorian Electrical Practice.* New York: Cambridge University Press, 2004.
Griffiths, Paul. *György Ligeti*, rev. ed. Contemporary Composers. Series editor: Nicholas Snowman. London: Robson, 1997.
Guthmann, F. "Ein neuer Taktmesser, welcher aber erst erfunden werden soll." *Allgemeine musikalische Zeitung* 9 (1806): cols. 117–19.
Hamburger, Michael, ed. and trans. *Beethoven: Letters, Journals, and Conversations.* London: Thames and Hudson, 1951.
Harding, Rosamond E. M. *The Metronome and It's [sic] Precursor.* Henley-on-Thames: Gresham, 1983.
Haspels, J. J. L. Automatic-Musical Instruments: Their Mechanics and Their Music, 1580–1820. PhD diss., Rijksuniversiteit te Utrecht, 1987.
Haupt, Günther. "J. N. Mälzels Briefe am Breitkopf und Härtel." *Der Baer: Jahrbuch von Breitkopf und Härtel* 3 (1927): 122–45.
Haynes, Bruce. *A History of Performing Pitch: The Story of "A."* Lanham, Md.: Scarecrow, 2002.
Helmholtz, Hermann von. *On the Sensations of Tone as a Physiological Basis for the Theory of Music*, trans. and rev. by Alexander J. Ellis. New York: Longmans, Green, 1895; repr., New York: Dover, 1954.
Jackson, Myles W. *Harmonious Triads: Physicists, Musicians, and Instrument Makers in Nineteenth-Century Germany.* Cambridge, Mass.: MIT Press, 2006.
Jungnickel, Christa, and Russell McCormmach. *Intellectual Mastery of Nature*, 2 vols. Chicago: University of Chicago Press, 1986.
Kielhauser, Ernst A. *Die Stimmgabel: Ihre Schwingungsgesetze und Anwendungen in der Physik.* Leipzig: Teubner, 1907.
Knecht, Justin Heinrich. "Ueber die Stimmung der musikalischen Instrumente überhaupt und der Orgel insbesondere." *Allgemeine musikalische Zeitung* 5 (1803): cols. 529–35.
Koch, Heinrich Christoph. *Musikalisches Lexikon.* Frankfurt: Hermann, 1802.
Kursell, Julia. "Wohlklang im Körper: Kombinationstöne in der experimentellen Hörphysiologie von Hermann v. Helmholtz." In *Resonanz: Potentiale einer akustischen Figur*, ed. Karsten Lichau, Viktoria Tkaczyk, and Rebecca Wolf, 55–74. Munich: Funk, 2009.
Leonhardt, Henrika. *Der Taktmesser, Johann Nepomuk Mälzl: Ein lückhafter Lebenslauf.* Hamburg: Kellner, 1990.
Levin, Thomas Y. "Integral Interpretation: Introductory Notes to Beethoven, Kolisch, and the Question of the Metronome." *Musical Quarterly* 77 (1993): 81–89.
Melde, Franz. *Akustik: Fundamentalerscheinungen und Gesetze einfach tönende Körper.* Leipzig: Brockhaus, 1883.

———. "Ueber die Erregung stehender Wellen eines fadenförmigen Körpers." *Annalen der Physik und Chemie* 109 (1859): 193–215; 111 (1860): 513–17.
Michaelis, C. F. *Mittheilungen zur Beförderung der Humanität und des guten Geschmacks.* Leipzig: Meißner, 1800.
———. *Moralische Vorlesungen.* Weissenburg in Franken: Verlag des oberdeutschen Addresse- und Industrie-Komptoirs, 1800.
———. *Ueber den Geist der Tonkunst, Erster Versuch.* Leipzig: Schäferische Buchhandlung, 1795.
———, and J. G. Schicht. "Aufforderung zur Festsetzung und gemeinschaftlichen Annahme eines gleichen Grundtones der Stimmung des Orchesters." *Allgemeine musikalische Zeitung* 16 (1814): cols. 772–76.
Morton, David. *Sound Recording: The Life Story of a Technology.* Westport, Conn.: Greenwood, 2004.
Ohm, Georg Simon. "Ueber die Definition des Tones, nebst daran geknüpfter Theorie der Sirenen und ähnlicher tonbildender Vorrichtungen." *Annalen der Physik und Chemie* 59 (1843): 513–65.
Olesko, Kathryn M. "Precision, Tolerance, and Consensus: Local Cultures in German and British Resistance Standards." In *Archimedes: New Studies in the History and Philosophy of Science and Technology*, vol. 1, ed. Z. Buchwald, 117–56. Dordrecht: Kluwer Academic Press, 1996.
Ouellette, Fernand. *Edgard Varèse*, trans. from the French by Derek Coltman. New York: Orion, 1968.
Pantalony, David. *Altered Sensations: Rudolph Koenig's Acoustical Workshop in Nineteenth-Century Paris.* New York: Springer, 2009.
———. "Rudolph Koenig's Workshop of Sound: Instruments, Theories, and the Debate over Combination Tones," *Annals of Science* 62 (2005): 57–82.
Powell, Ardal, ed. and trans. *The Virtuoso Flute-Player.* New York: Cambridge University Press, 1991.
Riehn, Rainer. "Beethovens Verhältnis zum Metronom." *Musik Konzepte 8 (Beethoven: Das Problem der Interpretation*, ed. Heinz-Klaus Metzger and Rainer Riehn). Munich: Edition Text und Kritik, 1979, 70–84.
Robel, Ernst. *Die Sirenen: Ein Beitrag zur Entwickelungsgeschichte der Akustik*, Teil I. Berlin: Gaertner, 1891.
———. *Die Sirenen: Ein Beitrag zur Entwicklungsgeschichte der Akustik.* Teil II: *Die Arbeiten deutscher Physiker über die Sirene im Zeitraume von 1830–1856.* Programm des Luisenstädtischen Gymnasiums, Beilage. Berlin: Gaertner, 1894.
Robinson, John. *A System of Mechanical Philosophy*, with notes by David Brewster, 4 vols. Edinburgh: Murray, 1822.
Savart, Félix. "Note sur la limite de la perception des sons graves." *Annales de chimique et physique* 47 (1831): 69–74.
———. "Notes sur la sensibilité de l'organe de l'oiue." *Annales de chimique et physique* 44 (1830): 337–52.
Schaffer, Simon. "Accurate Measurement Is an English Science." In *The Value of Precision*, ed. M. Norton Wise, 135–72. Princeton, N.J.: Princeton University Press, 1995.
———. "Late Victorian Metrology and Its Instrumentation: A Manufactory of Ohms." In *Invisible Connections: Instruments, Institutions, and Science*, ed. Robert Bud and Susan E. Cozzens, 23–56. Bellingham, Wash.: SPIE, 1992.
Scheibler, J. Heinrich. *Der physikalische und musikalische Tonmesser.* Essen: Bädeker, 1834.

Schicht, J. G. *Grundregeln der Harmonie, nach dem Verwechslungs-System entworfen und mit Beispielen erläutert*. Leipzig: Breitkopf and Härtel, 1812.

Schilling, Gustav, ed. *Encyclopädie der gesammten musikalischen Wissenschaften, oder das Universal-Lexikon der Tonkunst*, 6 vols. Stuttgart: Köhler, 1835–1838.

Schindler, Anton. *Biographie von Ludwig van Beethoven*. Münster: Aschendorff, 1840.

Schmalz, Eduard. *Erfahrungen über die Krankheiten des Gehöres und ihre Heilung*, 8 vols. Leipzig: Teubner, 1846.

Schmidgen, Henning. "The Donders Machine: Matters, Signs, and Time in a Physiological Experiment, c. 1865." *Configurations* 13 (2005): 211–256.

Schuller, Gunther. "Conversations with Varèse." *Perspectives of New Music* 3 (Spring–Summer 1965), 32–37.

Seebeck, August. "Beobachtungen über einige Bedingungen der Entstehung von Tönen." *Annalen der Physik und Chemie* 53 (1841): 417–36.

Serway, R. A. *Physics for Scientists and Engineers*, 2 vols. Philadelphia: Saunders College Publishers, 1996.

Sievers, G. L. P. "Pariser Musikalische Allerley." *Allgemeine musikalische Zeitung* 19 (1817): col. 302.

Sievers, G. L. P. "Pariser musikalische Allerley." *Allgemeine musikalische Zeitung* 21 (1819): cols. 588–93, 597–603.

Slonimsky, Nicolas. *Music since 1900*, 5th ed. New York: Schirmer, 1994.

Stadlen, Peter. "Beethoven und das Metronom." *Musik Konzepte 8, Beethoven: Das Problem der Interpretation,* ed. Heinz-Klaus Metzger and Rainer Riehn. Munich: Edition Text und Kritik, 1979, 12–33. See also Peter Stadlen, "Beethoven and the Metronome." *Music and Letters* 48 (1967): 333–349.

Steinitz, Richard. *György Ligeti: Music of the Imagination*. Boston: Northeastern University Press, 2003.

Stöckel, G. E. "Ueber die Wichtigkeit der richtigen Zeitbewegung eines Tonstücks, nebst einer Beschreibung eines musikalischen Chronometers und dessen Anwendung für Komponisten, Ausführer, Lehrer, und Lernende der Tonkunst." *Allgemeine Musikalische Zeitung* 3 (1800): cols. 657–66, 673–79.

Thayer, Alexander, Hermann Deiters, and Hugo Riemann, eds. *Ludwig van Beethoven's Leben* 5 vols. Berlin: Weber, 1901–11.

Ullmann, Dieter. *Chladni und die Entwicklung der Akustik*. Boston: Birkhäuser, 1996.

Vogel, Stephan. "Sensation of Tone, Perception of Sound, and Empiricism: Helmholtz's Physiological Acoustics." In *Hermann von Helmholtz and the Foundations of Nineteenth-Century Science*, ed. David Cahan, 259–87. Los Angeles: University of California Press, 1994.

Weber, Ernst Heinrich. *De pulsu, resorptione, auditu, et tactu. Cap. VI: De utilitate cochleae in organo auditus. De auditu*. Leipzig: Koehler, 1834.

———, and Wilhelm Eduard Weber. *Wellenlehre auf Experimente gegründet oder über die Wellen tropfbarer Flüssigkeiten mit Anwendung auf die Schall- und Lichtwellen*. Leipzig: Fleischer, 1825.

Weber, Gottfried. "Ein neuer Taktmesser, welcher aber erst erfunden werden soll." In *Allgemeine musikalische Zeitung* 8 (1806): cols. 117–19.

———. "Noch einmal ein Wort über den musikalischen Chronometer oder Taktmesser." *Allgemeine Musikalische Zeitung* 15 (1813): cols. 441–47.

Wehmeyer, Grete. *Prestißißimo: Die Wiederentdeckung der Langsamkeit in der Musik*. Hamburg: Roswohlt, 1993.

Wen-Chung, Chou. "Varèse: A Sketch of the Man and His Music." *Musical Quarterly* 52 (1966): 151–70.
Wheatstone, Charles. "Experiments on Audition." The Physical Society of London, ed., *The Scientific Papers of Sir Charles Wheatstone.* London: Taylor, 1879.
Wolfson, R., and J. M. Pasachoff, *Physics with Modern Physics for Scientists and Engineers* (Reading, Mass.: Addison-Wesley, 1999).

CHAPTER 9

CONVERSIONS: SOUND AND SIGHT, MILITARY AND CIVILIAN

CYRUS C. M. MODY

Introduction

Scholars in science and technology studies (STS) have long noted scientists' predilection for converting data into visual representations. Indeed, examining visual representations yielded many early STS keywords. Laboratory ethnographers, for instance, famously reinterpreted scientists as producers of "inscriptions" (Latour 1987). The conversion of Geiger counter "clicks" into "splodges" on a graph offered insights on inferential chains and interactions among subfields (Pinch 1985). Conversion to the visual provided a hook to mutually implicate vision and science as hallmarks of modernity (Crary 1990). Visual representations became raw materials for art historical analyses of science (Jones and Galison 1998) and histories of objectivity (Daston and Galison 2007).

Visual representations, however, are neither inexorable nor ubiquitous in science. Scientists still work with their ears, fingertips, and taste buds. Some scientists

---

Research for this chapter was supported by the National Science Foundation under Grant no. SES 0531184. Any opinions, findings, and conclusions are my own and do not necessarily reflect the views of the National Science Foundation.

prefer certain data in audible rather than visual form (Helmreich 2007). Alexandra Supper (this volume) has described the nascent "sonification" community's struggles for acceptance of this practice. Despite these struggles, sonification has found favor even among scientists whose work is primarily visual. I first ran across sonification, for instance, among scanning probe microscopists (Mody 2005), a group generally committed to visualization (hence micro*scopy*). Yet probe microscope technology lends itself equally well to visual and auditory outputs. The "microscope" is a small solid probe that scans and interacts with a surface; it assembles measurements of the probe-surface interaction into a data array that is usually rendered visually but sometimes in auditory or other forms.

Sonification has been part of probe microscopy since its invention in the 1980s. The technique's ability to image individual atoms, for instance, was discovered when its inventors heard—rather than saw—an unusual repetition in how a chart recorder printed data (Mody 2004, 107). Yet only a small minority of microscopists audibilize their data. Those who do offer justifications in phenomenological terms. Some consider sound better for perceiving change over time, while some combine visualization and sonification for a richer data environment. These reasons, however, are inseparable from sociological motivations. Sonification is usually practiced by high-end probe microscopists who build rather than buy their instruments. Sonification helps them present their work as distinctively fine-tuned and artisanal compared to the mass of probe microscopists. Prominent builders such as Jim Gimzewski use sonification to forge collaborations with artists (Roosth 2009), and other builders play their microscopes' auditory output as background music during public talks. Those who buy their microscopes are much less likely to use sonification this way.

Two Senses of "Conversion"

Let us group visualization and sonification under a wider category of "synesthetic conversion" from one sense to another. In probe microscopy, the same people who practice sonification often venture into other senses as well. For instance, when the inventors of the technique heard their chart recorder and realized they could image individual atoms, they turned the chart recorder strips into a three-dimensional, tactile sculpture of the atoms rather than a (more ordinary) two-dimensional micrograph. Other prominent microscope builders who audibilize data also sometimes use haptic feedback, where the microscope physically pushes against its operator (e.g., when the microscope nudges carbon nanotubes or pries proteins apart). As with sonification, tactile or haptic representations are associated much more with microscope builders and tinkerers than with those who buy a microscope for use with few modifications.

That is, those probe microscopists who view synesthetic conversion as an embodied strategy for perceiving data in nuanced ways also deploy synesthetic conversion in the politics of their research community. Sonification and haptic feedback distinguish developers of new microscope technology from the mass of microscope users. Synesthetic conversion provides unusual interdisciplinary bridges to fields such as art, computer science, and robotics, and it locates skill and authority in probe microscopy by showing whose data and technology are so subtle and sophisticated that multiple senses (rather than vision alone) are needed to understand or operate them.

In today's probe microscopy, synesthetic conversion is generally limited to the small-p politics of jockeying within a field. But synesthesia can be embroiled in big-p politics. To demonstrate how, this chapter traces the prehistory of probe microscopy back to Vietnam-era protest at Stanford University. At that time, Calvin Quate (the coinventor of the atomic force microscope [AFM], the most common kind of probe microscope) invented a closely related instrument, the scanning acoustic microscope. Acoustic microscopy and other synesthetic conversion technologies were at the forefront of Stanford researchers' response to calls for more socially responsible science. Synesthetic conversions—from sound to sight, sight to touch, sight to sound, touch to sound, and so on—were embroiled in the "reconversion" of American academic research from military funding and applications to civilian funding and an orientation to "human problems."

As today's probe microscopy shows, synesthetic conversion in science does not have to be freighted with ambitions for societal reform. Moreover, "reconversion" at Stanford and elsewhere was never limited to new synesthetic conversion technologies. However, the acoustic microscopy case shows that, at times, projects for reforming knowledge making are coproduced with scientists' conversions of phenomena into visual, tactile, haptic, or auditory form.

Boiling Campus, Anxious Engineers

In the late 1960s, "reconversion" was the widely used term for the turbulent, occasionally violent, debate about whether American academic scientists needed to forego defense research and take up civilian social problems. Often historians' stories about this period end with reconversion as the dissolution of the early Cold War "military-industrial-academic" arrangement. The last chapter of Bill Leslie's (1993) *The Cold War and American Science*, for instance, is "The Days of Reckoning: March 4 and April 3," the dates in 1969 when the reconversion debate erupted at, respectively, MIT and Stanford.

One need only visit the Stanford University Archives—the primary data source both for this chapter and for the Stanford portion of Leslie's book—to see how deeply reconversion affected that campus in the late 1960s and early 1970s. University administrators' memos, campus newspapers, and faculty members' private correspondence (including Quate's) all speak to the frantic need to make up for declining defense research funding and to either quash or appease campus protestors' demands for more civilian research. Yet Leslie demonstrates that most of the reconversion movement's long-term effects were cosmetic. Both MIT and Stanford ended classified research and divested themselves of some defense-oriented institutes, but most of the proposed reforms were abandoned. Postdivestment, centers such as the Stanford Research Institute and Draper Laboratory prospered and maintained closer ties to Stanford and MIT than reconversion proponents hoped.

Reconversion's consequences were not all short-lived, though. Matthew Wisnioski (2005) sees reconversion as part of a wider movement in which American scientists and engineers questioned their motivations and practices, founded STS programs, read works by the Frankfurt School, and collaborated with artists. Eric Vettel (2006) sees reconversion as triggering the emergence of the biotech industry. Both Leslie (2010) and Wisnioski show that, at MIT, some researchers did successfully move away from military funding and applications—though those who succeeded were skeptical their actions could be widely replicated.

Nevertheless, reconversion proponents heralded such success stories. As Holt Ashley, an aeronautical engineering professor in the Stanford School of Engineering, put it in a short-lived student-faculty publication dedicated to debate about reconversion, "Against the unregenerateness and glacial metamorphosis of some, I don't hesitate to place the valuable progress on socially significant projects led by Lusignan, Meindl, Anliker, Homsy, Pantell, McCarty, and DeVoto."[1] Of these, James Meindl offers the clearest instance of the entanglement of reconversion and synesthetic conversion and an explicit model for Quate's foray into acoustic microscopy.

An electrical engineer working for the Army Signal Corps, Meindl moved to Stanford in 1967 and began working with John Linvill, chair of the electrical engineering department (Lécuyer 2005). Since 1962, Linvill had been developing the Optacon, a device for scanning printed pages and converting text into mechanical vibrations that could be felt by a blind reader. Linvill's original work predated the lean budgets and political turmoil of the late '60s and was based partly on personal motivations (his daughter was blind). Meindl's arrival, though, coincided with the first reconversion murmurs at Stanford and triggered an acceleration of Optacon research and its extension into other civilian and socially conscious applications.

As a 1973 brochure put it, Meindl's lab's "central objective . . . is to prepare the student to use integrated circuit technology in an innovative manner in solving the problems of our society . . . particularly in the field of medical electronics."[2] "Problems of our society" was typical reconversion rhetoric. Synesthetic conversion was a recurring motif of the approach Meindl's lab took to solving those "problems of our society": conversion from sight-to-touch (the Optacon, or "OPtical to TActile CONverter"), sound-to-sight (ultrasonic soft-tissue imaging), and later sight-to-sound (an adaptation of the Optacon to read scanned letters aloud via synthesized voice) and touch-to-sound (an artificial ear project).

That synesthetic conversion was upheld by Stanford administrators as a model for reconversion can be seen in the annual reviews of Stanford's electronics research. These reviews summarized the results of funding from the military services and NASA. Thus, through 1968, defense applications headlined, with civilian applications tucked at the back. In the 1968 review, for instance, the Optacon was described briefly (in purely technical terms) on page 27. The 1969 review, however, came just four months after a student takeover of Stanford's Applied Electronics Lab and the university administration's decision to ban classified work on campus. Suddenly, the Optacon was promoted to page 1, where the report pointed out the following:

> Two nontechnical aspects of the [Optacon] reading-aid project are noteworthy. (1) Integrated circuits, principally developed to the present stage for space and military applications, are powerful tools for the solution of human problems, as this research project illustrates. (2) Such projects . . . [are suited to] channeling [the] interest of the imaginative graduate student to important social problems.[3]

The rhetoric here contains clear gestures to reconversion: "Space and military applications" are specifically contrasted with "human problems," and graduate students are imagined as above all interested in solving "social problems."

From Acoustic Wave Devices to Acoustic Microscopy

Linvill and Meindl's Optacon and other synesthetic conversion technologies were a bright light in a dark year for the Stanford administration. The Optacon offered a riposte to reconversion proponents and a guide to other faculty members, especially those needing new funding after Stanford lost $2 million when it canceled its classified research contracts. It is perhaps unsurprising that Calvin Quate was one of those who followed their example. His research was heavily funded by the military and therefore jeopardized both by the new Stanford policy and by cuts in federal defense research funding. Moreover, Quate was an ardent admirer of Meindl, and Linvill was his department chair.

Quate received his doctorate from Stanford in 1950, working on microwave traveling tubes. He continued that research for almost a decade at Bell Laboratories before taking a joint appointment in electrical engineering and applied physics at Stanford in 1961. In the '60s, he studied ultrasonic-electromagnetic interactions in crystals, particularly for signals processing. He became well known for research on acoustic wave devices, by which an electrical signal is converted into an ultrasonic wave, processed via interaction with a crystalline matrix or tiny interdigitated structures, and then reconverted into an electrical signal. The U.S. Air Force and U.S. Navy funded Quate to develop acoustic wave devices to pick radar signals out of background noise and recognize radar patterns corresponding to specific aircraft.

The idea that Quate's electroacoustic research could be adapted to microscopy was suggested by Rudolf Kompfner—another Bell Labs veteran—while they "were sitting around the swimming pool one day" in 1966 (National Science Foundation 1978, 36; Pierce 1983). Kompfner proposed that a thin film transducer could create an ultrasonic wave that could be focused with a lens onto a sample (figure 9.1). After interacting with the sample, the transmitted ultrasonic radiation would then hit another transducer (or reflected ultrasonic radiation would hit the original transducer) and be converted into an image.

The way Quate and Kompfner spoke of their turn to acoustic microscopy gestured toward an intertwining of synesthetic conversion and reconversion. For a few years after their poolside chat, they did little work in the area. As reconversion boiled over in 1969, though, both men put aside other commitments and focused on acoustic microscopy in a way that, consciously or not, aligned with the aspirations of reconversion proponents. In their justifications for this work, Quate and Kompfner made the case that acoustic microscopy would have enormous civilian benefits. They did so partly by investing the visual apprehension of invisible phenomena with nearly magical importance in world history:

> If one makes the crude assumption that the acquisition of new knowledge is in some way proportional to the extension of the scale of magnitude of the

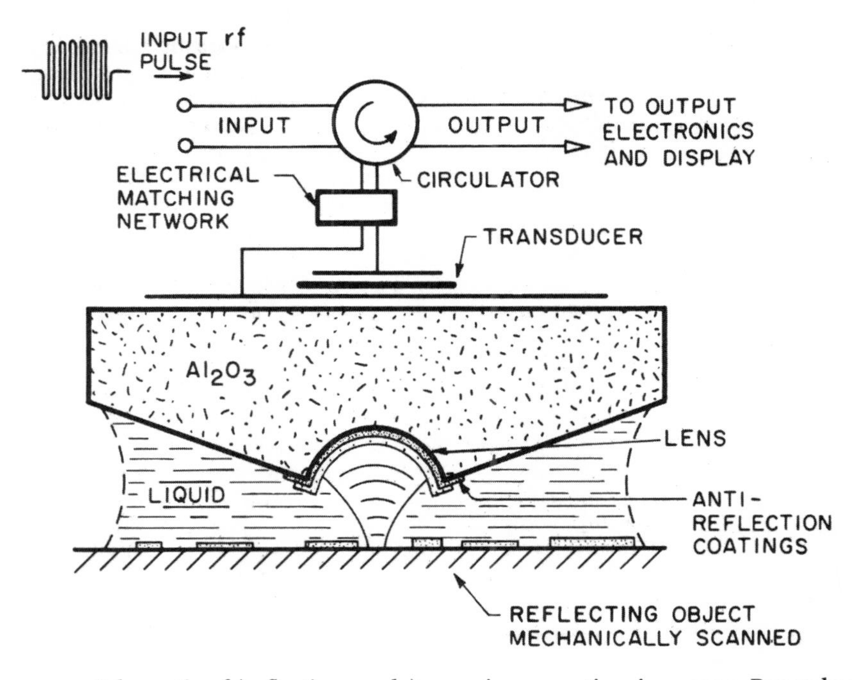

Figure 9.1 Schematic of (reflection mode) scanning acoustic microscope. Reproduced with permission of University of California Press, from Edwin S. Boatman, Michael W. Berns, Robert J. Walter, and John S. Foster, "Today's Microscopy: Recent Developments in Light and Acoustic Microscopy for Biologists," *Bioscience* 37 (June 1987): 384–94. Permission conveyed through Copyright Clearance Center.

> perceived objects . . . one might even be tempted to compare the total amount of information about nature at any time with the . . . useful magnification of the best microscopes. (Kompfner 1975, 619)
>
> It is hard to conceive of a modern laboratory involved with technology or the advancement of science that is without a microscope of some kind. It has been suggested that the number of instruments in a given country used to extend our "vision" beyond the limits of the unaided eye can be used as an indicator of the progress and advancement of that region.[4]

That is, for Quate and Kompfner, a society capable of making progress was necessarily one where human "vision" was extended "beyond the limits of the unaided eye" both in scale and in the types of radiation that could be perceived. Synesthetic conversion was therefore inseparable from societal "advancement."

For the first few years, the Office of Naval Research funded Quate's acoustic microscope work as part of an umbrella package to Stanford rather than an individual grant.[5] Before long, though, Quate disentangled acoustic microscopy from defense funding. I have no direct evidence that this was in response to reconversion pressure. There is no reason to think Quate ever had any personal qualms about defense funding for his research. Indeed, he was not cooperative with reconversion-minded students' investigations of defense-funded research at Stanford (SWOPSI, 1971, 215). It is probable that, like Meindl, Quate enthusiastically embraced "human problems" research while also viewing defense-funded research as both good science and good citizenship.

However, Quate's turn to acoustic microscopy eerily matched specific Stanford administration recommendations in the wake of the first protests against on-campus classified research in 1966. In particular, a memo from Linvill in 1967 detailed the electrical engineering department's steps toward reform:

> Stanford University can and should become more effective in studying and attacking the problems of today's society. Electrical Engineering, with its aim to bring technological tools to the solution of man's problems, is interested to join with other departments in working on these contemporary problems. . . . The usefulness of university engineering and science to our defense efforts since the beginning of World War II is clear and well documented. It is also clear that attention should be directed in a university to other problems. . . . [E]nvironmental studies, urban problems, problems of developing countries, etc.—are timely problems to which the university in many of its parts should be directed. These problems cannot be attacked within a single discipline. . . . Financial support for the research must be found outside the university, however, as has been the case in our DOD and space research. . . . Representative areas providing opportunities are: (a) education by teaching machines, (b) automation and cybernetics, (c) applications of solid-state electronics to the life sciences, and (d) application of computers in life and economic systems.[6]

Note that "defense efforts" are specifically contrasted with work on "the problems of today's society" and that the memo clearly encourages Stanford engineers to include more of the latter in their research portfolios.

Linvill's Memo

The acoustic microscope aligned with Linvill's memo in a variety of ways. Most important, it answered Linvill's call for "applications of solid-state electronics to the life sciences." At the time, Quate and Kompfner spoke of their microscope as having application only in biomedicine and not in areas of interest to the military. "[T]he original intention behind the Stanford work was to make an instrument which would achieve a resolution on the order of 0.1 μm for the purpose of studying biological objects, particularly the nuclei of cells" (Kompfner 1975, 627).

This initial exclusive emphasis on biological specimens is curious. The first suggestion for an acoustic microscope in the West was a translation (Devey 1953) of a Soviet article by a program manager at the Office of Naval Research (ONR). That article mentioned only metal and glass samples, not biological ones. Yet when Quate and Kompfner approached the problem, they were initially oblivious to non-biological samples. They were acutely aware that the initial funder for the acoustic microscope, the military's Joint Services Electronics Program, might object and ask whether Quate was "going to come up with an instrument for studying materials or an instrument for studying biological structures? And why is any of this related to a research program for electronics?"[7] Yet despite that potential objection (and their awareness of the 1953 ONR article), Quate and Kompfner initially downplayed acoustic microscopy's application to military-relevant materials, while highlighting its relevance to medicine.

Acoustic microscopy also aligned with Linvill's reconversion memo in that it enabled Quate to search for new "financial support . . . outside the university." Unlike his earlier defense-funded microwave work, Quate's acoustic microscopy research was funded by the John A. Hartford Foundation, National Science Foundation, National Institutes of Health, National Bureau of Standards, and IBM. Quate also licensed acoustic microscope patents to companies selling to the civilian market—something he had not attempted before. Those licenses were negotiated through Stanford's new Office of Technology Licensing (OTL—one of the first of its kind; see Smith-Hughes 2001; Colyvas 2007; Yi 2008, 182–3). The rhetoric of reconversion suffused justifications for the OTL. As a Stanford vice president put it, "I am pleased by the high societal value of inventions we have licensed [such as] a potential cure for viral infections and a potential ecologically safe insect control" (quoted in Colyvas 2007, 69–70).

As the quote indicates, the OTL concentrated on fields favored by reconversion proponents, especially biomedical and environmental applications. The OTL's initial contacts with Stanford engineers, therefore, emphasized inventions with humanistic and civilian application, especially ones that used synesthetic conversion to forge interdisciplinary links. John Chowning's patents for electronic sound synthesis and proprioceptive perception of sound, for instance, linked engineering to music (Nelson 2005). Similarly, Meindl's patents for ultrasonic blood imaging tied engineering to biomedicine. Meindl and Linvill's Optacon was also commercialized in the early '70s (Lécuyer 2005, 60).

The OTL viewed acoustic microscopy as yet another bridge between engineering and life science. In 1974, the office licensed American Optical to use Quate's patents to manufacture commercial microscopes. By that point, Quate had realized acoustic microscopy had applications in microelectronics manufacturing in addition to biomedicine. However, for the OTL, microelectronics had neither the "high societal value" nor the interdisciplinary cachet of life science. For that reason, OTL saw microelectronics as secondary to (indeed, a subset of) biomedical applications:

> A license agreement for commercial development of a microscope that uses ultrahigh-frequency sound waves to see into living cells and other materials has been signed. . . . By comparing [acoustic and optical microscopy], diagnosticians should be able to tell more about cell functions and disorders. Since ultrasonic waves can penetrate materials, they also can be used to show defects in microscopic integrated circuits. The new integrated circuit technology already has provided the world with life-saving heart-pacers, hearing aids, minicomputers, and many other benefits.[8]

For the OTL, electrical engineering was worth mentioning not for itself but for its contribution to biomedicine.

The OTL's attitude was common at Stanford at the time. As Linvill's 1967 memo and *The Grindstone*'s praise for Meindl's biomedical collaborations show, engineers working on biomedical topics were seen as "attacking the problems of today's society" in a way that those working on radar-signals processing were not. Linvill's declaration that biomedical, environmental, urban, and other civilian "problems cannot be attacked within a single discipline" was intended to make engineering expertise relevant to such problems and to stimulate Stanford's engineering faculty to pursue interdisciplinary collaborations that would bring diversified funding and a positive reception from reconversion proponents.

As Bill Leslie (1993) notes, the military had promoted interdisciplinarity among electrical engineers, physicists, and applied physicists at Stanford since the 1950s. Reconversion, however, brought calls for much wider-ranging collaborations between engineering and music, philosophy, economics, biology, and so on. Reconversion proponents believed narrowly discipline-based questions distracted engineers from solving "human problems" and shielded them from the moral consequences of their work. Working with other disciplines would allow engineers to become aware of a wider range of society's problems to which they could apply their knowledge.

Along those lines, Linvill's 1967 memo proposed to:

> introduce a Senior or Master's level program involving both engineers and students from the Humanities and Sciences for the purpose of discussion and formulation of problems of significance which society needs to have solved and to which technology can contribute as a partner with the humanities.[9]

The STS programs formed in this period—including Stanford's—also reflected this view of the healing power of interdisciplinarity. More generally, the number of degree-granting interdisciplinary programs at Stanford *doubled* from 1968 to 1969 (the largest relative or absolute increase in the school's history) and grew seven times more rapidly in the '70s than in the '60s (Nelson 2005).

Nor was Stanford alone in connecting interdisciplinarity to civilian reconversion. For instance, when the National Science Foundation was commanded in 1971 to fund research "to solve major problems such as pollution, transportation, energy and other urban, social and environmental problems and to initiate and expand applied research essential to technological advancement and economic productivity,"[10] it did so not through its traditional disciplinary units but through a program called Interdisciplinary Research Relevant to Problems of Our Society (renamed RANN—Research Applied to National Needs). In a time of declining federal research funding in general and defense research funding in particular, such interdisciplinary funding streams provided a lifeline. Quate, for one, sought funding from RANN for acoustic microscopy, as well as for power transmission line research.

From the Life Sciences to Microelectronics

So acoustic microscopy aligned Quate with reconversion aspirations by facilitating connections to life science, to new funders, and to the market (and hence to wider society). Those connections consisted in the microscope's ability to combine sound and vision and allow life scientists to see "the density, the elasticity and the viscosity [of cells]—properties which are far more vital to the functions of living tissue than the optical refractive index."[11] Getting nonengineers to "see" acoustically, though, was harder than Quate anticipated.

In the first place, he had difficulty weaning himself from defense funding partly because civilian agencies needed visual proof of synesthetic conversion before they would fund him. "'We had a concept, [but] we had no lens, we had no photographic film, we had no way of displaying the image,' Quate recalls. He applied for funding from the federal government, but was unsuccessful 'because there were no images, there were no results, there was just a theory'" (Jacobson 1984, 132).

Luckily for Quate, a Stanford Medical School colleague told him about the Hartford Foundation, a philanthropy that funded biomedical research. Hartford gave Quate $170,000 for two years starting in 1969, after which he had a prototype. From that first grant, Quate learned that the microscope would need to scan the lens relative to the sample (figure 9.1). Without scanning, the acoustic image would have to be received by an array of detectors, each kept in phase with the rest—a nearly insurmountable task at the time. With scanning, a single detector could capture a tiny picture of the sample at any moment; by moving over the sample, in time a complete two-dimensional image could be generated on an oscilloscope.

Quate then secured $300,000 from Hartford for the next three years by promising to collaborate with biomedical researchers. Thus, he began soliciting colleagues for biological samples to be imaged with the acoustic microscope. Eventually he could point to samples received from the Stanford departments of surgery, pathology, psychology, radiation oncology, hematology, anatomy, medical microbiology, and biology, as well as from the Mayo Clinic, University of California at Irvine, Albert Einstein School of Medicine, University of North Dakota, National Cancer Institute, U.S. Department of Agriculture, and the Faculty of Medicine in Montpellier, France.

Quate's pitch to collaborators and to funding agencies pivoted on comparing the resolving power of acoustic microscopy with more familiar optical and electron microscopes. The basic milestone Quate set himself was a comparative one:

> One of the earliest goals that we set for ourselves in the acoustic microscope program was to record an image using an acoustic wavelength smaller than that used in the optical microscopes. . . . Now it has happened—we have an instrument operating with an acoustic wavelength in water equal to 5450 Å!![12]

Later Quate described the event with even more excitement: "Never again will we have to state that the resolution is within 'a factor of two of the optical resolution.' Never again will we have to answer the query 'what is it good for?' "[13]

As I've written elsewhere (Mody 2000), proponents of a new technology use comparisons to other artifacts at their peril. Comparisons can create a logic that makes it difficult to exploit the new technology's unique abilities. In the case of acoustic microscopy, the technique's resolution matched the optical microscope's only after a decade of work, during which biomedical researchers provided Quate with samples but saw few reasons to adopt the new instrument.

Even after Quate achieved parity with optical microscopy, life scientists complained that (on the one hand) acoustic microscopy still had poorer resolution than electron microscopy, while (on the other hand) it was still less user friendly than optical microscopy. American Optical, in declining to manufacture commercial microscopes for life scientists, noted that "the key problem in encouraging broader use of the acoustic microscope was that it took on the order of one day from receipt of a specimen until a useable picture was obtained."[14] That is, one reason life scientists avoided the technique was that the synesthesia of acoustic microscopy was not effortless enough. Biomedical researchers could simply walk up to an optical microscope and peer through it to get the answers they wanted, whereas acoustic microscopy required patience. As Quate (1985, 134) put it, "We can compete in resolution, but we cannot compete with the immediacy of the optical images. When you look, the optical image is there. For an acoustic image of similar quality you must wait."

Biomedical researchers preferred the immediacy of vision to acoustic microscopy's mediation of sight and sound. To overcome the invidious comparison, Quate pointed out that acoustic microscopy could "see" things optical microscopes couldn't:

> If the observer had a choice it is not obvious that he would choose optical waves for the microscopic examination. . . . [F]or the most part the form, function, and growth patterns of cellular complexes do not depend on their optical properties. . . . Contrary to this the elastic properties are directly and intimately connected to the form and function of a cell. . . . In the new microscope acoustic waves in the form of propagating elastic waves—familiar as water waves in the oceans and sound waves in the air—are used in such a way that we can now monitor these elastic properties.[15]

Biomedical researchers were neither synesthetically nor cognitively familiar enough with ultrasonic radiation to be comfortable with acoustic microscopy. They knew what an optical micrograph of a cell meant, but images made with acoustic radiation required a difficult translation between senses:

> The training and background of such people as pathologists and histologists cannot be easily transferred in a way that will allow them to interpret the acoustic micrographs with the same facility that is now done with the optical photos. . . . [W]ork to date has been done in an electronic laboratory with people possessing a background in physics and electronics. No one in biology or in medical research has carried out a significant piece of research with this new instrument.[16]

Difficulties in interpreting acoustic micrographs stymied Quate's efforts to deepen collaboration beyond biomedical researchers' simply handing him samples.

Without those deeper collaborations, agencies balked at funding biomedical acoustic microscopy research. As an NIH reviewer told Quate in turning him down, "[T]his research will require a more definitive commitment and interaction with cell biologists, preferably one who can make a daily involvement."[17] By 1974, therefore, Quate was seeking alternatives to biomedical applications and the civilian agencies that funded them. This shift was aided by two changes in the political environment of American science. First, reconversion fervor had burned itself out. American universities were no longer overwhelmed by protests, sit-ins, teach-ins, or faculty strikes. Soul-searching publications like *The Grindstone* had petered out. While many Americans still wanted scientists to tackle society's problems in preference to the military's, few now vocally questioned the morality of researchers who took defense funding.

Second, the American public's laundry list of civilian problems that scientists should tackle began to converge with research problems relevant to the military. In the late 1960s reconversion proponents had pushed for research on environmental problems, urban housing, mass transit, and medical research. These were areas where applied physicists/electrical engineers like Quate were most out of their depth, as shown by his difficulties forging biomedical collaborations. After the oil crisis of 1973, those problem areas faded and were replaced by alternative energy research. The NSF's RANN budget, for instance, doubled from 1974 to 1975, due almost entirely to new funding for solar and geothermal energy research (WSF 1974). In these areas—particularly solar—physical scientists and electrical engineers could contribute without having to expand their horizons as broadly.

As the economy further deteriorated, a new problem area arose—national economic competitiveness, particularly with Japan in microelectronics. In the dynamic random access memory (DRAM) market—the microelectronics industry's barometer—the ratio of U.S. firms' share to that of Japanese firms went from 19:1 in 1971 to less than 5:1 in 1974 to 1.4:1 in 1977 (Macher, Mowery, and Hodges 1998). American policymakers, microelectronics manufacturers, and academic electrical engineers monitored this gap closing with alarm. The Japanese government's announcement in 1975 of a VLSI (very large-scale integrated circuits) crash program prompted calls for the U.S. government to aid domestic manufacturers by funding academic microelectronics research.

Here, applied physicists/electrical engineers like Quate could serve national needs without the complications of crossing many disciplinary boundaries. Moreover, in microelectronics the goals of military and civilian funding agencies coincided. The military wanted to stimulate domestic manufacturers to avoid dependence on offshore producers for advanced chips. Also, microelectronics research was perceived as more likely to have dual military-civilian applications than the biomedical and environmental research earlier favored by reconversion supporters. Thus, Quate could now propose to the armed services to use acoustic microscopy to inspect *both* microelectronic circuits and materials used in advanced airframes and naval vessels.[18] Perhaps the clearest sign that civilian and military interests aligned around microelectronics was a program that funded Quate from 1976 to 1978 on "Innovative Measurement Technology for the Semiconductor Industry"—run jointly by the National Bureau of Standards (part of the Department of Commerce) and the Advanced Research Projects Agency (the Pentagon's long-range research-funding arm).

Through that program, Quate had much more success finding collaborators than he had with biomedical applications. Samples of microelectronic devices for the acoustic microscope to inspect came to him from Hewlett-Packard, Avantek, IBM, Fairchild Semiconductor, and other companies. Many of the microelectronics researchers who sent samples found acoustic microscope images useful—much more so than biomedical researchers had. They were particularly pleased that the acoustic microscope could look through a thin film to see whether it completely adhered to its substrate, something other instruments could not do. Since applying thin films is an essential step in microelectronics manufacture, catching defective films could save firms millions.

For that reason, electronics firms—unlike biomedical researchers—built their own acoustic microscopes or bought them when they became commercially available: Bell Labs, Hughes Aircraft, IBM, Hitachi, DuPont, Westinghouse, Motorola, TRW, Intel, and others are all mentioned in this regard in Quate's correspondence.

This interest in turn meant that microscope manufacturers could foresee a market in a way that American Optical had not when the only customers were likely to come from biomedicine. With Quate's help, two firms, Leitz in West Germany and Olympus in Japan, developed commercial instruments for microelectronics customers.

Synesthesia, Conversion, and Awareness

With the return to electronics applications and defense funding, both reconversion and synesthetic conversion faded as drivers of Quate's research. To understand why, we need to delve into the reason these phenomena were paired in the first place. Here I will be extremely speculative. For now, the documentary record is too sparse to directly address this point. We can, however, approach the issue by asking, what was reconversion really about?

I would argue that reconversion aimed to bring researchers into a heightened state of awareness—of themselves and others and of connections between the two. Most narrowly, this meant making researchers aware of the consequences of their research for the people of Southeast Asia. It was on that narrow interpretation of reconversion that the debate became heated and occasionally violent. Nonetheless, many reconversion supporters sought a broader awareness. They wanted researchers not simply to be aware of society's problems (which many already were) but also to see themselves as agents that possessed the capacity to solve those problems.

Some reconversion proponents focused on institutional and political changes to facilitate this awareness. Lessening researchers' dependence on military funding, for instance, would make them more aware of civilian problems and of the agencies that fund scientists to solve those problems. Likewise, the creation of interdisciplinary centers would make researchers like Quate aware of "human problems" in the life sciences in such a way that they could envision how electrical engineering could solve those problems. Reconversion-oriented publications like *The Grindstone* pushed faculty to recognize their own political agency in effecting reconversion: "If the faculty truly desires to redirect the School, they must accept the inconvenience of lobbying the executive to change budget priorities, and then follow through by providing advice and testimony, if necessary, for the Congress."[19]

Yet reconversion was not limited to institutional and political reforms. The counterculture encouraged many scientists and engineers to follow a technological and synesthetic path to awareness as well. For them, awareness could be expanded by developing new technologies for integrating the human mind with a greater range of perception and/or for making the perceivers more aware of their place within society and the environment.

One epicenter for this approach was Quate, Linvill, and Kompfner's former employer, Bell Labs. Matthew Wisnioski has detailed efforts in the late '60s to carve institutional space at Bell Labs for collaborations with artists and other ways to humanize engineering practice. One result was the Experiments in Art and Technology (E.A.T.) organization that brought Bell Labs engineers together with artists such as Robert Rauschenberg. As Wisnioski (2005, 300–301) describes it, most of E.A.T.'s few successful engineer-artist collaborations involved synesthetic extensions of their audience's awareness, such as "a glass-enclosed cube with a high-intensity light beam, which illuminated a pile of dust that responded to acoustic vibrations of recorded heart rhythms" or a sculpture "sensitive to changes in its environment, [which] consisted of vibrating steel rods illuminated by strobe lights, which modulated to the sounds in the room."

Kompfner worked with one of E.A.T.'s organizers, Billy Klüver, and provided advice for John Cage's contribution to E.A.T.'s "9 Evenings: Theatre and Engineering" shows (Klüver 1988), in which performers' movements tripped photocells, triggering sound sources all over New York—a true synesthetic symphony. Around the same time, he wrote an unpublished essay expressing his own, complex views on music, technology, and cognition:

> *Music represents the human mind. It is the representation by the human mind of itself.* . . . In depicting the processes, the states and the modes of operation of the human mind music at first did it simply and in a primitive way, limited perhaps by the technology which was available for its expression. As western technology developed and surpassed anything that existed before on this planet, it became possible to express music in ever increasing complexity, and to depict ever increasingly complex mental phenomena, so that finally music has achieved the power to represent all of the human mind, and perhaps even more than that. Music describes what human mind has not yet reached, and may perhaps never reach; it goes beyond the compass of the human mind just as mathematics does when it operates in many dimensions.[20]

Quite possibly Quate knew about E.A.T. through Kompfner. If not, he certainly knew about Bell Labs' collaborations in electronic and computer-aided music with Stanford's John Chowning since he played a role in Chowning's tenure case. Quate and Kompfner's old boss, John Pierce (to whom Kompfner sent his essay on music), connected Chowning with Max Mathews, another Bell Labs musician-engineer. Pierce also encouraged Chowning to patent his FM synthesis work (Reiffenstein 2006). Later Quate was instrumental in bringing Pierce and Mathews to faculty positions in Chowning's Center for Computer Research in Music and Acoustics.[21] Though there is no evidence Quate himself participated in synesthetic artist-engineer collaborations, his closest colleagues did so while he worked on acoustic microscopy.

The other epicenter of technophilic-synesthetic extensions of awareness was the San Francisco Bay Area. At Berkeley, for instance, physicist Don Buchla was building particle accelerator components for the Atomic Energy Commission when the free speech movement pushed him toward more humanistic, civilian, and synesthetic applications of his talent for electronics (Pinch and Trocco 2002, 33–36). Initially that meant developing a transistorized hearing aid, then a proprioceptive sensor for blind people that "changed pitch according to its proximity to objects." Most famously, he developed the Buchla Box, an analog synthesizer distinguished from its rivals by Buchla's interest "in involving as many senses as possible" (Pinch and Trocco 2002, 49).

Similarly, Buchla's patron, Ramón Sender Barayón, coorganized the Trips Festival of 1966 with (Stanford alumnus) Stewart Brand. "In venues like the Trips Festival, the hippies of Haight-Ashbury sought to demonstrate the ability of technologies such as LSD, stereos, and stroboscopic lights to amplify human consciousness" (Turner 2006, 178) by mixing those technologies into a single, transcendently synesthetic experience. Brand, in turn, was closely associated with techno-synesthetic experimentation at the Stanford Research Institute (SRI). Willis Harman, Quate's colleague in the Stanford electrical engineering department and an SRI researcher, introduced Brand to LSD, and Brand subsequently connected SRI to the counterculture. Though SRI's classified contracts were a critical focus of reconversion protests at Stanford, many SRI researchers drew on countercultural aspirations in devising technologies for expanding consciousness to integrate an ever-larger sensorium.

At the time, for instance, SRI's best-known research was its "Electronics and Bioengineering Laboratory's" 1972 experiments on parapsychological synesthesia. Spoon bending, remote viewing, telepathy, and ESP attempted to expand awareness until mental states could convert into physical actions (Kaiser 2011). Today SRI is most famous as the birthplace of the mouse and the graphical user interface (GUI), which debuted at the so-called mother of all demos in 1968, for which Brand served as videographer. Both those technologies originated in the same techno-synesthetic aspirations as the Buchla Box, Trips Festival, and acoustic microscopy. Thierry Bardini (2000) argues that their inventor, Douglas Engelbart, explicitly conceived them as synesthetic devices that would enable "augmentation of the human intellect" through feedback between proprioception (movement of the mouse) and vision (seeing the cursor on the screen).

I have no evidence that Quate knew Engelbart (much less Buchla or Brand), but they were geographically, socially, and intellectually proximate. All three had multiple connections to Stanford's School of Engineering. More generally, the two organizations that defined Quate's career—Stanford and Bell Labs—were places where reconversion and synesthetic conversion intertwined. Quate's boss at Bell Labs, John Pierce, was one of the first prominent engineers to publicly criticize the military-industrial complex for "alienating engineering education from the civilian economy" (quoted in Leslie 1993, 252). He was also a staunch proponent of artist-engineer collaborations that fused multiple senses into one synesthetic experience. Likewise, Quate's Stanford department chair, John Linvill, outlined a clear program for reconversion two years before the campus erupted in protest—at the same time that he was developing the Optacon.

While administrators like Pierce and Linvill outlined political and institutional reforms, engineers in Bay Area and Bell Labs workshops and laboratories focused on new synesthetic technologies as a complementary or alternative path to reform. Most of these engineers were not politically active. Kompfner, Quate, Buchla, Engelbart, Chowning, Mathews, Meindl, Klüver, and others had all worked on Cold War technologies funded by defense agencies, usually with great pride. Yet some of this group also were attracted to a more humanistic, socially conscious engineering practice geared to "maximizing how much good I can do for mankind" (Engelbart, quoted in Bardini 2000, 8). Technologies for expanding awareness offered a less overtly politicized way to maximize the good they could do. As Fred Turner (2006, 108) says of Engelbart, he "worked to create an environment in which individual engineers might see themselves as both elements and emblems of a collaborative system designed to amplify their individual skills." Technologies for translating among visual, auditory, haptic, tactile, and proprioceptive sensations offered a way both to amplify individual skills and create collaborative systems linking engineers to "representatives from different tribes—e.g. sociology, anthropology, psychology, history, economics, philosophy" (Engelbart, quoted in Bardini 2000, 16).

Linvill may have seen projects like Engelbart's as the intended outcome of the institutional reforms he proposed. The first two "representative areas" of research Linvill called for in his 1967 memo were "(a) education by teaching machines [and] (b) automation and cybernetics." Engelbart's whole program was a kind of educational teaching machine, and he and Brand were closely associated with the nearby Portola Institute's work in instructional technologies (Turner 2006, 70). Both were also deeply influenced by cybernetics; like many, they saw it as a bridge from engineering to biomedicine, art, and music (Dunbar-Hester 2010). Likewise, Linvill and Quate were in close contact in shaping how acoustic microscopy could further Linvill's institutional reforms.

So acoustic microscopy resembled other synesthetic technologies at Stanford and Bell Labs in that it aimed to extend perception by fusing different senses, to connect engineering to more humanistic fields (i.e., the life sciences), and to answer administrators' calls for institutional reform. All of these synesthetic technologies, including acoustic microscopy, experienced some collapse or deflation of their original ideals by the mid-1970s. The E.A.T. project dissolved due to lack of funding and a disastrous exhibit for the 1970 World Expo in Japan. Engelbart's group imploded, and his most creative engineers moved to Xerox PARC. Brand founded the *Whole Earth Catalog* to provide information to the commune movement, only to see the communes evaporate by 1973.

A common theme of these unravelings was that new forms of liberated communication and synesthetic consciousness were too difficult to sustain—especially once less-challenging alternatives became available. For instance, Buchla's Box, which required its operators to unlearn the entire framework of Western music, lost out to the mini-Moog synthesizer, which (with its preset patches and familiar keyboard) could be played like an ordinary piano. Similarly, Engelbart's five-finger keyboard, which required learning an entirely new system for inputting information, lost out to the familiar QWERTY keyboard. The mouse and the graphical user interface, which Engelbart conceived as a means to challenge and stimulate users to a new stage of "coevolution," were co-opted by Apple for their unchallenging "user friendliness" (a concept Engelbart detested).

Something similar befell acoustic microscopy. Quate's original ideal of working closely with biomedical researchers to better understand living cells became mired in life scientists' reluctance to use an entirely new imaging technology and interpret an entirely new contrast mechanism—at least while more familiar imaging technologies sufficed. Thus, Quate turned away from biomedical applications and toward materials science and microelectronics. There he found an audience that already understood how the acoustic microscope worked and what its images meant and who even wanted to build new microscopes for themselves.

Quate's return to audiences in engineering and the physical sciences coincided with his palpable frustration with civilian agencies. He criticized RANN for its tendency to "revolution and abrupt changes in course" that demoralized NSF staff, and he rebuked the NIH for its inability to understand his difficulties collaborating with life scientists.[22] Defense agencies, meanwhile, showed great interest in acoustic microscopy and related technologies. Thus, Quate began work on an acoustic array imaging system that would allow the army to "see" at night and the navy to "see" under water (a macroscale equivalent of the array detector scheme he rejected for acoustic microscopy). By 1980, he and his colleagues had come to view synesthetic conversion as a way to signal their alignment with traditions in the military itself. As Tony Siegman, the director of the laboratory where Quate worked, asked:

> Have you guys ever thought seriously about heterodyning some of your rf [radio frequency] or microwave acoustic signals down to audio frequencies, and letting some human ears simply *listen* to them? From what I've been told of the near-miraculous ability of human sonar operators to recognize, and pull out of the noise, audible signals which defy any kind of electronic signal processing, perhaps this facility could be usefully employed in acoustic microscopy or NDT also. [Now there's a good basic research topic that should really be salable to ONR!].[23]

Conclusion: Basic Science to Nanoscience

Siegman's mention of "basic research" is telling, for it had been the *bête noire* of reconversion proponents. As one pro-reconversion student put it in *The Grindstone*:

> [S]ome engineers point out that they do basic research and if someone else uses it for something else, it is not their fault. This is a school of engineering—of applied science—not of physics. We are concerned with how to do things, not how the universe works. . . . The notion of "basic research" often acts as a smokescreen to hide what we are doing from others as well as to avoid facing the consequences ourselves.[24]

The final phase of acoustic microscopy, however, took place in a postreconversion environment where engineers no longer faced such criticisms for doing basic research.

The evolution of acoustic microscope design reflected that reality. In microelectronics, acoustic microscopy was successfully applied to nondestructive inspection of integrated circuits. Microscope manufacturers and microelectronics firms actively developed the technology for that application with only occasional input from Quate. In the life sciences, he accepted—perhaps grudgingly—that such applications would have to wait. Biomedical researchers provided him with samples but generally proved uninterested in acoustic micrographs of those samples. Though Quate initially justified acoustic microscope research on the basis of its routine use in clinical settings, biomedical researchers' indifference meant he made little progress toward that goal.

Quate therefore used their samples as test objects (Mody and Lynch 2010) to calibrate high-resolution versions of the microscope for basic biophysical research rather than clinical application. Quate's students of the early '80s focused on simply driving the microscope toward higher resolution. Cells and bacteria contained small features with which to demonstrate acoustic microscopy's improving resolution, but Quate made little attempt to generate new knowledge of those samples (figure 9.2). Partly this was because high-resolution acoustic microscopes operated in liquid nitrogen or liquid helium rather than water. In those conditions, cells and tissues would die, and any images generated would be too esoteric to be useful to life scientists. As one reviewer put it:

> the main advantage of acoustic microscopy, besides the ability to see below the surface, is the ability to observe biological specimens in their natural environment, or inorganic specimens without the need for etching or special preparation. The steps required to improve resolution seem to all require temperature changes or altered fluid couplant properties that would remove this advantage.[25]

Quate's research had become so basic that it had lost much of its relevance even to other basic researchers!

Ironically, the idea for this line of improvement arose from Quate's misunderstanding a biomedical colleague—the kind of misunderstanding that frustrated his original hopes for acoustic microscopy:

> "Gene Farber . . . said, 'Why don't you look at frozen tissues?' So we built a microscope that would look at frozen tissues." Actually, Farber was merely suggesting another use for the microscope. Tissue examined under an optical microscope is first frozen so that it can be cut. It is then placed on a slide where it warms to room temperature and stained to afford contrast for the viewer. [Says Quate,] "I thought they looked at it while it was frozen, so we build one at the liquid nitrogen level. Farber said, 'What the hell for?' But once you start looking at the properties of nitrogen, you soon realize it's a better instrument, it's a better resolving power." (Jacobson 1984, 135)

By 1982 Quate's students were almost exclusively operating the microscope in liquid helium-3, at the coldest temperatures in the universe.[26] Achieving acoustic

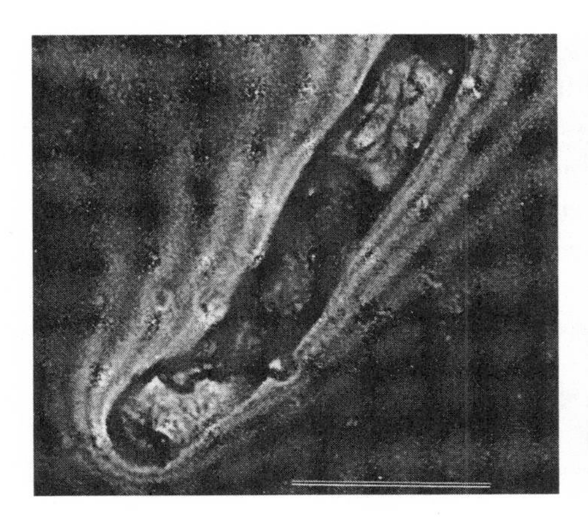
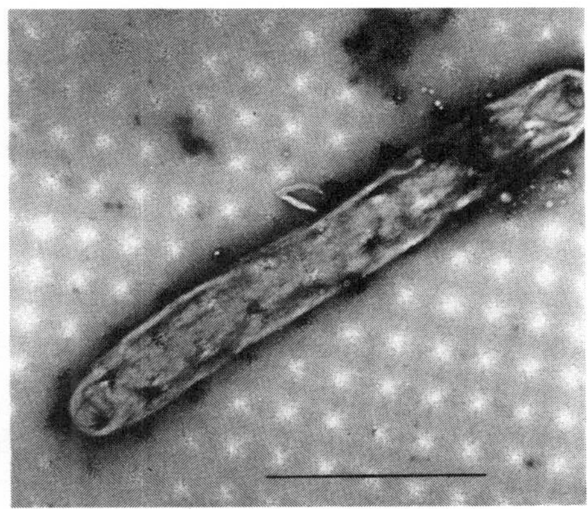

Figure 9.2 Cryogenic acoustic microscope image of myxobacteria (left) compared with transmission electron microscope image (right). The scale bar in both micrographs represents one micron (one-millionth of a meter). Reproduced with permission of University of California Press, from Edwin S. Boatman, Michael W. Berns, Robert J. Walter, and John S. Foster, "Today's Microscopy: Recent Developments in Light and Acoustic Microscopy for Biologists," ***Bioscience*** **37 (June 1987): 384–94. Permission conveyed through Copyright Clearance Center.**

microscopy's theoretical maximum resolution replaced biomedical collaboration and discovery as the end goal of Quate's research program.

So when he read an article (Schwarzschild 1982) that April about an entirely new microscope that used tunneling electrons to probe a metal or semiconductor surface, Quate abruptly moved into the area. Scanning tunneling microscopy (STM) was, in many ways, a natural extension of acoustic microscopy. Quate's students ported significant pieces of acoustic microscope technology to STM. The means of scanning an STM probe over a surface, for instance, resembled those for scanning the acoustic lens. The circuits for controlling the scan and outputting the microscopes' signals to a visual format were closely related. Quate was also confident that the STM could become a nondestructive testing tool for microelectronics manufacturing in the same way acoustic microscopy had.

The social relationships built around acoustic microscopy were also critical to Quate's success in STM. Indeed, Quate's introduction to STM came through an acoustic microscopy colleague—after reading that article, he obtained a meeting with the STM's inventors through Eric Ash, a British acoustic microscopist. As it happened, IBM, which had sponsored Quate's acoustic microscopy work, was also where the STM was invented—so Big Blue naturally funded Quate's STM research as well. Several of the students and postdocs who worked on acoustic microscopy had left Stanford for IBM in the early '80s; when these people learned their mentor had moved into tunneling microscopy, they followed. Moreover, once it became obvious that STM was not conducive to nondestructive testing, Quate convinced IBM to send one of the STM's coinventors to spend a year at Stanford so they could coinvent the atomic force microscope (AFM), a related instrument that *can* nondestructively test microelectronic components (Mody 2004).

Acoustic microscopy's deepest influence on Quate, though, may have been that it offered a new way to think about the objects of research. Before acoustic microscopy, Quate worked on *phenomena* (microwaves, surface acoustic waves) or areas of *application* (signals processing, nondestructive testing). In the late '70s, however, Quate began arguing that his expertise lay in a *size scale*—the submicron regime—that was critical to national security and economic competitiveness:

> We propose here a program of generic research in the field of acoustic microscopy . . . with a resolving power that approaches one-tenth of a micron. The realization of this goal will open new vistas in this field of "Microscience." . . . In the special issue on "Microscience" (*Physics Today*, November 1979) [and] . . . a report from the National Research Council ("Microstructure Science, Engineering & Technology") . . . the case is stated. . . . "Microstructure science, engineering, and technology are essential to . . . continued economic well-being."[27]

Expertise in "microscience" would not be pinned to any particular instrument but would range across all conceivable means of augmenting the senses:

> Unconventional methods for imaging in the microscopic world represent an emerging technology that permits us to go beyond what is possible with the optical or electron microscope. These technologies are based on other forms of radiation such as acoustic waves, tunneling electrons, ions and X-rays.[28]

Quate and his associates have lived this philosophy to the fullest. For several years in the '80s and '90s, for instance, his former postdoc, Kumar Wickramasinghe, was inventing a new microscope every six months.

As "microscience" morphed into "nanotechnology" in the '90s, veterans of the Quate group were therefore well positioned as the leaders in the field. For instance, when President Clinton gave a speech announcing the formation of the National Nanotechnology Initiative, the backdrop behind him was an STM image made by one of Quate's former acoustic microscopy students.[29] By focusing on size scale rather than phenomena or applications, Quate has retrieved some of the interdisciplinary aspirations that were stymied in the 1970s. When Quate first contemplated the STM, for instance, he waxed enthusiastic that it would ensure that "someday we should return to cells."[30]

Synesthesia, however, is largely missing from aspirations for nanotechnology. As I explain in the introduction, probe microscopists still listen to their data but with none of the utopian connotations of the late '60s. The moment when synesthesia and reconversion were two sides of the same reformist coin passed quickly. For a few years around 1970, though, these senses of conversion were intertwined in ways that shaped how (and why) scientists and engineers generated new knowledge. If we consider the role of the senses in research—and particularly the ways laboratory workers move from one sense to another—then we need to be mindful of the historical specificity of these conversions. What is simply the natural manipulation of data in one era can, in another, be the hook on which utopian reforms, unsettling convulsions, or appeals to new audiences are hung.

NOTES

SUA = Stanford University Archives
CQC = SUA, Calvin Quate Collection SC 347 (83-033), 1987 accession

1 Holt Ashley, "New Directions for Engineering Research," *Grindstone: A Forum for Controversial Issues of Special Interest to the Engineering Community* 1 (Feb. 22, 1971): 1–?, SUA, Collection Arch 3009 The Grindstone.

2 Integrated Circuits Laboratory, Integrated Circuits Technology: Opportunities for Graduate Study at Stanford University, 1973, SUA, Collection 3120/4 Electronics Labs.

3 Stanford Electronics Laboratories, Stanford University Electronics Research Review, 11, 12 (August 1969), SUA, Collection 3120/4 STAN.

4 Calvin F. Quate, Engineering Aspects of Acoustic Microscopy, 1979, CQC, box 1, folder Conferences 1979.

5 Calvin F. Quate, Improved Resolution in the Acoustic Microscope, April 1977, CQC, box 1, folder Contract ENG75-02028 NSF.

6 J. G. Linvill, memo to Committee for the Study of Stanford's Educational Program, Jan. 26, 1967, CQC, box 4, binder Electrical Engineering.

7 Calvin F. Quate, letter to John Dimmock, Feb. 9, 1977, CQC, box 1, folder Contract #N000014-75-C-0632 ONR (JSEP) correspondence 1974–1977.

8 Robert Lamar, Stanford University News Service press release, Jan. 15, 1975, CQC, box 1, folder American Optical Corporation.

9 Linvill memo to committee, 1967.

10 National Science Foundation, Appendix A: Background of RANN, 1976, National Archives, collection 307-130-37-16-(1–6) National Science Foundation, NSF Historian, box 30, folder RANN: Interviews.

11 Quate, Improved Resolution, 1977.

12 Calvin F. Quate, letter to J. Warren, Dec. 15, 1977, CQC, box 1, folder Contract—Hartford Foundation, Inc., Correspondence—Proposals (1970–1981).

13 Calvin F. Quate, letter to Eric Ash, Jan. 5, 1978, CQC, box 3, folder Correspondence 1978 January–June.

14 Neils Reimers, memo to file S73-46, May 20, 1976, re: Visit of Eli Snitzer, American Optical Director of Research, on May 3, CQC, box 1, folder American Optical Corporation.

15 K. Wickramasinghe, M. Hall, and C. Quate, Acoustic Microscopy for Biomedical Structures, 1977, CQC, box 1, folder Contract #N000014-75-C-0632 ONR (JSEP) correspondence 1974–1977.

16 Calvin F. Quate, memo to NSF/RANN re: Acoustic microscopy, Dec. 19, 1974, CQC, box 2, folder Contract APR75-07317, National Science Foundation, Apr. 1, 1975.

17 National Institutes of Health, report of Special Study Section on "Acoustic Microscopy for Biomedical Applications," July 26–27, 1978, CQC, box 2, folder NIH 25826 Renewal (January 1980).

18 Calvin F. Quate, letter to Capt. Stephen Wax, 1977, CQC, box 2, folder AFOSR 98 Correspondence 1978–1982.

19 Stanton A. Glantz, "Comments about Engineers for Engineering by an Engineer," *The Grindstone: A Forum for Controversial Issues of Special Interest to the Engineering Community* 1 (Nov. 30, 1970): 4–16. SUA, Collection Arch 3009 The Grindstone.

20 SUA, Rudolf Kompfner papers, SC 194 ACCN 86–125, box 1, folder "Music" personal reminiscences.

21 John R. Pierce, oral history conducted by Andy Goldstein, IEEE History Center, New Brunswick, N.J., Aug. 19–21, 1992.

22 Calvin F. Quate, letter to Dr. Richard C. Atkinson and letter to Dr. Suzanne Stimler, both Nov. 12, 1979, CQC, box 3, folder Correspondence 1979 July–December.

23 Anthony Siegman, memo to Quate et al., Nov. 12, 1980, CQC, box 4, folder GL [Ginzton Lab] Memoranda 1980–1982. Brackets and emphasis in original; "NDT" refers to nondestructive testing.

24 Glantz, "Comments," 1970.

25 National Science Foundation, referees' comments on ECS-8010786, NSF Automation, Bioengineering, and Sensors, Systems program, 1980, CQC, box 3, folder NSF 10786 Proposal January 1980.

26 John Foster, interview by author, Santa Barbara, Calif., May 15, 2009.

27 Calvin F. Quate, Research on Acoustic Microscopy with Superior Resolution, 1981, CQC, box 3, folder NSF 10786, Proposal December 1981.

28 Calvin F. Quate, flyer, UC Irvine School of Engineering, "Modern Microscopy," Nov. 28, 1984, CQC, box 1, folder Conferences 1984.

29 Dan Rugar, interview by author, San Jose, Calif., Mar. 14, 2001.

30 Calvin F. Quate, notebook, April–June 1983, SUA, Quate Collection SC 347 (04–117), 2004 accession, box 2.

REFERENCES

Bardini, Thierry. *Bootstrapping: Douglas Engelbart, Coevolution, and the Origins of Personal Computing*. Stanford: Stanford University Press, 2000.

Colyvas, Jeannette Anastasia. From Divergent Meanings to Common Practices: Institutionalization Processes and the Commercialization of University Research. PhD diss., Stanford University, 2007.

Crary, Jonathan. *Techniques of the Observer: On Vision and Modernity in the Nineteenth Century*. Cambridge, Mass.: MIT Press, 1990.

Daston, Lorraine, and Peter Galison. *Objectivity*. New York: Zone, 2007.

Devey, Gilbert B. "Ultrasonic Microscope." *Radio-Electronic Engineering* (February 1953): 8–9.

Dunbar-Hester, Christina. "Listening to Cybernetics: Music, Machines, and Nervous Systems, 1950–1980." *Science, Technology, and Human Values* 35 (2010): 113–39.

Helmreich, Stefan. "An Anthropologist Under Water: Immersive Soundscapes, Submarine Cyborgs, and Transductive Ethnography." *American Ethnologist* 34 (2007): 621–41.

Jacobson, Judith S. *The Greatest Good: A History of the John A. Hartford Foundation*. New York: The Foundation, 1984.

Jones, Caroline A., and Peter Galison, eds. *Picturing Science, Producing Art*. New York: Routledge, 1998.

Kaiser, David. *How the Hippies Saved Physics*. New York: Norton, 2011.

Klüver, Billy. "E.A.T. 9 Evenings: Theatre & Engineering, Variations VII by John Cage" (1988). http://www.9evenings.org/variations_vii.php (accessed January 6, 2010).

Kompfner, Rudolf. "Recent Advances in Acoustical Microscopy." *British Journal of Radiology* 48 (1975): 615–27.

Latour, Bruno. *Science in Action: How to Follow Scientists and Engineers through Society.* Cambridge, Mass.: Harvard University Press, 1987.

Lécuyer, Christophe. "What Do Universities Really Owe Industry? The Case of Solid State Electronics at Stanford." *Minerva* 43 (2005): 51–71.

Leslie, Stuart W. *The Cold War and American Science: The Military-Industrial-Academic Complex at MIT and Stanford.* New York: Columbia University Press, 1993.

———. "'Time of Troubles' for the Special Laboratories." In *Becoming MIT: Moments of Decision*, ed. David Kaiser, 123–43. Cambridge, Mass.: MIT Press, 2010.

Macher, Jeffrey T., David C. Mowery, and David A. Hodges. "Reversal of Fortune? The Recovery of the U.S. Semiconductor Industry." *California Management Review* 41 (Fall 1998): 107–36.

Mody, Cyrus C. M. Crafting the Tools of Knowledge: The Invention, Spread, and Commercialization of Probe Microscopy, 1960–2000. PhD diss., Cornell University, 2004.

———. "'A New Way of Flying': *Différance*, Rhetoric, and the Autogiro in Interwar Aviation." *Social Studies of Science* 30 (2000): 513–43.

———. "The Sounds of Science: Listening to Laboratory Practice." *Science, Technology, and Human Values* 30 (2005): 175–98.

———, and Michael Lynch. "Test Objects and Other Epistemic Things: A History of a Nanoscale Object." *British Journal for the History of Science* 43 (2010): 423–458.

National Science Foundation. "The Promise of Acoustic Microscopy: Aided by Gigahertz Sound, Acoustic Microscopists Add a Dimension to the Perception of the Very Small." *Mosaic* (March/April 1978): 35–41.

Nelson, Andrew J., "Cacophony or Harmony? Multivocal Logics and Technology Licensing by the Stanford University Department of Music." *Industrial and Corporate Change* 14(1) (2005): 93–118.

Pierce, J. R. "Rudolf Kompfner, May 16, 1909–December 3, 1977." In *Biographical Memoirs*, ed. Bryce Crawford Jr. and Caroline K. McEuen, 157–80. Washington, D.C.: National Academies Press, 1983.

Pinch, Trevor. "Towards an Analysis of Scientific Observation: The Externality and Evidential Significance of Observational Reports in Physics." *Social Studies of Science* 15 (February 1985): 3–36.

———, and Frank Trocco. *Analog Days: The Invention and Impact of the Moog Synthesizer.* Cambridge, Mass.: Harvard University Press, 2002.

Quate, Calvin F. "Acoustic Microscopy: Recollections." *IEEE Transactions on Sonics and Ultrasonics* SU-32 (March 1985): 132–35.

Reiffenstein, Tim. "Codification, Patents, and the Geography of Knowledge Transfer in the Electronic Musical Instrument Industry." *Canadian Geographer* 50 (2006): 298–318.

Roosth, Sophia. "Screaming Yeast: Sonocytology, Cytoplasmic Milieus, and Cellular Subjectivities." *Critical Inquiry* 35 (2009): 332–50.

Schwarzschild, B. M. "Microscopy by Vacuum Tunneling." *Physics Today* 35 (April 1982): 21–22.

Smith Hughes, Sally. "The First Major Patent in Biotechnology and the Commercialization of Molecular Biology, 1974–1980." *Isis* 92 (2001): 541–75.

Stanford Workshop on Political and Social Issues (SWOPSI). *D.O.D. Sponsored Research at Stanford.* Vol. 1, *Two Perceptions: The Investigator's and the Sponsor's.* Stanford: Stanford University, 1971.

Turner, Fred. *From Counterculture to Cyberculture: Stewart Brand, the Whole Earth Network, and the Rise of Digital Utopianism.* Chicago: University of Chicago Press, 2006.

Vettel, Eric J. *Biotech: The Countercultural Origins of an Industry*. Philadelphia: University of Pennsylvania Press, 2006.
Wisnioski, Matthew H. Engineers and the Intellectual Crisis of Technology, 1957–1973. PhD diss., Princeton University, 2005.
WSF. "NSF RANN Program Shifts Gears: Sun, Earth Core Energy Sources to Receive New Research Emphasis." *Environmental Science and Technology* 8 (August 1974.): 704.
Yi, Doogab. The Recombinant University: Genetic Engineering and the Emergence of Biotechnology at Stanford, 1959–1980. PhD diss., Princeton University, 2008.

CHAPTER 10

THE SEARCH FOR THE "KILLER APPLICATION": DRAWING THE BOUNDARIES AROUND THE SONIFICATION OF SCIENTIFIC DATA

ALEXANDRA SUPPER

Introduction: "The Ear Has Somehow Had a Bad Lobby"

This chapter deals with the sonification of scientific data, frequently defined as "the use of nonspeech audio to convey information" (Kramer et al. 1997). While our ears interpret information all the time, as many of the contributions to this handbook show, the usage of sound to represent scientific data remains contested. My focus in this chapter is thus on the strategies that the practitioners of sonification utilize in order to establish the legitimacy of sonification as a scientific method of data display.

The last two decades have seen the establishment of the *International Community for Auditory Display* (ICAD), as well as an increasing popularity of sonification in general—elsewhere in this volume, Sterne and Akiyama go as far as declaring that we live in "an age of unprecedented sonification" (Chapter 23, page 557). Sonification has been applied to a wide variety of data and phenomena, ranging from seismographic data to election results, from molecular structures to the electrical activity of the brain. The different sonifications not only represent a multitude of different kinds of data but also sound very diverse: While some sonifications lull the listeners with the sound of orchestral music, others might prompt them to dance to a techno beat, while yet others stay clear of any musical connotation and instead rely on abstract clicks reminiscent of a Geiger counter. A shared underlying assumption of these approaches has been that an auditory display and analysis of scientific datasets might not only be helpful for scientists who have a visual impairment but might—often as a complement to existing modes of representation (e.g., statistical tables, verbal descriptions or visualizations)—also yield a more thorough comprehension of certain scientific data and phenomena.

In the light of several centuries' worth of scientific tradition dominated by the sense of vision (Wise 2006), this is a bold claim to make. A thought experiment proposed by Kenneth Gergen, in an essay that discusses the centrality of visual language for the rhetorical accomplishment of objectivity, underlines just how bold:

> The language of objectivity is primarily a language of vision. A typical research description in psychology, for example, will speak of subjects, questionnaires, tachistoscopes, chimpanzees, and so on . . . . If the subjects were described in terms of smell, questionnaires in terms of taste, tachistoscopes in terms of touch, and chimps in terms of sound, the descriptions would rapidly be discounted—merely the personal and subjective experiences of the investigator—potentially biased and unreplicable. (Gergen 1994, 277)

Yet this chapter deals with a community of researchers who *do* attempt to describe and understand the world in terms of sound and who turn this thought experiment into a "breaching experiment" (Garfinkel 1967). Much like Garfinkel's experiments have unveiled social norms that are usually taken for granted by violating them, sonification violates established customs of scientific data display and thus makes evident how much is taken as self-evident about what constitutes an acceptable way of representing scientific data. However, the practitioners of sonification are not (necessarily) doing this as an ethnomethodological experiment in epistemology; rather, they are looking to establish the legitimacy of sonification as a scientific technique. In this chapter I highlight the strategies they employ in order to accomplish this balancing act of seeking acceptance for a scientific discipline that breaks with entrenched conventions of scientific analysis and display.

To be sure, science wasn't silent before the sonification movement came along and made it sound. In recent years, scholars of science and technology (Lachmund 1999; Sterne 2003; Mody 2005; Kursell 2008) have shown that sound is an important element of scientific and medical practice and that the monopoly of vision has never been quite as complete as assumed by many who have focused exclusively on visual aspects. Indeed, some studies of visualization have themselves undermined the belief in the dominance of vision by acknowledging the role of other senses in the production of scientific knowledge.[1]

Some studies of sound in science have investigated fields such as acoustics (Kahn 2002; Thompson 2002; Pantalony 2005), where sound is the very object of study. Others have given attention to the sounds of the laboratory in fields that are not explicitly dedicated to the study of sound (Schmidgen 2003; Mody 2005), pointing out that sound "is pervasive in laboratory life and impinges on experimental experience in surprising and often epistemologically significant ways" (Mody 2005, 193). In doing so, they have shown that science, if you move upstream from its public representations to the places where it is actually produced and takes shape, turns out to be less silent than often assumed. If science appears soundless, that is not because it never makes a sound but because the laboratory walls not only keep the unwanted sounds of the outside world out (Schmidgen 2008) but also trap the sounds of science inside the lab, out of earshot of the public.

In this chapter I cross over from the confines of the laboratory to the conference hall. Whereas the studies discussed earlier have looked at sound as an object of scientific inquiry or at its importance for the daily practices of scientists, I study a community that attempts to emancipate the sounds from their hidden existence inside the black box of science by using them as a tool in the analysis and representation of scientific data. In doing so, these researchers have to face conventions of scientific data display that are not exactly sympathetic to auditory representation—as one of my interviewees succinctly put it, "The ear has somehow had a bad lobby for a very long time" (FD, 3).[2]

The sonification researchers therefore have some crucial lobbying work to do for the ear. In this chapter I analyze the different strategies with which they attempt to make listening to data a legitimate way of dealing with scientific information. I understand these activities as "boundary work": In order to position their work and infuse it with legitimacy, these researchers engage in negotiations of the boundaries of their field, attributing it with certain qualities to establish its cultural authority and demarcating it from other endeavors. The concept of boundary work has been used to study the demarcation work with which science is distinguished from its "others," such as politics, engineering, and religion (notably by Gieryn 1995), as well as at the internal boundaries of science, those between disciplines (Amsterdamska 2005; Burri 2008). This chapter is concerned with both, investigating how the boundary between science and nonscience (notably art) is negotiated within the domain of sonification, as well as its positioning vis-à-vis other scientific disciplines.

I look at these processes from four different angles: In the first section, which focuses on the establishment of a core sonification community, I study debates about how to best define the field, as well as their implications for who may (or may not) legitimately speak for sonification. Subsequently, I discuss the community's search for a "killer application" and how expectations shape the community—which means following these expectations as they branch out from the community into commercial, academic, artistic, or popular spheres. I then zoom in on one particular borderline where (albeit peaceful) boundary conflicts loom large in the domain of sonification: that between science and art. Returning to (relatively) firm disciplinary academic ground in the last section, I get to the bottom of notions of "quality" and strategies for assessing it.

This chapter is based on empirical research conducted in 2008 and 2009. The methods range from the conduct of twenty-two semistructured qualitative interviews (listed in the appendix) with practitioners of sonification, mostly from within the International Community for Auditory Display (ICAD); a discourse analysis of ICAD conference proceedings and related journal articles, books, and dissertations; and a few short-term ethnographic studies. These episodes of participant observation include two weeks of self-experimentation in learning sonification while based in an institution that is active in the field, as well as attendance at two ICAD conferences, two sonification workshops, and several sonification-related concerts and talks.

"Sonification Is Defined As . . ."

The first ICAD conference was held in 1992 at the Santa Fe Institute in an effort to pool the previously scattered activities of various researchers working on issues related to sonification and auditory displays. Despite being labeled an "international conference," the character of this gathering, with its thirty-six (predominantly American) attendees, was closer to a workshop and a rather personable one at that: The participants' names are printed on the back of a T-shirt commemorating the event, and the proceedings volume (Kramer 1994) comes with a group picture. Gregory Kramer, conference organizer and founder, explains that, even prior to the first conference, it was never intended as a one-time congregation; he recalls thinking the following:

> This is a field, this could be, this should be, this will be, a field. . . . So calling it something as pompous as the International Conference on Auditory Display, when really it was a handful of researchers doing something, a handful of researchers not doing anything, and a handful of people who were kind of interested maybe—you know, it's a little bit of a reach, but I trusted it. (GK, 3)

This desire to establish a new field along with the conference is noticeable in the book publication, which contains not just a collection of conference papers and an accompanying CD but also a lengthy, historically informed background paper on auditory displays by Kramer and an annotated list of resources and publications. Throughout the book, it is acknowledged that a systematic and coherent terminology for the field has yet to be provided—see, for example, Kramer's own preface

(Kramer 1994, xxvi) or the contribution by Scaletti (1994, 224). Yet the book does propose some "working definitions" (Scaletti 1994, 224), which continue to be an important resource for the field today.

The next big step in codifying sonification was a report for the U.S. National Science Foundation. Prepared at a preconference workshop at the third ICAD (held in Palo Alto in 1997), the report provided some stock taking, as well as recommendations for a research agenda, and included what remains one of the most-cited definitions of sonification today: "Sonification is defined as *the use of nonspeech audio to convey information.* More specifically, *sonification is the transformation of data relations into perceived relations in an acoustic signal for the purposes of facilitating communication or interpretation*" (Kramer et al. 1997; original emphasis).

Several characteristics stand out in this definition: It specifies a process of transformation (something which wasn't already audible is turned into sound), it emphasizes the relational character (it's about expressing relations, not precise numerical values), and it indicates the purpose of sonification (communication or interpretation). The most explicit exclusion is that it categorically excludes speech displays; the other elements of the definition appear more elastic.

Compare this to a newer effort at defining sonification, proposed by Thomas Hermann at ICAD 2008. Hermann suggests that "a technique that uses data as input, and generates sound signals . . . may be called sonification, if and only if" (Hermann 2008, 2) a certain number of other criteria are met. Hermann's criteria specify that the sound reflects *objective* relations or properties in the input data, that the transformation is *systematic*, that it is *reproducible*, and that the same procedure is applicable to *different data.* Unlike in the previously quoted definition, then, the systematicity and reproducibility of sonifications are made explicit here: It's not enough that the sound tells us something about underlying data relations; that relationship must also be thoroughly systematic, and the output sound has to be the same, not merely similar, when the process is repeated. In other words, what we are witnessing is the establishment of a set of procedures to reduce subjective intervention in an appeal to "mechanical objectivity" (Daston and Galison 2007).

This new definition involves the drawing of a sharper, less permeable boundary between "scientific sonification" and "artistic sonification," or rather, it denies the existence of the latter category altogether and therefore declares the distinction redundant: "Being a scientific method, a prefix like in 'scientific sonification' is not necessary" (Hermann 2008, 3). Defining sonification in such a way also means to declare it as the jurisdiction—in Andrew Abbott's (1988) sense—of scientists, in particular, those scientists who have an intimate knowledge of the conventions of sonification research. While contributions by artists and composers are not ruled out categorically, they at least have to submit to an explicitly scientific logic.

This is not an insignificant point: Although the term *sonification* is not usually used by these artists, there is a rich tradition, especially within the domains of contemporary classical music and sound art, of transforming data into sound for musical purposes (Schoon and Dombois 2009). On the other hand, there are also scientists who, without any contact with the sonification community, use sound as a resource for popularization activities. For example, when giving popular talks, some asteroseismologists play "the sound of a star" in order to convey an idea of the stellar oscillations that they research. These scientists might regard these sounds as useful tools for popularization, but they are generally quite skeptical about the scientific accuracy and reliability of auditory displays and see no need to pay too much attention to the specifics of how these sounds were generated (CA, 7; JH, 5; DK, 14). For their purposes, such minutiae are of no importance: "I'm interested in the physics of these stars. I use the sounds to understand that. . . . But it's not a goal by itself" (CA, 6).

In other words, the researchers gathered at ICAD are not the only group of practitioners who transform data into sound. Proposals of a more rigid, scientific definition of sonification also serve to establish the ability to produce something that would qualify as a sonification as the exclusive competence of the core sonification community. The definition thus has implications not just for what counts as a sonification but also for who is considered qualified to make one.

It is perhaps no coincidence that such definitional efforts occur at the same time as a general trend to formalize the community. Kramer's (1994) volume is still the closest thing the field has to a handbook; there is no journal explicitly dedicated to sonification, while the annual ICAD conference proceedings serve as the most important outlet for publications; there are no chairs or dedicated university positions in sonification; it is still difficult to convince funding agencies of the merits of sonification research. Yet the last few years have seen increasing efforts to push for an institutionalization of the field: The structure of ICAD is becoming more formal (BW, 7), and a new handbook and a journal are in the works (BW, 7; TH2, 1f.). According to Bruce Walker, ICAD president, this means that the next few years will see an increase in definitional efforts:

> Now, as soon as you have a journal . . . you have to have an editorial statement and a mission statement for the journal. And that formalizes exactly what that journal is about. Now if that journal is the flagship journal for our community, then that will have the effect of kind of defining what we consider to be representative research, or representative papers. . . . So you will see more and more explicit definition of the field in the next two or three years. (BW, 7)

Clearly, debates over what sonification is (and is not) attract considerable interest within the ICAD community: After Thomas Hermann's 2008 talk proposing a new definition of sonification, the discussion went far beyond the few minutes reserved for responses at the end of the presentation and reached into various coffee breaks and dinner conversations. What was at stake was not so much the specific criteria Hermann proposed but rather the question of whether it is a good idea in the first place to try to narrow down the boundaries of the field. Some have argued that any definition should be flexible and fluid enough to accommodate future developments of the community rather than intervening in favor of any one particular conception of the community and its techniques. Stephen Barrass reasons that "the suggestion that sonification should be more scientifically rigorous is misguided,"[3] as it makes sense for particular approaches to sonification that exist within the field but not for other traditions that are informed by other kinds of (e.g., humanist) theories. To some, this might be an issue of principle; to others, one of timing: As one commentator at the ICAD 2008 session remarked, it was "too early" to restrict the boundaries of sonification, as only very few really convincing examples have yet been found. In the light of this situation, argued the commentator, it would be a mistake to narrow down the definition now. If an artist were to develop a really convincing auditory display that would effectively convey something about a dataset without fulfilling all of Hermann's criteria, it would be a loss for the field if this example could not count as a sonification. With these considerations, we are already addressing the issues that are the central topic of the next section: the capacity of expectations for the future to shape a community.

"That Might Be a Fairly Significant Killer Application"

The role of expectations becomes most manifest in the community's discussion about finding a "killer application" for sonification—precisely because, it seems widely agreed upon, this killer app still has to be found.[4] It seems significant that it is an "application" we are talking about here: Sonification deals in applied knowledge and is on the lookout for a service that could extend past the core community into commercial, scientific, artistic, or popular domains. Certain expectations are created as a resource for legitimizing research and mobilizing for funds, publicity, and support of other actors—all of which are functions of expectations discussed by Van Lente (1993) in an STS study on the dynamics of expectations in technological developments. Expectations are also drawn upon and created by actors in specific research communities in order to justify and push their field (Guice 1999) or a particular vision of their field (Hedgecoe and Martin 2003). It is this last aspect of expectations that I focus on here: How are expectations for sonification appropriated to shape a particular conception and vision of the community?

The term *killer application* is borrowed from informatics, where it "refers to an application program so useful that users are willing to buy the hardware it runs on, just to have that program" (Juolo 2008, 76). Juolo gives some historical examples of killer apps (e.g., early spreadsheet software, which showed such promise for bookkeeping and decision making that businesses bought computers just to be able to run these applications). Similarly, a videogame would be considered a killer app if it were so popular that thousands of gamers would buy a new console just to be able to play it.

For sonification, the term is used in a metaphorical sense: A sonification would be considered a killer app if it were so convincing that it would make people "buy into" the idea of sonification in general, contributing to its acceptance—not if it made practitioners run out by the million to buy new soundcards or speakers so they could hear the sounds better. Indeed, in many cases it will make sense to keep the cost of entry low by avoiding the need for new technical gadgets: For many scientists, engaging with sonification already requires a leap of faith, so it doesn't help to expect additional expenditures (e.g., new hardware or programming skills) from them. Consequently, Florian Grond, who works on a sonification platform for molecular structures and dynamics, realized after talking to a number of chemists that it would be better to provide this sonification utility in the form of a plug-in for a widely used, open-source software package for data visualization rather than to expect them to learn an entirely new programming language. This means to link the sonification application not only to existing tools but also to established modes of representing data and the skills required to work with them. Grond acknowledges that the chemists have a point in insisting on a combination with visual displays:

> And that's quite obvious, actually—because they have acquired expertise in [working with visualization], and if you told them now: "as of tomorrow, only listening and no more seeing," then they would throw away the expertise which they have acquired over many years. I wouldn't accept having to do that, either. (FG1, 11)

Here Grond illustrates a lesson that has been pointed out by scholars of science and technology studies: If a (research) technology is to succeed, it has to be adapted to and embedded into existing cultural practices in order to be appropriated into specific contexts of use (Bijsterveld 2004; Borck 2006). If the new technique is too foreign, too far removed from scientists' daily practices, they will be reluctant to give it a chance. This is perhaps one of the reasons that many sonification researchers focus their hopes of developing convincing displays on medical applications: Not only is there the potential of coming up with solutions that might prove beneficial to public health (EB, 8; TS, 14), but many perceive that they already have a foot in the door in the medical field (GB1, 1; TH2, 12), where existing listening practices have helped to prepare medical practitioners for the possibilities of auditory displays. While many sonification researchers complain about the lack of an education in listening skills, the medical field is one domain where students *do* learn how to listen (see Rice, this volume).

Thus, even though sonification breaks with some of the entrenched conventions of scientific data display, its success will depend on the ability to tie in with existing practices and skills. Of course, it also has to offer some clear benefits over existing approaches to data analysis and representation—it must carve out its own niche. The exact contours of such a niche are still unclear, as the demands are at times contradictory: For example, in order to be of use to users who have a visual impairment, who need to compensate for their lack of vision, a sonification might be at its best if it conveys something that a visual display is good at (TS, 14); to a person whose sight is intact, it might be more useful if it compensates for weaknesses in visual displays (MG, 13; TH2, 11). Different user groups might also have very different demands for accuracy versus aesthetic appeal (BW, 2). Another question is whether the sonification should be directed at a small community of specialists or appeal to a popular movement; while many projects have gone for the first strategy, there are also attempts to anchor sonification more firmly in popular culture (e.g., in social networking applications) (SB, 3). Others argue that "the only way to succeed will be to carve out a prototype or a convincing example and to demonstrate it on one disease, one symptom, one little corner where you can convince specialists" (GB2, 2).

It seems widely believed that most of the sonification work being done so far consists in "some very interesting prototypes" (EB, 3) but that there is very little in the way of generalizable knowledge, guidelines, or best-practice models—which arguably becomes more of a problem precisely as the community expands (EB, 3). Acknowledging that their field is still in its infancy, many practitioners of sonification rely on promises for the future to justify their work. Talk about "killer applications" seems to occur in the future tense.

Yet it is not projected into the very distant future; it could happen any day. One of my interviewees relates the following:

> And funnily enough, a fortnight ago, on Radio 4, a very mainstream BBC radio station, there was some physicists who had some sonifications of activity going on internally within stars. . . . You know, I thought that might be a fairly significant killer application, it certainly got quite a bit of publicity. (TS, 15)

However, when he contacted one of the scientists involved, the reaction was quite disappointing: The researcher replied that the sound display was intended as a publicity gag, not as a serious research component. Not much of an ally in the search for the killer app.

Yet this readiness to accept an example from outside of the ICAD community as a possible killer application seems symptomatic. We have witnessed this before in the previous section, when a commentator at ICAD warned against narrowing down the boundaries of the field, as this might result in excluding some very convincing sonification examples. To some extent, this openness is built into and explicitly fostered in the community—for example, with sonification concerts organized in the context of the conferences to stimulate involvement by composers or with the invitation of keynote speakers who are often not "the kind of person who would come to ICAD unless they were invited or unless the conference was in their city" and who can thus be introduced "to a field that they may not even realize is related to their work" (BW, 6). This openness is partly perceived as being in "the nature of the field" (GK, 4), but it's also explained by the potential of learning from each other (BW, 6), as well as a strategic choice to keep the area broader in order to build momentum, to get some "critical mass" (GK, 4). As we have already seen, this reasoning is rooted in expectations for the development of the community in the search for a killer app. Expectations thus not only guide the direction of research conducted in the field by favoring some types of research over others but are also used as a resource in shaping a conception of sonification that emphasizes openness, broadness, and self-reflexivity as characteristics of the community.

"What 'Art' Means for the Art System and What 'Science' Means for the Science System"

Drawing the boundaries of sonification often involves questions about the relationship between science and art, and I address these questions in more detail in this section. We have already seen how attempts to define sonification as a scientific (objective, reproducible) procedure constituted an effort to draw a sharper boundary between sonification as a scientific application and musical approaches of making data sound. However, we have also seen that these efforts of boundary work have not remained undisputed, as many consider the openness of the community an asset in the search for the "killer app."

This dispute might be interpreted as a struggle between different factions within ICAD, one in favor of doing only the most straightforwardly "scientific" sonifications and cutting the ties to the artistic community, the other stressing the value of artistic contributions and the need for openness. However, this interpretation would be too facile. It would be hard to reconcile, for example, with the performance of Thomas Hermann, who would have to be considered a spokesperson of this ostensible "scientific faction," at the 2008 *Wien Modern* festival for contemporary music, presenting—with Gerold Baier—a "live sonification of the human EEG," using the brainwaves of composer Alvin Lucier.

Rather than an issue of stable splinter groups within the sonification community trying to enforce their interests and impose their idea of how scientific/artistic sonification should be, I show that these boundaries are negotiated in context-specific ways. Take the example of a sonification talk given at a neurological workshop, reported by Gerold Baier: When he and his colleague presented their EEG sonification, one of the "big names" in neurology demonstratively left the lecture room and was soon followed by a number of other researchers. Even those who remained in the hall for the remainder of the lecture displayed reactions that Baier likened to those to free jazz performances during its prime: open mouths, signs of disapproval, confusion about what they were being confronted with. Those who were willing to engage with the presentation at all insisted on addressing it on musical rather than scientific terms—to the dismay of the presenters, who were there to discuss their results with epileptic data, not elements of composition. Yet, however undesired it may have been by the presenters, the association of their talk with music was not a complete accident, as the title of their talk (mentioning an "unpredictable concert") itself elicited musical connotations. When asked about this apparent contradiction—not wanting a paper to be associated with music, yet invoking music in its very title—my informant explained that the title was not chosen entirely voluntarily; the musical reference was proposed by the organizers and was for the presenters the label via which the material could be presented at the workshop at all (GB1, 1).

This is an interesting case because it combines two aspects of boundary work that are often discussed separately: Scholars of boundary work often focus exclusively on rhetorical demarcations of a field (Gieryn 1995), or they interpret interdisciplinarity in terms of a crossing of boundaries (Klein 1996). Here, however, we see that the two go hand in hand; Demarcation and boundary crossing occur simultaneously. As Willem Halffman points out, "boundary work has the double nature of dividing and coordinating" (Halffman 2003, 70), of demarcating as well as specifying conditions under which demarcations can be crossed. In the case of the earlier anecdote, it was the very promise of crossing the boundaries between science and music that opened up possibilities of presenting in this scientific venue in the first place. However, the association with music was not desired by the presenters themselves, who were quick to rhetorically demarcate themselves from being positioned on the artistic end of the science-art spectrum. While the researchers accepted and went along with the musical label to some extent (they did not turn down the offer of presenting at the workshop) because it provided opportunities for them, in the next step they tried to subvert this categorization.

This shows that the position of sonification in relation to science and/or art is not a given but becomes a balancing act that is dictated as much by strategic decisions and the desire for an audience as it is by the penchants of the practitioners. In other words, the categories of "science" and "art" do not refer to distinctly bounded domains; the distinction becomes permeable and mutable: subject to boundary work.

Such negotiations about the role of art (specifically, music) for sonification become visible in debates within the community about finding a language that allows for communication across different disciplinary backgrounds—an issue that has also been given attention in STS studies of collaboration, notably in Galison's (1997) discussion of pidgin and creole languages. While a pidgin refers to a contact language developed by native speakers of different languages in order to be able to communicate across specializations on delimited issues, a creole is more comprehensive and can be taken up as a native language for a new generation of researchers.

According to many in the community, sonification lacks such a comprehensive common language, although there seems to be a shared understanding that it will become necessary to develop one. Such an emerging language for sonification is considered to have "patchwork" (TH1, 7) qualities and should be suitable for analytic rather than emotional discussions of sound and data—a vocabulary that, so far, is considered to be largely absent (GB2, 5). In some instances, this missing vocabulary can be replaced by terminology borrowed from visualization—a practice that is eyed suspiciously by parts of the community, as evidenced by debates at ICAD 2009 about the need to break away from the appropriation of visual language and develop a language for sound.

To what extent does musical discourse come into play in developing such a language? A basic literacy in musical vocabulary, my interviewees seem to agree (TS, 11; EB, 4), is helpful for working with sonification. This does not mean that musicological terminology itself is a sufficient foundation for talking about sound, "especially as it is made for a different context. It is made, for instance, for a certain culturally grown concept of harmonic progressions" (GB2, 5), and using musical vocabulary therefore means dragging along the entire cultural context it sprang from. Indeed, some even regard a traditional musical training with very strict conventional ideas of harmonics and rhythmicity as a potential constraint for doing sonification (AdC, 12).

Terminology is just one of the things that sonification could learn from music. Musicians and composers—especially those who come from a contemporary and experimental musical tradition (AdC, 12; GB2, 5)—are also thought to be able to bring other assets to the table, such as the ability to listen for structures (GB2, 2), certain technological tools (GK, 12), the freedom to choose one's format of publication (FD, 3) and to experiment with ideas that are not a guaranteed success (AdC, 11), the diversity of different realizations of one idea (SB, 5), aesthetic considerations (GB2, 6; FD, 11; GK, 12), and the familiarity with a broad spectrum of different kinds of sounds and knowledge of the different possibilities they allow (AdC, 12). Besides, in a field that is chronically underfunded and barely established in the scientific domain, the possibility of crossing over into artistic territory opens up opportunities for funding and publicity. In the light of these considerations, it is no surprise that sonification frequently branches out into artistic contexts or invites participation from musicians (e.g., with the organization of concerts in the course of ICAD conferences). Clearly, a lot of boundary crossing is going on here.

However, boundary crossing goes hand in hand with demarcation, and the practitioners of sonification are often quite adamant about drawing a boundary between sonification and music. Thus, at a talk given in the context of a festival for contemporary music, one sonification researcher greeted his audience with these words: "I am glad to see that you dared to approach this topic even though it has nothing to do with music." Laughter ensued, but a message had been sent: Sonification does not equal music—even if its practitioners might take advantage of an opportunity to present their work in the context of a music festival. A similar rhetoric of demarcation can be found in quotes from my interviews: One interviewee emphasized that sonification—just like visualization—allows some aesthetic freedom but is "actually a technical procedure . . . it's rule bound. And a visualization is not yet a painting. And likewise, a sonification cannot automatically be compared to a musical composition" (FD, 3). While participants with musical backgrounds are numerous and welcome at ICAD, "if it's purely art or purely music, then it belongs in a different conference" (BW, 14). "People might still benefit from it; as a matter of fact in all likelihood [they] would if it's interesting stuff. But you wouldn't just take it in if it's a good piece of music and say it should be at ICAD" (GK, 13).

The question of where to draw the boundary also has to be faced by individual practitioners (e.g., in choosing appropriate contexts in which to present one's work). After all, the crossover into musical territory can secure funding and publicity but has some drawbacks as well. When I asked the media artist and researcher

Florian Grond whether he would make an art project out of his dissertation, he replied that he might consider it but would be "very cautious":

> If you approach chemists about sonification of chemistry, at first they always believe that it's art anyway. And then it's counterproductive, of course, if you end up using it as art because then it's just difficult to communicate why one thing was not art and the other thing suddenly is art. (FG1, 4)

Grond worries about muddling the message for the chemists he wants to collaborate with, but he is also concerned about communicating a clear message to funding agencies—after all, "you just have to be realistic: Being part of a system, with all the restrictions that system entails, simply also means that you can put things into practice because the system just distributes money in the best case" (FG1, 5). He is reluctant to tap into the kind of resources that might become available by presenting sonification as an art project, as this often "has a very, very decorative function that is mostly on the level of infotainment and actually of a justification for all the money being spent on these [large-scale research] projects. It's about public relations and all that" (FG1, 6). The demarcation here is not an essentialist distinction between science and art but a pragmatic realization that "by and by, you learn to detect what 'art' means for the art system and what 'science' means for the science system" (FG1, 2). To succeed in either one, you have to learn the rules of the respective system.

"Evaluating Ourselves to Death"?

"Bam-bam-bam-bam-bam." Over the phone, Bruce Walker sings me an auditory graph with rising pitch. I have asked him to name an example he likes to draw upon in order to convince his fellow psychologists of the value of sonification-related research. Walker's example is more than just the auditory graph itself: After letting them hear the graph and explaining that the sound represents money, he would ask his colleagues whether he was getting richer or poorer. Almost all of them would answer that, as the pitch goes up, the money increases. Walker would then tell them about a study in which hundreds of people, both blind and sighted, were asked this question. While the sighted respondents shared the interpretation of Walker's colleagues, the blind respondents thought the opposite. In the course of the research, it turned out that different mental models are at play:

> If I took a coin, and I dropped on a table, it would go "clink." If I took a roll of coins and dropped them on the table, it would go "clonk." And if I took a bag of coins and I dropped it on the table, it would go "thud." And for [the blind people], more coins, more money, more—volumetrically more of anything, makes a lower-pitch sound. (BW, 19)

Walker uses this anecdote to demonstrate to his psychologist colleagues that research on auditory perception yields interesting and unexpected results. Yet debates about the role of testing auditory displays are relevant not just in the psychology community but also within ICAD itself. Presumably few would deny the value of the research in auditory perception that Walker described; however, what is at stake is whether such research should be included with every sonification publication. Traditionally, this has not been the case; after the first ICAD conference, Gregory Kramer summed up as follows:

> Auditory display is a young field. As in any field, it takes time to establish a solid foundation of rigorous research. . . . Few of the papers at ICAD included the actual running of subjects along with the experimental design necessary for a more rigorous proof of display effectiveness. (Kramer 1994, xxix)

Recent literature studies of ICAD proceedings indicate that "there has been a trend toward increasing usage of [user or participant data] testing" (Bonebright and Miner 2005, 518). In Frauenberger's (2009) survey of the publications in the proceedings of ICAD 2007,[5] all but four of twenty-three papers included some sort of evaluation of the designs—but that includes subjective evaluation by the author or pilot studies promising further research, as well as the "experimental design" that Kramer had hoped for. One of my interviewees explained why he considered it important to have more quantitative user tests:

> You need some way to measure what you actually achieve when you're using sonification. It's not enough that you say this, listen, this really sounds better than yesterday. That's not the result. But if you can show that, when you have ten people doing this task, they do things 10 percent better when they're using the auditory display than when they're not using the auditory display—that's a result. (MG, 15)

Another respondent referred to such work as having "research components" as opposed to doing "show and tell":

> I always encourage papers at the conference that have evaluation components, that have research components, that are introspective or self-reflective. They look at what they've done, and they try to assess it and evaluate it and figure out, place it within some theoretical context. As opposed to the kind of papers where people are basically doing show and tell. . . . There will always be plenty of applications, but I think we always need to push for more of the science and evaluations. (BW, 3)

It's interesting to see what counts as a "research component" here: It has to do with placing one's work in a theoretical cadre and relating it the work of others, but it is also strongly associated with evaluation and experimental setups. Both respondents quoted earlier profess to exert (gentle) pressure for such "research components" when writing peer reviews for conference papers. This practice, however, is contested—as are the underlying ideas of what constitutes research. For instance, one respondent complained about the peer review practices at a previous ICAD conference:

> Many of the best sonification examples were curated out, peer-reviewed away. . . . There were a central stream and the poster sessions, and [many] good things

> were sent into the poster sessions because [the reviewers] had abstruse ideas about evaluability and intersubjectivity. So they said, if somebody makes a sound and did not make a series of user tests with 17, at least, I don't know, 17, 20 test persons, then we cannot accept this because that's not scientific. It's as if you would not have a graph printed if someone cannot prove that he let 17 people look at the graph to make sure they can see something in the graph. That is, I think, that's absurd. (FD, 13)

The comparison with visualization in discussing this phenomenon was made by another interviewee as well:

> The first visualizations of molecules were not evaluated; they were just made. And they were extremely functional. Certainly, there was some thought behind it and not just daubing something on paper, sure, but that's what I mean: Something was done; it was extremely functional. It is—I'm not sure if the medial representation, if you can call this a scientific method, but it is at the very least a valid scientific tool without having gone through an evaluation process. (FG1, 7)

This interviewee related the insistence on evaluations to a lack of involvement by the application sciences. In doing so, he took up a recurring theme in my interviews: the importance of intensive collaboration with specialists in the domains from which data are being sonified (EB, 7; TH1, 9), as well as the difficulties of finding such specialists who are willing to cooperate (FD, 17; GK, 8, 11). According to my respondent, the internal audience of ICAD attempts to compensate for this shortcoming by being extremely critical, which results in a danger of "evaluating oneself to death" (FG1, 7).

The difference between the perspectives discussed earlier partially lies in a different emphasis on what constitutes "research." The diversity of different backgrounds in ICAD manifests itself in the different motivations brought to the table: Depending on whether one's main research interest is to further understanding of human listening processes or to explore a particular dataset, the requirements of empirical testing might be dissimilar; testing for "display effectiveness" might not mean the same thing in both instances. Not everyone even aspires to the status of doing science to the same extent. After stressing that scientific strands within sonification indeed require testable hypotheses, Stephen Barrass states that the same is not true for all sonification work:

> In my view, sonification is a field of applied science and design. Design is a problem-solving process rather than a theory-proving process. It is multifaceted and specifically situated. It recognizes multiple solutions to the same problem and that the designer, context, and audience are critical elements. It is iterative and requires evaluations at incremental decision points. . . . Rather than a scientific experiment with large numbers of subjects, the designer needs to rapidly test alternatives with just a few subjects. (Barrass, personal email correspondence, July 10, 2009)

With such problems in mind, members of the ICAD community are developing alternative techniques of evaluations often based on qualitative methods.

Barrass's work toward a "sonification design meshodology" (not a methodology but "a mesh of methods") is an example, as is recent work by Brazil and Fernström (2009).

So, to some extent, the different takes on user testing can be ascribed to different disciplinary backgrounds, motivations, and notions of "research." They can also be regarded as diverse types of boundary work, that is, as different approaches to the legitimization of sonification as a viable technique. This becomes most explicit when Dombois argues that the insistence on user testing not only debases the papers that include no such evaluations but also undermines the accomplishments of the community as a whole: "It's like giving up something you have just gained, which is saying: Listening is an authority" (FD, 13).

The argument rests upon the comparison to visualization: Sonification is *just like* visualization (in this respect), and visualization is widely used without requiring user tests, so sonification should be self-confident enough to do the same or risk undermining its own accomplishment of establishing listening as a trustworthy technique. The reference to visualization is thus used as a rhetorical resource for boundary work. Sonification, like medical auscultation in Jens Lachmund's study, is not a purely auditory phenomenon but "an auditory [one] that, at least partly, [is] being shaped through the visual" (Lachmund 1999, 428). Any look at sonification papers or conference slides can confirm this: No less than in other disciplines, they are replete with pictures that range from visualized sound waves to diagrams and photos. (The accompaniment by sound files is less widespread, though often requested.) If the usage of graphical techniques to transcribe sound into images played a fundamental role in the establishment of acoustics as a science (Brain 2002; Kahn 2002), such visualization techniques continue to play an important role for sonification today. Moreover, the preceding analysis shows that visualization not only helps to illustrate sonification research but also becomes a tool for rhetorical demarcation. Visualization is presented as a competitor, as well as an ally of sonification research, and is invoked in order to make claims of scientific authority.

The approach favoring user tests, too, is involved with a quest for scientific credibility: Listening is an authority, and we're not just going to claim it; we're going to *prove* it. Especially with its emphasis on quantitative testing, it links up to common strategies of quantification as a means of establishing objectivity and overcoming distrust (Porter 1995). Both of these approaches are therefore concerned with different ways of establishing the scientific legitimacy of sonification.

Conclusions

This chapter has followed sonification through different contexts: Studying discursive formations and strategic demeanor within the sonification community, we have scrutinized the constitution and professionalization of an academic field dedicated to sonification; its attempts to find applications that would attest to its usefulness beyond this core community in a broader scientific, artistic, commercial, or popular sphere; the negotiation and balancing of an identity between science and art; and, finally, the disputes about notions of "quality" and how to best assess it. These episodes add up to a story of how an academic community that breaks with certain conventions of scientific data display attempts to establish the legitimacy of its activities.

One of my interviewees remarked that "the niche for [sonification] doesn't exist at all, not yet" (GB2, 4) Therefore, anyone doing sonification simultaneously has to "define a field, provide the listening education," and understand the scientific specialization from which the data are taken (GB2, 4). In order to create a niche in which sonification can be regarded as an acceptable way of analyzing and representing scientific data, much work needs to be done: transforming data into sound, making it sound good, developing and relaying listening skills, familiarizing oneself with different scientific specializations, building up a field. The creation of such a niche can be seen as a kind of "boundary work." According to Thomas Gieryn, such "boundary work" is part and parcel of any scientific field, as science "acquires its authority precisely from and through episodic negotiations of its flexible and contextually contingent borders and territories" (Gieryn 1995, 405).

This chapter has looked at such negotiations of flexible and contextually contingent borders in a field that sets out to cross boundaries as much as it draws them; it creates, maintains, reinforces, adjusts, transgresses, negates, and breaks down boundaries, often all in one move. The extent to which sonification is positioned as science or something else (e.g., music) itself becomes a part of the repertoire of boundary work, from which its practitioners selectively draw in order to position and legitimize their approach.

Indeed, perhaps the most salient characteristic of the boundary work of sonification is one that has been disregarded in previous studies of boundary work: The practitioners not only demarcate and cross boundaries but at times, if it is convenient for them, are also quite ready to submit to existing categorizations even if they may not agree with them—only to distance themselves again and question those categorizations at the next opportunity. For example, they may accept the label of "music" or "sound art" in certain instances in order to acquire funding or reach a particular audience even if they consider this a misrepresentation of their work, only to then reject the categorization again and insist that sonification is not music after all. Existing cultural attributions are thus simultaneously embraced and subverted; sonification both assimilates to and undermines existing conventions of scientific representation. However, as the debates about evaluation and quality assessments have shown, it is by no means clear how far sonification should go in its strategies of assimilation or subversion.

Sonification is thus sometimes positioned in conformity with scientific conventions such as objectivity and reliability; at other times, it is emphasized to be more cutting edge, avant-garde, and experimental than boring old disciplinary science—and often both at once. Which qualities of sonification are stressed and which are disregarded is very much contingent and varies from one context to another—the result of negotiations that both the field as a whole and the individual practitioners deal with. To some extent, it has to do with strategic choices related to research funds, audiences, and publication channels, but it is also a question of disciplinary identity and professional "jurisdictional control" (Abbott 1988). It has implications for the self-perception, the composition, the contours, and the size of the community: Who may speak for sonification, and who may not? If the goal is to establish sonification and gain acceptance, is it best to ensure its publicity and popularity (e.g., by being very open to artistic contributions) or to appear as a small but highly professional community of experts? This is a tightrope that has to be walked on many levels; it comes into play when designing sonifications (easy to use for laypeople or requiring highly specialized knowledge?), when assembling project teams, and when preparing conferences or founding journals. There is no consensus within the field of how to tackle these questions, and, indeed, they are more often than not dealt with in context-specific and self-reflexive ways.

As Olga Amsterdamska (2005, 20) has indicated, "the establishment of [boundaries between disciplines] creates the need to emphasize both the distinctiveness of a field of research and, simultaneously, its conformity with the prevailing, though changeable, standards or markers of scientificity." Such a pattern is certainly visible for sonification, which is positioned as both distinctive from and compatible with notions of scientificity. Studies of emerging disciplines have shown how these have struggled with defining their objects, facts, and methods and how they have differentiated themselves from or associated themselves with public health (Amsterdamska 2005), amateur science (Wilde 1992), or common sense (Derksen 1997); for sonification, a comparable reference point could be its positioning vis-à-vis not only music but also visualization. At the same time, what is specific for sonification is that it not only takes on one or a few disciplines as competitors but also challenges fundamental conventions about scientific analysis and data display. Since sonification is in principle applicable to data from any scientific discipline and its claims therefore potentially affect all of these fields, its boundary work, too, reaches out in many different directions.

As this analysis shows, the sonification community walks a rocky path on its way to have listening to data accepted as a legitimate scientific activity. It takes on many different fields simultaneously and has to face skepticism from as many sides. One of my interviewees remarked that scientists from other fields do not necessarily dismiss the possibility of interesting sonification applications; in classic "not in my backyard" fashion, they just tend to rule them out for their own field. Thus, physicists might be ready to concede the potential of sonification in neurology but not in physics (KV, 4). However, the success of sonification will depend on its ability to convince specialists of the promise of sonification for *their own turf*. The practitioners of sonification are well aware of these difficulties—hence, the widespread concern that the field should not only develop its own facts, objects, and methods but also come up with killer applications.

NOTES

1 See Pasveer's (2006) discussion of X-ray technologies in relation to medical listening practices, Myers's (2008) research on modeling in protein crystallography, which emphasizes the importance of gestures and incorporation, or Joyce's (2008) study of MRI, which concludes with a brief consideration of touch, smell, taste, and sound as sources of knowledge about the body.

2 I have translated this quote and several others from German. A table in the appendix indicates in which language each interview was conducted and therefore also whether translation has occurred.

3 Personal email correspondence, July 10, 2009.

4 This was especially apparent in discussions at the 2008 workshop on "Recycling Auditory Displays," organized by Frauenberger and Barrass and also mentioned in several interviews (MG, 13; TS, 15).

5 He took into account all eleven papers and twelve posters that presented a design for an auditory application, ranging from sonifications of scientific data to auditory games. "Papers describing purely artistic projects" were excluded since they follow a different rationale, where the issue of evaluation is posed differently (Frauenberger 2009, 59).

REFERENCES

Abbott, Andrew. *The System of Professions: An Essay on the Division of Expert Labor.* Chicago: University of Chicago Press, 1988.

Amsterdamska, Olga. "Demarcating Epidemiology." *Science, Technology, and Human Values* 30(1) (2005): 17–51.

Bijsterveld, Karin. "'What Do I Do with My Tape Recorder . . .?': Sound Hunting and the Sounds of Everyday Dutch Life in the 1950s and 1960s." *Historical Journal of Film, Radio, and Television* 24(4) (2004): 613–34.

Bonebright, Terry L., and Nadine Miner. "Evaluation of Auditory Displays: Comments on Bonebright et al., ICAD 1998." *ACM Transactions on Applied Perception* 2(4) (2005): 517–20.

Borck, Cornelius. "Between Local Cultures and National Styles: Units of Analysis in the History of Electroencephalography." *C. R. Biologies* 329 (2006): 450–59.

Brain, Robert M. "Representation on the Line: Graphic Recording Instruments and Scientific Modernism." In *From Energy to Information: Representation in Science and Technology, Art, and Literature*, ed. B. H. Clarke and Linda Dalrymple, 155–77. Stanford: Stanford University Press, 2002.

Brazil, Eoin, and Mikael Fernström. "Subjective Experience Methods for Early Conceptual Design of Auditory Displays." In *Proceedings of the 15th International Conference on Auditory Display*, ed. Mitsuko Aramaki, Richard Kronland-Martinet, Sølvi Ystad, and Kristoffer Jensen, 11–18. Copenhagen: Re:New – Digital Arts Forum, 2009.

Burri, Regula Valérie. "Doing Distinctions: Boundary Work and Symbolic Capital in Radiology." *Social Studies of Science* 38(1) (2008): 35–62.

Daston, Lorraine, and Peter Galison. *Objectivity*. New York: Zone, 2007.

Derksen, Maarten. "Are We Not Experimenting Then? The Rhetorical Demarcation of Psychology and Common Sense." *Theory and Psychology* 7(4) (1997): 435–56.

Frauenberger, Christopher. Auditory Display Design: An Investigation of a Design Pattern Approach. PhD diss., Interaction, Media, and Communication, Queen Mary, University of London, 2009.

Galison, Peter. *Image and Logic: A Material Culture of Microphysics*. Chicago: University of Chicago Press, 1997.

Garfinkel, Harold. *Studies in Ethnomethodology*. Englewood Cliffs, N.J.: Prentice Hall, 1967.

Gergen, Kenneth J. "The Mechanical Self and the Rhetoric of Objectivity." In *Rethinking Objectivity*, ed. A. Megill, 265–87. Durham, N.C.: Duke University Press, 1994.

Gieryn, Thomas F. "Boundaries of Science." In *Handbook of Science and Technology Studies*, ed. S. Jasanoff, G. E. Markle, J. C. Petersen, and T. J. Pinch, 393–443. Thousand Oaks, Calif.: Sage, 1995.

Guice, Jon. "Designing the Future: The Culture of New Trends in Science and Technology." *Research Policy* 28(1) (1999): 81–98.

Halffman, Willem. Boundaries of Regulatory Science: Eco/toxicology and Aquatic Hazards of Chemicals in the US, England, and the Netherlands. PhD diss., Science Dynamics, University of Amsterdam, 2003.

Hedgecoe, Adam M., and Paul Martin. "The Drugs Don't Work: Expectations and the Shaping of Pharmacogenomics." *Social Studies of Science* 33(3) (2003): 327–64.

Hermann, Thomas. "Taxonomy and Definitions for Sonification and Auditory Display." In *Proceedings of the 14th International Conference on Auditory Display*, ed. Patrick Susini and Olivier Warusfel, 1–8. Paris: IRCAM (Institut de Recherche et Coordination Acoustique/Musique), 2008.

Joyce, Kelly A. *Magnetic Appeal: MRI and the Myth of Transparency*. Ithaca, N.Y.: Cornell University Press, 2008.

Juolo, Patrick. "Killer Applications in Digital Humanities." *Literary and Linguistic Computing* 23(1) (2008): 73–83.

Kahn, Douglas. "Concerning the Line: Music, Noise, and Phonography." In *From Energy to Information: Representation in Science and Technology, Art, and Literature*, ed. B. H. Clarke and Linda Dalrymple, 178–94. Stanford: Stanford University Press, 2002.

Klein, Julie Thompson. *Crossing Boundaries: Knowledge, Disciplinarities, and Interdisciplinarities*. Charlottesville: University Press of Virginia, 1996.

Kramer, Gregory, ed. *Auditory Display: Sonification, Audification, and Auditory Interfaces*. Reading, Mass.: Addison-Wesley, 1994.

———, Bruce Walker, Terry Bonebright, Perry Cook, John Flowers, Nadine Miner, and John Neuhoff. "Sonification Report: Status of the Field and Research Agenda." International Community for Auditory Display, 1997.

Kursell, Julia. *Sounds of Science: Schall im Labor (1800–1930)*. Berlin: Max Planck Institute for the History of Science, 2008.

Lachmund, Jens. "Making Sense of Sound: Auscultation and Lung Sound Codification in Nineteenth-Century French and German Medicine." *Science, Technology, and Human Values* 24(4) (1999): 419–50.

Mody, Cyrus C. M. "The Sounds of Science: Listening to Laboratory Practice." *Science, Technology, and Human Values* 30(2) (2005): 175–98.

Myers, Natasha. "Molecular Embodiments and the Body-Work of Modeling in Protein Crystallography." *Social Studies of Science* 38(2) (2008): 163–99.

Pantalony, David. "Rudolph Koenig's Workshop of Sound: Instruments, Theories, and the Debate over Combination Tones." *Annals of Science* 62(1) (2005): 57–82.

Pasveer, Bernike. "Representing or Mediating: A History and Philosophy of X-Ray Images in Medicine." In *Visual Cultures of Science: Rethinking Representational Practices in*

Knowledge Building and Science Communication, ed. L. Pauwels, 41–62. Hanover, N.H.: Dartmouth College Press, University Press of New England, 2006.
Porter, Theodore M. *Trust in Numbers: The Pursuit of Objectivity in Science and Public Life.* Princeton, N.J.: Princeton University Press, 1995.
Scaletti, Carla. "Sound Synthesis Algorithms for Auditory Data Representations." In *Auditory Display: Sonification, Audification, and Auditory Interfaces*, ed. G. Kramer, 223–251. Reading, Mass.: Addison-Wesley, 1994.
Schmidgen, Henning. "Silence in the Laboratory: The History of Soundproof Rooms." In *Sounds of Science: Schall im Labor (1800–1930)*, ed. J. Kursell, 47–61. Berlin: Max Planck Institute for the History of Science, 2008.
———. "Time and Noise: The Stable Surroundings of Reaction Experiments, 1860–1890." *Studies in History and Philosophy of Biological and Biomedical Sciences* 34(2) (2003): 237–75.
Schoon, Andi, and Florian Dombois. "Sonification in Music." In *Proceedings of the 15th International Conference on Auditory Display*, ed. Mitsuko Aramaki, Richard Kronland-Martinet, Sølvi Ystad, and Kristoffer Jensen, 76–78. Copenhagen: Re:New – Digital Arts Forum, 2009.
Sterne, Jonathan. "Medicine's Acoustic Culture: Mediate Auscultation, the Stethoscope, and the 'Autopsy of the Living.' " In *The Auditory Culture Reader*, ed. M. Bull and L. Back, 191–222. Oxford: Berg, 2003.
Thompson, Emily. *The Soundscape of Modernity: Architectural Acoustics and the Culture of Listening in America, 1900–1933*. Cambridge, Mass.: MIT Press, 2002.
Van Lente, Harro. Promising Technology: The Dynamics of Expectations in Technological Developments. PhD diss., University of Twente, Enschede, 1993.
Wilde, Rein de. *Discipline en Legende: De Identiteit van de Sociologie in Duitsland en de Verenigde Staten 1870–1930*. Amsterdam: Van Gennep, 1992.
Wise, M. Norton. "Making Visible." *Isis* 97 (2006): 75–82.

Appendix: Overview of Interviews

Name	*Function*	*Date and Place*	*Language/ Remarks*
Conny Aerts	professor of asteroseismology at KU Leuven and RU Nijmegen	Leuven, March 6, 2009	English
Gerold Baier	lecturer in systems biology, University of Manchester	Bielefeld, February 21, 2008 Vienna, November 2, 2008	German/first interview unrecorded
Stephen Barrass	associate professor of digital design and interactive media, University of Canberra	Paris, March 26, 2009	English/ unrecorded
Eoin Brazil	PhD student at the Interaction Design Centre, University of Limerick	Limerick, February 29, 2008	English/Skype interview
Alberto de Campo	professor at Institute for Time-Based Media, Berlin University of the Arts	Berlin, October 16, 2009	German
Christian Dayé	PhD student in Department of Sociology, Karl-Franzens-University Graz	Graz, March 17, 2008	German
Florian Dombois	head of Institute for Transdisciplinarity (Y), Berne University of the Arts	Cologne, February 17, 2008	German

Appendix Continued

Name	*Function*	*Date and Place*	*Language/ Remarks*
Lars Graugaard	composer, flautist; responsible for artistic program of ICAD 2008 conference	Copenhagen, May 22, 2009	English
Florian Grond	PhD student, Bielefeld University (Ambient Intelligence Group); media artist	Paris, June 26, 2008 Bielefeld, October 13, 2009	German
Matti Gröhn	ICAD board member; visualization specialist, Finnish IT Center for Science	Espoo, July 1, 2009	English/phone interview
John Heise	astrophysicist, Netherlands Institute for Space Research (SRON), Utrecht	Utrecht, March 10, 2009	English
Thomas Hermann	ICAD board member; head of the Ambient Intelligence Group, Cognitive Interaction Technology Excellence Cluster (CITEC), Bielefeld University	Bielefeld, February 22, 2008 Bielefeld, October 14, 2009	German
Kristoffer Jensen	associate professor in medialogy, Aalborg University Esbjerg; conference chair of ICAD 2008	Copenhagen, May 22, 2009	English
Gregory Kramer	founder of ICAD; Metta Foundation, Portland, Oregon	German railway system, July 1, 2008	English
Donald Kurtz	professor of astrophysics, University of Central Lancashire	Preston, November 6, 2009	English/phone interview
Tony Stockman	ICAD board member; senior lecturer in computer science, Queen Mary, University of London	London, November 7, 2008	English
Katharina Vogt	PhD student, Institute of Electronic Music and physics department in Graz	Graz, August 11, 2009	German
Bruce Walker	associate professor at the School of Psychology and School of Interactive Computing; director of the Georgia Tech Sonification Lab; ICAD president	Atlanta, June 4, 2009	English/phone interview
Olivier Warusfel	head of room acoustics team of IRCAM, Paris	Paris, March 26, 2009	English

SECTION IV

SPEAKING FOR THE BODY: THE CLINIC

CHAPTER 11

INNER AND OUTER SANCTA: EARPLUGS AND HOSPITALS

HILLEL SCHWARTZ

Introduction

"Soon after the outbreak of war," wrote a ghostwriter in 1916,

> a strange apparition, wretched and frightening, appeared on the avenues of our cities. . . . His body shook as if overcome by a great chill, or he would stand stock-still in the middle of a quiet street, in thrall of experiences at the Front. We see others too, men who could move ahead only with the jerkiest of steps; poor, pale, and gaunt, they leapt as though a merciless hand gripped them by the neck.

The ghostwriter was Franz Kafka, proposing in the name of his chief at the Prague Workmen's Accident Insurance Institute the creation of "A Public Psychiatric Hospital for German-Bohemia" (Kafka 1916/2009; translation amended). Shellshock, said physicians, could be effectively treated only in calm residential settings, beyond the means of most (Lerner 2003; Schaffellner 2005). The Provincial Central Administration for the Welfare of Returning Veterans ought therefore to prepare vacant structures at Geltschbad, healthfully distant from the city, to receive the apparitions and give them the time, space, and quiet in which to return to themselves.

Kafka did not specify the quiet. It was as implicit in Geltschbad's clean air and green woods as it was explicit in millennia of Hindu, Egyptian, Greco-Roman,

Jewish, and Islamic prescriptions for the best milieu in which to heal the feverish and confused. Euro-American physicians, confronting a late nineteenth-century epidemic of nervous collapse, had taken this a step further. Accustomed to shushing nurses and patients during intent auscultations, reliant on surgical anaesthetics and postsurgical narcotics to mute most screaming, they had imposed a regime of strict silence on "neurasthenics" undergoing rest cures at home or in sanatoria. Such a regime harked back to the therapeutic quietudes of Quaker-run asylums and to the punitive-penitent quietudes imposed on prisoners under the Silent System (Berlin 1912; Kirkbride 1880/1973; Rosenberg 1987).

For patients neither nervous nor criminal, the sonicities had been fixed by generations of Gallican Sisters of Charity and more recently by German Lutheran deaconesses. Under the impress of vows of obedience that made of chit-chat a seductive jeopardy and clangor a mortal sin, nuns and lay sisters in medieval hospices and early modern hospitals had tended silently to worn-out pilgrims, the local sick, the decaying and dying. Attendants at almshouses, leprosaria, and workhouse infirmaries were held to a similar sacerdotal quiet, as would be the nurses trained by Florence Nightingale (1820–1910), a devout Unitarian who by the age of forty had become a reclusive invalid sleeping beneath prints of Raphael's *Sistine Madonna* and Murillo's *Virgin*. Her nursing school at St. Thomas's Hospital she regarded as akin to a convent, and each week she scoured the progress reports on her "probationers," who were assessed, among other virtues, as to their Quietness, for "A nurse who rustles . . . is the horror of a patient" (Bowers 2007; Cook 1913, vol. 1, 470–74, 486; McDonald 2004, 490–588, 750–56; Nightingale 1859, 27; 1990, 238, 385).

Rustling was no mere matter of inexpert bustle or crinoline fashion. Touring Rome's hospitals in 1848, Nightingale had taken a side trip to Santa Maria degli Angeli, a church whose design, by Michelangelo, impressed her as barren yet harmonious enough to merit an acoustic metaphor: "That noiseless (if you may use the word) growth of one part out of the other, which reminds one of the growth of the kingdom of heaven from a grain of sand, becoming a great tree, that want of bustle and glaring effect and impudently forcing itself upon one's notice . . . which is so like the works of God Himself" (McDonald 2004, 217, q221, 439, 455). Projected into the military hospitals of the Crimea (1853–1855), then into civilian wards of the Commonwealth and North America (Kalisch and Kalisch 2004), thence to missionary clinics in Africa, Oceania, and East Asia (Hisama 1966), Nightingale's sense of noiselessness was ecological, aristocratic, and metaphysical: a permeating "atmosphere" (Böhme 2000; Connor 2006) maintained on behalf of patients by nurses who should have within themselves the confidence to be still when nothing need be done. This confidence, which deaconesses at Kaiserswerth deepened through their *Stille halbe Stunde*, a daily half-hour of meditation, came to Nightingale through her wealth, her elite connections, and a "divine call" to work with "the sick poor," for whom a nurse's presence must be as commanding as it is reassuring (Bancroft 1890; Cook 1913, vol. 1, 485; Gill 2004).

Shortages of supplies, including ether, had plagued her twenty months in the Crimea, so the noise of men in pain in four miles of beds sixteen inches apart should have been grim, but Nightingale found that "As I went my night-round among the newly wounded that first night, there was not one murmur, not one groan[;] the strictest discipline, the most absolute silence & quiet prevailed." She enforced a similar sound-discipline among the British nurses and male orderlies under her command. ("I don't like the word discipline, because it makes people always think of drill and flogging," she would write, "but, if they would but associate it with the word disciple - - - - -!") When at last she returned home after bouts with fever, sciatica, and earache, she was on the verge of collapse; a breakdown in 1857 left her with chronic rheumatic pain and extreme sensitivity to sound (Nightingale 1990, 63–157; Goldie 1987; McDonald 2002, 468; Gill 2004, 437–45, probable brucellosis).

From her Rheocline Spring Bed, she wrote fourteen thousand letters and published *Notes on Nursing* (1859), *Notes on Nursing for the Labouring Classes* (1861), and *Notes on Hospitals* (1863), with sections on noise that drew as much from her own situation ("entirely a prisoner to my bed") as from her time in the Crimea. "Unnecessary noise, or noise that creates an expectation in the mind, is that which hurts a patient. It is rarely the loudness of the noise, the effect upon the organ of the ear itself, which appears to affect the sick," so a good nurse will act with purpose and "always make sure that no door or window in her patient's room shall rattle or creak; that no blind or curtain shall, by any change of wind through the open window, be made to flap." Those with brain injuries might be bruised by "mere noise," but all patients suffered intensely from "intermittent noise, or sudden and sharp noise." Worst were sounds that drew attention to themselves without arriving at closure: whispered conversations, the affectation of walking on tiptoe, "the fidget of silk and of crinoline, the rattling of keys, the creaking of stays and of shoes." Small unresolved noises. "Irresolution," which Nightingale detested personally and politically, "is what all patients most dread," for with it came strain, fuss, fatigue. "Unnecessary noise, then, is the most cruel absence of care, which can be inflicted either on sick or well" (McDonald 2002, 371; Monteiro 1974, 34; Nightingale 1859, 25–33; 1861, 35–37) (see figure 11.1).

"One thing more:—From the flimsy manner in which most modern houses are built, where every step on the stairs, and along the floors, is felt all over the house; the higher the story, the greater the vibration." Therefore, never install a patient where she will hear medicine carts rattling overhead or where he may feel "every step above him to cross his heart." This was important, for rich and poor alike avoided hospitals, whose high mortality rates would decline only with the recalcitrant adoption of systematic asepsis, whose antibacterial principles Nightingale herself disputed. She was all for cleanliness and quiet but had no truck with things malignly microscopic that might seduce a nurse into neglecting the grosser proprieties (Nightingale 1859, 32–33; 1861, 29, 42–43).

Victorian architects had meanwhile begun designing attic or annex sickrooms that afforded the better-off a quiet convalescence (Adams 1996; Bailin 1994). For Julia Stephen, a woman of the struggling upper-middle class who composed "Notes from the Sick Room" (1883) after years spent tending to a dying uncle, sister, mother,

PLEASE
don't wheel trolleys too fast:
don't let them bump into doors
or beds — or anything else:
don't let their wheels squeak:
don't let their contents rattle together

CLATTER
DOES MATTER

Figure 11.1 Cyril Kenneth Bird (1887–1965) created this and the posters in figures 11.2 and 11.3 for the King Edward's Hospital Fund for London around 1958. His first drawing for *Punch*, the English satiric magazine he later edited, was titled "War's Brutalising Influence" (1916) and appeared under the pseudonymn he used thereafter—"Fougasse," the French name for a dangerously unreliable mine. Reproduced by the kind permission of the family and estate of C. K. Bird.

and father, sleep was the nub of healing, quiet the nub of sleep: "[T]he room should be so gently hushed that the patient should feel able to drop off to sleep at any moment" (Stephen 1883/1987, 230). Wealthier families might benefit from ordinances that allowed their servants to strew straw on the street so as to damp carriage noise outside households tending to the ill, but the poor sick had to abide the thin walls of tenements, stertorous neighbors, street-lively neighborhoods. Poor, middling, female, black or brown, few had the luxury, at home or in hospital, of "a room of one's own," as later advocated by Stephen's rest-cured daughter, Virginia, who, like Kafka, appreciated how a veteran of the Great War might be wracked by the blast of a doorman's whistle or backfire from a passing lorry (Kingsdale 1989; Lee 1997; Woolf 1975).

Nervous, headachy, tubercular (he had a "cardiac neurosis"), Kafka was no stranger to hospitals and sanatoria, whether in his capacity as an insurance lawyer or in his own person, and wherever his digs, he found himself as tormented by noise as any apparition (Gilman 1995, 93–95; Kafka 1974). "The amount of quiet I need does not exist in the world," he wrote his sister Ottla, "from which it follows that no one ought to need so much quiet" (Karl, 1991, 85, 394–96). In his parable of 1917, "Silence of the Sirens," Odysseus at first eluded the Sirens by orneriness and Ohropax earplugs, bound to the mast by chains and sailing staunchly on. Homer had granted his hero the heroism of listening to the Sirens, whence the chains; during the Great War, heroic listening was no longer credible, whence the ear-peace of the Ohropax (Ohropax GmbH 2010). Then Kafka bethought himself of klaxons, modern sirens that warned of air raids or poison gas. It occurred to him that Sirens might be at their most irresistible when silent, seducing men from rough seas or deafening barrage by the promise of stillness (Kafka 1917/1971, 430–32; 1917–1919/1991, 19–20). After the war, after years of apparitions rising out of trenches, Kafka wrote *Der Bau*, about a burrow and its burrower: "The most beautiful thing about my burrow is the stillness. Of course, that is deceptive. At any moment it may be shattered and then all will be over." Listening for sounds of cave-in or incursion, the burrower hears an indefinable whistling, "audible everywhere, night and day," ominous (Kafka 1923/1971, 327, 353) (see figure 11.2).

Burrower and hero, burrow and mast: These are the figurations between which to hear out the history of the hospital and earplugs. Hospitals were, in this regard, sanctuaries from the noise of the world, their patients burrowed under blankets and trusting to devout nurses for protection against acoustic insult. Earplugs were sentries thrust into the meatus, inelegant guardians of an elegant labyrinth, an inner sanctum that was sound's lymphatic burrow and rebirth. Hospitals and earplugs shared a tactile origin—a healing touch; hands clapped to terrorized ears—but the more heroic the medical posture of physicians and the more technical the training of nurses, the more the sound history of hospitals would diverge from that of earplugs. When hospitals were primarily places for lyings-in, amputations, and recuperations, or (as lazarettos) for isolation and palliation, or (as asylums) for safe enclosure and rest cure, they were almost as passive in their acousti-cultural architecture as earplugs, though intended as much to contain screams from within as to

The nuisance value of sound often bears little relation to its intensity.
(Report on Noise Control in Hospitals, King Edward's Hospital Fund for London)

Figure 11.2 During World War I, Fougasse (C. K. Bird) had his back shattered by a shell at Gallipoli; thus, he had direct experience of hospital acoustics. Poster reproduced by the kind permission of the family and estate of C. K. Bird.

block noise from without (Leistikow 1967; Risse 1999; Stevenson 2000). When hospitals touted themselves as major research laboratories jealous and zealous of new procedures, they became industrial cities, with the attendant environmental problems. In short, hospitals became noisier as physicians became more invasive in their diagnostics and more aggressive in their treatments, while earplugs and muffs became more effectively defensive with the patenting of lighter-weight, sound-damping materials that fit more snugly. Only during the 1990s were earplugs put on the offensive against noise, by which time the sonic trajectories of hospitals and earplugs were again running in tandem.

Outer Sancta: The Hospital

The original tandem had been joined by anti-noise campaigners since the 1840s, for anti-noisites invariably enlisted infants and invalids in their briefs against dutiful daughters pounding out piano scales in upstairs apartments, maids beating carpets on balconies, organ-grinders at the curb. All who needed sleep to thrive or survive, all those with finely tuned minds or artistic sensoria—in brief, all those who, like the English philosopher Herbert Spencer, might be drawn to the defensive use of earplugs or muffs—deserved help from public authorities to eliminate offending sounds at their source (Duncan 1908, 79, 314). So anti-noisites gathered testimony from hospital patients, nursing mothers, harried businessmen, and ailing great-aunts who attributed colic, insomnia, irritability, and general debility to pestiferous sound (Bijsterveld 2008, 27–90; Schwartz 2011). Noisites, who professed to delight in the sonic robustness of civilization on the make, could scarcely attribute to infants and young mothers that "morbid sensitivity" to sound through which they—and the Bench—often dismissed noise complaints.

Anti-noisites also bemoaned the indefensibility (or "inadjustability") of the ear, which had no lid or lips and did not sleep (Hazlitt 1930; Schwartz 2003). Since plugging the ear with a finger, stuffing it with cotton wool, or enveloping it with muffs minimally buffered oncoming noises, and since sound was conducted through the bones of the skull as well as through the ear canals, there was only so much the sound-sensitive could do before leaving the field to the imperium of noisemakers.

Imperium: the command of space and time, which had been Nightingale's hardest-fought battle as she sought for her head nurses a paramountcy in the wards second only to physicians. Toward this end she promoted a pavilion style of hospital architecture that gave each head nurse her own demesne and the closest scrutiny of patients (Nightingale 1863). European hospitals had previously followed the model of army barracks, monastic hospices, and church naves, with beds arrayed in long columns, one or more aisles between, and with provisions for natural light, access to "necessaries," and ventilation adequate to feed the lungs while dispersing odors and diffusing those deadly miasmas that gave Nightingale the willies. Medieval and Renaissance architects, working for pious hospital patrons, had also been directed to situate a chapel so centrally that patients could see or at least hear the recital of the Mass, its ritual thaumaturgy no less vital than physick. Privacy, when and where accommodated, was ensured by curtains around bedsteads or, in cases of high privilege, by separate cells, but the new demand for privacy that arose among the middle classes in the mid-eighteenth century was not yet so strong that Nightingale thought it determinative when she laid out the wards of public hospitals in the mid-nineteenth (Abel-Smith 1964; Taylor 1991). The pavilion model, which she rather adapted than invented (Risse 1999), did presume an intimate supervision over fewer beds, but her emphasis on surveillance required an open floor plan, which was all the more reason to inculcate habits of quiet among nurses and orderlies (Billings 1895)—and to establish, as planned in 1863 for the "Nightingale" State Emigrant Hospital of Manhattan, a separate nurses' room and staff quarters, "thereby equalizing the temperature of the wards, and shutting off the noise of the halls, staircases and sanatoriums; affording likewise a moderate degree of conversation and amusement among the convalescents, without disturbing the weak, helpless, and suffering patients, who are unable to leave their beds" (Kennion 1868, pt. 3, 77). Where patients themselves were unruly, shrieking in pain, or unavoidably loud (with whooping cough, for example), she would have them assigned to their own open, mutually noisy wards. Private rooms she distrusted as warrens in which dirt could hide, miasma regroup, defiant patients conspire (McDonald 2002, 458, 468).

No more than three stories high (to ensure ventilation), pavilion hospitals expanded outward like army encampments. This could rarely be achieved in city centers, which was fine with Nightingale, who had grown up in a country manor and believed that hospitals belonged where air was sweeter and the quiet timeless. For fear of infecting or offending townspeople, medieval hospitals with their mortuaries had been built outside town walls; modern hospitals, divorced from morgues, were to be situated in the countryside for fear of cities disturbing their patients—a reversal neater in theory than in practice. Most new hospitals after 1850 arose where the need was greatest and patients nearest, in metropoli whose density constrained them upward (Abel-Smith 1964; Risse 1999). The new verticality was enabled by the advent of steel-frame construction, "safety" elevators, electric lighting, forced-air furnaces, and huge rooftop fans, whose ratcheting, flickering, thundering, screeching, and whoosh paradoxically added to the noise of cities, through whose avenues roared hospital ambulances with duplex sirens (Hornsby and Schmidt 1913).

Ambulances and their hospitals proliferated during the Belle Époque in response to an increasingly urban world beset by the derangements of density: road and rail accidents; skin, ear, and lung diseases from bad air; infections spread through the overwhelmed plumbing of tenements. As hospitals multiplied, they were parsed into medical specialties: contagious diseases, geriatrics, maternity, pediatrics, psychiatry, syphilis, tuberculosis. Anti-noisites exploited them all as wedges toward the passage of zoning laws that would keep industrial and commercial activity remote from the residences of good people. In 1905, New York did boast a section of city code that forbade peddlers from bawling their wares within 250 feet of a church during services, a court in session, and a hospital at all times, but this went largely unenforced, like other noise laws, for want of patrolmen and standard audiometrics. Lawyers and architects had had some luck forcing the relocation of hospital operating rooms from streetside rooms to walled-off interiors where screams could not be heard; hospital administrators had had no such luck in damping the clamor of nearby traffic (Gutton 2000, 88–90; *Kestner v. Homeopathic Medical and Surgical Hospital*, 245 Penn 326 [1914]). The eventual designation of hospital "quiet zones," with bold signage and stiff fines, would lay the groundwork for a policy of civic sonic set-aparts in other public contexts.

"Zones" had been on the lips of astronomers for a century, of geologists for eighty years, of railmen and diplomats for twenty, of timekeepers and trolley conductors for ten. While towns since Jericho had relegated smelly and noisy occupations to precincts just within or beyond their walls, and while Frankfurt and Berlin in the 1890s had passed ordinances specifying the functions of buildings in each ring of development around the city hub, American urban planners had still to defend the legality and morality of explicit zoning. Everywhere, however, cities were *ipso facto* zoned by the nature of the environing sounds, according to which districts were paved and with what—asphalt, brick, wood blocks, or cobblestone. Hospital quiet zones were in this sense a sound-*re*marking (Bijsterveld 2008, 243–53; Peterson 2003, 308–17; Rodgers 1998, chs. 4–5; Sandweiss 1997).

Julia Barnett Rice was credited across the Atlantic ecumene for making the *re*marking culturally persuasive. A graduate of the Women's Medical College of the New York Infirmary, she never practiced medicine. Instead, upon marrying the pianist, lawyer, and venture capitalist Isaac Rice, her career was more broadly that of "helping other people" and publishing *The Forum: A Magazine of Politics, Finance, Drama, and Literature*, a leading Progressive journal. Her run-in with noise, aside from raising six rambunctious children, came after she and Isaac in 1902 built a mansion overlooking the Hudson. If Manhattan's street traffic was loud enough to bump wealthy families off central arteries toward the banks of the Hudson or the East River, river traffic had its own clamjamfry with "the shrieks of passing steamers, the discordant notes of harbor craft, the puffing and wheezing of tugs, the din of escaping steam, clanging bells, howling men." (Van Dyke 1909, 32). Back from a tour of Europe in August 1905, Julia and her sleep-deprived children were assailed by ever-more strident whistles, steam sirens, and compressed-air horns from a flotilla of tugs. Not one to seethe in silence, she resolved to eliminate the noise, though cognizant that a campaign for quiet issuing from the mistress of Villa Julia would be subject to innuendos from the Left of ethnic prejudice or bourgeois maternalism and from the Right of hysteria or economic naïveté, objections voiced during prior anti-noise campaigns in London, Boston, and Chicago. She therefore deflected the debate from her person (Southern, Jewish) and position: not the wealthy matron at her high window but the indigent and infirm (immigrant, Black, Irish, Italian, Polish) in hospitals along the Hudson, and hundreds of sick poor children (ditto), none of whom could sleep while tugs blatted at two A.M. for crews in saloons (Gatlin 1913; Heinl 1908; Lomax 1996; *New York Daily Tribune*, November 29, 1905; *New York Times*, December 10, 1905; *New York Times Magazine*, January 14, 1906; Rice 1906 [both]; Seidler 1989; Sloane 2008).

Julia visited the Gouverneur and Bellevue hospitals on the East River, the Children's Hospital on Randall's Island, and the Metropolitan Hospital on Blackwell's Island, all of which lay near tugboat moorings. She called on the Board of Health, Police Department, and dock commissioner, armed with testimonials from superintendents of hospitals "representing the sufferings of 13,000 poor people" and letters from neurologists about the danger of shrieking whistles—at Bellevue, patients mistook the whistling for their death knells (Morse 1906; Rice 1906 [both]). The city's chief medical officer, Hermann M. Biggs, who had expanded its program of school medical inspections to test for hearing loss, was "thoroughly in sympathy" with Mrs. Rice. Less reclusive than Nightingale and more adept with reporters, Julia made sure that Biggs's own specialty, medical pathology, would be in play: "The bacilli which haunt the air are not worse enemies than the sound waves that quiver in the air and beat in upon our brains," wrote the editors at the *New York Daily Tribune* (November 26, 1906) after listening to Julia. "There are benign bacilli, but there are no benign noises." All Biggs could do was to ask corporation counsel about his jurisdiction, if any, over noise arising on the river. Counsel dithered.

An impasse, was this, or a signal moment? It is common nowadays for noise to be characterized as "liminal," as sounds that teeter on the edge of meaning, fitful of place and fretful of time. Liminality was not what Julia Rice had in mind when she campaigned against noise, nor was it what decades of New Yorkers understood when protesting doormen's whistles, itinerant brass bands, or church bells. They knew what the noises meant; they knew well where/who/what the noises came from; they knew very well how much earlier than dawn or later than dusk the noises arose. Indeed, knowing the time, place, and meaning of the noise was crucial, for only then could they argue, as they did, that the sounds came at the wrong time, in the wrong place, in excess, or as atavisms. Not liminal: obtrusive and inexcusable. A campaign against frustratingly unassailable whistles or sirens had to be an assertion that no engineered sound should escape human command. If tug whistles resembled "the wails of lost souls in torment, or children in terror, or wounded horses, or distant Scottish bagpipes, or the Banshee," one ought not grant them a legal limbo. True, the intensity of aural sensations had then no universal metric. Still, attention had to be paid and arguments made if banshees were not to haunt the recuperative sleep of the thousands in hospitals ventilated by invigorating riverside air (Lomax 1996; *New York Evening Sun*, November 27, 1907).

When propagated on interstate waters, noise turned out to be a federal issue requiring a revision of the U.S. Code to give municipal officials the authority to squelch local whistling—a revision that Congress, at Julia's instigation, did pass and Theodore Roosevelt signed into law. Julia meanwhile had been contemplating her next step: the formation of a Society for the Suppression of Unnecessary Noise

(SSUN), which would go after all sorts of noises and appeal to the public to relieve "the intense suffering of our sick poor from the noise-evil." This would entail a push for the creation of sound-protected areas, or quiet zones, around hospitals (*New York Daily Tribune*, December 4, 1906; *New York Times*, June 26, 1907; Rice 1906 [both]).

June of 1907 saw the first official signs ("NOTICE—HOSPITAL STREET") posted around each New York hospital. From Manhattan such signage spread to Chicago, London, Paris, Hannover, Amsterdam, and Milan. Patrolmen on hospital beats were to insist that horses be walked, newsboys rerouted, motorists fined for honking. Realizing that there could never be enough officers to make the signs other than cautionary, Julia got twenty thousand schoolchildren to compose pledges: no pounding of hospital fences with sticks, no roller-skating on hospital sidewalks. "I *promise not* to play near or around any hospital," wrote a child eager for membership in the Children's Hospital Branch of the SSUN and a smile from Mark Twain, figurehead of the campaign. "When I do pass I will *keep my mouth shut tight*, because there are many invalids there" (*New York Times*, June 26, 1907; February 27 and April 6, 1908; Rice 1908 [both]; Society for the Suppression of Unnecessary Noise 1908, 3.)

The Noise from Within

Alas, quiet zones around hospitals could not suppress the new noises arising from within. Self-noise, an issue of moment in twentieth-century physics and experimental psychology, became an equally serious issue for hospitals as they modernized. The greater the efforts of administrators to implement antisepsis with sterilizable metal or Pyrex apparatus, easily mopped floors, and impermeable counters, the louder and more reverberant became their hospitals. The more strenuous their effort to centralize systems of accounting, distribution, and communication, the noisier became their hospitals with devices of acoustic repercussion: the pounding of keypunch machines; the clattering of metal carts; the clunk of dumbwaiters; microphones and wall speakers fraught with feedback and interference (Govan, Anderson, and Reilly 1934). And the taller the hospital buildings, climbing upward to serve populations encouraged by modern medical "miracles" to enter diagnostic wards and maternity wings, the lower the ceilings and more blatant the HVAC. From the outside, hospitals could seem massively quiet; within, they were louder and shriller—or would seem so to the acutely ill and injured, who were becoming the primary clientele (Adams 2008; Hornsby and Schmidt 1913, 109) (see figure 11.3).

Shunted to nursing homes or back to families with cupboards of off-the-shelf analgesics, the chronically ill and/or debilitated would no longer occupy the majority of beds; instead, there was a more diverse, demanding hospital population, anxious for action and addicted to telephones and radios. Contrary to Nightingale's

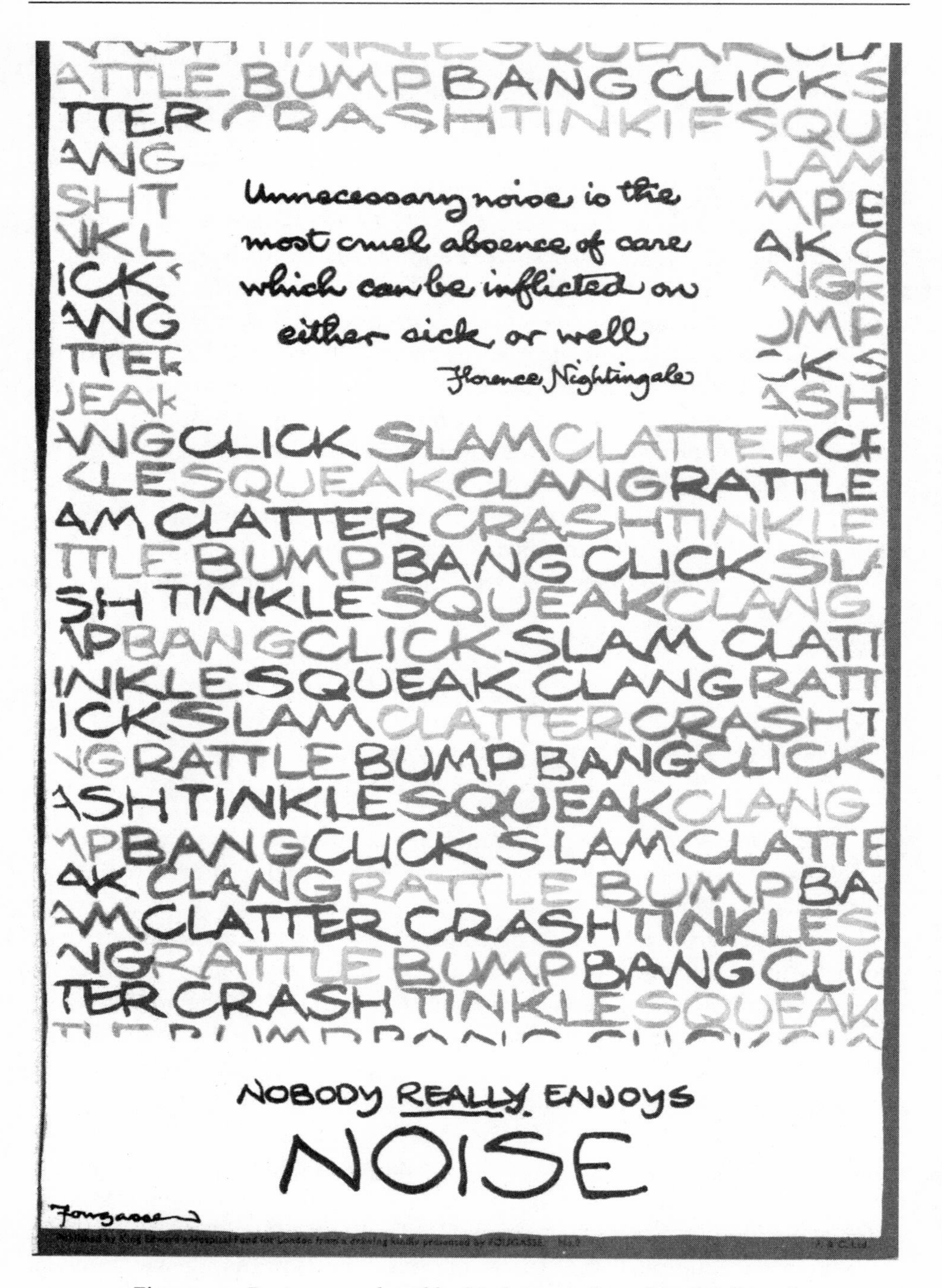

Figure 11.3 Poster reproduced by kind permission of the family and estate of C. K. Bird.

directives but at the insistence of well-heeled patients and donors, prosperous hospitals incorporated more semiprivate and private rooms. The better-off would no longer have to endure the rude hawking, coughing, snuffling, snoring, and shrieking common to open wards or tolerate noises they associated with lower classes, immigrants, and other races. Since the newly enclosed rooms were out of nurses' audiovisual range, hospitals had then to install bedside call buttons, monitors, and alarms, compounding the cacophony at and around nurses' stations (Coser 1963, 248; Goldwater 1932; Gorrell 1936; Griffin 1950; Underwriters Laboratories 1979).

Toileting, too, became more private, and no one had a cost-effective solution for the abominable acoustics of flushing, at the bowl or throughout the innards of a building, even where (as in England) drainpipes were bolted to outside brick. To isolate throbbing conduits from contact with other internal ducts was forbiddingly expensive, and industrial-strength hospital toilets were far louder than those at home. For bedridden patients, the clangor of new stainless-steel bedpans—slick, oblong, hard to stack, and often dropped—was more discombobulating than the shatter of earlier crockery.

It was a toss-up, depending on country, culture, and economic circumstances, as to whether rough wooden floors would be replaced by linoleum or terrazzo; creaking wooden bedsteads by adjustable metal; slamming shutters by flapping blinds; gas jets by incandescent or fluorescent lights; wooden bowls by Bakelite or tinplate; dust mops by vacuum cleaners. If they chose linoleum and Bakelite, hospitals could claim small victories with regard to noise control; terrazzo and tinplate meant pandemonium (Grumet 1993) (see figure 11.4).

At their most technically sophisticated, hospitals were intrinsically noisier: With each new device for diagnosis, monitoring, or treatment, the peskier would be the background hum and the louder or beastlier the foreground of (sometimes literally) bells and whistles. If physicians frowned upon faradic current generators and other Victorian electrotherapeutica, modern hospitals would proudly resound with X-ray machines, electric floor polishers, intercom paging, air conditioners, and mobile trays of metal and glass (Agnew 1974; Howell 1995; Kingsdale 1989; Rosenberg 1987; Stevens 1918; Kersbergen 1931 for images). In sum, the machines conduced to exactly that bustle deplored by Florence Nightingale, Julia Barnett Rice, Julia Stephen, and Virginia Woolf.

Should one put a gendered finger on the anti-noisite pulse as felt in the hospital, one might suppose that educated women were extending an age-old guardianship of household peace into institutional sick rooms. Or that, as domestic economists and "municipal housekeepers," modern women were seeking to effect a quietness that was elsewhere a sigil of efficiency. Or that the "civilizing mission" of women called for a quiet as restorative as that reconstructive surgery refined during the Great War, the ears tucked back and in. But insofar as anti-noisite impulses were gendered female—and many a male sanitary engineer, architect, and physician of the twenties and thirties was equally unhappy with hospital noise (Censorinus 1938; Dixon 1927; Horder 1938; League of Nations 1937; Neergaard 1929, 1932; Stevenson 1936; Tanon 1938; Thompson 2002; Tucker 1929)—it may be more valuable to consider quietude as invoking two experiences rare in most women's daily lives: uninterrupted time and insulated spaces. For women, more than men, the qualia of quiet resonated with the promise of self-healing and a room of one's own, which might be possible only in the hospital (Hofrichter 2006; Ibels 1916).

After World War II, campaigns for quiet within hospitals accelerated even as truck and motorcycle traffic exploded through quiet-zoned avenues. With ears sensitized by the battlefield experiences of physicians, the combat stress of veterans, and the psychiatric training of war nurses came not only more powerful (electric/electronic) hearing aids but also subtler measurement of the effects of acoustic aggravation on blood pressure, blood chemistry, adrenal excretion, and coronary and cortical activity (Ades 1953).

Armed with antibiotics that allowed for more intricate operations and better management of infection, surgeons and internists repositioned hospitals as spaces of urgency and potency whose quiet dramas attracted donors for new wings that housed noisier if otherwise more advanced technologies (Association des Hautes Études Hospitalières 1968; Falk and Woods 1973; Fife and Rappaport 1976; Gendre 1976; Golub 1969; Haslam 1970; Hinks 1974; King Edward's Hospital Fund 1958; U.S. Public Health Service 1963).

Of particular urgency and potency was obstetrics, which since the 1890s had been improving its skills at maintaining infants born prematurely. Where each Intensive Care Unit became a hospital within a hospital—Nightingale's legacy of I-See-You surveillance—each newborn intensive care unit (NICU) was that much more metonymic: The more transparently a preterm infant could be kept alive and thriving, the more appealing would be the hospital itself as an artificial womb (Russell 1977). Once ultrasound monitors became common and prospective parents could see that fetuses at six months were acoustically alive to the aerien world, investigators began to attend to the sense-surround of infant incubators and discovered that these were among the loudest of hospital environments, louder even than the bellows of iron lungs for those stricken with polio (Baker 1996; Vaizey 1986). The premature fetus and the neonate, like other patients, must suffer the indignities of a raucousness produced by the very devices that kept them safe—fiercely ventilating pumps, beeping monitors, the flap-slap of hinged glass (Falk 1974; Frank 1991; Kahn et al. 1998; Michaëlsson, Riesenfeld, and Sagrén 1992; Seleny and Streczyn 1969).

Inner Sancta: The Earplug

We are back to the sonic insecurities of the burrow, which were, inside-out, the sonic insecurities of earplugs. Like ear trumpets and other obstreperous hearing aids, earplugs invoked contradictory stigmata: Either the users were prematurely old and cranky or parading their hyperacuity in hopes of special dispensation; either

Figure 11.4 The colonel in this 1942 advertisement likely had a private room and high expectations of hospital quiet, which could be met, suggested the Celotex Corporation, through the installation of its sound-absorptive ceiling tiles. Reproduced by permission of the CertainTeed Corporation.

the users were misanthropic, holding people at a distance, or malingerers, conning them closer. Apart from such stigmata, there were practical dissuasions against the use of earplugs even in the noisiest of contexts, on battlefields, in boiler factories, at the power looms of textile mills (e.g., Holt 1882; Rodger 1915). Soldiers, sailors, engine mechanics, and artisans claimed for safety's sake, as well as for quality control, that, like musicians, they needed to hear the subtlest of sounds (Bijsterveld 2008; Boys 1915; Guild 1951). Further, it was a point of pride to men who labored among drop forges, printing presses, or metal lathes that their ears grew "stronger" over the years, and at home they seemed none the worse for wear. Besides, plugs themselves swiftly became the worse for wear: Gummy with wax or body oils, they attracted dust, dirt, and half-dead bugs that could infect the ear canal, producing the smelly effusions of otitis media, which, untreated, could cause hearing loss and, backing up into the brain (said some), meningitis or brain rot. Then there was the discomfort of plugs too tight, the uselessness of plugs too loose, their ineffective shielding from low-frequency, bone-shaking rumbles. Finally, although plugs had long been recommended to prevent cold salt water from getting into the ear during ocean swimming, there was little evidence that the use of earplugs in the long run prevented hearing loss since it appeared that, regardless of profession or circumstance, men and women (city men and women, at least) lost acuity as they aged (Bauer and Edgley 1939; Behar, Chasin, and Cheesman 2000, 26–30; Knudsen 1939; Knudsen, Gales, and Watson 1941; Warfield 1948).

Related to this finality was another, more covert reluctance: partial deafness. A majority of modern populations—industrial laborers, farmers driving tractors and combines, housewives in homes aswirl with flakes of lead paint, children who survived scarlet fever or suffered from rickets—already had a degree of deafness that would seem exaggerated by earplugs. That is, persons ignorant of (or undisturbed by) their hearing problems might be more deafened—or troubled—by the most innocuous of plugs or made more aware of the distressing tinnitus that occurs in half of all cases of noise-induced hearing loss (Schwartz 2011).

Inserting plugs was eargo a risky business, sociologically, otologically, psychologically. Where the loudness or shrillness of recurrent sounds was demonstrably deafening and disabling, with episodes of disorientation, loss of balance, and/or ringing in the ears—as with naval gunners, tank captains, artillerymen—then the use of "ear defenders" was less fraught, and orders were issued. During the Great War, when ear defenders were thought capable, perhaps, of reducing the incidence of shellshock or at least of tempering its sound-cued aftermaths of jerking and nightmare, plugs were sometimes requisite and frequently requisitioned, but most soldiers in the trenches did not cotton to the idea of putting mud-sopped plugs into ears rocked by hours of shelling that reverberated through their new metal helmets. Aside from gunners, those who did use plugs most regularly were pilots, who were expected nonetheless to return from the war with permanent hearing loss (Grant 1917; Horne 1914; Jaehne 1911; Trible and Watkins 1919).

How much hearing was lost due to the inconsistent or imperfect use of ear defenders during the Great War we will never know, as hearing tests at induction were perfunctory and at discharge were done only if pension claims were filed. Statistics in any case rarely convinced people to use earplugs even after electrical audiometers came into general use during the 1930s. By then, other technocultural habits were making people comfortable with having something about their ears. First, headsets, as worn by switchboard operators, by navy men listening for submarines, by secretaries at dictaphone desks, by home audiences sharing lines to early radios (Sterne 2003), and by schoolchildren listening for audible tones produced by audiometers. Next, telephone receivers themselves, to which one cuddled one's ear not only to hear more clearly but to block noises from outside. Next, miniaturized hearing aids made to fit almost invisibly within the meatus. Finally, commercial airplane travel in cabins so noisy that earplugs were *de rigueur* (Arnhym 1944).

During the Second World War, the most mechanized and industrial of wars, more men and women had good experiences with headsets and acquired also a healthy regard for ear defenders (Littler 1940; Knudsen 1939). Afterward, they had the benefit of a new war material, synthetic foam, that improved the fit and comfort of plugs and of headphones for portable radios. The strongest incentive toward the use of earplugs, however, came from having to tend to thousands of enlisted men who had returned from the war hard of hearing and had to be "reconditioned" or recompensed. Not that any statistician could prove that the use of earplugs would have prevented most cases of partial deafness but that otologists now had firmer evidence of noise-induced hearing loss due to long exposure to industrial-strength sounds and short but repeated exposure to high-frequency or percussive sounds (Luz, Decatur, and Thompson 1973; Prasher 1998). Although industrial trade associations and some researchers were still vouching for the marvelously recuperative powers of ears (Snow and Neff 1943), the military and the higher courts were increasingly inclined to acknowledge that hearing loss was not exclusively a matter of ageing, blunt-force trauma, or ototoxicity. Partial and compensable deafness could result from specifiable lengths of exposure to specifiable decibel minimums at specifiable frequencies and could be forestalled with noise-control programs that involved the erection of sound-deadening panels on the shop floor and requiring workers to wear earplugs (American Academy of Ophthalmology and Otolaryngology 1973; American Industrial Hygiene Association 1975; Associated Industries of New York State 1956; Employers Insurance of Wausau 1957; Robinson 1971).

Unions encouraged workers to wear the plugs so as to demonstrate that certain work environments were unfriendly to ears and to lay the basis for future compensation claims for cumulative hearing loss on the job. To the extent that workers during the fifties and sixties went along with blue-collar unions on this, earplugs were rather a pawn in bargaining over factory conditions and workers' compensation than a means of avoiding deafness (Frisch 1993, 115–17, 132–33; Mine Safety Appliances Company 1954; Braun, this volume, who also furnishes data on the effectiveness of plugs).

Though foam-fitted and less easily infected, earplugs still did no more than they had done for millennia, impassively blocking the passage of a range of vibrations through the ear canal. A more active approach, imagined for centuries, began to appear feasible during World War II as physicists and psychoacousticians studied masking techniques and noise cancellation in connection with the jamming of enemy radio, radar, and sonar. Masking could be sheerly electrical, smothering sounds under the white noise inherent in electrical devices and as easily amplified as in cheap battery-powered hearing aids, where static was rather a problem than a solution. Noise cancellation required sophisticated electronics in order to accomplish a series of instantaneous analyses of incoming sounds such that exactly inverse waves of different amplitudes could be produced to negate each successive offending vibration. Such "active noise control," first implemented for the quieting of commercial jet engines, could not be effectively applied to so small a feature as an ear until microprocessors had reached a third generation of development. Even then, noise cancellation was best achieved with a headset or pair of muffs about the size of those cosmetic powder puffs sewn into pilots' earflaps during the twenties on the advice of panjandrums of aviation medicine (Bauer 1926, 157).

Conclusion: Burrow and Mast

Active noise control made of the earplug something of an antenna—or a Homeric mast—for sensing and silencing sounds that threatened the integrity of the cochlea, while the well-insulated, musically accomplished headset protected the absorption of the listener. No more a burrower, the noise-canceling plug or muff stood like a guardian automaton at the entrance to a classical labyrinth, programmed to repel all intruders. If this was rather more wily than heroic, so was the program of care in late twentieth-century hospitals, where noise cancellation was technically still a Nightingale dream (Busch-Vishniac et al. 2005; McCarthy, Ouimet, and Daum 1991) and nurses had begun recommending earplugs to their patients (Haddock 1994). Active noise control meant rather a repudiation of hospitals as burrows and a vigorous effort to discharge as soon as possible the most obstinate of noisemakers—the loudly demented, the chronic coughers, the second opinionators (Mizrahi 1986; Stevens 1989; Strumpf and Tomes 1993). The motive for rushing patients out the door was to some extent the rediscovery that hospitals themselves—due to incestuous HVAC systems and a democratic openness to visitors—were significant sources of secondary infection. To a greater extent, rapid discharge was motivated by an ideology of urgency. Except for lucrative cosmetic or "optional" procedures, hospitalization now implied swift intervention involving invasive tests and soundings, complex arrays of drugs, and/or surgery, an expertise in which hospitals prided themselves and, in some countries, profited highly therefrom (Betsky 2006; Miller and Swensson 1995; Rosenfield 1969).

Convalescence from such traumatic doctoring could be relegated to auxiliary institutions where the need for quiet was not played out against a calculus of mortal peril. So hospitals, too, began to be re-presented as antennae and masts. No longer refuges for the weary, decaying, or dying, they were "centers" for exquisite detection and Oddyssean mainstays for hairsbreadth escapes (Knipping and Kenter 1967, 14–38; Verderber and Fine 2000).

As embodiments of two distinct modes of dealing with noise, the individual and the institutional, earplugs and hospitals meet up time and again at precisely that intersection at which Kafka put his hapless heroes, subject to the perpetual buzzing of *The Castle* and an ominous, indefinable whistling.

REFERENCES

Abel-Smith, Brian, with Robert Pinker. *The Hospitals, 1800–1948: A Study in Social Administration in England and Wales*. London: Heinemann, 1964.

Adams, Annmarie. *Architecture in the Family Way: Doctors, Houses, and Women, 1870–1900*. Montreal: McGill-Queen's University Press, 1996.

———. *Medicine by Design: The Architect and the Modern Hospital, 1893–1943*. Minneapolis: University of Minnesota Press, 2008.

Ades, Harlow W., chair, Committee on Hearing and Bio-Acoustics for the U.S. Air Force and National Research Council. *An Exploratory Study of the Biological Effects of Noise (BENOX) Report* (Chicago: University of Chicago Press, 1953).

Agnew, G. Harvey. *Canadian Hospitals, 1920–1970*. Toronto: University of Toronto Press, 1974.

American Academy of Ophthalmology and Otolaryngology (AAOO). *Guide to the Conservation of Hearing in Noise*. Rochester, NY: Author, 1973.

American Industrial Hygiene Association (AIHA). *Industrial Noise Manual*, 3rd ed. Akron: Author, 1975.

Arnhym, Albert A. *Comfortization of Aircraft*. New York: Pitman, 1944.

Associated Industries of New York State. "A Third Industrial Noise Conference." *American Industrial Hygiene Association Quarterly* 17 (1956): 17–60.

Association des Hautes Études Hospitalières. *La Lutte contre le bruit*. Paris: Masson, 1968.

Bailin, Miriam. *The Sickroom in Victorian Fiction*. New York: Cambridge University Press, 1994.

Baker, Jeffrey P. *The Machine in the Nursery: Incubator Technology and the Origins of Newborn Intensive Care*. Baltimore: Johns Hopkins University Press, 1996.

Bancroft, Jane M. *Deaconesses in Europe and Their Lessons for America*. New York: Hunt and Eaton, 1890.

Bauer, Louis H. *Aviation Medicine*. Baltimore: Williams and Wilkins, 1926.

Bauer, William W., and Leslie Edgley. *Your Health Dramatized: Selected Radio Scripts*. New York: Dutton, 1939.

Behar, Alberto, Marshall Chasin, and Margaret Cheesman. *Noise Control*. San Diego: Singular, 2000.

Berlin letter. "Quiet at Health Resorts." *Journal of the American Medical Association* 59 (1912): 2271.

Betsky, Aaron. "Framing the Hospital: The Failure of Architecture in the Realm of Medicine." In Cor Wagenaar, ed., *The Architecture of Hospitals*, 68–75. Rotterdam: NAi, 2006.

Bijsterveld, Karin. *Mechanical Sound: Technology, Culture, and Public Problems of Noise in the Twentieth Century.* Cambridge, Mass.: MIT Press, 2008.
Billings, John S., and Henry M. Hurd. *Suggestions to Hospital and Asylum Visitors.* Philadelphia: Lippincott, 1895.
Böhme, Gernot. "Acoustical Atmospheres." *Soundscape Journal* 1(1) (2000):14–18.
Bowers, Barbara S., ed. *The Medieval Hospital and Medical Practice.* Aldershot: Ashgate, 2007.
Boys, C. V. "Ear-Guards for War-Noise." *Literary Digest* 51 (August 28, 1915): 403.
Busch-Vishniac, Ilene, James E. West, Colin Barnhill, Tyrone Hunter, Douglas Orellana, and Ram Chivukula. "Noise Levels in Johns Hopkins Hospital." *Journal of the Acoustical Society of America* 118 (2005): 3629–45.
Censorinus. "Pleasant and Unpleasant Noises in Hospitals." *Quiet* 2(7) (1938): 10–11.
Connor, Steven. "Atmospherics." 2006. http://www.stevenconnor.com/atmospheres.htm (accessed January 17, 2010).
Cook, Edward T. *The Life of Florence Nightingale.* 2 vols. London: Macmillan, 1913.
Coser, Rose Laub. "Alienation and Social Structure: Case Analysis of a Hospital." In Eliot Freidson, ed., *The Hospital in Modern Society*, 231–65. Glencoe, Ill.: Free Press, 1963.
Dixon, Ronald F. "Psychology Studies Noise: Suggestions for Its Elimination." *Trained Nurse and Hospital Review* 78 (1927): 253–55.
Duncan, David. *The Life and Letters of Herbert Spencer.* London: Appleton, 1908.
Dundas-Grant, J. "The Organs of Hearing in Relation to War." *Royal Institution of Great Britain Proceedings* 22 (April 2, 1917): 91–99.
Employers Insurance of Wausau, Accident Prevention Department. *Industrial Noise and Hearing Protection.* Wausau, Wis.: Author, 1957.
Falk, Stephen A. "Noise Pollution: Neonatal Aspects." *Pediatrics* 54 (1974):46–79.
———, and Nancy F. Woods. "Hospital Noise Levels and Potential Health Hazards." *New England Journal of Medicine* 289 (1973): 774–81.
Fife, Daniel, and Elizabeth Rappaport. "Noise and Hospital Stay." *American Journal of Public Health* 66 (1976): 680–81.
Frank, A., P. Maurer, and J. Shepherd. "Light and Sound Environment: A Survey of Neonatal Care Units." *Physical and Occupational Therapy in Pediatrics* 1, 2 (1991): 27–45.
Frisch, Michael, with photographs by Milton Rogovin. *Portraits in Steel.* Ithaca, N.Y.: Cornell University Press, 1993.
Gatlin, Dana. "Mrs. Isaac L. Rice." *American Magazine* 75 (February 1913): 34.
Gendre, Pierre N. "Le Bruit à l'hôpital." Thèse pour le doctorat en médecin, Université de Rennes, UER Médicales et Pharmaceutiques, 1976.
Gill, Gillian. *Nightingales: The Extraordinary Upbringing and Curious Life of Miss Florence Nightingale.* New York: Random House, 2004.
Gilman, Sander L. *Franz Kafka: The Jewish Patient.* New York: Routledge, 1995.
Goldie, Sue M., ed. *"I Have Done My Duty:" Florence Nightingale in the Crimean War, 1854–56.* Manchester: Manchester University Press, 1987.
Goldwater, S. S. "Preparing the Building Program for a General Hospital." *Modern Hospital* 38 (March 1932): 49–55.
Golub, Sharon. "Noise and the Nurse." *RN* 32 (May 1969): 40–45.
Gorrell, John. "Planning Hospital Communications." *Modern Hospital* 46 (March 1936): 87–91.
Govan, James, C. R. Anderson, and H. E. Reilly. "Twin Problems in Construction: Insulation and Acoustics." *Modern Hospital* 42 (March 1934): 52–56.

Griffin, Noyce L. *Telephones in the Hospital*. Washington, D.C.: Public Health Service, 1950.
Grumet, Gerald W. "Pandemonium in the Modern Hospital." *New England Journal of Medicine* 328 (1993): 433–37.
Guild, Elizabeth. "Acoustic Trauma in Aircraft Maintenance Workers." *Journal of Aviation Medicine* 22 (1951): 477–90.
Gutton, Jean-Pierre. *Bruits et sons dans notre histoire*. Paris: PUF, 2000.
Haddock, J. "Clinical Sleep: Reducing the Effects of Noise in Hospital." *Nursing Standard* 8(43) (1994): 25–28.
Haslam, P. "Noise in Hospitals: Its Effect on the Patient." *Nursing Clinics of North America* 5 (1970): 715–24.
Hazlitt, Henry. "In Dispraise of Noise." *Century Magazine* 120 (January 1930): 4–6.
Heinl, Robert D. "The Woman Who Stopped Noises." *Ladies' Home Journal* 25 (April 1908): 19.
Hinks, M. Dorothy. *The Most Cruel Absence of Care: Report of a Follow-Up Study of Noise Control in Hospital*. London: King's Fund Centre, 1974.
Hisama, Kay K. "Florence Nightingale's Influence on the Development and Professionalization of Modern Nursing in Japan." *Nursing Outlook* 44 (1966): 284–88.
Hofrichter, Linus. "Hospital Rooms: Where They've Been, Where They're Going." In *The Architecture of Hospitals*, ed. Cor Wagenaar, 418–22. Rotterdam: NAi, 2006.
Holt, E. E. "Boiler-Maker's Deafness and Hearing in a Noise." *Transactions of the American Otological Society* 3 (1882): 34–44.
Horder, Thomas Jeeves. "Human Reactions to Noise." *Journal of the Royal Sanitary Institute* 58 (1938): 713–21.
Horne, Jobson. "Gun Deafness and Its Prevention." *Lancet* (August 15, 1914): 462–64.
Hornsby, John Allen, and Richard E. Schmidt. *The Modern Hospital*. Philadelphia: Saunders, 1913.
Howell, Joel D. *Technology in the Hospital: Transforming Patient Care in the Early Twentieth Century*. Baltimore: Johns Hopkins University Press, 1995.
Ibels, Louise C. *Une journée à l'hôpital: 20 lithographies en couleur*. Paris: Chachoin, 1916.
Jaehne, Arthur. "Untersuchungen über Hörstörungen bei Fussartilleristen." *Zeitschrift für Ohrenheilkunde und für die Krankheiten der Luftwege* 62 (1911): 111–34.
Kafka, Franz [as Eugen Pfohl]. "Public Psychiatric Hospital for German-Bohemia" and "German Society for the Establishment and Maintenance of a Public Veterans Psychiatric Hospital for German Bohemia in Prague." 1916. Trans. Eric Patton with Ruth Hein. In *Franz Kafka: The Office Writings*, ed. Stanley Corngold, Jack Greenberg, and Benno Wagner, 336–45. Princeton, N.J.: Princeton University Press, 2009.
———. "Das Schweigen der Sirenen / The Silence of the Sirens." 1917. Trans. Willa Muir and Edwin Muir. In *The Complete Stories*, ed. Nachum N. Glatzer, 430–31. New York: Schocken, 1971.
———. "Das Bau / The Burrow." 1923. Trans. Willa Muir and Edwin Muir. In The Complete Stories, ed. Nachum N. Glatzer, 325–59. New York: Schocken, 1971.
———. *I Am a Memory Come Alive: Autobiographical Writings*, ed. Nahum N. Glatzer. New York: Schocken, 1974.
———. *The Blue Octavo Notebooks*, ed. Max Brod and trans. Ernst Kaiser and Eithne Wilkins. Cambridge, Mass.: Exact Change, 1991 (written 1917–1919).
Kahn, D. M., T. E. Cook, C. C. Carlisle, D. L. Nelson, N. R. Kramer, and R. P. Millman. "Identification and Modification of Environmental Noise in an ICU Setting." *Chest* 114 (1998): 535–40.

Kalisch, Philip A., and Beatrice J. Kalisch. *American Nursing: A History*, 4th ed. Philadelphia: Lippincott, Williams, and Wilkins, 2004.
Karl, Frederick R. *Franz Kafka, Representative Man*. New York: Ticknor and Fields, 1991.
Kennion, John W. *The Architects' and Builders' Guide*. New York: Fitzpatrick and Hunter, 1868.
Kersbergen, L. C. *Geschiedenis van het St. Elisabeths of Groote Gasthuis te Haarlem*. Haarlem: Enschedé, 1931.
Kestner v. Homeopathic Medical and Surgical Hospital, 245 Penn 326 (1914).
King Edward's Hospital Fund for London. *Noise Control in Hospitals*. London: Author, 1958.
Kingsdale, John M. *The Growth of Hospitals, 1850–1939: An Economic History in Baltimore*. New York: Garland, 1989.
Kirkbride, Thomas. *On the Construction, Organization, and General Arrangements of Hospitals for the Insane*, 2nd ed. New York: Arno, 1973 [1880].
Knipping, H. W., and H. Kenter. *Amerika: Aerzte über Reisen in den Amerikas: Spitalarchitektur, Konquista-Barock, magische Medizin*. Stuttgart: Schattauer, 1967.
Knudsen, Vern O. "Ear Defenders." *National Safety News*. 1939. Clipping in Vertical File, "Noise." Volta Bureau, Washington, D.C.
———, Robert Gales, and Norman Watson. Letter to Harvey Fletcher, April 10, 1941, and Report to National Defense Research Committee. In Box 5, f.7, Vern O. Knudsen Papers, Charles E. Young Library, Dept. of Special Collections, UCLA.
League of Nations. Health Organisation. Housing Commission. *The Hygiene of Housing. Vol. 3, Report on Noise and Housing*. Geneva: Author, 1937.
Lee, Hermione. *Virginia Woolf*. New York: Knopf, 1997.
Leistikow, Dankwart. *Ten Centuries of European Hospital Architecture*. Ingelheim am Rhein: Boehringer, 1967.
Lerner, Paul F. *Hysterical Men: War, Psychiatry, and the Politics of Trauma in Germany, 1890–1933*. Ithaca, N.Y.: Cornell University Press, 2003.
Littler, T. S. "Effect of Noises of Warfare on the Ear." *Nature* (August 17, 1940): 217–19.
Lomax, Elizabeth M. R. *Small and Special: The Development of Hospitals for Children in Victorian Britain*. London: Wellcome Institute, 1996.
Luz, George A., Richard A. Decatur, and Robert L. Thompson. *Psychological Factors Related to the Voluntary Use of Hearing Protection in Hazardous Noise Environments*. Fort Knox: U.S. Army Medical Research Laboratory, 1973.
McCarthy, Donna O., Mary E. Ouimet, and Jane M. Daum. "Shades of Florence Nightingale: Potential Impact of Noise Stress on Wound Healing." *Holistic Nursing Practice* 5(4) (1991): 39–48.
McDonald, Lynn, ed. *Florence Nightingale's European Travels*. Waterloo, Ont.: Wilfrid Laurier University, 2004.
———. *Florence Nightingale's Theology*. Waterloo, Ont.: Wilfrid Laurier University, 2002.
Michaëlsson, M., T. Riesenfeld, and A. Sagrén. "High Noise Levels in Infant Incubators Can Be Reduced." *Acta Pediatrica* 81 (1992): 843–44.
Miller, Richard L., and Earl S. Swensson. *New Directions in Hospital and Healthcare Facility Design*. New York: McGraw-Hill, 1995.
Mine Safety Appliances Company. *Noise: Pertinent Questions and Answers*. Pittsburgh: Author, 1954.
Mizrahi, Terry. *Getting Rid of Patients: Contradictions in the Socialization of Physicians*. New Brunswick, N.J.: Rutgers University Press, 1986.
Monteiro, Lois A., ed. *Letters of Florence Nightingale*. Boston: Boston University Press, 1974.

Morse, Edward Sylvester. Papers 1858–1925, Box 88, f.7, Phillips Library, Peabody Essex Museum, Salem, Mass., including letters from Supt. E. C. Dent (January 11, 1906), Dr. M. S. Gregory (undated), and Hermann M. Biggs (December 21, 1905) to Mrs. Isaac L. Rice.

Neergaard, Charles F. "Controlling Hospital Noise." *Architectural Forum* 57 (November 1932): 449–50.

———. "Sound Proofing the Hospital." *Architectural Record* 66 (August 1929): 174–86.

New York Daily Tribune. "Crusade on Noises" (December 4, 1906), 8: 3.

———. "Fight on Harbor Noise." (November 29, 1905), 5: 4.

———. "Noise." (November 26, 1906), 6: 3–4.

New York Evening Sun. "A Benefactor of the City." (November 27, 1907).

New York Sun. "A Great Cry of Less Noise!" (October 20, 1908).

New York Times. "Anti-Noise Society Reviews Progress." (February 27, 1908): 2.

———. "Letting the Children Help." (April 6, 1908): 7.

———. "Pass Quiet Zone Ordinance." (June 26, 1907): 6.

———. "Woman Starts a War on Tooting River Tugs." (December 10, 1905): 8

New York Times Magazine. "What One Public-Spirited Woman Can Do—Mrs. Isaac L. Rice's Campaign." (January 14, 1906): SM3

Nightingale, Florence. *Ever Yours, Florence Nightingale: Selected Letters*, ed. Martha Vicinus and Bea Nergaard. Cambridge, Mass.: Harvard University Press, 1990.

———. *Florence Nightingale on Hospital Reform*, ed. Charles E. Rosenberg. New York: Garland, 1989, from 3rd ed. of 1863.

———. *Notes on Nursing for the Labouring Classes*. London: Harrison, 1861.

———. *Notes on Nursing: What It Is, and What It Is Not*. London: Duckworth, 1859.

Ohropax GmbH. www.ohropax.de/2-1-history.html (accessed September 2010).

Peterson, Jon A. *The Birth of City Planning in the United States, 1840–1917*. Baltimore: Johns Hopkins University Press, 2003.

Prasher, Deepak. "Factors Influencing Susceptibility to Noise-Induced Hearing Loss." In *Biological Effects of Noise*, ed. Deepak Prasher and Linda Luxon, 125–31. London: Whurr, 1998.

Rice, Mrs. Isaac L. "The Anti-Noise Society." *New York Times* (December 23, 1906): SM4.

———. "Children's Band for Quiet." *New York Times* (January 26, 1908): 5.

———. "Children's Hospital Branch of the Society for the Suppression of Unwanted Noise." *Forum* 39 (1908): 560–67.

———. "An Effort to Suppress Noise." *Forum* 37 (1906): 552–70.

Risse, Guenter B. *Mending Bodies, Saving Souls: A History of Hospitals*. New York: Oxford University Press, 1999.

Robinson, D. W., ed. *Occupational Hearing Loss*. London: Academic Press, 1971.

Rodger, T. Ritchie. "Noise Deafness: A Review of Recent Experimental Work, and a Clinical Investigation into the Effects of Loud Noise upon the Labyrinth of Boiler-Makers." *Journal of Laryngology, Rhinology, and Otology* 30 (1915): 91–105.

Rodgers, Daniel T. *Atlantic Crossings: Social Politics in a Progressive Age*. Cambridge, Mass.: Belknap, 1998.

Rosenberg, Charles E. *The Care of Strangers: The Rise of America's Hospital System*. New York: Basic, 1987.

Rosenfield, Isadore, with Zachary Rosenfield. *Hospital Architecture and Beyond*. New York: Van Nostrand Reinhold, 1969.

Russell, Carolyn M., ed. *Planning and Design for Perinatal and Pediatric Facilities*. Columbus, Ohio: Ross Laboratories, 1977.

Sandweiss, Eric. "Paving St. Louis's Streets: The Environmental Origins of Social Fragmentation." In *Common Fields: An Environmental History of St. Louis*, ed. Andrew Hurley, 90–106. St. Louis: Missouri Historical Society, 1997.
Schaffellner, Barbara. *Unvernunft und Kriegsmoral: am Beispiel der Kriegsneurose im Ersten Weltkrieg*. Vienna: Lit, 2005.
Schwartz, Hillel. "The Indefensible Ear." In *The Auditory Culture Reader*, ed. Michael Bull and Les Back, 487–501. London: Berg, 2003.
———. *Making Noise: From Babel to the Big Bang and Beyond*. New York: Zone, 2011.
Seidler, Edward. "An Historical Survey of Children's Hospitals." In *The Hospital in History*, ed. Lindsay Granshaw and Roy Porter, 181–97. London: Routledge, 1989.
Seleny, Frank L., and Michael Streczyn. "Noise Characteristics in the Baby Compartment of Incubators." *American Journal of Diseases of Children* 111 (1969): 445–50.
Sloane, David C. "A (Better) Home Away from Home: The Emergence of Children's Hospitals in an Age of Women's Reform." In *Designing Modern Childhoods*, ed. Marta Gutman and Ning de Coninck-Smith, 42–60. New Brunswick, N.J.: Rutgers University Press, 2008.
Snow, W. B., and W. D. Neff. *Effects of Airplane Noise on Listening with Headphones*. New York: Columbia University Department of War Research, 1943.
Society for the Suppression of Unnecessary Noise. *Annual Reports* 1 (1908).
Stephen, Julia Duckworth. "Notes from Sick Rooms." 1883. In *Stories for Children, Essays for Adults*, ed. D. F. Gillespie and E. Steele, xviii–xxi. Syracuse, N.Y.: Syracuse University Press, 1987.
Sterne, Jonathan. *The Audible Past: Cultural Origins of Sound Reproduction*. Durham, N.C.: Duke University Press, 2003.
Stevens, Edward F. *The American Hospital of the Twentieth Century*. New York: Architectural Record, 1918.
Stevens, Rosemary. *In Sickness and in Wealth: American Hospitals in the Twentieth Century*. New York: Basic, 1989.
Stevenson, Christine. *Medicine and Magnificence: British Hospital and Asylum Architecture, 1600–1815*. New Haven, Conn.: Yale University Press, 2000.
Stevenson, R. Scott. "Nursing Homes, Sick Rooms, and Noise." *Quiet* 1(2) (1936): 26–28.
Strumpf, Nevill E., and Nancy Tomes. "Restraining the Troublesome Patient." *Nursing History Review* 1 (1993): 3–24.
Tanon, L. "Sur les méfaits du bruit." *Bulletin de l'Académie de Médecine*, sér. 3, 120 (1938): 377–80.
Taylor, Jeremy. *Hospital and Asylum Architecture in England, 1840–1914*. London: Mansell, 1991.
Thompson, Emily. *The Soundscape of Modernity: Architectural Acoustics and the Culture of Listening in America, 1900–1933*. Cambridge, Mass.: MIT Press, 2002.
Thompson, John D., and Grace Goldin. *The Hospital: A Social and Architectural History*. New Haven, Conn.: Yale University Press, 1975.
Trible, G. B., and S. S. Watkins. "Ear Protection." *U.S. Naval Medical Bulletin* 13(1) (1919): 48–60.
Tucker, W. S. "Noise and Hearing." *Nineteenth Century and After* 104 (February 1929): 246–55.
Underwriters Laboratories. *Standards for Hospital Signaling and Nurse Call Equipment*, 2nd ed. Chicago: Author, 1979.
U.S. Public Health Service. Division of Hospital and Medical Facilities. *Noise in Hospitals*. Washington, D.C.: Department of Health, Education, and Welfare, 1963.

Vaizey, John. *Scenes from Institutional Life and Other Writings*. London: Weidenfeld and Nicolson, 1986.
Van Dyke, John C. *The New New York*. New York: Macmillan, 1909.
Verderber, Stephen, and David J. Fine. *Healthcare Architecture in an Era of Radical Transformation*. New Haven, Conn.: Yale University Press, 2000.
Warfield, Frances. "I Know What It Means to Be Deaf." *Saturday Evening Post* (March 13, 1948): 18, 63.
Woolf, Virginia. *The Flight of the Mind: The Letters of Virginia Woolf*. Vol. 1, *1888–1912 (Virginia Stephen)*, ed. Nigel Nicolson. London: Hogarth, 1975.

CHAPTER 12

SOUNDING BODIES: MEDICAL STUDENTS AND THE ACQUISITION OF STETHOSCOPIC PERSPECTIVES

TOM RICE

INTRODUCTION

IN 1999 I conducted a short study of patients' experiences of the sound environment at the Edinburgh Royal Infirmary (Rice 2003). For many patients the sounds of the ward had become symbolic of the tedious routine of hospital life. The cacophony of voices, footsteps, coughing, cries of pain, rattling trolleys, bleeps from medical equipment, and so on served to reinforce and exacerbate feelings of powerlessness among many patients. Nurses to whom I spoke tended to have a different perspective. They generally interpreted the noise of activity on the ward as a positive sign, indicative of order and the smooth running of care. They were more alert to signals given by call buttons and to potential emergencies indicated by particular tones from medical equipment and unusual sounds or quietness from the beds. Spending time on the wards, I also observed doctors going about their work. I was interested to see them listening to patients' bodies using stethoscopes. The instruments enabled the doctors to shut out the ambient noise of the ward and to immerse themselves instead in the acoustic space created by the patient's body. The hospital, it seemed, was a space in which different forms of listening and layers of acoustic knowledge were constantly in play (see Schwartz, this volume).

In order to broaden and deepen my understanding of how sound is implicated in hospital life I carried out a year of ethnographic fieldwork between 2003 and 2004 based at St. Thomas' Hospital in London. I was particularly eager to explore stethoscopic listening during this period. Aware of the importance of auscultation in assessing the health of the heart, I approached consultant cardiologist Dr. John Coltart, then head of cardiothoracic services. He told me that he considered his ears to be one of his most important clinical tools and agreed to act as my supervisor. He invited me to participate in the classes he taught, in which medical students learned to listen.

This chapter gives a brief survey of a number of techniques used to appropriate body sounds as a diagnostic resource in the Western medical tradition. The work of Reiser (1978), Lachmund (1998, 1999) and Sterne (2003) indicates that auscultation was instrumental in bringing about important changes in the way in which doctors engaged with patients in Western medicine. It allowed them to begin to make diagnoses without having to rely on patients' illness narratives, creating scope for the formation of independent clinical judgments and permitting listeners a degree of physical and perceptual distance and detachment. It also required specialist knowledge, the possession of which granted doctors new social status and distinction. The stethoscope's application in examinations of the heart and lungs is now frequent and routine in many modern clinical settings, and the instrument has become the archetypal symbol of the Western doctor.

Though a well-established technique, auscultation has by no means been a universal or homogenous practice. Lachmund (1999) explores two distinct (though clearly related) approaches to auscultation and the interpretation of sounds as diagnostic signs that emerged in Paris and Vienna during the nineteenth century. In doing so he draws attention to the need to understand the practice of auscultation (and forms of scientific work more generally) as historically located, embedded in contextually specific networks of social and spatial relations. With Lachmund in mind, this chapter explores some of the techniques through which students are taught to focus their hearing and analyze sounds received through the stethoscope. It positions auscultation as a way of knowing through sound that is in the process of transmission between generations of medical practitioners.

Sterne, Reiser, and Lachmund document early problems identified with auscultation. They describe difficulties encountered in creating a robust taxonomy for body sounds and in codifying sounds in ways that guaranteed consistent and reproducible interpretations between listeners. While the difficulty of establishing consensus over the character and significance of subtle heart sounds remained a subject of discussion among doctors at the cardiothoracic unit at St. Thomas', auscultation at the time of my research also formed part of a repertoire of increasingly sophisticated diagnostic techniques. The work for which the stethoscope was used was considered by some to be better performed using other kinds of equipment that drew on alternative skill sets (cardiac ultrasound scanning, for example). This chapter, then, locates auscultation not only in relation to discussions of the perceptual and linguistic problems which the practice generates, but by reference to debates over the value and relevance of auscultation in the twenty-first century. In doing so, it echoes Lachmund's call for the particularities of stethoscopic listening to be understood within their specific social and historical context.

The students with whom I worked generally approached auscultation with enthusiasm. However, the technique also represented a new and sometimes disconcerting way of engaging with the body. In becoming attuned to heart sounds, students were required to encounter the bodily interior in a novel acoustic light. Their reactions to being asked to apply a newly acquired auditory "gaze" not only to hospital patients but also to friends and family members highlighted the intriguing and sometimes disturbing experiential subtleties and complexities of an apparently simple and straightforward diagnostic technique. Outside the clinical setting, their new knowledge of sounds as diagnostic signs could have a destabilizing effect by repositioning family members (and even the students themselves) as diagnostic objects. Clinical knowledge did not always remain within fixed or tightly delineated communities of practitioners but, so to speak, bled out into interactions beyond the hospital. The chapter therefore points to the conceptual and imaginative, as well as the diagnostic power of the stethoscope.

A History of Listening

During one of the first classes I attended during fieldwork, Dr. Coltart asked his students, "Who invented the stethoscope?" None of us knew. "I'll give you a clue," he added. "It wasn't Mr. Steth." Then Dr. Coltart went on to tell us how one day in 1816 a young doctor named René Laënnec was walking through the Tuileries Garden in Paris. The case of one of his patients, a young woman suffering from a heart condition, was playing on his mind. He had been unable to learn anything about her problem from the accounts she gave. In those days doctors used the patients' description of their symptoms as the basis for diagnosis, so Laënnec's inability to glean anything from the young woman's account represented a serious problem. While he was wondering what to do, Laënnec saw a group of children playing around a log that was sitting on top of a pile of trash. The children at one end of the log were pressing their ears to the wood. They seemed to be able to hear the knocks and scratches made by the children at the other end. This was Laënnec's "Eureka!" moment. He returned immediately to his patient at the Necker Hospital. Rolling a book into a tight cylinder so that it resembled a log, he pressed it to her heart and found he could hear her heartbeat clearly.

I heard Dr. Coltart repeat this story to other groups of students, and, indeed, it can be found in several texts (e.g., Marks 1972; Welsby, Parry, and Smith 2003). Others writing on the invention of the stethoscope are less specific as to what took place. Reiser states that Laënnec simply "recalled the well-known acoustic phenomenon: that sound was augmented when it traveled through solid bodies, as when a scratch noise made at one end of a piece of wood can be heard at the other end" (1997, 828). He does not specify what caused Laënnec to make this recollection. Fleming goes as far as to declare that "[t]here is . . . no documentary evidence to support the attractive traditional story that it was the sight of children at play, scratching one end of a log of wood and listening at the other in the courtyard of the Louvre, which first gave Laënnec the idea of mediate auscultation" (1997, 88). However, while Fleming aims to discredit the myth, he in fact only reproduces the "attractive traditional story" by referring to it in such detail. Despite a lack of substantiating evidence and even counterclaims or suggestions that the stethoscope might have been invented in ancient Egypt (e.g., Rackrow 2009, Martinet et al. 1998), the story of Laënnec's invention of the stethoscope was well established in the clinical setting in which I conducted my research. It was being passed on, reproduced in new generations of doctors. The story was an accepted and generally unscrutinized creation myth for a now ubiquitous piece of medical technology.

Although Laënnec is widely credited with the invention of the stethoscope, medical practitioners have evidently used the sounds of the body as clues to the health of its interior for centuries. Mangione indicates that references are made to the diagnostic significance of breath sounds in the Ebers papyrus (c. 1500 BC) and the Hindu Vedas (c. 1400–1200 BC) (2000, 295). In the fourth century BC, Hippocrates described what came to be known as "immediate auscultation," which involved pressing an ear to the patient's chest in order to listen to the internal sounds (Mangione 2000, 295; Marks 1972, 19–20). Doctors continued to employ this technique as the centuries progressed, and it was still in use in the early nineteenth century. Fleming (1997) and Lachmund (1999) argue that the practice was not completely abandoned until much later. In the medical community in which my research took place, Hippocrates was widely credited with being the first to recognize and teach listening as a way of discerning the health of the interior organs. However, Reiser suggests that it was Robert Hooke who first grasped the huge scope and diagnostic potential of auscultation:

> There may be . . . a Possibility of discovering the Internal Motions and Actions of Bodies by the sound they make, who knows but that as in a Watch we may hear the beating of the Balance, and the running of the Wheels, and the striking of the Hammers and the grating of the Teeth, and Multitudes of other Noises; who knows, I say, but that it may be possible to discover the Motions of the Internal Parts of Bodies, whether Animal, Vegetable or Mineral, by the sound they make, that one may discover the Works perform'd in the several Offices and Shops of a Man's body, and thereby discover what Instrument or Engine is out of order. (Hooke 1705, 39–40).

The idea that the "Motions of the Internal parts or Bodies" might be discerned by their sounds suggests that the actions and movements of each organ could be rendered intelligible to the ear, producing a rich and detailed source of diagnostic information.

A further technique for the use of sounds in diagnosis is thought to have emerged in 1761, when Leopold Auenbrugger hit upon the idea of percussion. He was an innkeeper's son and as a child had learned to test the fullness of barrels by thumping them (Porter 1997, 256). Switching kegs for rib cages, Auenbrugger noted that, if struck with a finger, healthy and unhealthy chests produced different sounds: "a healthy chest sounded like a cloth-covered drum . . . a muffled sound or one of high pitch indicated pulmonary disease" (Porter 1997, 256). Porter writes that Auenbrugger's work attracted little attention when it first appeared. It was not until after the charismatic physician Jean Corvisart published a French translation of it in 1808 and began to apply percussion himself that the technique gained general acceptance and began to be used routinely (Porter 1997, 308).

Percussion remains an important technique in the clinical examination. The students with whom I studied were taught that, as Auenbrugger had attempted to demonstrate, percussive sounds are resonant over healthy, aerated lungs and dull in spaces where there is fluid. They would tap the chest using the middle finger of one hand to knock on the second phalange of the middle finger of the other. As part of the chest examination students were taught to percuss the lungs by tapping at an upper, middle, and lower point on each side of the patient's back and on top of the shoulder. Percussion was also applied to the abdomen to determine whether abdominal distension, if present, involved solids or liquids (producing dull sounds) or gas (producing resonant ones).

René Laënnec apparently knew of and admired Auenbrugger's work on percussion. He recognized, however, that "we frequently stand in need of a more constant and certain sign than that furnished by percussion" (Laënnec 1846, 3). Laënnec was also familiar with immediate auscultation, but he considered the technique unsatisfactory in many ways. Because it involved direct contact between the ear and the patient's chest, its best use in his own experience was as a tactile means of discerning the pulsations of the heart. Little could actually be heard with any clarity. Laënnec also felt it to be "alike inconvenient to the physician and the patient; its disagreeableness alone often renders it almost impracticable in hospitals" (Porter 1997, 2). Sterne gives careful attention to the cultural dynamics of disgust at the time Laënnec was practicing, but it is easy to understand that, at a time when many patients would have arrived at the Necker hospital in the advanced stages of tuberculosis, pressing one's ear to the chest may have been an unpleasant task (2003a, 115–17). Laënnec also wrote of immediate auscultation that "it can hardly be proposed to females in general, and in some the large size of the mammae presents an insuperable obstacle to its adoption" (1846, 2). Not only was immediate auscultation often impracticable on women; for a Frenchman in the early nineteenth century it required an unacceptable level of intimacy.

As indicated earlier, it is unclear exactly where Laënnec's inspiration for the stethoscope came from. Though Sterne argues that he may well have crafted the story with the benefit of hindsight, Laënnec relates how, sitting at the bedside of a young female patient whose gender and obesity made immediate auscultation useless, he rolled a small stack of paper into a cylinder and pressed it to his patient's heart (Sterne 2003a, 102). He was pleasantly surprised to find how well his first impromptu stethoscope worked: "I . . . was not a little surprised and pleased, to find that I could thereby perceive the action of the heart in a manner much more clear and distinct than I had ever been able to do by the immediate application of the ear" (Laënnec, cited in Reiser 1997, 828–29). Listening with what he later called the "stethoscope," "pectriloque," or "cylinder" became known as *l'auscultation mediate*: auscultation mediated by an instrument (Lachmund 1999, 424).

Laënnec's method in developing auscultation was to listen to the chests and particularly the lungs of patients admitted to his hospital, carefully noting what he heard. Many of these patients were suffering from tuberculosis, and Laënnec would continue to listen as the disease progressed. When the patient died, he would conduct a postmortem, systematically referencing the sounds he had heard to pathological changes found during dissection (Porter 1997, 263). Laënnec began to identify acoustic signs that he believed were characteristic and therefore diagnostic of tuberculosis. Ultimately, Laënnec's approach enabled him to identify the presence of abnormalities in the bodies of patients who were *still alive*. It was no longer necessary for patients to die (though they very frequently did) in order for anatomization to take place. Auscultation created scope for the "autopsy of the living" (Sterne 2003b). In this respect, Reiser states that auscultation marked "a new age in diagnosis" (1997, 831).

In describing the adoption of stethoscopic listening by other practitioners it is important to avoid reproducing what Latour describes as "diffusionist" narratives, whereby instruments spread through society as if by their own volition or as though moving through a static medium (Latour 1987, 136). Nor was it simply a question of "technology transfer," by which the stethoscope was taken to other countries. As Maulitz, referring to the introduction of the stethoscope to England, points out, "[t]he process was rather one in which *experience*, from the dissection table and the hospital wards, flowed through the careers of multitudinous young Englishmen as they made the journey out [to Paris] and back" (1987, 136; original emphasis). Auscultation *as a skill* was carried between communities of medical practitioners. Also, Furst writes that "it would be erroneous to jump to the conclusion that the stethoscope was rapidly hailed and assimilated into general practice." Instead, the "adoption of the stethoscope was patchy" (1998, 57). Lachmund (2009) and Nicolson (2004) indicate that Laënnec's ideas were variously accepted, contested and subjected to reevaluation by communities of doctors in major centres of medical practice across Europe. At the same time, though auscultation was eventually widely adopted it has never been an entirely homogenous or completely standardized practice. I recognize that in what follows, I am describing some of the specifics of auscultation as taught to students at a particular socio-historical juncture—a teaching hospital in London between 2003 and 2004. As I go on to explain, the technological sophistication of medicine at this time has had important implications for the way in which auscultation is taught and learned, and for the value which is assigned to the practice.

The Body as a Soundscape

Of the methods of listening just described, all except immediate auscultation have retained a place in the repertoire of skills and techniques taught to medical students at the time my fieldwork was conducted. These techniques, and in particular mediate auscultation, allow the body to be constructed as a dynamic acoustic space. Schafer (1977) is credited with having coined the term *soundscape* to refer to the specific acoustic character of a *space* or *place*, but it is clear that the body, too, has a soundscape of its own. It is a "sounding cavity" (Gell 1995, 240). It "sounds and resounds" (Rée 1999, 53).

Some psychoacousticians engage with sound originating within the body as an acoustic problem. Barany (1938) refers to the noise produced by the flow of air, blood, and the creaking of joints and muscles attached to the skull. Tonndorf writes that "trans-lational" waves of sound may be conducted through the body from within (1972, 233). It is argued that the ear filters out or minimizes these sounds. Thus, for instance, the auditory ossicles (the *malleus*, *incus*, and *stapes*) are separated by fluid-filled membranes that reduce bone-conducted vibration. Sounds from inside the body are screened out so as to make the outside world audible. Ackerman suggests that, were this not to occur, even the sound of a person's own blood flow "would be as deafening as sitting in a lawn chair next to a waterfall" (1990, 189). Using the same watery metaphor, Carpenter and McLuhan suggest that the ears filter out "the continuous Niagara of sound . . . in the circulation of the blood" (1960, 68; see Helmreich, this volume, on water sounds).

Both Ackerman and Carpenter and McLuhan imagine the dominant sonic characteristic of the body to be one of "flow." Perceived through the stethoscope, too, the soundspace of the body is characterized by flow and by recurrent patterns of movement: of blood around the body and through the vessels and chambers of the heart, of breath in and out of the lungs, of matter and gases through the gut. Of course, auscultation also allows the detection of sounds produced by, for instance, contact or friction in the body. A pleural rub, for example, which can occur in patients with pneumonia, is created when thickened, roughened surfaces rub together as the lungs expand and contract. The sounds that constitute the heartbeat are made by the snapping closure of the heart valves. Broadly speaking, however, as Sterne suggests, auscultation creates a "hydraulic hermeneutics, charting the motions of liquids and gases through the body" (2003b, 203). Feld (1996) describes how the Kaluli of Papua New Guinea have a specialized knowledge of the sonic space created by waterways flowing through their (visually inaccessible) rainforest environment. There is a sense in which, like the Kaluli, doctors have a specialized sonic knowledge, an "acoustemology" of patterns of flow through the visually obscured space of the bodily interior (Feld 1996).

Flowing blood describes the corporeal surfaces and spaces across and through which it moves. Turbulence and variation in velocity create an audible hemodynamics that can convey detailed information about the efficiency and condition of, for instance, the heart valves. Anthropologists have detailed South American contexts in which the blood is said to "speak" directly to the healer, revealing the condition and needs of the heart (Nash 1967). The healer "hears" it or "listens to what it wants" (Nash 1967, 132–33). The blood "speaks" or "talks" (Tedlock 1982). Reiser writes that when stethoscopic listening was new to medicine, doctors were concerned that they might appear ridiculous because they behaved "[a]s if the disease itself were a living being that could communicate its condition" (1978, 37). On several occasions I heard teaching doctors who had asked students to auscultate and give a diagnosis exclaim that a particular sound was "screaming" or "shouting" the correct answer—as though it was actively communicating. But these colloquialisms, which effectively personifed the body sounds and lent them an intentioned voice, represented a deliberate departure by teaching doctors from a formal medical discourse whereby the sounds of the body were framed more neutrally as incidental by-products of corporeal processes and events.

The heart's most obvious "sound event" is the heartbeat (Schafer 1977). Most people know this sound through its being at the edge of their awareness of their own bodies (Rée 1999, 51; Ackerman 1990, 178). However, recordings of the heartbeat are also frequently used in popular music and television and film sound tracks. Pink Floyd's *The Dark Side of the Moon*, for example, begins and ends with the heartbeat, while the Minimoog synthesizer has a chart showing the necessary oscillator configurations required to produce versions of this sound. In film sound tracks a gradual speeding up of the heartbeat is often used to build tension or expectancy, while in medical dramas it is employed to evoke a sense of jeopardy—of a patient's life hanging in the balance, for instance. The heartbeat has become a sonic icon of human life. However, auscultation requires the listener to internalize a new semiotics of the heartbeat whereby its character and constituent sounds become diagnostically associated with physiological states and events. Like cars as described by Krebs (this volume), hearts and bodies more generally are positioned as mechanisms in which a range of problems and inefficiencies are articulated through sound. At the same time, like car mechanics, medical students must develop diagnostic listening skills over the course of their professional apprenticeship. However, whereas for the trainee mechanics Krebs describes diagnostic listening skills might be acquired through tacit learning, for the medical students with whom I studied these skills became the focus of overt and targeted learning exercises.

Learning to Listen

At the hospital, Dr. Coltart, my supervisor, suggested I attend the classes at which he introduced a group of third-year medical students to auscultation. The time taken to qualify as a doctor in Britain varies depending on a person's route through the education system, but the students in whose classes I became a participant observer expected their medical training to take five years in total. For the first two years, the preclinical years, they had learned through lectures and tutorials; now they were embarking on the first of three clinical years that would involve contact with actual patients. This year's practical teaching was subdivided into three "rotations," each lasting for three months and focusing on a specific area of the body: the abdomen, head or chest. Dr. Coltart's classes were part of the chest rotation, which included cardiovascular and respiratory medicine. Owing to Dr. Coltart's specialization in cardiology, heart sounds became the particular focus of attention during his classes.

Auscultation of the heart tends to be used in the context of a more general cardiovascular examination that usually includes the taking of a clinical history, a series of careful observations from the foot of the bed, and the feeling of pulses, as well as other small checks and procedures, all of which shape the listener's expectations of what will be heard. Indeed, *The Oxford Handbook of Clinical Medicine* states that auscultation is:

> generally, but wrongly, held to be the essence of cardiovascular medicine at the bedside. A caricature of cardiology ward rounds is of the anxious junior gabbling through the history, while noting his chief's fingers twisting his stethoscope, impatient to "get down to the main business" of listening to the heart—thereby blotting out all talk in favour of a few blissful minutes communing with the "lub" and the "dub." This is absurd . . . if you spend time listening to the history and feeling pulses, auscultation should hold few surprises: you will often already know the diagnosis. (Longmore, Wilkinson, and Torok 2001, 39)

Stethoscopic listening, then, is rarely used on its own but fits within a web of interconnected techniques for examining patients. Signs detected by means of the technique often confirm diagnoses formulated on the basis of other observations.

However, while auscultation forms just part of a repertoire of diagnostic practices and techniques available to the doctor, at St. Thomas' it was also regarded as an important skill in its own right. This, I suggest, is not only an index of the value of sounds as diagnostic signs but also a reflection of the symbolic importance of auscultation. As indicated earlier, the act of placing the diaphragm on the patient's chest might be described as the archetypal gesture of modern clinical expertise. It articulates the doctor's power, knowledge, and skill relative to the patient (Hansen 1997, 78; Rice 2010a). The stethoscope is the doctor's "symbol of office," and listening provides an opportunity for doctors to express what, following Bourdieu, might be described as their key professional "dispositions": concern, responsibility, and so on (Marks 1972, 16; Bourdieu 1980, 54). Furthermore, the chest and especially the heart are popularly regarded in the West as the seat of life. A person listening to the heart and having medical power in relation to it is accorded particular prestige among both medical professionals and patients.

At the first chest rotation session I attended, Dr. Coltart began explaining that the heartbeat is made up of two main sounds. These are known as the first and the second heart sounds (often abbreviated as S1 and S2) or simply the "lub dub" (I have also heard them referred to as the "lub dup" and "lup dup"). The first heart sound, or the "lub," is caused by the closure of the mitral and tricuspid valves. The second heart sound, or the "dub," by the closure of the aortic and pulmonary valves (see figure 12.1).

In addition, Dr. Coltart explained that, in a normal heart, blood flows around the heart and through the valves smoothly, so that the only sounds that can be heard through a stethoscope are the closing snaps of the valves that produce the "lub" and the "dub." Sometimes, though, a physiological abnormality can mean the blood flow becomes turbulent, and this turbulence creates what are known as "heart murmurs." For instance, if a valve becomes stiff (perhaps through calcification) so that blood is forced through a narrower opening than is normally made by a valve (this restriction of blood flow is known as *stenosis*), turbulence will be created and a murmur produced. If the valve ceases to close properly or becomes floppy through tissue damage, blood will often flow back through it (this is known as *regurgitation*),

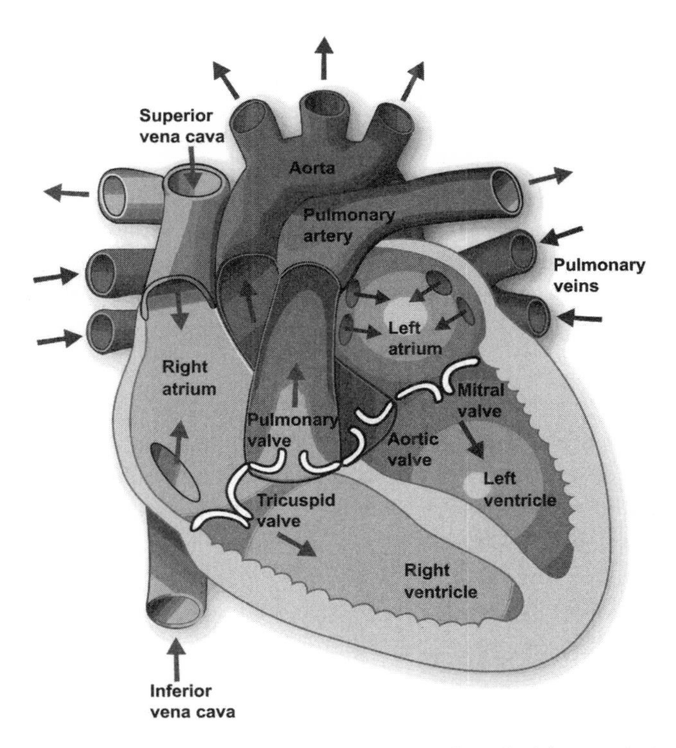

Figure 12.1 A simple diagram of the heart. Reproduced with permission of the Texas Heart Institute (www.texasheart.org).

and a murmur will again be produced, though at a different stage in the cardiac cycle and with a different sound. Valve problems are not the only causes of murmurs. Holes in the septum (the muscle wall separating the two sides of the heart), for instance, can cause large volumes of blood to be forced back and forth between the chambers, producing a murmur known as a *shunt.* Some murmurs appear to have no physiological basis at all and are caused by unexplained turbulence in the blood flow. These are known as "innocent" or "flow" murmurs. In general, though, heart murmurs are linked to malfunctioning valves.

By indicating the manner in which they correspond to the underlying physiology of the heart, Dr. Coltart set out a basic interpretative framework for heart murmurs. He worked on the assumption that auscultation serves to "give light to the bodily interior" (Draper 2002, 777). It allows the doctor to visualize what is taking place inside the heart, drawing on an existing "body" of knowledge gained through dissection (Foucault 1973). Listening acts as an extension of the clinical gaze, so that the "murmur of mitral regurgitation," for example, might be described as a "quasi-visual depiction" of a malfunctioning valve (Lachmund 1999, 432). Despite the closely interconnected nature of auscultation with forms of anatomically orientated clinical gazing, however, the technique demands the practical acquisition of auditory skill, and after his introduction to the theoretical underpinnings of the auscultation of the heart, Dr. Coltart told us that most of our time with him would be spent actually learning to hear and identify murmurs.

AN "EARS ON" APPROACH

Describing the trips made by English doctors to study auscultation in Paris, Maulitz writes: "The year in France was, far from a passive period of observation, a veritable *tour de main*" (1987, 136; italics in original). The notion of a "*tour de main*" captures the distinctly practical and skill-orientated nature of the education the doctors received. The auditory training auscultation requires means this period of education must often have also been a *tour d'oreille.* For the students with whom I worked, stethoscopic listening certainly involved an "ears on" approach (Lachmund 1999, 440).

Sessions for the chest rotations took place on the cardiothoracic wards. Dressed in white coats and clutching our notebooks and stethoscopes we (the students and I) would follow Dr. Coltart to the bedside of a patient he judged to be a suitable case study for us. We would form a semicircle around the bed, with Dr. Coltart at the head on the patient's immediate left. He would then introduce us to the patient briefly and tacitly draw our attention to the importance of being pleasant and polite in professional interactions. With the patient's consent he would then demonstrate the cardiovascular examination, going through each stage and drawing our attention to any relevant diagnostic signs. Among the checks involved were inspection of the patient's hands and fingers, the taking of pulses in the wrist and neck, noting the coloration of the cornea and condition of the teeth, and palpation of the chest for abnormal vibrations.

After demonstrating these examination stages, Dr. Coltart would proceed to auscultation; listening quickly and with well-practiced skill he indicated the key points on the chest at which sounds produced by the heart valves are most easily discerned. In doing so he essentially produced in our minds a map of the surface of the chest, where each point indexed an underlying valve. Having given this demonstration, he would instruct each of us to auscultate in turn. Like the other students, I found it was initially very difficult to make any assessment of the sounds detected. I was unsure what to listen for, and fumbling with or misplacing the stethoscope often made it impossible to hear anything apart from vague and confusing or sudden, sometimes painfully loud, sounds.

Correcting our listening postures ("You need to keep your head down and forward") and our handling of the device ("You've got your earpieces in the wrong way around"), Dr. Coltart advised us to listen first of all for the "landmark" sounds, namely the "lub" and "dub" that constitute the heartbeat. He suggested we nod our heads with each beat or tap out the rhythm with a foot, lending the sounds a kind of kinesthetic amplification. He also instructed us to listen while feeling the patient's pulse not only to anchor the sounds of the heart in tactile sensation but also to check for delays between the heartbeat and the rise of the pulse as these could be of clinical significance.

As we became more confident at detecting the first and second heart sounds, Dr. Coltart instructed us to begin to listen for heart murmurs and other abnormal heart sounds. He suggested we employ a kind of temporal system of focus, spending a few seconds listening to the sound as a whole, then another few seconds concentrating on the "lub" and "dub," then a further few seconds listening for additional sounds. He also advised us to mark the first and second heart sounds in our heads using imaginary lines. If additional sounds were present, we should shade in the spaces on either side of the first or second heart sounds in which they occurred. From this visualization we would be able to work out the position of the murmur relative to the cardiac cycle. Many students found this technique of visualization helpful, but Dr. Coltart was eager for us to become so familiar with different kinds of murmurs that we would cease to need to visualize them; instead there would be an automatic auditory recognition.

In addition, Dr. Coltart emphasized that our ability to auscultate would improve only with practice. He suggested that we revisit patients whose murmurs we had found tricky and go around other wards asking the nurses in charge if they knew of patients in their care with interesting murmurs to which we might listen. St Thomas' was a teaching hospital, which meant that patients had been alerted to the importance of education in the institution, and students could quickly gain access to them. There was almost invariably a range of patients with different heart sounds for the students to listen to, and these patients generally agreed to repeated examinations (for a discussion of patient involvement in auscultation classes see Rice 2008).

The students with whom I worked considered auscultation to be a solitary, even isolating perceptual experience (Rice 2010b). This is partly because stethoscopic listening, like the iPod listening described by Bull (this volume), is privatized. The sounds of the body are narrowcast—channeled through earpieces—not broadcast into public space in the way that music played over a stereo might be. The students could never be sure that they were hearing the sound Dr. Coltart wanted them to be hearing or whether they were making some mistake in their placement of the stethoscope. Technologies have been developed to get around this problem. Teaching stethoscopes with as many as ten pairs of earpieces leading off a single central diaphragm (ensuring that everyone is listening to the same sound) are available, and recently digital stethoscopes have been produced that allow heart sounds to be amplified, recorded, and played back. However, while they are available to purchase, I did not see either of these devices in use at St. Thomas'. At the time of writing, the iPhone application iStethoscope Pro became available. This software allows heart sounds to be recorded and played back into shared auditory space via the iPhone's speakers. Filters can be applied to make the desired frequencies of the heart sound clearer and to minimize background noise. The application may find use in the teaching of auscultation in the future precisely because of its capacity to make heart sounds easy to share between listeners.

To get around the problem of the subjective isolation of listening, Dr. Coltart and his colleagues provided students with a heart murmur CD. Professionally produced by a major stethoscope manufacturer, this set of recordings included voice-overs naming each individual heart sound and describing its particular sonic characteristics. The CD enabled the students to listen collectively during classes and ensured that they were hearing the intended sound. Whenever they detected subtle or more unusual sounds, they were also able to refer to the recorded examples as a kind of acoustic template. However, while the recorded sounds were helpful, they tended to be both louder and clearer than those the students were able to hear through their stethoscopes. The recordings also lacked the background noise of the wider ward, the tactile realism of the stethoscope-user interface, and the contextual detail of the doctor/patient interaction. Though after a few days the students were able to complete the heart sound quiz at the end of the CD, finding and recognizing the same murmurs in real life was more challenging and would be achieved only gradually by practicing on real patients.

The teaching of auscultation also presented a problem of vocabulary. Standing at the bedside, Dr. Coltart often tried to describe heart sounds for his students. "It's that long, low, rumbling sound I want you to be getting," he might say, or "Can you hear that harsh, rasping, almost squeaking sound?" But the adjectives were usually inadequate and prone to misinterpretation. He also resorted to mimicry to capture the quality of murmurs. "It's the ush, ush, ush I want you to listen to," he might say, or the "lup-dup-shh, lub-dup-shh." His use of these nonsense words often had a comic effect, particularly because of their contrast with the serious tone and formal terminology that generally characterized clinical discussions.

The problem of providing reliable descriptions of sounds evidently preoccupied René Laënnec. He employed a range of similes, describing one lung sound, for example, as "a tinkling similar to that of a small bell just ceasing to ring, or of a fly buzzing in a china vase" (1846, 320). Others resembled, for instance, snoring, the cooing of a wood pigeon, or the rubbing of a bass string (Marks 1972, 71–72). Lachmund (1999) writes that Skoda, a key figure in the development of the "Viennese approach" to auscultation, also made use of analogies between body sounds and acoustic experiences from the everyday soundscape, though to a lesser extent than Laënnec.

Both Laënnec and Skoda also struggled to produce functional classification systems for the lung sounds they detected. Lachmund (1999) describes how Skoda conscientiously critiqued and reworked an (over)elaborate and fine-grained taxonomy put forward by Laënnec, producing a simpler and more straightforward system of his own. In the taxonomy of heart murmurs used and taught at St. Thomas's, the sounds were described by reference to four parameters: tone, volume (on a scale of I to VI), place in the cardiac cycle, and location on the chest. Thus, it became possible to refer, for instance, to "a loud, grade IV, ejection-systolic murmur in the aortic area." A murmur meeting these criteria would usually indicate aortic stenosis. Use of this system was evident in numerous referral letters, suggesting that it had attained a high degree of standardization in communities of clinical practice in Britain.

This system of classifying heart sounds did not mean that there were no longer difficulties in establishing consensus on the nature and significance of heard sounds. Auscultation is inevitably interpretive or hermeneutic. Findings made by using the technique are contestable. *The Oxford Handbook of Clinical Medicine* states that "[t]he first and second sounds are usually clear"; however, "[c]onfident pronouncements about the other sounds and soft murmurs may be difficult. Even senior colleagues disagree with one another about the more difficult murmurs" (Longmore, Wilkinson, and Torok 2001, 80). During my fieldwork I saw a number of instances in which a doctor would have his assessment of a heart murmur corrected by a more senior member of his team or in which a third doctor was brought in to offer an opinion on a murmur whose significance two other doctors could not agree on. Importantly, however, systems for communicating about and classifying body sounds have evidently been *good enough* to make auscultation viable and applicable as a diagnostic technique in a range of Western medical contexts. By developing their practical and sensory understanding of auscultation alongside a more technical appreciation of the structure and significance of heart murmurs, the students with whom I worked gradually increased their proficiency at detecting and interpreting sounds. They became more effective participants in and bearers of what Sterne calls "medicine's acoustic culture" (Sterne 2003b).

A Dying Art?

In the medical culture in which my fieldwork took place the stethoscope was considered by some to represent something of a throwback to the "old world."

Though still in routine use, a number of doctors referred to the instrument as "a dying technology" or to auscultation as "a dying art." The difficulties of consistent interpretation and reliable communication over heart sounds (as indicated earlier, a problem throughout the history of the use of the technique) were considered by some to be unacceptable flaws. Though in the right hands (and ears) a highly sensitive instrument, the stethoscope was felt by some to have reached its diagnostic limits. Investigations could be done with greater thoroughness and accuracy and with less scope for interpretive variation by more powerful technologies. Echocardiography, for instance, also known as cardiac ultrasound, can be used to produce detailed images of the interior of the heart, showing the valves and making their functioning clearly visible in real time. It can also create images of hemodynamic flow and allow the thickness of the heart's muscle walls to be measured with considerable accuracy.

The students with whom I worked knew they would be obliged to demonstrate a degree of skill in auscultation in order to pass exams that would allow them to progress through medical school and eventually qualify. But they were also conscious of the diagnostic power of echocardiography. They could imagine that, once qualified, listening would be necessary in making an initial assessment of whether a patient's heart was normal. However, they felt that, should any abnormality be detected, they would almost certainly be expected to send the patient for an echocardiogram. They suspected they might never have to make fine distinctions between subtly different sounds. Auscultation might effectively become a form of triage for them.

During fieldwork I attended several teaching sessions with a young woman named Zani, a doctor from Pakistan who was attending St. Thomas' in order to qualify to practice in Britain. She pointed out that in Pakistan many hospitals are comparatively underresourced and tend to be less technologically sophisticated than those in the West. Echocardiography was not widely available and was too expensive for routine use. As a consequence, auscultation was considered to be a highly important component of both medical training and practice. Zani believed the training in stethoscopic listening she had received in Pakistan was more rigorous than that undergone by most students in Britain. She also felt it was important that she be self-assured and confident in her listening skills. Clearly the emphasis placed on expertise in stethoscopic listening varies considerably according to cultural locale. I noticed that Dr. Coltart told several groups of students about the time he had spent at Stanford University, where he had had a visiting professorship. He related how at this prestigious medical school he had impressed his colleagues with his bedside diagnostic abilities and in particular his skill at auscultation. Physicians there, he said, did not receive such extensive training in stethoscopic listening as those in the UK. They were generally more heavily dependent on test results than their counterparts in Britain. His remarks resonate with comments in a number of written sources (Mangione and Nieman 1997; Kirsch 1998; Babu 1999).

It is easy to see that in auscultation, the "network" of relationships between people and technologies used in making a diagnosis is relatively simple (Latour 1987, 204).

The doctor "enlists" the stethoscope and engages his own skill, knowledge, and powers of judgment in examining a patient, sometimes also conferring with other doctors (Latour 1987, 178). In echocardiography, a wider network of machines and technicians becomes involved in diagnostic decision making. The reconfiguration of people and technology this wider network requires may be resource intensive and expensive to operate, only becoming viable in comparatively wealthy settings. Where it is mobilized, however, diagnostic agency is dispersed, and skill sets are reconfigured. The capacity to auscultate may no longer be demanded of the physician as a priority.

Yet even in contexts where echocardiography is in routine use, it seems hasty to state categorically that the stethoscope is "dying" or that auscultation is "a dying art." Echocardiography and auscultation do not necessarily cover precisely the same diagnostic territory or have identical purposes. For instance, auscultation may be useful and expedient in the first phase of examination. It may provide a means of distinguishing serious problems from trivial ones. It may also offer an efficient means of repeatedly checking or monitoring the health of, for instance, a heart valve over time, while echocardiography creates more detailed diagnostic information where time and resources allow and is likely to be of particular value where patients are being prepared for surgery or other procedures. For Dr. Coltart it was important that his students should be independent and resourceful and not rely too much on the wider networks of equipment and technical knowledge that echocardiography requires. Where he determined the orientation of classes, auscultation was emphasized as a key skill and it seems unlikely that the students he was training will soon definitively pronounce the stethoscope dead. Indeed, they are more likely, through their own practice, to testify to its continuing life. Nonetheless, the students also indicate that auscultation has both a variable status and a value that are contingent upon the availability (or otherwise) of more powerful medical technologies and those with the expertise to operate them.

The Auscultation of Everyone

At every available opportunity Dr. Coltart told his students that they should listen to people's chests. This meant not just hospital patients but people everywhere. He joked that even when we were sitting on the Tube (London's rapid transit system) we should ask the person sitting next to us if it would be all right to listen to their heart. From many of its rooms St. Thomas' Hospital has remarkable views out over the River Thames, the Houses of Parliament, and Westminster Bridge. On one occasion Dr. Coltart made a sweeping gesture toward the bridge and asked us to imagine what would happen if we were to stop all the people walking across it and listen to their hearts: "Think of all those interesting sounds. There would be hundreds of people walking around with murmurs who didn't even realize they had them!" For Dr. Coltart the population of London could be understood as a vast collection of medical cases within and among whom lurked many murmurs.

The group would often kill time between teaching sessions drinking coffee in the hospital café. After the session in which Dr. Coltart had suggested that we listen to all Londoners, we began to discuss how strange this thought was. Tom explained:

> You suddenly realize that they are inside everyone, these heart sounds . . . and the stethoscope is like a window to that. If I just walked around the street and listened to everyone . . . I would be bound to hear murmurs that people never knew they had, and suddenly they've got a disease. Suddenly there's something wrong with them.

Harjit thought similarly: "I had never really thought about people in that way before. I could go around with my stethoscope, and I would pick up heart murmurs in people everywhere, not just in the hospital but in the street." A student named Dave described how, listening to a CD of heart murmurs as he traveled to the hospital by bus, he was surprised to find himself imposing the murmurs he was hearing onto people he saw in the street as if their (unhealthy) heart sounds were amplified or he could hear them from a distance. Making the population at large subject to (albeit hypothetical) medical scrutiny, then, the students effectively collapsed any distinction between patients and nonpatients. Adopting a kind of stethoscopic macrofocus, they evoked the project of "surveillance medicine," with "all persons . . . becoming patients" (Armstrong 1995, 397).

In *Medicine, Rationality, and Experience,* Good explores the manner in which Harvard Medical School students are led to "think anatomically" through their training and how as a consequence they internalize "an alternative way of seeing" (1994, 73). Good suggests that the students carry and apply the gaze they acquire at medical school beyond the campus. After conducting dissections and autopsies, for instance, they begin to mentally anatomize the living people they encounter in everyday life. Similarly, the students with whom I studied found their acquired gaze and, in this particular instance, their acoustic perspective on the body applying beyond the hospital.

Another of Dr. Coltart's teaching strategies was to encourage his students to practice listening on the people close to them, as well as people in and outside the hospital. They should auscultate their boyfriends and girlfriends, as well as their family members. This brought the stethoscopic focus to a domestic or family rather than a societal level. The change proved worrying and difficult. There was always a possibility that the students might hear something they didn't want to. As Mary explained:

> When you're in hospital, you want to hear murmurs. You want to hear loads of noises. But I was listening to my Dad's chest, and I was thinking "Please, let there be just normal heart sounds, no murmurs." I was worried because I know he smokes. And I was thinking "Come on Dad! Be normal! Be normal!" Then I was listening and I thought "Thank God I can't hear anything! Thank God there are no murmurs."

Rishi told me about the time he and his identical twin brother, who was also studying medicine in the same year, listened to their father's chest:

> I remember my brother listening to my dad's chest. He's had an MI [myocardial infarction, commonly known as a "heart attack"] in '95. Anyway, my brother was listening to my dad's chest, and he heard a diastolic murmur. I came into the room, and I listened and heard it, too. My brother and me both agreed it was a diastolic murmur. It worried my Dad—enough to make him go to the GP [general practitioner].

Several other students found their family members were unwilling to undergo the examination at all. They were too concerned about what the stethoscope might reveal. The students were also uncomfortable putting close relatives under their clinical gaze. They disliked momentarily reformulating kinship relationships as doctor/patient relationships. They found they could not maintain the position of clinical/social distance and detachment that characterized their relationship to the people they saw during teaching sessions.

Sinclair writes that the medical students he studied were generally encouraged to develop and test their anatomical knowledge by relating it to their own bodies (1997: 153). However, when it came to auscultation they were told "Don't listen to your own heart; it's confusing!" (1997: 203). During my fieldwork Dr. Coltart actively encouraged us to listen to ourselves, a technique known as "autoauscultation." He said it would give us a chance to familiarize ourselves with normal heart sounds (assuming our bodies would be clinically normal and reaffirming a dialectic between the "healthy" doctor or medical student and the "sick" patient) and would enable us to think through the relationship of the heart sounds to the cardiac cycle. Auscultation could be properly practiced only *on a person*, and as we ourselves were more readily available teaching material even than patients it made sense to listen to our own hearts.

Still, bringing a stethoscopic focus to the individual self could be frightening; as Tom put it, "You just never know what you're going to hear." He continued: "My Dad has had a few heart problems, and I'm a bit of a hypochondriac . . . I don't really want to turn this ear that I'm being trained to use on myself." In fact, both Tom and another student named Alistair found they had quite bad sinus arrhythmia. Sinus arrhythmia can often be observed in young men. The heart tends to speed up perceptibly during inspiration and to slow down again during expiration. Although not a sign of disease or abnormality, it can be disconcerting, particularly if an entirely regular heart rhythm is expected. Harjit experienced a similar fear of what she might hear: "I came back from class and was sitting at my desk, and I thought, 'Wait, I've never listened to *my* heart before.' I was uncertain about whether I should or not. I was very scared that I might hear something bad." None of the students I studied detected an abnormality, though it later became known that a female medical student in a different class had heard a murmur and become very distressed. She had gone crying to her teaching doctor and had never been entirely convinced by his reassurances that hers was a flow murmur and therefore "innocent." It is easy to understand why she was reluctant to believe him given the strong emphasis placed on murmurs as indicators of valve disease. Autoauscultation then, and the diagnostic anatomization of ones self it involves, was responsible for cases of what Sinclair refers to as 'medical student disease', a form of hypochondria which frequently results from the student's application of their clinical knowledge to their own bodies.

For those who heard their heart sounds as normal, autoauscultation became a reassuring, even pleasant experience. Some students would try running on the spot while listening to themselves, seeing how fast they could make their hearts beat and how loud they could make them sound. Rishi's friend Ambrose said that he would spend large amounts of time listening to his own heartbeat and would lie in bed, sometimes falling asleep with the stethoscope still in his ears. He found the sound comforting and speculated that this might be because the sound of one's own heartbeat created acoustic conditions similar to those experienced in the womb. It is clear, though, that diagnostic techniques and the perspectives they create do not remain restricted to the hospital setting. Instead, they spill out of the clinical context and create informal perceptual engagements and interactions. Students encounter not only "the body," unassociated with a specific owner or linked to a socially distant patient, but also socially close and familiar bodies through medical methods of perception. Turning the stethoscope back on themselves, they also encounter their own lived and experienced bodies —what Draper (2002) calls the "my body"—in new and potentially unsettling ways.

Conclusion

I have introduced the practice of cardiac auscultation as a specialized appreciation of the soundscape of the body and briefly described the manner in which medical students begin to learn the auscultation of the heart. The training in auscultation that medical students receive grants them a particular acoustic perspective on the sonic space of the bodily interior, amplifying it, bringing it to the center of their attention. Importantly, however, it is not only the patient body that is acoustically configured through auscultation. Medical students must also change their own bodies in the course of their training. They are obliged to learn new bodily techniques—types of practical sense and sensitivity—in order to become competent auscultators. They must take on unfamiliar frames of auditory attention and analysis.

Writing on fetal ultrasound, Draper (2002) describes the manner in which the introduction of this technology allowed women to see images of their unborn babies. These images were shared with the technician operating the scanner and could also be distributed to partners, family, friends, and so on. Draper argues that in this process the uniqueness of the woman's pregnancy experience is disrupted. The fundamental axis of pregnancy shifts from one of haptic hexis (characterized by notions of "touch," "feeling," and "being") to an engagement with a visual image. I suggest that Draper slightly overstates her case; after all, the ultrasound scan does not erase the hapsis of pregnancy altogether. Nonetheless, she does make the important point that techniques of medical perception have the capacity to affect radically the way in which people engage with their own bodies. As the example of autoauscultation demonstrates, the stethoscope can create new senses of bodily awareness and possibility organized through sounds. At the same time, as was shown when the students began to listen to their friends and families, it can produce or exacerbate anxieties over what constitutes a healthy body and how a healthy body *should* sound.

Historically, auscultation has been involved in the wider medical project of deconstructing the body by positioning its organs, tissues, and processes as "scientific work objects" (Sharp 2000, 298). More recently, echocardiography might be construed as another means by which scientists technologically shape and manipulate sound (albeit ultrasound) in order to propagate and reproduce this clinical-perceptual approach. At the beginning of this chapter I briefly described the soundscape of the hospital wards at the Edinburgh Infirmary. One feature was the electronic tones produced by medical equipment (in particular the electrocardiograph). This sound was also very noticeable in the wards of the cardiothoracic unit at St. Thomas' Hospital, where the monitoring of patients' hearts was particularly intensive. Sound, then, plays an important role in medical projects of surveillance, as well as anatomization. I would suggest that medical techniques like echocardiography and electrocardiography and potentially even tools such as the iStethoscope Pro described earlier, are dictating new chapters in the story of medicine's sonic mediation of information about bodily spaces, surfaces, states, and events. Interestingly, an "ears on approach" will be crucial not only for medical students, but also for those involved in science and technology studies as they seek to understand new acoustic technologies and their significance for the production of medical and scientific knowledge.

REFERENCES

Ackerman, Diane. *A Natural History of the Senses*. New York: Vintage, 1990.

Armstrong, David. "The Rise of Surveillance Medicine." *Sociology of Health and Illness* 17(3) (1995): 393–404.

Babu, Ajit N. "Death of the Stethoscope." *American College of Physicians—Internal Medicine*. http://www.acpinternist.org/archives/1999/03/letters.htm. 1999.

Barany, E. "A Contribution to the Physiology of Bone Conduction." *Acta Otolaryngol.* 26 (1938): 1–223.

Bourdieu, Pierre. *The Logic of Practice*. Cambridge: Polity, 1980.

Carpenter, Edmund, and McLuhan, Marshall. "Acoustic Space." In *Explorations in Communication*, ed. Edmund Carpenter and Marshall McLuhan, 65–70. Boston: Beacon, 1960.

Draper, Janet. "It Was a Real Good Show: The Ultrasound Scan, Fathers, and the Power of Visual Knowledge." *Sociology of Health and Illness* 24(6) (2002): 771–95.

Feld, Stephen. "Waterfalls of Song: An Acoustemology of Place Resounding in Bosavi, Papua New Guinea." In *Senses of Place*, ed. Steven Feld and Keith H. Basso, 91–135. Santa Fe: School of American Research Press, 1996.

Fleming, Peter R. *A Short History of Cardiology*. Amsterdam: Rodopi, 1997.

Foucault, Michel. *The Birth of the Clinic: An Archaeology of Medical Perception*. New York: Vintage, 1973.

Furst, Lilian. R. *Between Doctors and Patients: The Changing Balance of Power*. Charlottesville: University Press of Virginia, 1998.

Gell, Alfred. "The Language of the Forest: Landscape and Phonological Iconism in Umeda." In *The Anthropology of Landscape: Perspectives on Place and Space*, ed. Eric Hirsch and Michael O'Hanlon, 232–54. New York: Oxford University Press, 1995.

Good, Byron J. *Medicine, Rationality, and Experience: An Anthropological Perspective*. New York: Cambridge University Press, 1994.

Hansen, Helen P. "Patients' Bodies and Discourses of Power." In *Anthropology of Policy: Critical Perspectives on Governance and Power*, ed. Chris Shore and Susan Wright, 68–81. New York: Routledge, 1997.

Hooke, Robert. "A General Scheme, or Idea of the Present State of Natural Philosophy." In *The Posthumous Work of Robert Hooke*, ed. Richard Waller. London: Smith and Walford, 1705.

Kirsch, Michael. "The Death of the Stethoscope: Murmurs of Discontent." *American College of Physicians—Internal Medicine*. http://www.acponline.org/journals/news/dec98/stetho.htm. 1998.

Lachmund, Jens. "Between Scrutiny and Treatment: Physical Diagnosis and the Restructuring of 19th-Century Medical Practice." *Sociology of Health and Illness* 20(6) (1998): 779–801.

———. "Making Sense of Sound: Auscultation and Lung Sound Codification in Nineteenth-Century French and German Medicine." *Science, Technology, and Human Values* 24(4) (1999): 419–50.

Laënnec, René-Théophile Hyacinth. *A Treatise on Mediate Auscultation and on Diseases of the Lungs and Heart*. Trans. a member of the Royal College of Physicians. London: Bailliere, 1846.

Latour, Bruno. *Science in Action: How to Follow Scientists and Engineers through Society*. Milton Keynes, UK: Open University Press, 1987.

Longmore, Murray, Ian Wilkinson, and Estee Torok. *The Oxford Handbook of Clinical Medicine*. New York: Oxford University Press, 2001.

Mangione, Salvatore. *Physical Diagnosis Secrets*. Philadelphia: Hanley and Belfus, 2000.

———, and Linda Z. Nieman. "New Doctors Have Dangerously Poor Stethoscope Skills." *Journal of the American Medical Association* 278 (1997): 717–22.

Marks, Geoffrey. *The Story of the Stethoscope*. Folkestone, UK: Bailey and Swinfen, 1972.

Martinet, X., J. L'Helgouarc'h, L. Roche, I. Favoulet, and P. Cougard. "Laënnec, ré-inventeur du stéthoscope?" *Presse Medicale* 27(30) (1998): 1534–35.

Maulitz, Russell C. *Morbid Appearances: The Anatomy of Pathology in the Early Nineteenth Century*. New York: Cambridge University Press, 1987.

Nash, June. "The Logic of Behaviour: Curing in a Maya Indian Town." *Human Organisation* 26 (1967): 132–40.

Nicolson, Malcolm. "Having the Doctor's Ear in Nineteenth Century Edinburgh." In *Hearing History: A Reader*, ed. Mark M. Smith, 151–186. Athens and London: University of Georgia Press. 2004.
Porter, Roy. *The Greatest Benefit to Mankind: A Medical History of Humanity from Antiquity to the Present*. London: Fontana, 1997.
Rackrow, Eric. "A Brief History of Physical Diagnosis." http://www.antiquemed.com/invention.html. 2009.
Rée, Jonathan. *I See a Voice: A Philosophical History of Language, Deafness, and the Senses*. London: Flamingo, 1999.
Reiser, Stanley Joel. *Medicine and the Reign of Technology*. New York: Cambridge University Press, 1978.
———. "The Science of Diagnosis: Diagnostic Technology." In *The Companion Encyclopedia of the History of Medicine*, ed. William F. Bynum and Roy Porter, 826–51. New York: Routledge, 1997.
Rice, Tom. "'Beautiful Murmurs': Stethoscopic Listening and Acoustic Objectification." *Senses and Society* 3(3) (2008): 293–306.
———. "Soundselves: An Acoustemology of Sound and Self in the Edinburgh Royal Infirmary." *Anthropology Today* 19(4) (2003): 4–9.
———. "The hallmark of the doctor: the stethoscope and the making of medical identity." *Journal of Material Culture* 15(3) (2010a): 287–301.
———. "Learning to listen: auscultation and the transmission of auditory knowledge". *Journal of the Royal Anthropological Institute* 16(1) (2010b): S41–S61.
Schafer, R. Murray. *The Tuning of the World*. New York: Knopf, 1977.
Sharp, Lesley A. "The Commodification of the Body and Its Parts." *Annual Review of Anthropology* 29 (2000): 287–328.
Sinclair, Simon. *Making Doctors: An Institutional Apprenticeship*. New York: Berg, 1997.
Sterne, Jonathan. *The Audible Past: Cultural Origins of Sound Production*. Durham, N.C.: Duke University Press, 2003a.
———. "Medicine's Acoustic Culture: Mediate Auscultation, the Stethoscope, and the Autopsy of the Living." In *The Auditory Culture Reader*, ed. Michael Bull and Les Back, 191–217. New York: Berg, 2003b.
Tedlock, Barbara. *Time and the Highland Maya*. Albuquerque: University of New Mexico Press, 1982.
Tonndorf, Juergen. "Bone Conduction." In *Foundations of Modern Audition Theory*, ed. Jerry V. Tobias, 206–35. New York: Academic Press, 1972.
Welsby, P. D, G. Parry, and D. Smith. "The Stethoscope: Some Preliminary Investigations." *Postgraduate Medical Journal* 79(938) (2003): 695–98.

CHAPTER 13

DO SIGNALS HAVE POLITICS? INSCRIBING ABILITIES IN COCHLEAR IMPLANTS

MARA MILLS

INTRODUCTION: THE NEURAL-COMPUTER INTERFACE

IN 1998 roboticist Hans Moravec published a millennial prediction about the imminent convergence of humans and machines—based on evidence from the evolution of sound technologies:

> In a few decades, people may spend more time linked than experiencing their dull immediate surroundings . . . Linked realities will routinely transcend the physical and sensory limitations of the "home" body. As those limitations

I would like to thank Michael Chorost and Charles Graser for their generous feedback and thoughtful comments on drafts of this chapter. I am also grateful to Susan Burch and Michele Friedner for their assistance with specific concepts.

> become more severe with age, we might compensate by turning up a kind of volume control, as with a hearing aid. When hearing aids at any volume are insufficient, it is now possible to install electronic cochlear implants that stimulate auditory nerves directly. Similarly, on a grander scale, aging users of remote bodies may opt to bypass atrophied muscles and dimmed senses and connect sensory and motor nerves directly to electronic interfaces. Direct neural interface would make most of the harness hardware unnecessary, along with sense organs and muscles, and indeed the bulk of the body. The home body might be lost, but remote and virtual experiences could become more real than ever. (Moravec 2000, 169)

Electroacoustics has been at the forefront of signal engineering and signal processing since "the transducing 1870s," when the development of the telephone marked the first successful conversion of a sensuous phenomenon (sound) into electrical form and back again (Hunt 1954, 37). By the second half of the twentieth century, acoustics research centers in the United States, such as Bell Telephone Laboratories, the Harvard Psychoacoustic Laboratory, and Bolt, Beranek and Newman, had made central contributions to the digital coding of signals and to computer networking; the "overarching themes," as John Swets has argued, were "information processing and man-machine integration."[1]

In 1984, after FDA (U.S. Food and Drug Administration) approval of the 3M cochlear implant for adults, neuroprosthetics entered the commercial sphere.[2] The cochlear implant (CI) delivers electrical signals directly to the auditory nerve. With approximately 200,000 users today, these devices remain the most common neural-computer interfaces in the world.[3] The current technology includes an external microphone and a speech processor—a tiny computer with software that can be upgraded. The processor variously transduces, samples, and codes environmental sound in order to transmit it to the auditory nerve through up to twenty-four electrodes, which functionally replace the thousands of hair cells in the inner ear.

Concurrent with their entrenchment in futurist discourse, cochlear implants quickly entered the canons of bioethics and disability studies, raising questions about the definition of impairment, the feasibility of pediatric informed consent, and the cost-effectiveness of neuroprostheses. Many bioethicists have taken up the Deaf culture or linguistic minority critique of implantation, which situates this technology in the long history of eugenicist attempts to promote oralism through the medical eradication of deafness and through pedagogical bans on sign language (Beard 1999; Berg, Alice, and Hurst 2005; Crouch 1997; Levy 2002; Sparrow 2005). Despite the prominence of the cochlear implant in disability studies, bioethics, and science fiction, however, it has inspired little research in science and technology studies (STS). Stuart Blume, a sociologist of science and parent of a deaf child and a hard-of-hearing child, conducted the most substantial fieldwork in the 1990s on the reception of implants in France, the United States, England, Sweden, and the Netherlands (Blume 2000). Blume detected "two very different accounts of cochlear implantation":

> One is a tale of medicine's triumph, akin to many other such tales: a tale of courageous pioneers, of the wonders of medical science and technology.

> The other is in a genre which has emerged only in the past two decades and which highlights the subordination of medicine to surveillance, social control and normalisation. This a tale of the oppression of the deaf: of hearing society's inability to accept deaf people for what they are.[4]

Yet more than one identity group is accommodated within the category of deafness. It includes members of Deaf culture, self-defined as linguistic minorities who sign (and whose "disability" is largely an effect of the built environment and social stigma); late-deafened adults, who tend to claim disability from hearing loss; and oral deaf and hard-of-hearing individuals of all ages.[5] Without rejecting Blume's assertions about the Deaf response to cochlear implants, it is possible to write a third history of this technology—a history that includes the active participation of late-deafened volunteers in research and development and at the same time depicts their distinctive patterns of stigmatization and exclusion. In *The Artificial Ear: Cochlear Implants and the Culture of Deafness*, published in 2010, Blume largely maintains his focus on Deaf culture and the debate surrounding pediatric implantation. He concludes, "The demands of Deaf community leaders and advocates had little or no effect either on development of the cochlear implant or on the beginnings of local implantation practices. The experience of deaf people was not accepted as essential or even as relevant" (Blume 2010, 197).[6]

Reflecting on the participation of cochlear implant users (including himself) in trials of sound-processing software, technology theorist Michael Chorost offers a different perspective on the relevance of deaf experience to the making of this technology:

> Even without being able to write code themselves, implant users do have a crucial impact on how the code is written. When engineers write new code, they have to test it on implant users to see if it helps them hear better. They also have to find out if implant users like it and can get used to it. To do that they need to recruit articulate users and convince them to offer their time. It's a highly collaborative process and is integral to how the field makes new advances. (Michael Chorost blog, comment posted January 5, 2006)[7]

In this chapter I follow some of the trajectories by which the autoexperiments, field notes, and laboratory tests of early users have left traces in the hardware, as well as the software, of cochlear implants. A species of "the co-construction of users and technologies" genre in STS, this chapter also considers the distant but intimate relations between lead and end users (Hippel 1986; Bijker 1995; Oudshoorn and Pinch 2003).

On the one hand, the anatomy and phenomenology of experimental research participants exert a subtle influence on the experiences of users downstream. STS scholars have noted that "scripts"—defined by Madeleine Akrich as "the representations of users" embedded within technology—often materialize during the research and development, clinical trial, or testing phases of technical development.[8] Designers do not simply "project" users into cochlear implants; from surgery to speech processing, these devices are inscribed with the competencies, tolerances, desires, and psychoacoustics of early users.

On the other hand, the recommendations of test subjects have as often been expunged as built into cochlear implants. In what I am calling *cross-purpose collaboration*, social norms, medical ideals, and commercial interests have vied with the needs and preferences of deaf people in the construction of implant technology. In 1980 Langdon Winner asked whether artifacts had politics; for present-day electronic and digital media, politics can be found at the level of signals. Specifically, CI signal processors embody a range of cultural and economic values, some of which are deliberately "scripted" into design, others of which accrete inadvertently. These scripts include the privileging of speech over music, direct speech over telecommunication, nontonal languages over tonal ones, quiet "listening situations" over noisy environments, and black-boxed over user-customizable technology.[9] All technical scripts are "ability scripts," and as such they exclude or obstruct *other* capabilities. Due to the complexity and opacity of electronic technology, these constraints often prove impossible for users to circumvent. The "home body" is thus not lost with this new medium; practices of listening are radically materialized. Users, moreover, experience their devices across the corporeal registers of hearing, vision, and tactility.

Bionic Rhetoric

According to Bonnie Tucker, a deaf legal scholar, "The hatred with which Deaf culturalists view cochlear implants is expressed in the ASL sign for the cochlear implant, which includes a two-fingered stab to the back of the neck, indicating a 'vampire in the cochlea' " (Tucker 1998, 9). Although the sign language community is more diverse than is often acknowledged—and increasingly includes bicultural users of cochlear implants—existing animosity toward these devices derives from countless disappointments in Western science and medicine, accumulated over the last two centuries. Harlan Lane, a hearing author who earned a MacArthur Award for his philosophical histories of signing Deaf culture, has detailed the repeated scientific victimization of deaf individuals—from Jean-Marc Gaspard Itard's application of leeches and electricity to his students' ears in the nineteenth century, to Alexander Graham Bell's prohibitions on deaf intermarriage, to Nazi sterilizations and executions (Lane 1993). Harry Lang, a deaf professor at the Rochester Institute of Technology, tells of the more subtle losses that have attended scientific "progress" for deaf people:

> A glimpse into history also provides some understanding of why there is so much emotion attached to technological advances. "Advances" in voice telephony led to a ninety-year delay in access to the telephone for deaf people. "Advances" in adding the sound track to silent movies led to more than forty years of lost access to films. For hundreds of years, deaf people, viewed as "disabled," have been treated with chemical and electrical "cures," sent up for airplane dives, and subjected to a multitude of other medical fixes. (Lang 2002, 91–92)

Equally sobering, cochlear implants have been correlated to sign language death and "cultural genocide," especially since 1990, when the FDA approved them for children (Wrigley 1996; Ladd 1985). In 2004, Australian linguist Trevor Johnston, whose parents are deaf (and sign), published a demographic survey in the *American Annals of the Deaf* (with the bittersweet title "W(h)ither the Deaf community?"), which predicted an end to Australian Sign Language (Auslan) within "half a lifetime" as the result of "improved medical care, mainstreaming, cochlear implants, and genetic science" (Johnston 2004, 370). *Sign Language Studies* dedicated its Winter 2006 issue to the international "comments" spurred by this article; authors reported comparable situations in Norway and—with a more gradual timeline—the United States. Teresa Blankmeyer Burke, professor at Gallaudet University, insisted upon the resilience of sign language and Deaf culture; she noted that if Johnston's forecast turned out to be true, however, it would constitute a novel "instance of scientific progress directly threatening a linguistic community" (Burke 2006, 175).

Stuart Blume has examined the establishment of "clinical feasibility" and respectability for cochlear implants throughout the 1970s and 1980s. He concludes that most Deaf people were initially indifferent; their response "was quite unlike that of actual or potential AIDS patients, who (at least in the United States) stressed the right to earliest possible access to what might prove a life-saving drug" (Blume 1997, 33).[10] Blume acknowledges that *deafened* individuals "deluged" electroacoustic and otological researchers with inquiries; deafened advocates in fact urged physicians in France, the United States, and Australia to develop the first implants and offered themselves for surgery (Blume 2010, 31, 33). Blume argues that physicians subsequently deployed "bionic rhetoric" and sensational performances by early volunteers to convince further test subjects that implants were beneficial. While Blume does not explain this ostensible drop-off in interest among deafened people, I suspect that the invasiveness of the experimental procedure and the limited definition of the first cochlear implants were deterrents, not to mention the outspoken skepticism of many others in the medical community. Moreover, after the 3M/House implant gained FDA approval in 1984, "the market grew far more slowly than had been anticipated," an outcome due in part to the tremendous expense of the device (Blume 1997, 38; see also Blume 2010, 51; Zeng 2007).

The rhetorical promotion of cochlear implants as "bionic ears" (as opposed to imperfect prostheses) unintentionally generated a counterrhetoric among Deaf activists. In the 1990s, Blume concluded that the force of their counterrhetoric was such that this "stigmatized and relatively powerless group [became] a significant actor in the process of technological change" (Blume 1997, 46). The contributions of Deaf actors were, in Blume's account, inhibitory—discouraging implantation rather than affecting the design and fabrication of these devices. However, with the global escalation of implant adoption (due to factors ranging from improved technology to lowered age limits for legal implantation), Blume (2010) has reevaluated the protests of the Deaf community as disappointingly ineffectual. He registers a nearly impassable ethical predicament: When hearing parents choose implants and mainstream (oral) education for their deaf children, they diminish the population of native signers. One solution, Blume suggests, entails "rewriting kinship"—rethinking the imperative of common ability within the family and imagining linguistic communities external to family or nation.[11] At the most basic level, some of this work could occur through changed counseling practices in otolaryngology clinics.

Single Channel

For the most part, the first implantations were conducted with late-deafened individuals; those born deaf were not considered good trial candidates because they could not compare the electronic "ear" to prior experiences of hearing. From the outset, a tension has existed between normalizing therapeutics and the unexpected effects of the technology. Implants provide a novel mode of auditory perception, and, at once, they often socially disable those who wear them. Although cochlear implants seem to promise the replacement of a lost sense, by audiological standards implant users continue to have a hearing impairment.

Direct electrical communication with the human auditory nerve dates to 1957, when Charles Eyriès and André Djourno implanted "Monsieur G." at l'Institut Prophylactique (Arthur Vernes) in Paris. The patient—an engineer—had lost hearing in both ears after a surgical procedure.[12] Eyriès was a practicing otolaryngologist, Djourno a trained physician who had turned to basic research in medical electricity. Djourno studied electrocution and electroshock in animals and cadavers, funded by Électricité de France, which was interested in potential countershock applications. To this end, he constructed a number of induction coil implants ("microbobinages") for muscle and organ telestimulation.

Earlier, in 1953, Djourno had met a deaf man who "had considered suicide" and "said that it was utterly unbearable, this condition of no longer hearing . . . he said: 'I would prefer any noise . . . even if it's far from a real sound.' " Djourno stimulated the man's ears temporarily with electrodes, and he heard a few sounds. In a published article soon thereafter, Djourno speculated that the implants he had designed might be appropriate for treating deafness.[13]

Djourno and Eyriès claimed that Monsieur G. had similarly "requested that the impossible should be tried in order to correct—even to the most limited degree—his total deafness. He was so insistent in his desire that we decided—in spite of the possibility of a failure, as we informed the patient—to embed an induced coil" (Djourno, Eyriès, and Vallancien 1957).[14] They implanted one of Djourno's coils through the skull to the eighth nerve on February 25, 1957. During postoperative testing, Monsieur G. heard a number of high chirps and whistles (Seitz 2002, 81). He participated in "reeducation" four times a week for several months; when delivered speech signals through a microphone, he perceived low tones as the tearing of "jute cloth," higher tones as that of "silk."

Despite the incredible acoustic distortion, Djourno and Eyriès reported:

> He very much likes to have the apparatus working for the pleasure of hearing people come and go, slamming the door, or listening to conversation going on around him. Turning off the apparatus plunges him into a silence which he finds unbearable for a few minutes. In contrast, an unfortunate bump against the microphone results in a violent noise which deafens him for several seconds. (Djourno and Eyriès 1957, 1417)

When the implant ceased functioning after some months, they repeated the procedure. This, too, failed, at which point Eyriès abandoned the project (see figure 13.1).[15]

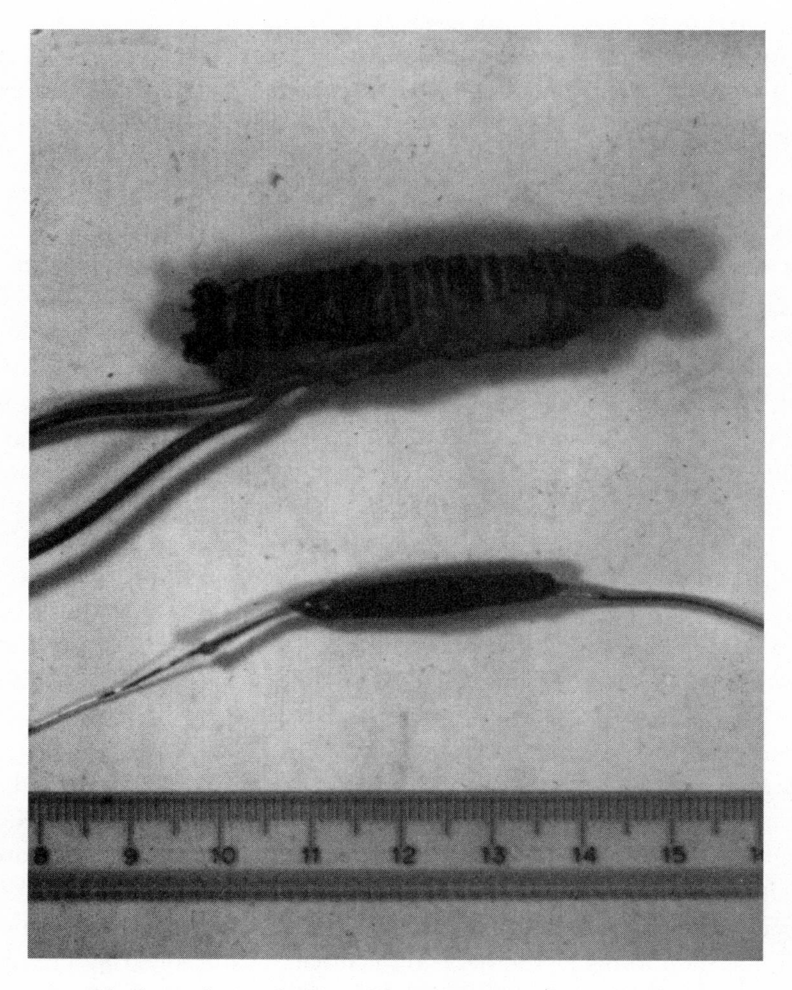

Figure 13.1 The lower "microbobinage" is the implant, the upper is the (external) signaling coil, which attached to a microphone transmitter. Designed by Djourno (France, 1950s). Photograph courtesy of the John Q. Adams Center for the History of Otolaryngology, Head and Neck Surgery, Alexandria, Va.

Djourno performed one final implantation, collaborating with otolaryngologist Roger Maspétiol. This second patient, N.T.L., was a Vietnamese girl who had lost her hearing after taking streptomycin for tuberculosis. She, however, was "reluctant" to undergo the surgery ("it was her father who had committed her to this operation").[16] Afterward, she disliked the "low moaning sound" produced by her implant. She also disliked wearing the external transmitter, even after they attached it to a headband so her hair would conceal it. "She spoke impeccably," Djourno commented, "yet complained a lot all the time because Indochinese men and women do not have the same accent or manner of speaking. She said 'It's a catastrophe, because when I begin to speak, it sounds like a man.' "[17] Over the course of two years N.T.L. visited the clinic only occasionally. When she and her family returned to Vietnam, they never contacted the doctors again. Djourno had by then begun planning a multichannel implant, based on the Fourier analysis seemingly performed by the human ear, but in 1959 the institute stopped funding this line of research (Seitz 2002, 84).

From the outset, then, implants and their electroacoustic signals had "politics." Implantation began as a response to *deafening*—a condition at once physical and social—caused, in these first cases, as a by-product of modern medicine. Whereas the adult engineer voluntarily immersed himself in a world of strange and even painful sounds, this capacity for self-determination was denied to the child. The assistive device easily became a stigma symbol, drawing attention to an otherwise invisible disability. Moreover, linguistic politics transferred instantly to this technology for communication: Which sounds counted as linguistic? How did tones correlate to gendered social norms? Should language be prioritized over other kinds of sounds? These early experiments raised the question of whether auditory cues, no matter how uncanny or uncomfortable, were in fact preferable to "unbearable" silence. In an interview near the end of his life, Djourno acknowledged that implants were not appropriate for everyone: "There are deaf [people] who do not want to be equipped with devices, who are happy with sign language, lip reading; they consider devices as a bother . . . I knew a family like that, [who used] a very complicated and perfected sign language, they fared much better than with any kind of electric stimulation . . . To impose a single solution on all the deaf, that's nonsense."[18]

In the United States, early volunteers also faced the difficult choice between impairment in sonic social settings, or surgical pain, atypical audition, and visible stigma. At the same time, a number of these volunteers were technical experts and enthusiasts. William House, who designed the 3M device (the first to be granted FDA approval), began his research in 1957, when a patient gave him a newsclipping about Djourno and Eyriès.[19] House and his brother owned a private ear institute in Hollywood, founded in 1946 by their father, Howard.[20] In early 1961 House collaborated with brothers John Doyle and James Doyle, a neurosurgeon and an engineer, respectively, to run preliminary tests on E.K., a deaf patient. House and the Doyles placed an electrode in E.K.'s inner ear, and the man—himself an engineer at a plastics plant—was able to describe the distinct sounds resulting from inputs such as pulses and square waves.[21] E.K. then tested a single-electrode implant for three weeks, followed by an insert of five gold wires that had to be surgically removed after two weeks due to allergic reaction.[22] Through these bodily demonstrations, it became evident that wires required insulation to prevent electrode failure; gold, moreover, was not a tolerable material.[23]

John Doyle soon disclosed these experiments to the press, and House dissolved their partnership.[24] House recalls, "We began to be deluged by calls from people who had heard about the implant and its possibilities. The engineer who had constructed the implant exercised bad judgment and encouraged newspaper articles about the research we were doing" (House and Urban 1973, 505). *Space Age News* was a particularly enthusiastic venue for these reports; by and large, cochlear implantation was unpopular with the medical establishment throughout the 1960s.[25]

When Dr. F. Blair Simmons of Stanford entered into cochlear implant research shortly after House and the Doyles, Blume notes, "The American Otological Society rejected presentation of this work at their 1965 meeting, while an application for funding to the National Institutes of Health (NIH) was turned down" (Blume 1995, 101–102). Simmons conducted an exploratory multichannel implant surgery on a human volunteer in 1964, but then turned to animal research to investigate such factors as surgical approach and ideal number of electrodes.[26] Simmons' move from clinical to basic research garnered respectability—and, by the 1970s, NIH funding—for further work on the electrical stimulation of the auditory nerve.

Robin Michelson, who began his career in private practice and then moved to the University of California-San Francisco (UCSF), similarly became interested in the possibility of cochlear implantation through clinical work, which he combined with animal research in the mid-1960s. Working with an engineer at Beckman Instruments, Michelson implanted several patients with single-channel devices around 1970 (Michelson 1971). He subsequently collaborated with a UCSF team that included Michael Merzenich and Robert Schindler to conduct basic research toward the development of multichannel implants (this research would lead to the Clarion model from Advanced Bionics). The UCSF group insisted that this animal research eventually be paired with psychoacoustic studies of human subjects. (Loeb et al. 1983, 252).

According to a report written by Caroline Hannaway of the NIH Office of History, House resisted animal studies and continued to feel "that at some point risks had to be taken. He believed the reluctance of scientists to pursue work involving human beings, and what he perceived as the greater readiness of granting agencies such as the NIH to support projects involving animal experimentation, to be a response to the Nuremberg trials" (Hannaway 1996, 6). House worked in a clinical context rather than a university one, and his early research preceded many of the current laws regarding human experimentation, such as the requirement of ethical review. Extensive records of this research are available in the archives of the American Academy of Otolaryngology-Head and Neck Surgery, including rare first-hand accounts from test subjects regarding their auditory sensations, personal experiments, and proposals for future technology. House's patient Charles Graser received one of the first portable cochlear implants, a trial device assembled by electronics technician Jack Urban.[27] Graser, who is still alive, has permitted me to quote from the journals of his field-tests.

A high school social science teacher and ham radio operator, Graser drove an oil rig in the summers to earn extra money to support his family. In 1959 his truck caught fire, and he was severely burned; over the course of several months in the hospital, he lost his hearing from the mycin drugs he was given. As a man in his forties, he had seen sign language only a few times; he once mentioned that he found it "beautiful," but his family, friends, and employment moored him in the English-language world. As a patient at the House Ear Institute in 1961, he first learned of the implant experiments; that summer he wrote to William House and volunteered to enroll.

House, at that point, was reluctant to proceed too quickly or to publicize his work. Graser wrote to him every six months until the end of the decade, inquiring about the possibilities for another experiment. Not until 1968, with advances in surgical plastics and miniaturized transistors—as well as the development of other medical electronics components in the growing field of artificial pacemakers—was House willing to make another attempt (House and Urban 1973). That year, he invited Graser to participate in an exploratory surgery with local anesthetic. He tried a temporary implant in several locations to determine its ideal placement and number of wires. He also tested the maximum intensity of the stimulus, noting, "When an intensity of two volts was introduced, he [Graser] responded by jumping, indicating that the pain threshold had been reached."[28] This new round of human experimentation met with more public criticism, most notably from Dr. Nelson Kiang of the Massachusetts Institute of Technology, who called the work "premature" (Kiang 1973, 512).

In 1970 Graser received a permanent "button" implant behind his right ear. Having five input wires, it divided the signal into separate bandwidths. Over the next four years Graser spent thousands of hours with House and Urban in the lab, testing different circuits, carrier waves, and modulation schemes. House remarked that "many of these devices took months to construct and proved worthless after a few hours of testing" (House and Urban 1973, 505). Graser, he attested, was a genuine collaborator (see figure 13.2):

> As an ex–ham radio operator he was a sophisticated listener and could fully describe the different signals presented as stimuli . . . He is an ambitious and goal-oriented individual who is tenacious in his desire to maximize the use he can derive from the implant . . . In addition, he is both articulate and an excellent observer. C.G. has been able to communicate to us much valuable information concerning his experience with the implant and has made thoughtful suggestions concerning ways of upgrading the system.[29]

Graser, on the other hand, was eager to have greater jurisdiction over the experimental process. In his own log, he recorded the following: "This electronic cochlea testing does bother you. It's like having someone say, 'Have a seat in the electric-chair while I fiddle with controls.' It may not hurt, but it is sometimes frightening

Figure 13.2 House (left) and Graser (right), c. 1974. Photograph courtesy of the John Q. Adams Center for the History of Otolaryngology, Head and Neck Surgery, Alexandria, Va.

in its intensity and your inability to control it. Even to begin with, you don't know how much you can take."[30] He was finally allowed to field-test a portable implant in 1972 even though House and Urban were not certain it was safe. There was a risk of infection; moreover, fluorescent lights, electrical wires, and highway radar traps caused interference.

Graser was aware that this prototype was never expected to provide "normal" hearing. In his field notes he commented, "You would probably describe my current progress as changing from profoundly deaf to just hard-of-hearing, but difficulty hearing and comprehending is in a completely different league from silence. For instance, tonight I can finally hear the bell that indicates that I am at the right hand margin, as I type this letter." In many respects, this early implant was a radically different and limited kind of ear: "I used to be a radio operator, and sometimes I would get a distant signal that I couldn't really hear. It sounded dim and garbled. That's the way this sounds. It's definitely an electronic sound." Yet in some ways, Graser's acoustic sensitivity exceeded that of his wife: "I will be startled by a brief exhaust sound of a car going by outside the house. Barbara doesn't even hear the car."

Graser began tinkering with his processor at home and recorded his findings about battery lifespan, microphone type and placement, and signal modulation. "I am constantly experimenting with the device," he wrote to Dr. House (House and Urban 1973, 510). He painstakingly documented the transformations of his domestic soundscape: "Walking on the floor in the house is not just padding sounds, but is more of a hammering with an echo for each step. Sound is too sputtery." "You can hear water run into the sink. Almost too noisy." His first transmitter had multiple dials for control over features such as carrier amplitude and frequency, modulation, microphone sensitivity, and high-filter cutoff. Graser recommended against standardizing the design: "I would have as many manual controls on the instrument as possible so that each patient could customize sound, much as if they were using a short wave receiver."[31] His field tests led to a number of concrete and lasting improvements: carrier waves that received less interference from environmental electricity; microphones worn at the head rather than in the pocket (where they picked up too much "clothing noise"); continued miniaturization, so the processor could also be worn behind the ear.[32] Like other lead users, Graser also contributed physically (if not deliberately) to surgical procedure; the determination of suitable implant materials; protocols for minimizing the destruction of hair cells; and evidence for the ability of the inner ear to withstand electrical stimulation over a period of many years.

In 1972 an electrode short damaged Graser's skin, causing House to work toward a fully implanted stimulator to replace the "button" model. As part of this restructuring, House decided to convert to a single-channel implant. Graser's "skin began to show evidence of retracting and reacting to the external button. Some leaking and shorting of the electrodes was observed. House secured all wires into a single bundle to prevent loss of the whole system. This event made it urgent that the electronics be converted into a single electrode system."[33] Graser received a fully implanted induction coil (as opposed to "hard wires") at his left ear in 1972, and in 1974, as a result of "secretions and debris around the button causing the wires to fail," his right ear was finally reimplanted.[34]

Blume regards William House as critical in establishing an international clinical reputation for cochlear implants. Before FDA approval of the single-channel implant (and before the FDA became responsible for medical devices), a very few volunteers participated in clinical experiments—each potentially having a large impact on design. Throughout the early 1970s Charles Graser served as House's primary evidence in publications and presentations.[35] In those years Nelson Kiang rebuked House for assuming that his trials with Graser were valid: "Enthusiastic testimonials from patients cannot take the place of objective measures of performance capabilities" (Kiang 1973, 512). Nevertheless, Blume contends that House's "early successes with a simple implant aroused the interest of clinicians in many countries, as well as of potential manufacturers" (Blume 2010, 173). Furthermore, as Hannaway points out, when the National Institute of Neurological and Communicative Disorders and Stroke (NINCDS) "sought the first objective evaluation of cochlear implantees in the United States" in 1975, it sent just thirteen individuals with single-channel devices—Charles Graser among them—to the lab of Robert Bilger at the University of Pittsburgh. "Bilger was a known skeptic about cochlear implants, and, as the result of the investigation, he was converted to being a modest supporter. So were the others who had previously opposed cochlear implants" (Hannaway 1996, 23).

Not all of Graser's recommendations survived in commercial versions of the technology. In the most significant enduring design shift, House began to build portable models with a single control dial, "eliminating the necessity for frequent fine adjustments"—in other words, eliminating the personal control and auditory customization that Graser so appreciated.[36] This black-boxing of the technology compounded disability, implying a lack of technical facility among users, as well as an obligatory dependence upon physicians and medical engineers. Moreover, although Graser strongly preferred receiving signals at both ears, bilateral cochlear implants would largely be refused by insurance companies for the next thirty-five years. Here, the drive toward normalization crossed purposes with the economics of health care (see figure 13.3).

Signal Processing

House eventually partnered with 3M Corporation to produce a commercial device; in 1984 their single-channel model was the first cochlear implant to receive FDA approval. By that time, investigators in Australia, France, Austria, England, and elsewhere in the United States had taken up cochlear implants, mostly with an interest in marketing a multichannel device. At the University of Melbourne, Graeme Clark (whose father was deaf) began work on a multichannel implant in the 1970s; he claims that his subjects *solicited him* after a 1977 news brief in the Melbourne *Herald*. Clark arranged for an ethics committee to oversee his research, following the 1975 Declaration of Helsinki amendment regarding biomedical experiments on humans. Working with a single patient implanted in 1979, Clark concentrated on the problem of signal processing: what information to extract from the sonic environment for transmission down a limited number of channels; the rates at which to stimulate the auditory nerve; the placement of electrodes:

> In theory, the coded signals should simulate in the auditory nerve fibers the temporal and spatial patterns of action potentials seen when sounds excite the normal cochlea . . . Therefore, it is necessary to know if speech processors can present speech as coded signals with a more limited number of stimulus channels and still adequately simulate the physiology. Alternatively, the speech processors should extract only the essential speech information that can be processed by the auditory nervous system via a relatively small number of stimulus channels. (Clark 1992, 95)

Clark tested a number of processors on this first patient, who had been deafened as an adult due to a head injury, and he evaluated each strategy according to the man's self-reported ability to perceive speech. Clark ultimately decided, in phonetics terms, to build a processor that extracted fundamental frequency and one to two speech formants.

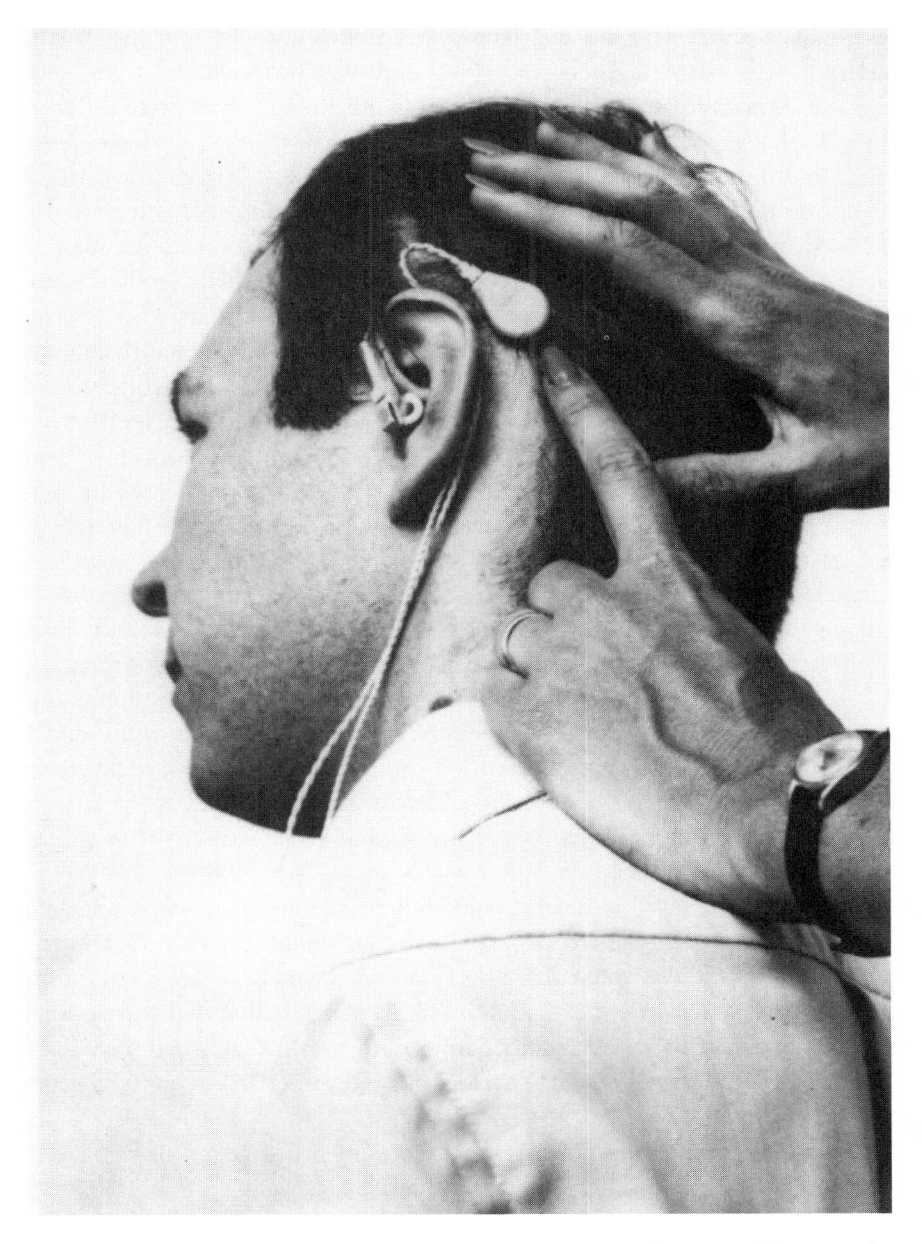

Figure 13.3 Early single-channel implant with fully implanted electrode. (The superfluous pointing finger objectifies and diminishes the user of this device.) Photograph courtesy of the John Q. Adams Center for the History of Otolaryngology, Head and Neck Surgery, Alexandria, Va.

After implanting a second person in 1979, Clark determined to proceed with a commercial device, necessitating a series of clinical trials on deafened adults. With no government funding, Clark could afford to give free implants to only his first six patients. Subsequent "volunteers" paid a minimum of $10,000 to participate in experiments that their health insurance would not cover. In 1985 the Australian (Nucleus) twenty-two-channel implant was approved by the FDA.

In the United States, researchers at UCSF, MIT, the Research Triangle Institute, and the University of Utah—often in collaboration with one another—generally departed from the feature extraction approach in the 1980s. Rather, they attempted to build multichannel speech processors, both analog and digital, that simply filtered and compressed incoming sound waves, with a focus on the frequencies of speech. Although the implants developed during this time varied in their numbers of electrodes, they shared many other hardware features. Users of experimental implants often tested multiple speech processors (eventually instantiated as software), and processing schemes in turn shared emerging psychoacoustic data. In *Rebuilt: How Becoming Part Computer Made Me More Human*, a "scientific autobiography" of obtaining an implant, Michael Chorost acknowledges his own debts to this second generation of user-researchers, who helped establish how to increase the number of electrodes; how to prevent crosstalk between those channels; and how best to dissect a signal. He specifically recognizes Michael Pierschalla, who participated in the clinical trials of Ineraid—the third implant to receive FDA approval—at the Massachusetts Eye and Ear Infirmary and the Research Triangle Institute. "His feedback integrally shaped most if not all of the software that is widely used today, not just in my own implant but in those of other manufacturers as well . . . Just as the DNA of all my ancestors lives on in me, a bit of his sensibility lived on in my software. My own body bore the stamp of his intelligence and generosity" (Chorost 2005, 106).[37] Signal processing continues to be the major site of cochlear implant development today, and it develops with a nonlinear incrementalism.

Pierschalla had lost his hearing suddenly at the age of twenty, an effect of Cogan syndrome. He studied at the Rochester Institute of Technology and became an artist and a noted furniture craftsman. While living in Cambridge, Massachusetts, he read a news article about local cochlear implant research. A friend wrote on his behalf to Dr. Donald Eddington of the Massachusetts Eye and Ear Infirmary; by June 1985 Pierschalla had obtained an experimental six-electrode Ineraid implant.

Pierschalla lamented his estrangement from the hearing and the sign language worlds even as he embraced his implant:

> I came to my hearing loss late, too late to abandon my hearing ways, too old to change the culture of my nativity, and too resistant to losing the remnants of my identity. Thus, I tried to tell them I was like a man without a home. Looked on as deaf by those who hear, and as Hearing by the native Deaf, and so for years have existed between those worlds, knowing much of both, but embraced and understood by neither, longing for the seamless brotherhood of either one, but incapable of owning that due to circumstances . . . Finally, one day in the laboratory, a man turned on a switch and my head lit up with sounds and alien noise and before long I went home and for the first time in years, I called my momma on the phone and we both cried like babies, and things have been very different after that.[38]

Pierschalla participated in laboratory as well as field tests of this device: independently trying out different microphones from Radio Shack; documenting the recharge times of his batteries; recording environmental sounds (subways, busy streets) to play back in the laboratory. While watching a film, he would estimate speech intelligibility for off-screen voices and under conditions of theater noise, multiple talkers, and "background" music. Lecturing at the Museum of Fine Arts in 1992, he reflected on his interactions with physicians and scientists: "As a craftsman and an armchair philosopher, I've sometimes wondered where the tool ends and the art begins. In the same way, when I'm in the lab with these scientists, we at our opposite ends of the same wire, I wonder where the observer ends and the subject begins. And I don't have an answer."[39] On multiple occasions, he co-presented cochlear implant findings with otolaryngologists at medical conferences.

Shortly before Pierschalla's death in 2002, science writer Victor Chase interviewed him over the phone. Chase noted the man's transition from *patient* to *expert*:

> Pierschalla's knowledge and success was widely known within the cochlear implant community. As a result, in 1995 Med-El, a cochlear implant manufacturer based in Innsbruck, Austria, asked him to establish its North American office in Boston, as an organization of one. He accepted, and shortly thereafter he relocated the company's office to Research Triangle Park, North Carolina. He then worked on obtaining FDA approval for sale of the Med-El implant in the United States. (Chase 2006, 26)

In 1995 Pierschalla served as a member of the second NIH Consensus Conference on Cochlear Implants in Adults and Children. The panel concluded that the population of implant users was steadily growing, that the device seemed to be successful in children, and that "cochlear implants should be made available to adults with severe hearing loss," as well as those who were "profoundly" deaf (Hannaway 1996, 53). Along with this medical consensus, one by-product of the many years of advocacy by late-deafened implant users was thus a renewed medical encroachment upon the Deaf world.

Chorost worried that his own perception—and, in turn, his perspectives—might be unduly influenced by the anatomy and predilections of the test subjects whose "scripts" he had inherited. He quickly learned that the prosthesis "doesn't change your values in the slightest"—although it does embody an "epistemology," complete with certain biases (Michael Chorost blog, comment posted March 2, 2009). The drive of early investigators toward *speech* processing, for instance, is inscribed into his implant. Thus, music sounds "flat and dull . . . flutes and soprano saxophones sounded as though someone had clapped pillows over them . . . oboes and violins had become groans" (Chorost 2005). In an article written for *Wired* magazine ("My Bionic Quest for *Boléro*") Chorost discusses his own participation in numerous software upgrades and engineering experiments in order to expand the frequency range of his sixteen channels. He further suggests that by flying "to labs around the country with [his] own agenda—to try out their software specifically with music," he's "gotten them to focus on music sooner than they might have otherwise" (Michael Chorost blog, comment posted January 5, 2006).

Philip Loizou, an electrical engineer at the University of Texas, Dallas, notes that speech-processing strategies also contain biases against environmental sounds, noisy environments (often due to unrealistic laboratory testing situations), and tonal languages (which convey information through pitch rather than simply formants/timbre) (Loizou 2006).[40] Furthermore, the endemic electromagnetic interference between implants and wireless or cordless phones indicates the primacy of face-to-face communication for implant design—not to mention the long-standing disregard for hearing aid and implant users within the telecommunications industry.[41] Finally, Chorost reminds that the design of cochlear implants is of course constrained by economics. Although his initial device allowed him to switch between two speech processing strategies, one analog and one digital, an upgrade reduced him to just one, "because audiologists simply don't have the time to fit all of their patients with two different kinds of software." (Chorost 2005, 189).

Deaf Futurism

Narratives that depict the history of cochlear implants as a binary conflict between Deaf culture and normative biomedicine additionally obscure the radical aspirations of a minority of deaf implant users. In a 1999 *Hastings Center Report*, G. Q. Maguire and Ellen McGee contended that cochlear implants had "set the stage" for neuroenhancement:

> Three stages in the introduction of such devices can be delineated. The earliest adopters will be those with a disability who seek a more powerful prosthetic device. The next step represents the movement from therapy to enhancement. One of the first groups of nondisabled "volunteers" will probably be in the professional military, where the use of implanted computing and communication devices with new interfaces to weapons, information and communications could be life-saving. The third group of users will probably be people involved in information-intensive businesses who will use the technology to develop an expanded information transfer capability. (Maguire and McGee 1999, 9)

This linear evolutionary trajectory assigns deaf people to the primitive state and suggests a structural need for individuals with disabilities to serve as test beds for new technologies. It ignores the possibility that *any* person might desire "enhancement" rather than therapy from a prosthetic device.[42]

A small subset of "deaf futurists" has always maintained an investment in technical innovation and posthumanism. Unlike the futurist art movements of the early twentieth century—which tended to reject the poor, the nontechnological, and the unfashionable, along with the "traditional"—minority futurisms counter histories of exclusion in science and engineering.[43]

Often unrecognized, deaf scientists and technophiles number among the leading theorists of implant futurity. Internet founder and Google vice president

Vinton Cerf, who wears two hearing aids, previously asserted an enduring relationship between deafness and computer networking. "In creating the Internet with my colleagues," he has stated, "in part I wanted to help people with hearing loss as well as other communication difficulties" (Better Hearing Institute 2007).[44] Years later, for the *Time* magazine special report on "visions of the 21st century," Cerf predicted web-enabled implants: "The speech processor used today in cochlear implants for the hearing impaired could easily be connected to the Internet; listening to Internet radio could soon be a direct computer-to-brain experience!" (Cerf 2000, 103).

Cerf's wife, Sigrid, whom he met at a hearing aid center, acquired an implant in 1996. In a short story written for *The Little Magazine*, he lists the ways "the little deaf girl" had grown up and begun experimenting with technology:

> Her aggressive approach to hearing led her to obtain and in some cases invent assistive methods to augment the basic speech processor/implant combination. She had patch cords made to connect her speech processor to the armrest of airplane seats so she could hear the movie sound track *directly* rather than through audio headphones. She obtained books on tape and listened to them too through a patch cable connecting the sound output of the tape recorder directly into the auxiliary audio input of her speech processor. She had numerous microphones made on wires ranging from six feet to sixty feet in length so she could put a microphone close to the speaker at lectures or at the dinner table. She obtained FM transmitters with microphones built in, so she could put the transmitter on a lectern and then, using an FM receiver, listen to the speaker—again using direct audio input into the speech processor. She obtained infrared receivers to pick up the sound track in movie theatres which are equipped to transmit this signal to receivers for the hearing impaired.
>
> She had patch cords made to plug into her mobile telephone and used magnetic telephone coils with patch cords for use with ordinary telephones. The list goes on and on. (Cerf n.d.)

Sigrid herself has intimated that she would one day like to link to the Internet via her implant (Hamilton 2002).

Similarly, Mike Chorost comes out as a cyborg in his autobiography (which was endorsed by Manfred Clynes). Part human and part machine, he describes himself as at once limited and endowed with new capacities. His implant is an object lesson about the imperfections of the technology, the ways normalization can be desired alongside enhancement, and the fact that "posthumanism" can coexist with impairment. He marvels about patching into his Walkman and "hearing music that never actually exists as sound. This could be evidence of a profound transformation in how human beings take in information from the world around them." He continues, reflecting on the inevitable domestication of his device, "Or, it could just be a cozy domestic scene: a cyborg and his cat" (Chorost 2005, 58).[45]

In a piece for *The Futurist* magazine, Chorost muses, "While my friends' ears will inevitably decline with age, mine will only get better." He warns of the potential costs to privacy and security of networked implants: "Neural devices such as cochlear implants are computers and can be wirelessly networked. People's bodies and brains could become visible to the global network in ways we can only dimly imagine now, much the way people could barely imagine the impact of computer networking back in the age of Apple IIs" (Chorost 2006, 68).[46] At the same time, Chorost criticizes the escalating barriers to users' control. Present-day implants allow users to adjust volume, microphone sensitivity, and "private tones" (which give cues about the functioning of the device), as well as to choose among three or four "customized" programs based on highly generalized listening environments (i.e. "quiet," "noisy," and "focus"). Parents and teachers can use a remote control to set these programs at a distance for children wearing implants. For adults and children alike, choice of speech-processing software is limited and software settings (i.e. amount of stimulation per electrode) must be set by a clinician.

Responding to a blog entry by Cory Doctorow titled "Deaf hacker rewrites implant-firmware so he can enjoy music again," Chorost counsels:

> I didn't actually rewrite the software. To do that I'd need a degree in electrical engineering, an insider's knowledge of the code, and an understanding of how electricity interacts with body tissues. If I broke into Advanced Bionics in the dead of night, fired up the computers, hooked up the interface gadgetry to my processor, and started changing code at will, I could cook my inner ear or electrocute myself. Much more likely, my processor would crash and I'd slink out in total humiliation. I'd be deaf until I went back to my audiologist to have the software re-uploaded . . . Cory is right that the more control a person can have over his or her prosthesis, the better. I would love it if Advanced Bionics built a gadget that let me change some of my parameters on my own. While I have a lot of control over the device through my audiologist, it's control that takes weeks and months to unfold and explore. My audiologist and I experiment with various parameters during fitting sessions, and it always frustrates me that we only have a few hours in each session to try to find the best values for 20 or more variables. I would love to have the freedom of playing with them on my own. That would considerably accelerate the process. But it is possible in principle to hack one's own implant. (Michael Chorost blog, comment posted January 5, 2006)

The complexity of electronic objects such as cochlear implants means that most users will not be able to modify them at will. Moreover, the drive to standardize medical devices means that most examples of "deaf futurism" exist in theory rather than being instantiated in the material-semiotic register of signals.

Conclusion

A comprehension of the politics of circuitry and software enables, on its own, a small measure of perceptual control. Along these lines, media theorist Vilém Flusser once argued that the very *constructedness* of his hearing aids provided an advantage over "normal" hearing. Having fled to Brazil in 1940, Flusser managed a radio and transistor factory before turning to the philosophy of communication. In a manuscript titled "Hoerapparate," he suggested that "the hearing aid is freedom"—and he wondered "why there has not been more philosophical writing on the subject." Deafness plus a hearing aid meant that he had "ear-lids"; he could choose when to be immersed in the world's noises and voices. More important, he felt that his hearing aid helped him *to see and hear better*: on the one hand, to be aware of the manufacture and obstructions present in all communication; on the other hand, to understand the programming behind his own auditory perception:

> If you are listening to the world, you will notice that sounds are "instrumentized." Not a white buzzing that comes to the ears, but an orchestrated swinging. A programmed noise. Therefore it must be supposed that between you and the world there is a sound-sieve turned on, a hearing aid. The unpleasant, even unacceptable thing about this apparatus is that one cannot see it. Therefore one cannot know who programmed it. If its program is coming from the world out there or from you yourselves, for instance through the way your ear is built. Even good old Kant puzzled his head on this subject. My own hearing aid is visible. One knows who programmed it, a Japanese company. And this finally is an advantage I have in comparison to you. I can, better than you are able to, see through my hearing aid. And therefore hear better than you.[47]

In much the same terms, Chorost recognizes the epistemological value of the cochlear implant. After an extended analysis of Donna Haraway's "Cyborg Manifesto," he weighs in against "the blithe assumption that one's sensory organs deliver a truthful representation of the universe...people with normal ears are not off the epistemological hook, because their "software" was written haphazardly by millions of years of evolution and has no greater claim to reality" (Chorost 2005, 147).[48]

The history and consequences of communication engineering, particularly at the level of electrical signals and machinic filters, must be considered more broadly within sound studies. Electroacoustic devices—with sleek casing, miniaturized circuitry, and confidential corporate histories—increasingly resist "seeing-through." Yet in the case of cochlear implants, the desires of early users, the conflicting demands of mainstream medicine and economics, and the mediated features of electrical listening—in other words, the politics attendant upon communication—can be found embedded in the design of electroacoustic objects.

NOTES

1 At BBN, Swets notes, researchers "thought of computers as symbol processors—for example, theorem provers and pattern recognizers—rather than as number crunchers . . . these psychologists would lend what they knew about human perception, thinking, language, and motor control to the design of computers that would augment or supplant human behavior in, for example, libraries, process control, and robotics" (Swets 2005, 15, 18). Paul Edwards traces the Harvard PAL transition "from wartime work on human-machine integration to postwar concerns with information theory to the computer as a metaphor for the human mind." The PAL scientists investigated problems such as communication between humans and complex military machines (e.g., submarines, airplanes); the transmission of speech in battlefield noise; the jamming of enemy communications; training in listening and

articulation for military personal. These scientists "made no distinction between the technology of hardware and the technology of language and listening" (Edwards 1996, 210, 214).

2 Approval followed for children over the age of two in 1990 and for twelve-month-olds in 2000.

3 These users are mostly wealthy individuals or citizens of countries with socialized medicine, as the procedure costs approximately $30,000 (Spelman 2006). On poverty and the global distribution of implants, see Zeng (2007). The NIDCD website maintains current statistics on cochlear implant use: http://www.nidcd.nih.gov/health/hearing/coch.asp.

4 "It is an unequal battle since the two accounts differ greatly in their authority. The medical understanding draws on and reflects the authority of science and the promise of medicine, both of which have become fundamental components of modern industrial culture. The deaf perspective draws on and reflects the experience of a traditionally marginalised and stigmatized group" (Blume 1999, 1265–66).

5 Brenda Jo Brueggemann has recently suggested that the deaf/Deaf distinction has been ineffective and might even be unrealistic. In this chapter, however, I follow these conventions of capitalization when referring to the audiological ("deaf") versus linguistic minority ("Deaf") definitions of deafness (Brueggemann 2009, 9–15).

6 In this book, Blume includes a short section on "adult implantees" and the potential for their experiences to complicate the Deaf/hearing binary, as well as any reductive audiological definition of deafness: "The boundary between deafness and hearing is a complex region, marked by values, memories, histories, and commitments, and to be crisscrossed in many ways." See pages 163–70.

7 To the contrary, Blume insists that patients are "typically not seen as competent interlocutors in the innovation process" (Blume 1997, 32).

8 Trevor Pinch and Nelly Oudshoorn elaborate as follows:

> [I]n the design phase technologists anticipate the interests, skills, motives, and behavior of future users. Subsequently, these representations of users become materialized into the design of the new product. As a result, technologies contain a script (or scenario): they attribute and delegate specific competencies, actions, and responsibilities to users and technological artifacts. (Oudshoorn and Pinch 2003, 9)

See the chapters by Oudshoorn and Blume and Rose regarding the significance of clinical trials or research and development to user configuration.

9 Helen Nissenbaum asserts that "computer and information systems can embody values," and these must be considered in tandem with their "social effects" (Nissenbaum 2001, 120).

10 Blume surmises that the initial disinterest of the Deaf community led implant manufacturers to direct marketing pressure *toward* the hearing parents of deaf children. The marketing focus has since expanded to include late-deafened seniors, by far the largest population with severe hearing loss (Blume 2010, 116–117, 144).

11 Here he draws on the work of Rayna Rapp and Faye Ginsburg (Blume 2010, 162).

12 Bernard Frayssend to Dr. House, June 11, 1978, "Djourno/Eyriès Publications" Folder, Cochlear Implants, 1961–1995 Collection, John Q. Adams Center. Drawing on this collection, Marc Eisen (also an otolaryngologist) has written two short articles on the early French implant (Eisen 2003, 2006).

13 Phillip Seitz, "Interview with André Djourno and Danièle Kayser 1/12/1994," 612-OH-11, Cochlear Implants, 1961–1995 Collection, John Q. Adams Center. See also Seitz (2002).

14 As Eisen narrates, Djourno and Eyriès ran into each other at the medical school morgue, where Eyriès was looking for a potential nerve transplant for Monsieur G. Djourno convinced him to try an implant instead (Eisen 2006, 3).

15 Eisen explains that Eyriès would have preferred to contract an engineering firm to manufacture the implant (Eisen 2003, 503). Blume offers a different interpretation of the conclusion to this experiment; he says the patient "decided that he had had enough: it was not worth the investment of so much time and emotion" (Blume 2010, 32).

16 Phillip Seitz, "Interview with André Djourno and Danièle Kayser 1/12/1994," 612-OH-11, Cochlear Implants, 1961–1995 Collection, John Q. Adams Center.

17 Ibid. Djourno also claimed that the implant allowed her to "hear her own voice for the first time in eight years" (Djourno and Vallancien 1958, 555).

18 Phillip Seitz, "Interview with André Djourno and Danièle Kayser 1/12/1994," 612-OH-11, Cochlear Implants, 1961–1995 Collection, John Q. Adams Center.

19 William House, "Cochlear Implants: The Development of an Idea," February 1976 (typescript), p. 9, 921-HSB-2, Cochlear Implants, 1961–1995 Collection, John Q. Adams Center.

20 For the story of Howard's own unconventional approach to medicine, see Hyman (1990).

21 The year before, James had designed similar equipment to record the sounds produced by the auditory nerve of a person with tinnitus. At the time he realized, "Well, gee, if we can do this, we ought to be able to reverse the process and make the person hear that's deaf." Even though the biophysics of hearing had not been clearly determined, James felt that "as an engineer you always look for a practical way of making something work, whether it's the absolute truth or not." Phillip Seitz, "Interview with John and James Doyle (8/22/93)," 612-OH-9, Cochlear Implants 1961–1995 Collection, John Q. Adams Center.

22 A woman, S.S., was implanted at the same time, but her electrode had to be removed almost immediately due to poor insulation.

23 In 1973 E.K. volunteered for another round of experiments, this time with new materials.

24 John Doyle claimed his slides had been stolen from a meeting of the American Medical Association (Schmeck 1962). The Doyle brothers continued with this research for several years until lack of funds forced them to abandon the project (Doyle et al. 1963).

25 "California Electronics Firm Readies 'Artificial Ear' Implant," *Space Age News* 3(18) (1961): 1; "Electronic Firm Restores Hearing with Transistorized System in Ear," *Space Age News* 3(21) (1961): 1. *Radio-Electronics* carried many of the earliest reports of the French implant. For instance, "News Briefs: Electronic Ears," *Radio-Electronics* 29 (December 1958): 6.

26 These first experiments by Simmons did not result in take-home implants.

27 Credit for the first functional portable implant remains controversial within the otolaryngology community. Although House is often recognized for developing the first successful take-home implant technology, Michelson's experimental devices, described in a 1971 publication, were close competitors.

28 William House, "Cochlear Implant: Hope for the Nerve-Deafened Person: A Decade of Progress" (typescript, n.d.), p. 4, 921-HSB-2, Cochlear Implants, 1961–1995 Collection, John Q. Adams Center.

29 Ibid., 27, 29.

30 Charles Graser Papers, 921-HSG, Cochlear Implants, 1961–1995 Collection, John Q. Adams Center. For a similar sentiment two decades later, see Chorost 2005, 163.

31 For all quotes in this section, see Charles Graser Papers, 921-HSG, Cochlear Implants, 1961–1995 Collection, John Q. Adams Center.

32 Today, some processors are again available as body-worn devices, marketed for their relative invisibility.

33 William House, "Cochlear Implants: The Development of an Idea," February 1976 (typescript), p. 28, 921-HSB-2, Cochlear Implants, 1961–1995 Collection, John Q. Adams Center. House and Urban had, moreover, not been able to eliminate crosstalk between electrodes.

34 They found that the silver wire had not corroded after four years. William House, "Cochlear Implants: The Development of an Idea," February 1976 (typescript), p. 36, 921-HSB-2, Cochlear Implants, 1961–1995 Collection, John Q. Adams Center.

35 C.R., who was implanted at the same time, had late-stage syphilis and moved away from the Los Angeles area.

36 William House, "Cochlear Implant: Hope for the Nerve-Deafened Person: A Decade of Progress" (typescript), p. 8, 921-HSB-2, Cochlear Implants, 1961–1995 Collection, John Q. Adams Center.

37 In Cambridge, the cochlear implant program was a joint effort of the Infirmary, Harvard Medical School, and M.I.T., coordinated by Dr. Donald Eddington. Another physician described Pierschalla as "more than a test subject . . . he was a valued colleague, an integral contributor, and a friend" (Chorost 2005, 106).

38 Michael Pierschalla, 1990, typescript in the author's possession.

39 Ibid., 4.

40 Biases against tonal languages within signal processing have long been of concern in the hearing aid industry. See as one example McCullough, Tu, and Lew (1993).

41 See Goggin (2006, ch. 5). "Politics" also reside in the medical determination of implant candidacy: psychological criteria, as well as willingness to maintain a relationship with the clinic for implant maintenance, are often taken into consideration. As a last example, experimental devices were designed with more or less flexibility for upgrades.

42 For a more thorough discussion of the ways the same technology can be used for either treatment or enhancement, see Parens (1998, 1–2).

43 A different iteration of this concept can be found in Afrofuturism, defined by sociologist of science Alondra Nelson as instances of "sci-fi imagery, futurist themes, and technological innovation in the African diaspora" (Nelson 2002, 9). The heterogeneity of the deaf community means that examples of "deaf futurism" vastly exceed cochlear implants. Members of Deaf culture who reject cochlear implants, for instance, may embrace a range of "deaftechs" from wireless relay services to videophones to motion capture to ASL vlogs. More important, Brenda Jo Brueggemann calls attention to the fact that the number of Gallaudet students with cochlear implants "has virtually doubled itself each year" in recent times; the relationship between Deaf culture and this technology is not inevitably one of opposition (Brueggemann 2009, 16–17).

44 In 1978 Cerf reported on "The Electronic Mailbox: A New Communication Tool for the Hearing Impaired" to the readers of *American Annals of the Deaf.* At that time, the U.S. Department of Health, Education, and Welfare (in collaboration with Bolt, Beranek, and Newman) was testing email on the deaf population of Framingham, Massachusetts (Cerf 1978, 771).

45 Chorost's current implant model allows him to use standard earbuds and skip the patch cord step. Mike Chorost, in discussion with the author, June 17, 2009.

46 On this note, Chorost's forthcoming book examines the emerging technology of optogenetics. On controlling genetically altered neurons with light see Chorost (2007).

47 Vilém Flusser, "Hoerapparate" [Hearing Aids], trans. Silvia Wagnermeier. Vilém Flusser Archive, Berlin University of Arts. Thanks to Siegfried Zielinski for alerting me to this document and Silvia Wagnermeier for providing the translation.

48 Chorost disagrees with Haraway's expansive definition of "cyborg," preferring to apply the term only to cases of cybernetic technology—that which "exerts control over the body" (Chorost 2005, 41).

REFERENCES

Beard, Marion. "Signs of the Times: Cochlear Implants May Not Be the Best Way to Help Deaf Children." *New Scientist* 162 (1999): 52.

Berg, Abbey L., Herb Alice, and Marsha Hurst. "Cochlear Implants in Children: Ethics, Informed Consent, and Parental Decision Making." *Journal of Clinical Ethics* 16 (2005): 239–50.

Better Hearing Institute. "Vinton Cerf, One of the Internet's Inventors, to Partner with Better Hearing Institute." 2007. Last modified January 3. http://www.betterhearing.org/press/news/pr_vintonCerf.cfm.

Bijker, Wiebe. "Sociohistorical Technology Studies." In *Handbook of Science and Technology Studies*, ed. Sheila Jasanoff, G. E. Markle, J. C. Petersen, and T. Pinch, 229–56. London: Sage, 1995.

———. *The Artificial Ear: Cochlear Implants and the Culture of Deafness*. New Brunswick, N.J.: Rutgers University Press, 2010.

———. "Cochlear Implantation: Establishing Clinical Feasibility." In *Sources of Medical Technology: Universities and Industry*, ed. Nathan Rosenberg, Annetine C. Gelijns, and Holly Dawkins, 97–124. Washington, D.C.: National Academies Press, 1995.

———. "Histories of Cochlear Implantation." *Social Science and Medicine* 49 (1999): 1257–68.

———. "Land of Hope and Glory: Exploring Cochlear Implantation in the Netherlands." *Science, Technology, and Human Values* 25 (2000): 139–66.

———. "The Rhetoric and Counter-Rhetoric of a 'Bionic' Technology." *Science, Technology, and Human Values* 22 (1997): 31–56.

Brueggemann, Brenda Jo. *Deaf Subjects: Between Identities and Places*. New York: NYU Press, 2009.

Burke, Teresa Blankmeyer. "Comments on 'W(h)ither the Deaf Community?' " *Sign Language Studies* 6 (2006): 174–80.

"California Electronics Firm Readies 'Artificial Ear' Implant." *Space Age News* 3(18) (1961): 1.

Cerf, Vinton. n.d. "The Little Deaf Girl." *The Little Magazine* 3. http://www.littlemag.com/listen/vintoncerf.html (accessed February 13, 2011).

———. "The Electronic Mailbox: A New Communication Tool for the Hearing Impaired." *American Annals of the Deaf* 123 (1978): 768–72.

———. "What Will Replace the Internet?" *Time* (June 19, 2000).

Chase, Victor. *Shattered Nerves: How Science Is Solving Modern Medicine's Most Perplexing Problem*. Baltimore: Johns Hopkins University Press, 2006.

Chorost, Michael. "Hacking My Own Ear: What Would It Take?" *Michael Chorost blog*, January 5, 2006; http://www.michaelchorost.com/archived-news/.
———. "Making Deaf Ears Hear with Light." *Technology Review* (August 10, 2007); http://www.technologyreview.com/biotech/19206/?a=f.
———. "The Mind-Programmable Era." *Futurist* 40 (2006): 68.
———. "My Bionic Quest for *Boléro*." *Wired* 13 (2005) (accessed October 12, 2010); http://www.wired.com/wired/archive/13.11/bolero.html.
———. "Questions from Steve Potter's Class." Michael Chorost blog, March 2, 2009; http://www.michaelchorost.com/.
———. *Rebuilt: How Becoming Part Computer Made Me More Human*. Boston: Houghton Mifflin, 2005.
Clark, G. M. "The Development of Speech Processing Strategies for the University of Melbourne/Cochlear Multiple Channel Implantable Hearing Prosthesis." *Journal of Speech-Language Pathology and Audiology* 16 (1992): 95–107.
Cochlear Implants, Manuscript Collection, 1961–1995. John Q. Adams Center for the History of Otolaryngology—Head and Neck Surgery, Alexandria, Va.
Crouch, R. A. "Letting the Deaf Be Deaf: Reconsidering the Use of Cochlear Implants in Pre-Lingually Deaf Children." *Hastings Center Report* 27 (1997): 14–21.
"Direct Wire to Deaf Inner Ear." *Medical World News* (December 21, 1962).
Djourno, André, and Charles Eyriès. "Prosthèse auditive par excitation électrique à distance du nerf sensorial à l'aide d'un bobinage inclus à demeure." *La Presse Médicale* 65 (1957): 14–17.
Djourno, André, and B. Vallancien. "De l'excitation électrique du nerf cochleaire chez l'homme, par induction à distance, à l'aide d'un micro-bobinage inclus à demeure." *Comtes Rendus de la Société de Biologie* (Paris) 151 (1957): 423–25.
———. "L'interrogation électrique du nerf cochléaire." *Comtes Rendus de la Société de Biologie* (Paris) 152 (1958): 555–56.
Doyle, J. B., D. H. Doyle, F. M. Turnbull, J. Abbey, and L. House. "Electrical Stimulation in Eighth Nerve Deafness." *Bulletin of the Los Angeles Neurological Society* 28 (1963): 148–50.
Edwards, Paul. *The Closed World: Computers and the Politics of Discourse in Cold War America*. Cambridge, Mass.: MIT Press, 1996.
Eisen, Marc D. "Djourno, Eyriès, and the First Implanted Electrical Neural Stimulator to Restore Hearing." *Otology and Neurotology* 2 (2003): 500–506.
———. "History of the Cochlear Implant." In *Cochlear Implants*, 2nd ed., ed. S. B. Waltzman and J. T. Roland, 1–10. New York: Thieme, 2006.
———. "The History of Cochlear Implants." In *Cochlear Implants: Principles and Practices*, 2nd ed., ed. John K. Niparko, 89–94. Philadelphia: Lippincott, Williams & Wilkins, 2009.
"Electronic Firm Restores Hearing with Transistorized System in Ear." *Space Age News* 3(21) (1961): 1.
Flusser, Vilém. "Hoerapparate." Vilém Flusser Archive, Berlin University of Arts, Berlin.
Goggin, Gerard. *Cell Phone Culture: Mobile Technology in Everyday Life*. New York: Routledge, 2006.
Hamilton, Tyler. "'Father of the Internet' Ponders Mars." *Toronto Star* (November 15, 2002).
Hannaway, Caroline. *Contributions of the National Institutes of Health to the Development of Cochlear Prostheses*. Bethesda, Md.: NIH Office of History, 1996.

Hippel, Eric von. "Lead Users: A Source of Novel Product Concepts." *Management Science* 32(7) (1986): 791–805.

House, William F., and Jack Urban. "Long-Term Results of Electrode Implantation and Electronic Stimulation of the Cochlea in Man." *Annals of Otology* 82 (1973): 504–17.

Hunt, Frederick V. *Electroacoustics: The Analysis of Transduction, and its Historical Background.* Cambridge, MA: Harvard University Press, 1954.

Hyman, Sidney. *For the World to Hear: A Biography of Howard P. House, M.D.* Pasadena, Calif.: Hope, 1990.

Johnston, Trevor. "W(h)ither the Deaf Community? Population, Genetics, and the Future of Australian Sign Language." *American Annals of the Deaf* 148 (2004): 358–75.

Kiang, Nelson. "Discussion." *Annals of Otology* 82 (1973): 512.

Ladd, Paddy. "Oralism's Final Solution." *British Deaf News* (Autumn 1985): 5–6.

Lane, Harlan. "Cochlear Implants: Their Cultural and Historical Meaning." In *Deaf History Unveiled: Interpretations from the New Scholarship*, ed. John Vickrey Van Cleve, 272–91. Washington, D.C.: Gallaudet University Press, 1993.

Lang, Harry G. "Book Review: *Cochlear Implants in Children: Ethics and Choices* by John B. Christiansen and Irene W. Leigh." *Sign Language Studies* 3 (2002): 90–93.

Levy, Neil Louis. "Reconsidering Cochlear Implants: The Lessons of Martha's Vineyard." *Bioethics* 16 (2002): 134–53.

Loeb, G.E., C.L. Byers, S.J. Rebscher, D.E. Casey, M.M. Fong, R.A. Schindler, R.F. Gray, and M.M. Merzenich. "Design and fabrication of an experimental cochlear prosthesis." *Medical & Biological Engineering & Computing* 21 (1983): 241–254.

Loizou, Phillip. "Speech Processing in Vocoder-Centric Cochlear Implants." In *Cochlear and Brainstem Implants*, ed. A. Moller, 109–43. Basel: Karger, 2006.

Maguire, G. Q., Jr., and Ellen M. McGee. "Implantable Brain Chips? Time for Debate." *Hastings Center Report* 29 (1999): 7–14.

McCullough, June, Clara Tu, and Henry Lew. "Speech-Spectrum Analysis of Mandarin: Implications for Hearing Aid Fittings in a Multi-Ethnic Society." *Journal of the American Academy of Audiology* 4 (1993): 50–52.

Michelson, Robin. "Electrical Stimulation of the Human Cochlea: A Preliminary Report." *Archives of Otolaryngology* 73 (1971): 317-323.

Moravec, Hans. *Robot: Mere Machine to Transcendent Mind.* New York: Oxford University Press, 2000.

Nelson, Alondra. "Introduction: Future Texts." *Social Text* 20 (2002): 1–15.

Nissenbaum, Helen. "How Computer Systems Embody Values." *Computer* 34 (2001): 120.

Oudshoorn, Nelly, and Trevor Pinch. *How Users Matter: The Co-construction of Users and Technologies.* Cambridge Mass.: MIT Press, 2003.

Parens, Eric, ed. *Enhancing Human Traits: Ethical and Social Implications.* Washington, D.C.: Georgetown University Press, 1998.

Schmeck, Harold, Jr. "Persons Deaf since Birth Hear Pure Sounds in a Doctor's Tests." *New York Times* (November 28, 1962), 32.

Seitz, Phillip. "French Origins of the Cochlear Implant." *Cochlear Implants International* 3 (2002): 77–86.

Sparrow, Robert. "Defending Deaf Culture: The Case of Cochlear Implants." *Journal of Political Philosophy* 13 (2005): 135–52.

Spelman, Francis. "Cochlear Electrode Arrays: Past, Present, Future." *Audiology and Neurotology* 11 (2006): 77–85.

Swets, John. "The ABCs of BBN: From Acoustics to Behavioral Sciences to Computers." *IEEE Annals of the History of Computing* 27 (April–June 2005): 15–29.

Tucker, Bonnie P. "Deaf Culture, Cochlear Implants, and Elective Disability." *Hastings Center Report* 28 (1998): 6–14.

Wrigley, Owen. *The Politics of Deafness*. Washington D.C.: Gallaudet University Press, 1996.

Zeng, Fan-Gang. "Cochlear Implants: Why Don't More People Use Them?" *Hearing Journal* 60 (2007): 48–49.

SECTION V

EDITING SOUND: THE DESIGN STUDIO

CHAPTER 14

SOUND AND PLAYER IMMERSION IN DIGITAL GAMES

MARK GRIMSHAW

INTRODUCTION

THE perceptions arrived at from different senses often combine either to provide further information about the environment or to confirm the information already provided by one sense. This is particularly the case with sight and hearing, which may, in some cases, be one perceptual phenomenon. A ventriloquist's dummy or a movie screen have no sound-producing capabilities, yet sound from elsewhere is perceptually located on the moving mouth or parts of the film image. Even when the sensations of sight and sound are temporally separated, they may still combine to make one perceptual phenomenon; a gunshot might be heard some time after a muzzle flash is seen, but the firing is perceived as one event. However, notwithstanding the perceptual effects of combined sensory phenomena, removing all sensory information but that derived from my hearing and combining it with general experience and knowledge, I would still be able to describe in great part the environments in which I may be situated. Even when the other senses are available, I often rely solely on sound to inform me. For example, from the sounds of vehicles passing unseen on the road outside, I can, to varying degrees of accuracy, ascertain the direction they are traveling in, their speed, the type of vehicle, and the density of traffic. From this latter deduction, I can further deduce the time of day, for example. Furthermore, as a single taste can spawn a monumental novel, unseen sound sources can provoke images; founded upon experience, specific denotations derived from a visualization of the sound-producing object or activity or, typically more varied, connotations that may be communal in their meaning or quite individual. Conversely, seen but unheard objects and activities can be mentally sounded; the sight of a coin spinning ever more slowly on a table top will lead to a mental sound object closely synchronized to the object's material, form, and action. At the very least, assuming both senses are functioning normally, if vision and hearing are not perceived as one perceptual event, there is a strong correlation between them.

This is a chapter about sound in digital games—games played on gaming consoles and home computers— and how the design of sound for such a medium contributes to player immersion in the game world, especially in worlds designed to be immersive. This statement encapsulates an assumption: that sound in such digital game worlds contributes to the immersion of the player or, at the very least, that the game designer uses sound in an attempt to facilitate such immersion. Before this assumption can be tested, a number of questions raised by that assumption must be answered: *What is sound in the digital game? What is the relationship between sound and image? Is there a difference in the use or perception of sound in the real world and in virtual worlds?* and *What is meant by immersion?* To illustrate the answers to these questions, sound use in first-person perspective games, of which the most notable subgenre is the first-person shooter (FPS) game, will be used. Such games have an immersive premise; on-screen, the player is typically presented with a pair of arms receding perspectivally into a representation of a three-dimensional visual space; the implication is that the pixellated arms are the player's own arms extending into and interacting with the game world. To parallel this, the player is positioned in a field of sound as a first-person auditor, and that field contains sounds that change their location as the player's character moves and have a dynamically processed reverberation (particularly in modern games) to approximate the effects of the materials and spaces in the game world. There are many game genres beyond those using the first-person perspective, but the latter offer a greater range of immersive possibilities as afforded by sound; indeed, not all digital games can be described as immersive at all. However, the study of immersion in those games that do invite it is important from an STS perspective that seeks to understand the effects of game sound experience on technology and design and vice versa. Immersion is actively pursued in the design of some games, yet its attainment is little understood, and still less understood are the long-term implications of increasing periods of immersion in virtual worlds.

Throughout the chapter I appear to be suggesting that real world and virtual world are distinct contexts, polar opposites, and ne'er the twain shall meet. While this serves the purpose of brute analysis, the *reality* is somewhat less clear cut, and I beg the reader's indulgence in my use of the artifice of polarity. It is a means to an end, and that end, as will be clear at the end of the chapter, is to demonstrate that, while differences do exist, such absolute distinctions are a nonsense where sound use and player immersion are concerned.

Sound in Digital Games

My hands clutch the machine gun nervously as I survey the scene before me. I am amid the ruins of a French village. A single bell tolls in the church tower behind me while pigeons flap and coo around it. An aircraft roars overhead, and I have just passed a sign creaking disconsolately on one rusty hinge and offering *vin rouge*. Somewhere, a record of Edith Piaf crackles through the empty streets, stuck on one lingering phrase. A tank burns in the distance, and the sound of muffled explosions and gunfire comes from the left and ahead behind a pile of rubble. The static of radio occasionally updates me on my comrades' status while I edge forward cautiously. Footsteps gradually impinge on my consciousness, and I whirl around, expecting to engage with the enemy. Nothing. All of a sudden I see a grenade arc gracefully through the air toward me, landing with a metallic, clattering bounce off the wall in front of me. As I jump back, the grenade explodes, and I take a hit in my leg, causing me to involuntarily grunt in pain and limpingly seek cover. Edith sings of regret to the accompaniment of the lonely bell. A head appears over the wall, and I instinctively empty a clip of my machine gun at it; blood spatters my face, and I hear a scream that satisfyingly mingles with the musical tinkle of empty, ejected shells. Silence. I stop to bandage my wound.

Analyses of sound and sound use in digital games—of which the scenario just now sketched is typical— usually take as their starting point principles defined in cinema theory. As an example, the notion of diegetic and nondiegetic sound has been imported from film criticism (Curtiss 1992; Chion 1994, for example) to game sound theory by a number of authors (such as Grimshaw 2008a; Jørgensen 2006). For games, following definitions in cinema, a simple assertion is that diegetic sound is the sound that derives from the internal logic of the game world and that nondiegetic sound is all other game sound such as the musical score and menu interface sounds—this chapter deals with diegetic sound as just defined and does not concern itself with nondiegetic sound. Similarly, the notion of acousmatic sound has been borrowed following its development from electroacoustic composition by film theorists such as Chion (1994). In the case of digital games, as for film, acousmatic sound is off-screen sound—a problematic definition for some (Metz [1985] disputes the on-screen/off-screen distinction) while others argue that technologies such as surround sound blur distinctions between on-screen and off-screen [Chion 1994, 129–31])—but having the benefit of denoting sound that derives from a source not seen on the screen.

No semantic classification is ever black and white, and there are several gray areas in the digital games application of terminology such as *diegetic* and *acousmatic* that are best illustrated by a brief discussion of the mechanics and technology of game sound. In the first instance, the soundscape of a digital game is different each time the game is played. Film soundscapes may change with censorship, dubbing, reproduction equipment, and interaction with the external acoustic environment. Essentially, though, the sound track on the medium made available to distributors and viewers is fixed at the point of production. This is not the case with digital games, which, as a highly interactive activity, come not with a fixed sound track on the distribution medium but with discrete audio objects used as required during gameplay to create a compound soundscape that is unique at each playing. Early games used real-time sound synthesis, and games in the future may well return to synthesis as a means of sound creation. Current games (as of 2009) with few exceptions make use of audio samples that take advantage of the large capacity of modern digital-storage media. These audio samples, be they intended as diegetic or nondiegetic, are sounded at the player's command or by the game engine in response to the current game state (which itself is dependent upon the actions of the player or, in multiplayer games, players). The individual, unprocessed audio samples for any particular game may remain the same,[1] but the resultant soundscape is different at each playing of the game because the player's actions in creating that soundscape are different at each playing. Where the intended soundscape of film is fixed at the point of production, digital game soundscapes are created anew at the point of reproduction.

Given these differences then, it is no surprise that, while similarities may exist between film sound and game sound, there are also significant differences that warrant care in the transfer of sound theories and concepts from film to digital games. Defining a film's musical score as nondiegetic is fraught with difficulty not least because of an ambiguity that is often exploited by the director. For example, the musical score, apparently nondiegetic,[2] that unmasks itself as diegetic through a panning shot that reveals a radio or an orchestra (Count Basie's Orchestra anachronistically in the desert in *Blazing Saddles* [Brooks 1974] is an extreme case in point). Furthermore, where the film's musical score typically follows the action onscreen, perhaps heightening the emotional impact or aiding in the intended interpretation of the scene (thus, in some cases it may be interpreted as having diegetic aspects to it), the technological genesis and the interactive nature of digital games make the boundary between diegetic and nondiegetic sound even more porous. All digital games are predicated upon the actions of the player, and there are many instances in which sections of the game's musical score are played only in response to certain player actions. *Rez* (Mizuguchi 2001) is an extreme example where the musical score derives almost entirely from the player's actions. Notwithstanding these conceptual debates, the design of many digital games (especially FPS games) does recognize a distinction between diegetic and nondiegetic sound through the inclusion of separate volume controls for sound effects and music. The "hunter and the hunted" premise of such games requires that the player be particularly attentive to game-world sounds—headphones aid further in excluding sounds external to the game—and thus it pays to turn the music off.

Acousmatic sound is similarly problematic as a concept particularly when applied to digital game sound. As Stockburger (2003) states, the player has kinesthetic control over the display (or not) of sound sources. Whereas in film, such visualization and acousmatization of sound is controlled by the director, in digital games, particularly FPS and other first-person perspective games, it is under the control, for the most part, of the player. Where an acousmatic sound is heard from behind the position of the player's character, the player can maneuver the character within the game world toward the source of the sound so that that source becomes displayed on-screen.

The Relationship between Sound and Image in the Game

I have already hinted at a strong relationship between sound and image not only in the scenario I sketched at the start of this chapter but also, first, in the suggestion that sound and image might be perceived as one event and, second, through the use of terms such as *visualization* and *seen sound source*. Upon hearing a sound (where the source remains unseen), a person will use experience to attempt to create a mental image of the sound source; a form of visualization exploited by directors and sound designers in the horror genres of both films and digital games. Gaver (1993) goes further in suggesting that sounds inherently betray the broad definitions of their sound sources; large objects tend to produce lower-pitched sounds than smaller objects, and particular events, such as bouncing, have a characteristic sound pattern. Such aural signatures, Gaver proposes, might be the basis for the synthesis of caricature sounds (by analogy with visual caricature as opposed to representative photography), a notion to which I return later.

The game's sound sources displayed on the screen are, as in film, not the actual sound sources. They are apparent sound sources, and all sounds and their sources are, quite literally, offscreen. This is because of the displacement of image-reproduction hardware compared to sound-reproduction hardware. Sound is not reproduced by means of the screen but is reproduced by headphones or loudspeakers that are often quite distant from the images displayed on the screen.[3] Disregarding the effect of the reflection of direct sound, the physical distance between sound (re)production and (apparent) sound source is one of the major differences between sound-image phenomena in the natural world and sound-image phenomena in the virtual worlds of digital games. As I write in my room, the hum of my fridge really does emanate from the fridge I see before me, and the scratch of my pen arises directly from the moving point of contact between pen and paper. Nevertheless, the sensations that arise from disjunct sound and image sources can be perceived as one event. This perceptual co-location of sound and apparent sound source has been described as synchresis or synchrony in cinema (Anderson 1996; Chion 1994) but is also known in acoustics and psychoacoustics as the audio-visual proximity or ventriloquism effect. In part, the conjunction has to do with experience (it becomes obvious that the music heard from loudspeakers is actually the music played by the orchestra on-screen); in part it has to do with the close synchronization between sound events and moving images (if an object on-screen starts to move, then it may well be "producing" the sound that started simultaneously); and in part it has to do with correct mental visualization of sounds (as in the orchestra mentioned earlier but also on the presumption that fundamental parameters of sounds—frequency, intensity, timbre—betray the general physical outlines and material properties of their sources). It is this perceptual cohesion between the results of physically disparate sound- and image-reproduction technologies that plays a major role in player immersion in the game world. Paradoxically, it is the distortion of reality inherent in the separate reproduction systems that is a contributing factor to allowing the player to respond to the game world as if it were real.

Game Technology—Real World, Virtual World

The differences and similarities between real-world sounds and sounds as mediated through digital games are best understood from the vantage points of both what is technically feasible and what the game designer's intentions are. If we assume that, in reality, we are immersed in a range of environments, then one can also assume that many digital games attempt to simulate, if not emulate, this immersion, particularly in first-person-perspective games.

Sitting at my desk as I write this, for the purposes of analysis I can assume I am immersed in the environment I perceive around me. Furthermore, I can identify several hierarchies and classes of environment; the environment of my room is part of the larger environment of my house, which itself is part of the larger suburban environment, and distinct visual and sonic environments (indeed, tactile and olfactory environments) can also be identified. From my particular perceptual location, these environments occupy different shared or overlapping spaces. Furthermore, positioning myself and others within those environments enables me to describe them as ecologies; in sound terms, the acoustic ecologies and acoustic communities described by Schafer (1994) and Truax (2001). Assuming a human is immersed in such environments,[4] then the game designer (of FPS games, for example) attempts to create similar ecologies by immersing the player in the environments of the game world. There are, however, several problems in establishing this immersion, and they all relate to the technology used in digital games.

Some of the differences and similarities between sound in the real world and sound in digital games have already been discussed as regards their technical origin (e.g., the workings of synchresis in combining distinct sensory events in different modalities into one perceptual event). Sound emanates from the vibrating object that is the cause of the sound; digital games, however, must propose apparent sound sources on the screen physically separate from the real vibrating sound sources (e.g., loudspeakers or headphones). Similarities usually arise from the simple fact that the reproduction of sonic and visual artifacts in the game world takes place in the real world; whether game-world sound or real-world sound, sound still reaches our ears through vibrating air, where it is processed by the same sensory and perceptual organs and faculties and interpreted using experience from both worlds. Likewise, the sound fields of FPS games are designed to be all-encompassing and omnidirectional[5] as opposed to the restricted visual field; FPS games provide a more highly restricted field of view than in reality, and, barring the use of virtual-reality headsets or goggles, this field is presented to the player as merely part of the larger, real visual field.

Limitations of digital-game technology impose other differences than a separation between real-sound source and apparent-sound source. In an FPS game, the player is presented with a first-person perspective on the visual game world. Similarly, the player is positioned within the game's sound field as a first-person auditor (Grimshaw 2008a, 83). This is analogous to the way a human hears the real world. No digital game, however, other than experimental games using head- or eye-tracking devices, will play and process sound according to the position of the player's head or body. In real-world environments, dynamic sound sources move around the listener, but listeners, too, can alter the relative position, the intensity, and the timbral properties of sound through their own head or body movement. In digital games, the player must move the character in a certain direction or rotate the character's entire body in order to effect a change in the sound properties listed earlier. In effect, the entire visual and acoustic environments of the game world are rotated or otherwise acoustically processed as per the sophistication of the game engine being used, relative to the first-person point of audition (which is the same as the first-person point of perspective). This point, regardless of whether the player is wearing headphones or is the focal point of loudspeakers, is located at the player's head. Visually and aurally, the player is at the center of the game world, and, in this, there is no difference between viewing and listening to the real world. In FPS digital games, though, while images seem to move around the *character's* point of view, sounds always move about the *player's* head.

In reality, the number and variety of sounds are, for all practical purposes infinite. Various drops of water have subtle sound differences depending upon the physical properties of the drop, the position of the listener, the environment, and the growing size of the pool they form. Digital games are not yet able to sonically compete with nature.[6] The reason for this is twofold: a limit to the storage capacity of digital-game equipment and a limit to the processing power of the equipment. Sound design, for FPS games particularly, is primarily an atomistic process rather than a holistic one. The sound designer has no way of knowing, for example, how many footstep sounds, at what speed, and on what surface will need to be heard during a game and so, ideally, should provide an infinite number of audio samples in order to match the potential variety of such sounds. The number of footsteps, speed of movement, and surface on which to move are choices within the gift of the player,[7] and so, individual audio samples, tailored to any possible combination, should be provided to be triggered when required. However, the storage capacity of the game's distribution medium limits the number of audio samples that can be provided. Furthermore, audio samples will be buffered in random access memory (for rapid access), which not only has other requirements made of it but also has a much-reduced capacity compared to hard drives and DVDs. There is also both a hardware and a software limit to the number of audio samples that can be played simultaneously. Complicating matters further is the fact that there might not only be a requirement for a variety of footsteps in a typical FPS game but also a requirement for a range of gunshots, voices, ambient noises, thuds, bangs, explosions, screams, and so on such that, in addition to the game software itself and the image files for screen display, no current game-storage medium has the capacity needed. The number of audio samples delivered with the game is, therefore, reduced to such an extent that no game yet can truly emulate the sound of a sonically rich, real-world environment.

Game designers attempt to ameliorate these limitations by a variety of methods. For example, a single audio sample might comprise several sounds, particularly if the sample consists of ambient sounds and is intended to be looped.[8] Audio compression[9] might be used to reduce the storage requirement of each sample without overly compromising the sound quality. Variety in repetitive sound classes (such as footsteps) is mimicked to some extent by the random use of a small set of similar samples. Increasingly, audio samples will be processed upon playback; for example, a reverberation will be applied that approximates the acoustic properties of the spaces depicted on-screen relative to the first-person point of audition. In game consoles that use dedicated audio digital-signal processor (DSP) chips, the real-time processing achieved can be quite complex but will always require controlling from the central processor unit (CPU). This processor must also be used for other tasks within the game, many of which will be prioritized. In attempting to limit the amount of processing required of the CPU for reverberation or instructions to DSP chips, the result can only be a rough approximation of the dynamic reverberation artifacts a human will usually hear upon moving through a typical real-world space.

Audio samples might be synthesized, or they might be recordings of real-world objects that are then synchronized with representations of those objects within the game world, whether those objects are displayed on-screen or not. They might be those recordings processed and used in a potentially schizophonic manner (Schafer 1994, 89–91), applied to a context different from that when originally recorded. For example, an authentic recording of a SPAS shotgun might be used every time the player fires that model in an FPS game. The type of realism that FPS games in particular attempt, where game objects' parameters and motions are modeled on real-world physical properties and behaviors, makes use of authentic audio recordings in an attempt to create a sense of realism. The use of authentic audio samples recorded in the real world and played back in the virtual world raises several important questions, the answers to which will help understand immersion in FPS games. For instance, how far should one pursue authenticity in digital games? Too much realism—a realism of theme that comprises "plausibility of characterization, circumstance and action" (Corner 1992, 100) combined with authentic audio samples, photorealistic imagery, and accurate physical modeling—potentially results in a simulation or even an emulation rather than a game. Realism in the FPS game, pursued to its logical conclusion (the death of players rather than characters), would not be permitted under any current legislature; thus, rather than realism per se, codes of realism are used that are often based on convention within the genre or are imported as schemas (Douglas and Hargadon 2000) from our experiences elsewhere. A second question is based upon the notion of schizophonia; namely, are the sounds heard in the game dissociated from their cause and context? Prima facie, the answer would be yes; sound has been plucked from a real-world sound source and uprooted from its environmental and causal context and placed in a virtual world to act as the sonic surrogate to mute pixels. Certainly, in realism FPS games, there would seem to be an irreconcilable tension between the desire for realism and authenticity of sound and Schafer's claim that such a situation is schizophonic, producing a synthetic soundscape rather than an authentic, causally real soundscape.

The answer to such questions of authenticity and schizophonia may be found by examining the significance and level of sound indexicality in the context of the game world (rather than the context of the recording). Film theorists have long pointed out that cinematic sound is not only schizophonic but, very often, is also quite unrelated to the apparent sound source shown on-screen. In a reference to the synchresis already discussed here, Lastra (2000, 147, 207), writing about tin-sheet thunder and coconut-shell horse hooves, states that "fidelity to source is not a *property* of film sound, but an *effect* of synchronization" and views the stockpiling of analog and digital audio samples for use in film audio dubbing as a production method to be used for the construction of a representation of reality. This bears a similarity to Chion's (1994, 108) suggestion that the use of such sound conventions, no matter how divorced from reality, creates the "impression of realism [and such sounds become] our reference for reality itself." The notions of new references for reality and of constructed representations of reality, codes of realism, may equally be applied to digital games. As already stated, however, games are highly interactive, and the resultant soundscape for games is not preconstructed and supplied on the distribution medium but, rather, is constructed in real time according to the actions of the player (or players in a multiplayer game). The game engine, therefore, acts as a sonification engine by converting the nonaudio data, the player's actions, and the player-derived game state into sound (Grimshaw 2008b, 119–20). In this case, the authentic audio samples in realism FPS games superimpose the indexicality of the real-world context with an indexicality derived from the sonification processes of the game.[10] The recording of a SPAS shotgun is heard not as the original recorded sound source in its real-world context but becomes, through the synchresis of player action, game image and audio samples, the sound of the player firing the game's shotgun. The construction of another reality in FPS games using sounds that are indexical to the player's actions in the game world rather than being strongly indexical to the external world is a strong motivating factor in the facilitation of player immersion in the game.

Immersion

I have previously proposed that humans are immersed in their real-world environments and that this immersion is what computer game designers attempt to replicate, especially in first-person-perspective game worlds such as those found in FPS games. I have carefully elided the question of what exactly is *meant* by immersion but, having now laid the groundwork for understanding some of the processes at work in FPS game sound, I can no longer avoid attempting to answer that question—it is, after all, the focus of this chapter. The short answer to the question *what is immersion?* is that no one knows. Grau (2003, 13) may describe the phenomenon as a mix of "diminishing critical distance . . . and increasing emotional involvement," but this does not explain how immersion occurs, its processes, nor does it account for the possibility of physical immersion. Am I not physically immersed in the sound environments of both reality and of digital games at the "*center* of auditory space" (Ihde 2007, 207)? I have lost track of the number of conference speakers I have heard who have owned up to their ignorance (before proceeding to claim how such and such a technology will aid in creating more immersive digital-game environments), and I, too, guiltily hold up my hand. Immersion has become the holy grail of first-person-perspective game design, the touchstone by which the quality of such games is measured. Yet, many questions remain unresolved; theoretical answers lack experimental data that would help formulate the procedural rules required to create truly immersive digital-game worlds. This is not to say that no one has attempted to define immersion in digital games both theoretically and/or experimentally, but such work, especially the latter form, is at an early stage and thin on the ground. This section briefly discusses such work as preparation for a discussion on how sound is *used* for facilitating player immersion in FPS games.

Early discussion describing psychological states similar in many ways to the descriptions of immersion listed later do exist. The philosophy and ideal of absorption held by eighteenth-century French art critic Diderot has been summed up as an obliteration of a beholder's presence in front of the painting and the transportation of "the beholder's physical presence [to] within the painting [whereby beholder and painting become] a closed and self-sufficient system" (Fried 1980, 131–32).[11] Immersion in virtual environments,[12] whether the virtual environments of digital games or of other applications, has been explained by a variety of related terminology. Thus, concepts such as *presence*, *being there*, *involvement*, *engagement*, and *flow* are used in part, whole, or combination to describe states or processes of immersion. Discussing the technology of virtual environments, presence is used to describe either "a sense of being able to touch and manipulate a virtual object" or, more germane to this chapter, "a sense of being and acting inside a virtual place" (Reiner and Hecht 2009, 183). In the latter sense, presence is the "direct result of perception rather than sensation [and] the mental constructions that people build from stimuli are more important than the stimuli themselves" (Fencott 1999). Perception rather than sensation is at the root of Slater's (2002) Gestalt-derived theory of presence in virtual environments, in which the brain chooses from among a set of hypotheses in order to determine where one feels present—in the real world or in the virtual world. A switch between hypotheses is termed a "break in presence" by Slater. Brenton and others (2005), in a discussion that uses theories of presence to derive a perceptual basis for the uncanny valley theory,[13] suggest that we instead superimpose these hypotheses and that our sense of presence relates to the dominant hypothesis.[14]

A number of scholars have defined frameworks or rule sets for player immersion in digital games that prescribe the conditions for immersion, where immersion relates to player engagement or involvement in the game. Kearney and Pivec (2007) state that immersion provides the motivation—which they relate to Csíkszentmihályi's concept of flow—for the player to repeatedly engage with the game, and they use a study of eye movement and blink rate to determine when a state of immersion is being experienced; theoretically, the less movement and the lower the blink rate, the more immersive the game experience is. Their requirements for an immersive state in the game include an emotional involvement in the game, an altered sense of time, and a lowered awareness of the player's surroundings during gameplay. The latter state, in particular, may be related to Brenton et al.'s theory of the superimposition of presence hypotheses rather than Slater's switch of hypotheses, which requires a *lack* of awareness of the surroundings once the immersive switch has taken place; the player is still required to use the game's hardware to engage with the game since such hardware is not a part of the game world and is still able to respond to real-world alarm signals, for example. Calleja (2007) appropriates Goffman's metaphor of the frame in order to construct an involvement model to explain immersion in digital games. In order to foster what Calleja terms *incorporation* in the game (a term he prefers to *immersion*), the player fluidly switches between six frames of involvement: tactical; performative; affective; shared; narrative; and spatial. The process of internalizing these frames is related to flow but, Calleja explains, cannot be equated with it as flow is a description of activity rather than a description of the environment in which such activity takes place. I return to the concept of incorporation later when I discuss the acoustic ecology of FPS games.

Following Fencott's analysis of presence, McMahan (2003, 75–76) suggests that immersion is facilitated in digital games through the design of *surities* and *surprises* into the game world. The former are cues (Fencott's term) in which elements in the game confirm and conform to the player's expectations; in FPS games, these might be the ability to navigate (the character) around the apparently three-dimensional spaces of the world or an appropriate use of paraspaces (Parkes and Thrift 1980) in which, for example, there is a set of weaponry that is contemporaneous with the game's premise. Surprises are either an aid to navigation around the game world or provoke an action from the player.[15] These are an indication of the interactivity of digital games, in which the player is able to engage with the game spaces and objects; they might be visual signs that indicate routes through the game level or guns that allow the player to deal harm to enemy characters. For McMahan, a defining structure of immersion is realism: a consistent social realism (similar to Corner's plausibility of theme) and a perceptual realism, the use of perspective in the visual design of the game world, for example. She further states that, for immersion to occur, the actions of the player "must have a non-trivial impact on the environment" and, following from the recognition of active interactivity as an element of immersion, that "immersion is not . . . wholly dependent on audio- or photo-realism" (McMahan 2003, 68–69). Hidden within this last statement is the implicit suggestion that immersion in game worlds *is* dependent upon audio- and photo-realism and that anything less risks a loss of immersion, a break in presence. Here, realism, in the case of sound, is assumed to be represented by the authenticity of audio samples in the game; as I have already noted in the context of cinema, a variety of scholars have stated that such authenticity is not a prerequisite for a sense of realism. In the context of sound in digital games, I return to this point later.

Building upon Pine and Gilmore's (1999) work on experience, Ermi and Mäyrä (2005) claim that "immersion [in digital games] means becoming physically or virtually a part of the experience itself." Furthermore, they distinguish between three forms or states of immersion: sensory; challenge based; and imaginative immersion. Again, I return to these ideas later, particularly the sense that Calleja's incorporation is related to the notion of sensory immersion and of being physically a part of the game's sound world, when I discuss the role of sound in digital-game immersion. The idea of sensory immersion, as opposed to imaginative or challenge-based immersion, is fundamental to Carr's (2006, 69) category of perceptual immersion, where the player's senses are monopolized by the game world.[16] A final statement on immersion comes courtesy of Garcia (2006, 23), who suggests that "in the most immersing environments reminders of the structural level of the game are gone." To bring the discussion back to Diderot, the unity of beholder and painting in a "self-sufficient system" or the unity of player and game world in an autopoietic, self-organizing system (Grimshaw 2008c) "is the strongest magic of art" (Diderot, quoted and translated in Fried 1980, 130).[17]

Three themes can be abstracted from the preceding overview of immersion theories: presence; active engagement; and codes of realism. They are not mutually exclusive, and indeed there is the strong suggestion that all three must be achieved to some degree to provide the conditions for immersion in digital games. However, common to the three themes is the interplay, perhaps even tension, between sensation and perception. Is presence, for example, a matter of being physically within a concrete reality, the phenomenological realm of the senses, or is it a matter of a mental construct, a Platonic realm of thoughts and ideas? Is immersion effected only when all of the elements of the digital game have the prerequisite levels of verisimilitude and/or veridicality, the "semblance of truth" required for Coleridge's (1817) "willing suspension of disbelief," and, if so, what is that level? Can someone with no experience of the virtual environments of digital games and, therefore, without the weight of prior knowledge and expectation, become immersed in those worlds? The following section delves into the role of FPS game sound in immersion in an attempt to answer just such questions.

The Role of Sound in FPS Game Immersion

Here I focus on the relationship between digital game sound and the player and how such sound is designed to achieve a perception of immersion and, indeed, whether such immersion is achieved. It is important to note that immersive game worlds do not attempt this immersion through sound alone; the designer's arsenal contains a battery of sensory and interactive game technology for just such a purpose. Haptic feedback in FPS games is at an early experimental stage. It is more advanced (but still primitive) in other digital game genres such as racing games on gaming consoles, particularly where the handheld controller, for example, might vibrate in response to crashes or uneven road surfaces. In the context of FPS games, therefore, I deal solely with image as it relates to sound.

In addition to hearing, vision forms a large part of the experience of FPS games, particularly where the display on-screen is designed to lead to player immersion in the game world. As already noted, this immersion is initiated (in FPS games) with the game character being seen only in part. A pair of arms clutching a weapon recedes from the bottom of the screen into the space of the game world—a virtual prosthetic extension of the player, bridging reality and virtuality. The identification of the player with the character is further strengthened by the control the player has over these arms, the ability to move the character around the game world, and the player's ability to use that character to interact with objects and other characters in the environment. As the character responds to the player's control, it is the player who perceives that they are navigating in and interacting with the game world, and, in FPS games, responses to misfortune are always in the nature of "I got killed" rather than "my in-game character was destroyed."

As previously mentioned, sound can be used in conjunction with image to make sense of the real-world environments in which we are situated; it can also be used in the absence of image by making greater use of experience and imagination. In the FPS scenario sketched earlier in the chapter, I used sound in both ways to engage with the game. Even though the *vin rouge* sign itself does not move, its creaking sound and the direction it arrives from allow me to imagine that it does, especially as the image itself is now out of view. The images depicting the ruined village may make use of perspective in which there is a scaling of size toward a vanishing point but in which the entities themselves are flat, two-dimensional objects displayed on a flat, two-dimensional screen. The distant muffled explosions, the locational properties of the sounds I hear, and their reverberant characteristics create of that village a three-dimensional world that, in its sound environment at least, approaches somewhat the level of detail of sound environments of the real world.

Much of the sound I hear in the FPS game is acousmatic sound. Sound from the game world surrounds me, unlimited in its directionality; my visual window onto that world is severely restricted, as in reality, and even more so in that the computer screen and its images form only a small part of what I can see. Acousmatic sound in the FPS game allows me, with the benefit of an experience that derives both from reality and from previous playings of the game, to visualize sound. Footsteps behind me indicate the presence of someone or something in both reality and virtuality; in this game's world, they further connote a potential threat. Distant sounds of battle combined with radio messages allow me to imagine situations that team members might be experiencing and to follow the progress of the game beyond what the screen offers. This, then, is the sonic world of the FPS game, which works in conjunction with image and imagination to engage the player in the game world.

I previously noted three themes that are apparent in scholarly discussion of immersion: presence, active engagement, and codes of realism. The presumption is that all three, in varying degrees and combinations, are required to define and/or create immersion. It might be supposed that presence in digital games (i.e., a sense of being and acting inside the game world) is analogous to immersion—Brown and Cairns (2004, 3) equate what they call "total immersion" with presence—but this definition does not take account of the requirement to have McMahan's nontrivial effect on that world for immersion to occur. Acting inside a virtual world is not a strong enough term to describe the player's ability to effect change in the world, indeed, to create that world. Limiting our discussion to sound, the soundscapes of FPS games may well require a "discerning Subject" to be present (Böhme 2000, 15), but all of the sounds in that soundscape are not merely sounded by the game engine's sonification processes in response to the player's very presence in the game world but are also changed in direct response to the player's actions. This is because of the player's engagement with the interactive affordance offered by the technology of the game: As the player navigates through the FPS game world and plays the game, the soundscape changes accordingly. Movement creates the sound of footsteps, firing the sound of gunfire, and environmental or ambient sounds fade in and out as the player moves through the game's spaces. The player's imagination is engaged by acousmatic sounds, and the sounds of other players' characters in a multiplayer game indicate that those players are also present in and engaged with the same world; players' acoustic environments dynamically overlap and part company.

What level of realism or authenticity of audio samples is required for immersion? Jørgensen (2006, 13), writing in the context of digital games, states that "realistic audio samples . . . will make the audio world more immersive." If, by "realistic," she means authentic, we have already seen that her assertion is not necessarily the case. What is the authentic sound of the zombies in *Left 4 Dead* (Valve Corporation 2008)? What is the authentic sound of the BFG[18] in *Quake III Arena* (id Software 1999) and the various imaginary monsters in *Half-Life 2* (Valve Software 2004)? Would the authentic, realistic absence of sound in a game played out in the vacuum of space make it more of an immersive experience?

Shilling, Zyda, and Wardynski (2002), using the realism FPS game *America's Army* (MOVES Institute 2002), measured players for electrodermal and heart-rate responses in order to test emotional arousal in the presence of the game's authentic sounds. While not an experiment to measure immersion per se, the authors assumed that "emotional arousal has a positive impact on [the] sense of immersion in virtual environments" and, following the results of their experiment, observed that immersion is crucially enhanced by precise synchronization of sound and the action displayed on the screen; this is objective (but not conclusive) evidence, in the context of digital games, for the effect of synchresis.

Even in the presence of nonauthentic sounds, however, game players experience immersion. This was the case in a psychophysiological experiment on diegetic sound and nondiegetic music in the FPS game *Half-Life 2* (Grimshaw, Lindley, and Nacke 2008). This game has a range of authentic audio samples of footsteps on a variety of surfaces, for example, and has reasonably accurate, real-time modeling of reverberation according to the volumes and materials of the immediate game space; it also includes a variety of decidedly nonauthentic vocalizations from the game's imaginary creatures.[19] Players' subjective responses gathered after gameplay categorically indicated perceptions of immersion in the game world (and these perceptions increased with the addition of sound), while the results of electromyography (EMG) and galvanic skin response (GSR) measurements taken during gameplay indicated heightened arousal when game sound was heard (as opposed to when it was muted).[20]

Darley (2000, 16–17) defines realism, in the context of digital media such as games and computer-generated films, as the degree of resemblance to real-world objects; while not discussing audio realism, photography is given as the yardstick for images. Photo-realist images, he claims, are indexical to their real-world counterparts, and, by this definition, authentic audio samples are indexical to real-world sounds and, by extension, to their sound sources. Following this logic, where digital visual realism derives solely from the indexicality of the photograph, if used in a digital game, audio samples are phono-realist regardless of their use and context. Yet realism may also derive from theme and action as discussed earlier, and, although authentic, phono-realist audio samples are widely used in FPS games, a sense of realism also derives from the way they are used in the game. Cinematic sound design has the mantra "see a sound, hear a sound"; this is extensively used in FPS games, but the dictum in this case can be expanded to "do a sound, hear a sound." The sonification of the player's in-game actions is the realism required, in part, for immersion in the game world rather than, necessarily, the use of authentic audio samples. In simulating the processes of acoustic environments of the real world within virtual worlds, game designers provide not only indexical, real-world sounds, sometimes in addition to more fantastical sounds, but also a simulation of sound genesis and behavior in the game world that is similar to that found in the real world. In this, the active relationship between the player and sound may be likened to the acoustic ecologies found in nature. Immersion is to be arrived at not only through the inclusion of authentic sound objects from the real world but also from the use and context of such sounds and other sounds. This is an immersion based primarily on contextual realism rather than object realism, verisimilitude of action rather than authenticity of sample.

The game's soundscape, its acoustic environment, and the player together form the acoustic ecology of the FPS game. Like Truax's (2001, 66) concept of the acoustic community "in which acoustic information plays a pervasive role in the lives of inhabitants," the FPS game's acoustic ecology has an effect upon the actions of the player (or players in a multiplayer game). Footsteps approaching from behind might cause the player to turn to meet a potential threat or to escape a particularly vicious in-game character (Grimshaw 2009). Psychophysiological data provide more detailed evidence of the effect on the player's physiology and emotions while in that acoustic environment. However, the concept of an acoustic ecology goes further than that of an acoustic community in suggesting that the human is a fundamental component of that ecology who not only responds to acoustic information but also generates acoustic information. This generation of acoustic information and, indeed, of the acoustic environment itself is evidence of the immersion of the player (or players) in the game world and of the sonically concrete reality of that game world. Calleja's six frames of involvement to explain his notion of the incorporation of the player into the game world is missing a seventh frame. While "incorporation" and the performative frame hint at it, Ermi and Mäyrä's concept of sensory immersion provides the physicality lacking in Calleja's model. Gene Youngblood (1970, 206) wrote that "The notion of 'reality' will be utterly and finally obscured when we reach that point [of generating] totally convincing reality within the information processing systems . . . We're entering a Mythic age of electronic realities that exist only on a metaphysical plane." Although he was discussing the future and the potential of visual computing hardware, it is not stretching the imagination too far to suggest that the immersion of the player in the game world through the incorporation of the player as part of the game's acoustic ecology leads to a reality that is different from that of the real world—one where the player, through presence and a realist active engagement with the acoustic environment, is truly immersed in a mythic electronic reality, that of the game world.

Conclusions

Throughout this chapter I have used a variety of polarizing sets of terminology. With regard to one particular set, from a superficial reading of the chapter, the reader might depart with the supposition that there is a real world and a virtual world and that immersion is simply a matter of stepping from one space into the other. This would be wrong. Although I have used such polarity to clarify my arguments, I have also dropped hints throughout the chapter pointing to a different state of affairs; the immersed player still utilizes real-world objects to interface with the game world and is attentive to real-world alarms, for instance. In "reality," real world and virtual world are two poles of a continuum, along which the player is able to be transported, importing conventions and experiences and expropriating and interpreting meaning from either world. In the context of digital games, the waters of this apparent divide are about to become murkier still.

William Whittington, describing sound and animation production practices at Pixar Animation Studio (this volume), discusses the future of such sound design as encompassing the synthesis of sound generated in response to the image and extends this to the possibilities for real-time sound synthesis in digital games (a possibility discussed elsewhere, such as in Grimshaw 2009). Sounds would be synthesized according to player action, game context, and game architecture. Let us take this idea a step further. As demonstrated throughout this chapter, there is plenty of evidence as to the effect of sound upon player affect and emotion. Turn this around: What about the effect of affect and emotion upon sound? Primitive consumer headsets that measure EMG and electroencephalography (EEG) are already available to interface with home computers and gaming systems. The work my colleagues and I are involved in asks the question, *Can we use EMG and EEG output from the player to process or synthesize sound in a game world such that that sound itself alters the player's affect and emotion in a predictable manner?* Can, for example, the player be made more frightened in a survival horror game through the specific alteration or creation of sound if the game engine "senses" the player is not frightened enough?[21] Can the stress, pitch, and rhythm patterns of a nonplayer character's speech be changed in response to the player's psychophysiological state? Can the sound of a monster be ripped out of the hidden recesses of the player's terrors? Such a topic is one for another chapter (and continuing empirical work), so I conclude by asking how, should such a vision of all-encompassing, real-time sonic biofeedback be reached, would we then define immersion? What are the implications for the already muddy distinctions between real world and virtual world? Would—and could—immersion be attained through a precise and calibrated manipulation of the player's psychophysiological state, and would there be distinctions such as real world/virtual world, or would the gamer, instead, inhabit a *blurred world* immersed neither here in reality nor there in virtuality but in a new form of space somewhere in between?

NOTES

1 However, it requires little skill to replace these audio samples with others if they are stored on a read-write medium.

2 I discount concert films, music video, and other forms where the musical score is intended to be diegetic.

3 New sound-panel technology may change this in future; screens may also act as loudspeaker transducers able to reproduce sound that is sourced, with varying levels of accuracy, from the images on-screen.

4 The notion that human + environment = ecology makes that assumption explicit.

5 In the sense that sound will arrive at the player's ears from a variety of game-world directions.

6 As previously mentioned, modern digital games, in particular first-person-perspective games, typically use audio samples.

7 Within the boundaries of the game's level design.

8 For example, a repeated audio sample of birdsong.

9 Audio compression is a reduction in the dynamic range of the sound and thus a reduction in the digital storage required.

10 See the Gestalt-based discussion given later on the superimposition of presence hypotheses and the domination of one over the other.

11 I am indebted to Hillel Schwartz for directing me to Fried's fascinating book.

12 See Helmreich (this volume) for a discussion of immersion in sonified aquatic environments.

13 The notion that, with robots and synthetic characters, fear can be the response as the visual representation of the character becomes more humanlike.

14 This might help explain the ability to engage with and to foreground the virtual world of a digital game while simultaneously being aware of the real world in the background.

15 To a certain extent, Schafer's keynote, signal, and soundmark terminology may be related to McMahan's concept of surities and surprises.

16 This is an unfortunate conjunction of perception and sensation given the distinctions noted previously, particularly by Fencott (1999). Carr's other category is psychological immersion, where the players become "engrossed through their imaginative or mental absorption" (2006, 69).

17 This is not the place to continue the sometimes heated debate on whether digital games are an art form, but champions of that position who are looking for academic ammunition could do worse than triangulate between Diderot, absorption/immersion, and digital games/art.

18 Big Fucking Gun.

19 Accepted that defining these sounds as "vocalizations" potentially classes them with other, ostensibly more authentic vocalizations taken from the real world. In this case, they have a level of authenticity because of a shared causality and indexicality, which is why they are recognized as vocalizations.

20 As yet unpublished results using other physiological data gathered during the same experiment intriguingly point to gender differences, both subjective and objective, in the perception and experience of immersion.

21 What would the ethical considerations of such biofeedback be?

REFERENCES

Anderson, Joseph D. *The Reality of Illusion: An Ecological Approach to Cognitive Film Theory*. Carbondale: Southern Illinois University Press, 1996.

Böhme, Gernot. "Acoustic Atmospheres: A Contribution to the Study of Ecological Acoustics." *Soundscape* 1(1) (2000): 14–18.

Brenton, Harry, Marco Gillies, Daniel Ballin, and David Chatting. "The Uncanny Valley: Does It Exist and Is It Related to Presence?" In *Workshop on Human-Animated Characters Interaction*, Edinburgh, September 6, 2005.

Brooks, Mel. *Blazing Saddles*. Warner Bros. Pictures, 1974.

Brown, Emily, and Paul Cairns. "A Grounded Investigation of Game Immersion." In *Proceedings of the Conference on Human Factors in Computing Systems*, 1297–1300. New York: ACM, April 24–29, 2004.

Calleja, Gordon. "Revising Immersion: A Conceptual Model for the Analysis of Digital Game Involvement." In *Proceedings of Situated Play*, 83–90. DiGRA. University of Tokyo, September 24–28, 2007.

Carr, Diane. "Space, Navigation, and Affect." In *Computer Games: Text, Narrative, and Play*, ed. Diane Carr, David Buckingham, Andrew Burn, and Gareth Schott, 59–71. Cambridge: Polity, 2006.

Chion, Michel. *Audio-Vision: Sound on Screen*. Trans. Claudia Gorbman. New York: Columbia University Press, 1994.

Coleridge, Samuel Taylor. *Biographia Literaria*. 1817.

Corner, John. "Presumption as Theory: 'Realism' in Television Studies." *Screen* 33(1) (1992): 97–102.

Curtiss, Scott. "The Sound of Early Warner Bros. Cartoons." In *Sound Theory, Sound Practice*, ed. Rick Altman, 191–203. New York: Routledge, 1992.

Darley, Andrew. *Visual Digital Culture: Surface Play and Spectacle in New Media Genres*. London: Routledge, 2000.

Douglas, Yellowlees, and Andrew Hargadon. "The Pleasure Principle: Immersion, Engagement, Flow." In *Proceedings of the Eleventh ACM on Hypertext and Hypermedia*, 153–60. New York: ACM. San Antonio, Texas, May 30-June 3, 2000.

Ermi, Laura, and Frans Mäyrä. "Fundamental Components of the Gameplay Experience: Analysing Immersion." In *Proceedings of Changing Views: Worlds in Play*. DiGRA. Toronto, June 16–20, 2005.

Fencott, Clive. *Presence and the Content of Virtual Environments* (1999). http://web.onyxnet.co.uk/Fencott-onyxnet.co.uk/pres99/pres99.htm (August 4, 2005).

Fried, Michael. *Absorption and Theatricality: Painting and Beholder in the Age of Diderot*. Berkeley: University of California Press, 1980.

Garcia, Juan M. "From Heartland Values to Killing Prostitutes: An Overview of Sound in the Video Game Grand Theft Auto Liberty City Stories." In *Proceedings of Audio Mostly 2006*, 22–25. Piteå, Sweden, October 11–12, 2006.

Gaver, William. W. "How Do We Hear in the World? Explorations in Ecological Acoustics." *Ecological Psychology* 5(4) (1993): 285–313.

Grau, Oliver. *Virtual Art: From Illusion to Immersion*. Cambridge, Mass.: MIT Press/ Leonardo Books, 2003.

Grimshaw, Mark. *The Acoustic Ecology of the First-Person Shooter: The Player Experience of Sound in the First-Person Shooter Computer Game*. Saarbrücken: Mueller, 2008a.

———. "The Audio Uncanny Valley: Sound, Fear, and the Horror Game." In *Proceedings of Audio Mostly 2009*, 21–26. Glasgow, September 2–3, 2009.

———. "Autopoiesis and Sonic Immersion: Modelling Sound-Based Player Relationships as a self-organizing system." In *Proceedings of the Sixth Annual International Conference in Computer Game Design and Technology*. Liverpool, November 12–13, 2008c.

———. “Sound and Immersion in the First-Person Shooter.” *International Journal of Intelligent Games and Simulation* 5(1), ed. Q.H. Mehdi, Leon Rothkrantz, and Ian Marshall, 119–124. 2008b.
———, Craig A. Lindley, and Lennart Nacke. “Sound and Immersion in the First-Person Shooter: Mixed Measurement of the Player’s Sonic Experience.” In *Proceedings of Audio Mostly 2008*, 9–15. Piteå, Sweden, October 22–23, 2008.
id Software. *Quake III Arena*. Activision, 1999.
Idhe, Don. *Listening and Voice: A Phenomenology of Sound*. Ohio: Ohio University Press, 1976. (2nd ed., Albany: State University of New York Press, 2007).
Jørgensen, Kristine. “On the Functional Aspects of Computer Game Audio.” In *Proceedings of Audio Mostly 2006*, 48–52. Piteå, Sweden, October 11–12, 2006.
Kearney, Paul R., and Maja Pivec. “Immersed and How? That Is the Question.” Unpublished paper presented at Game in’ Action. Gothenburg, Sweden, June 13–15, 2007.
Lastra, James. *Sound Technology and the American Cinema: Perception, Representation, Modernity*. New York: Columbia University Press, 2000.
McMahan, Alison. “Immersion, Engagement, and Presence: A New Method for Analyzing 3-D Video Games.” In *The Video Game Theory Reader*, ed. Mark J. P. Wolf and Bernard Perron, 67–87. New York: Routledge, 2003.
Metz, Christian. “Aural Objects.” In *Film Sound: Theory and Practice*, ed. Elisabeth Weis and John Belton, 154–61. New York: Columbia University Press, 1985.
Mizuguchi, Tetsuya. *Rez*. Sega, 2001.
MOVES Institute. *America’s Army*. Monterey, Naval Postgraduate School, 2002.
Parkes, D. N., and N. J. Thrift. *Times, Spaces, and Places: A Chronogeographic Perspective*. New York: Wiley, 1980.
Pine, B. J., and J. H. Gilmore. *The Experience Economy: Work Is Theatre & Every Business a Stage*. Boston: Harvard Business School Press, 1999.
Reiner, Miriam, and David Hecht. “Behavioral Indications of Object-Presence in Haptic Virtual Environments.” *Cyberpsychology and Behavior* 12(2) (2009): 183–86.
Schafer, Raymond Murray. *The Soundscape: Our Sonic Environment and the Tuning of the World*. Rochester, Vt.: Destiny, 1994. (Originally published as Raymond Murray Schafer, The Tuning of the World. New York: Knopf, 1977).
Shilling, Russell, Michael Zyda, and E. Casey Wardynski. “Introducing Emotion into Military Simulation and Videogame Design: America’s Army: Operations and VIRTE.” In *Proceedings of Game On*, 151–154. London: GameOn November 30, 2002.
Slater, Mel. “Presence and the Sixth Sense.” *PRESENCE: Teleoperators and Virtual Environments* (MIT Press Journal) 11(4) (2002): 435–39.
Stockburger, Axel. “The Game Environment from an Auditive Perspective.” In *Proceedings of Level Up*. DiGRA. Utrecht Universiteit, November 4–6, 2003.
Truax, Barry. *Acoustic Communication*, 2nd ed., Norwood, N.J.: Ablex, 2001.
Valve Corporation. *Left 4 Dead*. Valve Corp., 2008.
Valve Software. *Half-Life 2*. Electronic Arts, 2004.
Youngblood, Gene. *Expanded Cinema*. London: Studio Vista, 1970.

CHAPTER 15

THE SONIC PLAYPEN: SOUND DESIGN AND TECHNOLOGY IN PIXAR'S ANIMATED SHORTS

WILLIAM WHITTINGTON

Introduction

In 1986 a newly formed, high-tech startup company named Pixar began producing animated shorts in order to challenge its designers to develop computer applications that showcased the design possibilities of its new technology—the Pixar Image Computer, which could render photorealistic images albeit in a rudimentary way. A few years earlier, these same engineers worked with George Lucas and Industrial Light & Magic (ILM) to develop special-effects software and sequences for Hollywood studio features such as *Young Sherlock Holmes* (1985) and *Star Trek II: The Wrath of Khan* (1982), for which they created the "Genesis effect" (named for a science probe that generated a planet out of "lifeless rock"). Eventually, Pixar transformed from a computer company to become a leading studio in the field of 3-D computer-generated animation, and its efforts marked Hollywood cinema's transition into the digital age. Much like the impact of cinema's transition to sound in the late 1920s, this new focus on computer-based production processes shifted industrial practices, established new studio styles, and fostered a new form of filmmaking. This revolution was not entirely image based, however. Concurrently, film sound was undergoing a digital transformation of its own with the introduction of new sound formats such as Dolby Digital and Digital Theater System (DTS) and the rise of the sound design movement, which was driven in part by new portable recording and mixing devices, multichannel exhibition, and changes in the mode of sound production.

Using a variety of theoretical perspectives from digital culture, sound studies, and traditional film studies, I argue that as the techniques of computer-generated animation developed at Pixar, sound design became an integral aspect of this new mode of storytelling and overall filmic design. Foregoing an exclusively onomatopoetic approach to sound, defined by the *bangs*, *booms*, *zooms*, and *honks* commonly found in traditional cartoons, sound designers at Pixar worked with producers, directors, animators, and software engineers to establish an unprecedented unification of sound and computer imagery by borrowing live-action production techniques and reworking traditional animation strategies for sound use. In regard to technology, Pixar's sound designers adopted new portable recording devices for the collection of sound effects, unified these "raw sounds" with the aid of sound samplers like the Synclavier and, later, software like Pro Tools, and mixed their efforts in the newest multichannel formats typically reserved for feature productions. More important, sound designers like Ben Burtt (*Star Wars* series, *WALL-E*) and Gary Rydstrom (*Terminator 2: Judgment Day*, *Saving Private Ryan*, and *Monsters Inc.*), who worked on many of the initial shorts, fostered audio strategies that emphasized sound perspective, spectacle (localization of effects and the establishment of offscreen space and environmental effects), and "hyperrealism" (a technique Rydstrom would later adapt to films such as *Jurassic Park* and *Titanic*, which were also heavily laden with computer graphics, or CG). These technological innovations and new aesthetic approaches quickly established the sound-image relations in Pixar films as cinematically credible and viable for filmgoers. In short, the sound designs "sold" the images. The short films produced at Pixar are important in regard to this trend because they established the aesthetic and production patterns that formed the studio's house style, which eventually migrated into the company's successful feature films such as *Toy Story* (1995), *Finding Nemo* (2003), and *Up* (2009). In the short films, which make up the case studies for this chapter, sound is not only a dominant formal element, but also an important thematic one, principally as it relates to the notion of play. Many of the shorts explore childrearing and children's games as a form of play, the interaction of sound and images as play, and the notion of play associated with music performance and the voice. From *Luxo Jr.* (1986) to *Jack-Jack Attack* (2005), filmgoers are immersed in a kind of sonic playpen, surrounded by innovative sound designs and computer images that have reshaped our notions of cinema and animation in the digital age.

Sound Design's Expanding Role in the Era of Digital Animation

Animation is unique within the mode of Hollywood film production, above all in relation to sound. Typically, the overall sound track for any film consists of dialogue, music, and effects (Foley, hard effects, and ambiances), and within Hollywood cinema, a hierarchy has formed around issues of the voice, predominantly focusing on narrative intelligibility or who *tells* the story.[1] For this reason, the dialogue tracks form the scaffolding around which the entire sound track for a live-action feature is designed. Traditional animation, however, lacks production recordings and sometimes even dialogue, which is the case in a number of the Pixar shorts. This lack offers both creative freedoms and challenges for sound designers, who must not only develop the specific sounds for the film but also establish the overall rhythm of the film, which supports the story beats, character actions, and plot points. In the 2001 Academy Award–winning short *For the Birds*, director Ralph Eggleston and sound designers Tom Meyers and Jory Prum build comedic tension around the squeaks, squawks, and pecking of a gang of birds who want their club to remain exclusive. In a rhythmic chant of squeaks (recorded from a collection of "squeaky toys"), the birds egg on two of their flock to peck at the feet of a gangly outsider, who dangles precariously upside down from the telephone wire on which they are all perched (Amidi 2009, 30). Gravity and a well-timed moment of silence on the sound track provide the crucial comedic punch line as the birds achieve their goal but are launched upward (and featherless) with the sharp sound effect of a "boing." Sound and image work together rhythmically to comment on the absurdity and perils of a mob mentality without ever resorting to the use of dialogue. It is important to note that the choice of squeaky toys to create the character sound designs accesses nostalgic notions of childhood play and the tactile sensations and emotional delight of squeezing these air-filled toys. The visual design of the plump little birds reinforces the concept, and, even within the narrative action, the larger bird puffs and honks at the air that is filled with tiny feathers from the now naked flock. It is a comedic gesture that affirms the themes of air and flight, breathing and exhaling, and silence and squawking. In this way, play and peril are linked through sound.

Since its introduction in the 1970s, the term *sound design* has been multifaceted in its application and definition within cinema and sound studies. One of the initial definitions of sound design comes from the process of designing specific sound effects like those noted earlier. Animation has always been highly adept at creating and applying individual sounds for sonic punctuation and comedic effect, borrowing techniques from radio and the theater. As sound historian Robert L. Mott has noted in relation to early Disney animation, "When Mickey Mouse hit a baseball over the fence, the sound of the hit was provided by a wooden block, and the slide whistle sounded as the baseball went soaring into the air" (Mott 1990, 83).

These sound effects displaced realistic sounds to create broad comedic gestures or slapstick, which Charlie Chaplin once called "playful pain" (Mott 1990, 80). These early sound constructions have since become cliché. While sound designers today often apply the same strategies of timing and comedic intents, the sounds are much different in their material manufacture, as well as spatial encoding and aesthetic design. This new approach is currently supported by sound-editing software like Pro Tools, which has consolidated sound editing, mixing, and previewing into a virtual work environment. Within the program, the sound session for a particular sequence is laid out horizontally, providing a graphic representation of the sound wave forms for each individual effect, which can then be manipulated by using various sound filters (high- or low-frequency pass filters, for example) or processors. This digital system differs from the previous magnetic film system in that effects can be layered without significant noise buildup, sound sets can be created and stored within the digital environment for easy recall, and composite sound effects can be previewed in stereo or even multichannel formats without delay. Gone are the days when sound personnel would have to wait until the final mix to hear the composite sound effects.

In the Pixar short *Luxo Jr.*, a baby lamp pops a ball, which deflates with a comedic whistle of air. Unlike the earlier Disney example, the sound-effects design is multilayered. It includes the squeak of the plastic as the ball rolls on the hard wood surface, the pop of plastic, the expulsion of air through a plug, the deflating whistle, and the shifting of the coiled springs of the tiny lamp as it rides the ball down. In addition, the effects are recorded in close perspective to their sources using a portable Nagra in order to anchor them with a sense of image and sound credibility. This is not to suggest that the sound designer just gathers one version of each of these effects and combines them for the desired result. Sound design is a process of ongoing experimentation. Gary Rydstrom notes, "I have a PowerPoint to explain the design process that I use when I lecture about sound. The first slide reads, *Record a sound, try it, find it doesn't work*. The next card reads, *Select another sound, try it, find it doesn't work*, and so on. I keep doing this until I get to the final card that reads, *Run out of time*" (Rydstrom 2010). The design process in the digital age is one of nearly infinite choices that are limited only by aesthetic imagination and, of course, postproduction economics. While the comedic outcomes remain similar between the old and the new animation forms in the *Luxo Jr.* short, the sensibilities of design and their organization have shifted to create a heightened cinematic reality that resembles live-action sound.

The development of overall aesthetic sensibility for computer-generated animation is in part due to the involvement and sensibilities of specific sound personnel, considerations that are linked to another definition of the term *sound design*. Historically, the term *sound design* has also been applied to "the planning and patterns of the overall sound track" by a specific individual (Whittington 2007, 2–3). In many cases, the sound designer becomes akin to the director of the sound track, making sure that the various film reels exhibit thematic and formal unity. This is accomplished by the control of sonic aspects such as the use of leitmotifs, the design of effects for emotive impact, and attentiveness to mix practices. The key sound designers at Pixar have been Ben Burtt, Gary Rydstrom, and Tom Meyers (*Pitch Black*, *Armageddon*, and *Up*). In the Academy Award–nominated short *Lifted*, the film's director, Gary Rydstrom, makes playful reference to being a sound designer by comically drawing parallels between the job of the mixer and the job of a young teenaged alien learning to abduct humans from a hovering spacecraft. In this homage to Steven Spielberg's *Close Encounters of the Third Kind* (1977), an alien trainee named "Stu" sits behind a vast console of toggle switches (an analogy to a sound-mixing board), trying to retrieve his target human from a farmhouse. All the while, he is being graded on his very spotty performance by a gelatinous and impassive instructor (a stand-in for a producer/director). The overall sound-design strategy is an exercise in cross-cutting, as Stu's frustrated actions, articulated by the sounds of his hands raking the console and a Pong-like computer ping, lead to the abducted man being flung about the interior of the home. The sound design illustrates this through various thumps, body hits, and crashes. Causality is the source of the comedy, but the analogy to the perilous process of Hollywood filmmaking cannot be missed. Sound designers are always under the scrutiny of someone, whether it be the producer, the director, or other sound personnel. Finally, the instructor intervenes at the console and reverses the chaos by returning the abductee to his bed. The lesson of this exercise in computer animation is that at times sound and image design are about interplay. As in the performance of a duet, the comedy can be drawn from either side of the design equation. For example, when the abducted man gets stuck in the tree, the visual track reveals a static wide shot of the tree and a long beam of light emanating from the flying saucer. At the same time, the sound track offers the "brutal" sounds of the crunching of leaves and tree branches, as well as body parts, as the man gets stuck, clearly offering filmgoers an example of Chaplin's "painful play" (Amidi 2009, 39).

For Rydstrom, if the sound design is recorded and organized well, it "merges into the image, brings it to life," and the constructions are "not cartoon-y" but rather "reality-based" (Kenny 2004, 1–2). This is the foundation of his "hyperrealistic" sound-design style, a philosophy that draws inspiration from hyperrealism in visual culture, which is interested in our perception of the "real," a property that is both the subject and objective in hyperrealistic art. In the field of sound design, Rydstrom creates stylized constructions that access familiar sounds (raw sound "events"), yet he recombines these effects to create seamless sound impressions that are viscerally dynamic, anthropomorphic, and preoccupied with codes of heightened cinematic realism often found in live-action film. His goal is not to create an exact duplicate or simulation of a sound event, but rather to provide a lie that tells a dramatic *truth*. In unpacking this lie, we find cinematic codes of spatial and temporal design, genre expectations, and the engagement of an array of psychoacoustic properties of sound. Visual equivalents of these properties (mainly an attentiveness to movement, space, and temporality) have been a goal of animated Disney films from the very beginning. Digital theorist Andrew Darley notes, "Precedents can be found in Disney Studio's attempts, from the late 1930s, to mobilize *certain* of the existing aesthetic codes of classical narrative cinema (live action) and to integrate them in a more rigorous fashion than had hitherto been the case within drawn cartoon form" and in doing so provided a "heightened realism" (Darley 1997, 19). Rydstrom adapts this approach to sound design.

Early in their careers, many of the key sound designers at Pixar were associated with San Francisco Bay Area film-production companies and sound houses such as Sprocket Systems, later redubbed Skywalker Sound. They have collaborated with various directors such as James Cameron, Francis Ford Coppola, John Lasseter, George Lucas, and Steven Spielberg. These sound-conscious filmmakers encouraged and in many instances demanded innovations in relation to sound technology, mixing, and design. As a result, the duties of sound-effects recording, editing, Foley, and mixing became less constrained in terms of the division of labor within the mode of production. This shift arose in part as independent production units adopted collaborative strategies that grew out of production approaches from film schools and very low-budget productions in which every member of the crew was required to perform one or more tasks. According to Rydstrom, this environment fostered an important "mentoring" network for a core group of sound designers, in which attentiveness to sound experimentation and technological innovation was the norm, unlike the hierarchical system within the classical Hollywood mode of production (Kenny 2004, 1). John Lasseter, Pixar's chief creative officer and one of its most important directors, agrees with this assessment: "As a manager, it is my task to abolish hierarchies" (Reis 2009, A6). This sense of collaborative creativity in the field of sound was carried over to the newly formed Pixar Animation Studio and integrated well into its creative environment, which was attempting to blur the lines between filmmaking and computing.

Technological Innovation and Animation in Transition

The exchanges between technological development and aesthetic design are at the core of the Pixar philosophy, and they have deeply influenced the image and sound styles that have emerged at the studio. According to Lasseter, "Art challenges technology, and technology inspires the art. That's it in a nutshell—the way we work at Pixar" (Milsom 2008, 08:50 min.). One of the key pieces of sound technology adopted very early by Pixar sound personnel was the Synclavier, a synthesizer that was modified with a Winchester diskette add-on and floppy drive (and later a hard drive) to function as a sound sampler that could hold a library of "raw" materials for instant access and manipulation. According to Rydstrom, "With sampled sounds in RAM, you can instantly pitch-bend it and layer it and play it without using any processing time" (Kenny 2004, 3). In his short *Lifted*, Rydstrom worked with a similar yet far more advanced software-based system to build the frustrated vocalizations of the alien trainee, "Stu," using sampled growls and vocalizations from his own dog, Sparky. In the short film *Tin Toy*, the Synclavier technology brought together all of the sounds for the one-man band "Tinny," including his horn, drum, and cymbals. During the production process, the Synclavier allowed the consolidation of the duties of a sound recordist, editor and mixer, which supports the aims (and definition) of the sound designer. Among sound scholars and practitioners, there is consensus that technological developments such as the Synclavier and Digidesign's Pro Tools formulate an essential part of the definition and process of sound design. In fact, sound designer and editor Walter Murch (*THX1138*, *The Conversation*) first associated the term with his experience using the quadraphonic exhibition technology during the development of the sound track for *Apocalypse Now* (1979). As he used the multichannel technology to map out the sound fields established by the various speakers in the motion picture theater, he would drape or "design" sound effects and music like a set designer in order to establish an immersive experience for the filmgoers.

Historically, sound technology has never been fixed or standardized within the mode of production, however. From the very earliest attempts at film sound synchronization, the development and use of sound advances have sprung from a complex set of drivers, including economic competition, licensing and patents, exhibition quality control, and, perhaps most important of all, the particular needs of a specific film production. Innovation often results from logistical and production challenges that arise from a particular story that is being told. Similarly, innovation within the field of animation has not been a fixed or formal process. The needs of animated stories have often driven the development of specific audio and visual technologies and techniques. In the 1940s, for example, Disney developed the multiplane camera setup, an animation stand that allowed cels to be divided into layers to simulate depth of field, and "these [cels] could be moved frame by frame at varying rates toward or away from the camera, giving a powerful illusion of gliding through a three-dimensional space" (Thompson and Bordwell 1994, 261). This technique was developed for use on films such as *Pinocchio* (1940) and *Bambi* (1942), and it arguably set the stage for the development of 3-D exhibition technologies, which have become part of the production and exhibition strategies of the current computer-animation cycle. In regard to sound technology specifically, Disney also worked with RCA at this same time on the development of a new multichannel sound presentation for *Fantasia* (1940), which featured what many describe today as the precursor to "surround sound." The goal was to separate various orchestral instruments to match the movements of the characters within the animated musical sequences of the film. This effect was achieved by using two separate but interlocked optical tracks. In one configuration scheme, the speaker array consisted of fifty-four speakers throughout the theater and provided one of the first multichannel experiences for filmgoers. While the system was never adapted as an industry standard, it did present the possibility of a new cinematic-sound experience that would inspire companies like Todd AO and later Dolby to pursue multichannel exhibition possibilities (Blake 1984, 20). The coupling of image and sound developments presented a unique opportunity for Disney to reshape not only animation as form but also the visceral experience of cinema itself for filmgoers by creating new immersive environments for audiovisual play.

In their move to lead Hollywood cinema into the digital era, Pixar brought this approach up to date by embracing the power of computing for both image and sound production. They also coupled these developments with new exhibition formats from multichannel to 3-D presentation but always within the context of the character-based stories they wanted to tell. The blending of technological concerns and design considerations is never easy, however. As Steve Jobs has noted, "Pixar did an impossible thing . . . It blended the creative culture of Hollywood with the high-tech culture of Silicon Valley . . . The best scientists and engineers are just as creative as the best storytellers, just in different ways . . . The Pixar culture, which respects both, treats both as equals" (Paik 2007, 295). Central to Pixar's success has been the ability to forge not just a shared vocabulary between the two groups but a successful workflow as well. It is also important to point out that, at Pixar, technology is valued as a necessary component of the design process, not simply a high-tech item or piece of software to be purchased.

To present a more complete understanding of the interplay between computer graphics and sound, it is necessary to specifically address the background related to the integration of computer-generated images into the process (and cultural understanding) of animation and live-action filmmaking. In my previous work on sound, I have discussed the integration of multichannel technology into the Hollywood studio system, and I have noted that within the film industry, technological changes have encountered significant challenges as concerns are often raised regarding economic viability, issues of standardization related to exhibition and production practices, fears of diminished quality control, and lack of widespread testing of potential success (Whittington 2007, 28). The introduction of computer-generated imagery software and hardware by Pixar into the production process faced all of these concerns. Pixar's cofounder and president, Dr. Ed Catmull, has noted that during the period of early development in the 1970s and 1980s, both "artists and studios disliked and feared computers. . . . But even in that environment, a few small groups had the vision to use the computer for picture making. At that time, we thought a lot about what it would take to make the process economical and practical" (Street 1995, 79). Ironically, Disney, in particular, was eager to integrate computers into their highly successful 2-D animation production units—but also wary. Fearing a critical backlash about declining quality due to technological expedience, Disney did not publicly admit to using computers until the DVD release of *Beauty and the Beast*. However, during the 1980s, Disney worked closely with Pixar to create CAPS (Computer-Assisted Production System), a "digital ink-and-paint system that employ[ed] a sophisticated multiplane camera within a digital environment" (Street 1995, 79). The system shifted the labor-intensive process of inking and painting of cels to the digital realm. Ultimately, the Disney executives knew that computers would streamline the production process for animation;

however, their fears about computer technology were realized somewhat when computer-animated productions eventually supplanted 2-D animation techniques and practices in popularity.

While computer-generated images transformed the field of animation, the technology also had an equally profound impact on live-action films and their cultural reception by filmgoers in terms of expectations of credibility and verisimilitude of audiovisual design. As part of their technological focus, Pixar researchers began developing new software, such as Motion Doctor, MenV (Modeling Environment) and notably RenderMan, in order to create virtual characters and environments and to integrate or "composite" computer-generated images with live-action footage. With this innovative software, artists could engage in "image processing," specifically modeling photorealistic images, characters, and sets and incorporating realistic surfacing, lighting effects, and motion blurs that simulated the way photographic film captured action in motion. Early examples of the software's importance can be seen in the water-pod effects in *The Abyss* (1989), the liquid-metal effects in *Terminator 2: Judgment Day* (1991), and the design of the dinosaurs in *Jurassic Park* (1993). The software continues to be the foundation of most special-effects software today, and this technology arguably enabled the cinematic adaptation of books and graphic novels that Hollywood producers once considered unfilmable—from the *Lord of the Rings* series (2001–2003) to *Watchmen* (2009). Not surprisingly, one of the early adopters of the software was George Lucas, who pressed the technology into use to create his densely layered *Star Wars* prequels. As many critics have noted, this intervention raised the important question, In this new age of computer-generated images, what *is* animation?

Between the use of computer graphics for specific special effects and its use to create virtual sets, characters, and action, a blur between the categories of the live-action film and animation began to occur. In turn, this breakdown shifted the cultural reception of various image-sound relations, mainly related to "special effects." Michele Pierson, the author of *Special Effects: Still in Search of Wonder*, argues: "If an effect is only *special* in relation to something else—something that it isn't—how do viewers decide what is a special effect in this context? Does the scope for the kind of transmutation of the visual field that might make an effect special even *exist* once a film begins to be made over in the mode of an animated feature?" (Pierson 2002, 152–53). Initially, filmgoers and critics often contained the "work" of computer-generated images within the category of spectacle or special effect and addressed specifically the verisimilitude and authenticity of particular constructions. In short, filmgoers and critics were asking, "Does this special effect *look* like an effect?" And for the longest time, because of the rudimentary nature of the computer graphics (which were heavily support by sound), this was an easy question to answer.

However, this question lost it authority as the integration of live-action and computer-generated images became more refined and seamless and as computers have made their way into every aspect of the mode of production from color timing to film printings. As a result, the clear distinction between live-action and computer-generated animation has fallen away. Contemporary blockbuster films such as *Spiderman* (2002), *The Hulk* (2008), and *Avatar* (2009) regularly combine both forms, and the result has been a reshaping of cultural expectations around image-sound relations for these types of films. The debates have moved from questions about credibility to questions of immersion, visceral spectacle, and emotional resonance, as they support a new "cinema of sensation." This shift has also been supported by the rise in console video games, which offer various game environments that regularly blur the lines between live-action and animation without significant objection or notice from gamers. Historically, Pixar's films benefited greatly from this blurring of the lines. Specifically, computer-generated animation is no longer marginalized into the category of cinematic gimmick or novelty; rather, it has established itself as a unique form of filmmaking. As director Brad Bird argues, "Oftentimes people call [computer] animation a genre, and that's completely wrong. It's a medium that can express any genre" (Corrlis 2004, 80).

In addition to creating a new form of filmmaking, computer-graphics technology has even challenged the hierarchies and divisions of labor that once separated the areas of visual design such as lighting, set design, cinematography, and costuming. Today, one artist with a computer workstation and the right software can perform all of the tasks of set designer, costumer, cinematographer, and even lighting gaffer if needed. In many ways, the collapse of these duties and the merged production processes mimic the way in which sound design emerged as a new mode of production. It is historically important to note that sound design leads the way in theory and practice in this era of visual culture. I would even go so far as to argue that sound design is *the* crucial factor that has fostered the new perceptual gestalt of computer-generated animation in the digital age—a model of understanding that computer-generated images cannot fully achieve alone. Despite the efforts of computer images to seamlessly bring together collages of unexpected elements—animated or live-action—the current historical poetic of cinema in the digital age demands that the sound design present equally refined constructions to establish cinematic verisimilitude. In the article "A Back Story: Realism, Simulation, Interaction," Andrew Darley provides a systematic history of visual digital culture and even specifically addresses Pixar's position in the development of visual "techniques (programs) . . . for the various phases and procedures involved in three-dimensional modeling, animation, and rendering," but, like many others, he does not consider the role of sound or music (Darley 2000, 20). One of my aims in this chapter is to offer a counter-history that includes sound design in the discourse of the digital media.

But this raises the question, How do these computer-generated images and sound design come together to be "true?" As various case studies of Hollywood studios in the past have shown, new production processes and shifts in technology are often contained within aesthetic and stylistic perimeters set by the studios or production units. These factors become part of the development of a "house style," which filmgoers eventually recognize and come to expect from that studio.

Pixar's films established not only a new and innovative visual style but also a sonic style that has defined its house style.

Pixar Studio Style: Character Animation, Storytelling and Sound Design

From the very inception of Pixar, the founders were determined to focus their efforts on creating character-based films by using computer animation. According to *Toy Story* director, John Lasseter, "The animation was the groundbreaking thing about this movie; but our intention was for the audiences to get so caught up in the story and characters that they would forget the animation" (Street 1995, 91). This move was in part facilitated by an acceptance of the limitations of the computer technology. It is an axiom within filmmaking that if filmgoers become invested in the plight of characters in a story, they will forgive all manner of technical limitations—and early computer animation faced many limitations. The processing power and memory of the early hardware was severely limited. Specifically, because processing speeds were slow, longs hours were required to render a single image, and, once rendered, the image files presented significant storage and access problems. More important, the software was not adept at creating "organic" images, particularly in regard to modeling animals with fur or human figures (Corliss 2004, 80). For these reasons, the initial short films, like *The Adventures of André and Wally B* (1984) and *Luxo Jr.* (1986), featured characters that were "geometric" and relatively smooth in their visual design (Corliss 2004, 80). In order to mask the rudimentary nature of these images, the filmmakers at Pixar employed a host of cinematic strategies from an emphasis on character-driven stories and genre elements specifically drawn from the tropes of comedy and drama to experiment with formal aspects of the medium, such as rhythmic editing and camera moves, innovative music placement, and hyperrealistic sound design, which established the Pixar house style.

Their first short, *The Adventures of André and Wally B*, is a foundational example of this new style. The film, which was created for the SIGGRAPH (Special Interest Group on Graphics) conference in 1984, features a character named André, whose blissful afternoon is disturbed by a friendly and playful bumblebee, Wally B. In an unexpected homage, the character's names are derived from the 1981 Louis Malle film, *My Dinner with Andre*, and the short also gives a nod to the existential questions raised in the Malle film by referencing the playful nature of connections between people that are both pleasurable and painful (Amidi 2009, 14). In this short, a collaboration of animation styles—both old and new—are evident. The plot is a simple chase in which the anxious André distracts Wally B in order to run away, thus avoiding a painful sting. However, Wally B zooms after his new friend and delivers an inevitable offscreen sting. Floriane Place-Verghnes, author of *Tex Avery: A Unique Legacy*, places this idea in historical context: "The chase is a recurrent theme in the Averyan corpus and, generally speaking, in the cartoon industry of the late 1940s" and is primarily "an element of acceleration of the rhythm which can lead to total madness" (Place-Verghnes 2006, 137). In this way, the chase calls up the codes of the madcap comedy, a genre category defined by comedians such as Buster Keaton, Charlie Chaplin, and Laurel and Hardy.

While the sound ratchets up the comedic pace through an ever-increasing tempo related to both the music and the sound effects, the characters are the center of attention. Sound design is crucial in creating the integrity of the characters and the comedy of this short, as it both humanizes the characters and lends live-action credibility to their actions. One moment in the film is particularly telling: When André awakens, he *comes alive* with the sound of a yawn, a scratch of his belly, and a yowl that shivers through his frame. The sound design has done what the computer images cannot: It has rendered the character's internal structures (specifically through the breath and vocalizations from his lungs and mouth), his surface texture (through the sound of his hand on his belly), and the integrity of his body (the elasticity of his frame and spine through his throaty and quivering vocalization). The sound designer for this short, Ben Burtt, based his sound-design philosophy on a balance between how an effect might sound in the physical world with the overall dramatic needs or "truth" of the cinematic construction, and both of these factors establish the credibility of this character within the computer-generated environment (LoBrutto 1994, 142).

Lest we forget, this is also a chase comedy, and some elements of traditional cartoon sound and music strategies are blended in as well to transcend the rudimentary image design and to update familiar cinematic gestures as a form of acknowledgement and homage. In particular, Wally B's introduction is characterized by the sound of his flapping wings and buzzing, but these sounds are superseded by the self-reflexive placement of the music from "The Flight of the Bumble Bee." It is clearly the sonic setup for a joke, which will end with a sting. During the chase, sound and picture editing play a crucial role as the camera angle shifts to a bird's-eye view, and Wally B's flight sounds undergo a metamorphosis into the sound of a propeller-driven dive bomber dropping down to deliver its payload. This sound design calls on a long history in animation, in which sounds of the natural transform into the mechanical, and the humor is evoked as the "character loses his fluidity and becomes stilted" (Place-Verghnes 2006, 137). The audio and visual denouement of the short calls on another staple of animation—the role reversal, which is illustrated when an offscreen André (just stung) hurls his hat at Wally B and hits him on the stinger, knocking him out of frame with an onomatopoeic "boing." The evocation of the history of animation style and the new attentiveness to sound-design elements such as sound perspective, localization of sound on- and offscreen, and codes of live-action verisimilitude establish a new gestalt of understandings and expectations for filmgoers. As Ben Burtt notes, "It's forging those connections between familiar sound and illusionary sound that I think is the basis of the success for a lot of the sound that sound designers have put into these [Pixar] movies" (Milani 2008, 1).

Pixar's style expands on animation tradition by embracing anthropomorphism and recasting it to assist in the design of character "speech," which often mimics human sound patterns and behaviors and even combines human sounds as familiar anchors. In the short film *Luxo Jr.*, Gary Rydstrom offered director and animator John Lasseter an example of this new sound-design strategy:

> I wanted to give the lamps in *Luxo Jr.* character through sound. I told John that I'd come up with these voices. He'd never imagined they'd have voices and was wary of the idea. But I experimented with taking real sounds—a lot of it as simple as unscrewing a light bulb or scraping metal. Every once in a while, a sound would be produced that would remind you of sadness or glee . . . It felt like the birth of something new, even then. (Kenny 2004, 2)

This design approach established an emotional vocabulary for the characters through sound effects, thereby offering an analogy to language. In the film, a tiny Luxo lamp plays with a beach ball while being supervised by a parent lamp. After a mishap that pops the ball, the adult lamp admonishes the pint-sized Luxo with a snap of attention and a wag of its head. The sound design for the adult lamp focuses on punctuating the neck or joint movements of the apparatus, specifically the sound of the snap of the joint and its silence, then a wagging and clicking of metal, an audiovisual gesture that thematically supports the lamp's role as guardian and responsible adult. The visual design reinforces this idea as well by incorporating a light source that shines down on the tiny lamp like a spotlight on the bad behavior. Luxo Jr. deflates with a musical whine accompanied by a deep sigh of metal as if stifling tears. In terms of narrative storytelling, the short taps into the archetype of parent-child dynamics, while also offering a comment on the roller coaster of emotions that young children experience from moment to moment, following the ecstasy and joy of play to the deepest sadness from a sense of the loss of a beloved toy. In this way, the story is thematically about the play of emotions.

According to Lasseter, this short film solidified Pixar's philosophy about film sound and storytelling: "Gary's brilliant work made those lamps so real, so believable. It taught me that sound has an incredible ability to be a partner in the storytelling of a film, and ever since then Pixar has put a lot of emphasis on thinking of the sound as we develop our stories" (Paik 2007, 72). It is, therefore, not surprising that the Luxo Jr. lamp has become the unofficial mascot of the company and even forms the "I" in the company's logo before all of its feature films.

Character-based design strategies are evident in nearly every aspect of the filmmaking process at Pixar from behavior modeling to set design, yet the philosophies that drive these constructions can be traced back to sound design. As John Lasseter notes:

> Our philosophy for the set [design] came from Gary Rydstrom, our sound designer . . . He taught me long ago that, in doing sound effects, if a ball bounces, you don't just record the sound of a ball bouncing—because when the sound

> effect is cut in, it won't sound like it should. You have to make it bigger. To create the bark of a dog in *Toy Story*, Gary combined dog sounds with tiger sounds to make it bigger and more impressive. That's the philosophy we used in the look of the film. We went beyond reality, caricaturizing to make it more believable. (Street 1995, 83)

This philosophy encouraged artists to consider both the images and the sounds as characters in the film. One of Rydstrom's most important innovations at Pixar was to establish this sense of "hyperrealism," in which sound and images take on multiple layers of meanings and emotion by concentrating not simply on aspects of film form but also on emotional intents, expectations, and intertextual connections such as historical homage. For Rydstrom, this often meant displacing a sound from its original context, augmenting it, then reinserting into a new sound-image pairing to create a heightened effect like the tiger sounds inserted into the bark of a dog. However, rather than entirely stripping the sound of it previous meanings, Rydstrom was borrowing various qualities of the realistic or raw effect and effectively splicing them into a new creation that filmgoers would crane their ears to recognize and accept. For Rydstrom, sound design was the key to establishing the credibility of computer animation: "It was clear to me that this form was a whole new thing that would require a whole new approach to sound" (Paik 2007, 72).

The philosophies of hyperrealism became firmly established at Pixar in the 1989 Academy Award–winning short *Tin Toy*, which became the inspiration for *Toy Story*. The plot centers on "Tinny," a mechanical version of the one-man band, and his first encounter with a curious baby. Terrified at the prospect of being drooled on, Tinny eludes baby "Billy" and hides under a couch, which is where all wide-eyed toys seem to go to elude destructive children. But when the baby stumbles during the chase and begins to cry, Tinny emerges to enchant the child with his musical abilities. The moment is fleeting, however, as the baby quickly moves on to something more enticing—the discarded packaging for the toy. It is a humorous commentary on the attention span of children. It is also the perfect film to exploit the thematic notion of sound design and music as a form of play.

By today's standards, the computer-generated images in this short appear somewhat crude as the baby's oddly textured skin makes it appear more monstrous than perhaps intended, but both the character-based storytelling and the sound design transcend these deficiencies. Once again, there is no dialogue in the film; rather, the images and sound evoke the sense of humor and peril. In dramatic terms, the filmmakers cleverly devised Tinny to be his own worst enemy. His very nature is sonic by the fact that various musical instruments are woven into his jacket. His movements are, therefore, musical—so any chance of a stealthy retreat under the couch is impossible. This type of connection between movement and sound in animation has become known in animation as "Mickey Mousing," whereby the image and sound elements are connected in self-aware synchronization, and, of course, the process is named after the famous Disney character.

While this comedic idea is drawn from traditional animation, it has been completely redesigned for computer-animation. Using the Synclavier, Rydstrom created a complex database of sounds, including gears, cymbal crashes, drum hits, and horns, as well as eyelid flutters, shivers, and breathing. The instruments were not simply just recorded toy versions but they included also a series of real instruments of different scales, which were recorded with a resonant dynamic range and stacked as needed. In establishing the sound design pattern, Rydstrom composed the sound in layers and engaged them at various tempos, so they not only realistically fit the movement and actions of the character but also serve as score, which heightens the frenetic sense of peril. According to Rydstrom, "The complication for sound in *Tin Toy* was to make it sync up with the animation . . . John [Lasseter] didn't animate the cymbal and drums with the idea of what music he would play" (Amidi 2009, 24).

Along with representing the musical exterior of the character, the sound design also illustrates the inner life of the character through the emotive use of sound. In particular, when the baby falls and begins to cry, all eyes turn toward Tinny, and the accusation is registered through character reaction and sound. With an immediate click of his cymbal, a blink, and the clicking of his neck gear, Tinny turns and, in surprise, realizes what he has done. Shame overcomes him, and he lowers his head; his accordion deflates with a sound like a mournful groan. In these instances, sound provides the emotional language that renders the inner states of the character, which in turn tells the filmgoers how to feel.

In pursuit of an overall unified design, Rydstrom sets these sounds in relief against a bed of immediately recognizable vocal and ambient effects, always balancing traditional animation strategies and live-action expectations. In particular, Rydstrom recorded and edited the vocalizations of a real baby to support the rudimentary images of the "monster child" in the film rather than employing an actor to perform these sounds, which was standard practice in animation at the time. The child's coos and giggles are recorded and edited to mimic a documentary aesthetic style and even include an unscripted sneeze, which was animated into the character design. Rydstrom further heightened the live-action qualities of the environment by implying offscreen space by means of sound effects. Through an open doorway, the sound design implies the audio from a television set as someone channel surfs briefly before finally tuning into to familiar game show *The Price Is Right*. The recording features a sense of spatial encoding. The television effect is recorded from the perspective of the room on-screen, implying distance, and the chatter is also compressed and muted to give it a sense of the size of the room and the television speaker. Both of these sound constructions serve to create a psychoacoustic gestalt of a living space that is unseen by the camera, and the rapid editing of content implies the presence of a person who is changing channels, perhaps a parent just home from a trip to the toy store who is taking a break from child care. The balance of this pattern of effects creates a heightened sense of cinematic reality, effectively smoothing over the technical wrinkles in the computer-generated animation. Rydstrom reiterates the importance of *Tin Toy*: "I think it's the most sound-intensive movie per square inch that I've ever done" and for Pixar revealed the importance of sound design in "shaping the content of the film" (Paik 2007, 72).

In any discussion of Pixar's house style, it is essential to discuss the role of both music and the voice; however, these topics could merit another chapter entirely. Instead, I briefly explore how Pixar's approach has reshaped the functions of both of these elements to fit their design philosophies. In the 1989 short film *Knick Knack*, director John Lasseter paid homage to Warner Bros.-style cartoons by presenting the story of a snowman who is trapped in a snow globe but is eager to join the frivolity of a shelf full of warm-weather knickknacks. The visual design evokes a "retrovibe" that draws on trends present in "1950s' modernism" (Amidi 2009, 25). However, the music in the film is not orchestral as was traditionally the case in the early Warner Bros. animated films; rather, it features the instrument of a single voice, specifically that of performer Bobby McFerrin. McFerrin, who is perhaps best known for his hit "Don't Worry, Be Happy." He typically employs his entire body in a composition, using his voice to follow the melodic line while employing his chest to beat out the rhythm section and when necessary, using editing and re-recording techniques to fill in the gaps. His vocal style and his technique are akin to animation because they draw on the strategy of anthropomorphism to recast the voice in place of the various orchestral instruments. True to the Pixar's sound style, this composition, which draws on 1950s' bebop becomes an idiosyncratic character within the piece by providing a human rhythm that activates the computer-generated images. Specifically, the various characters, including a cactus, a pyramid, and a pink flamingo, "come alive" and sway to the beat of the music, which is situated as both score and source.

Within the hierarchy of image-sound relations, Pixar's use of music in this short flips the audio paradigm to resemble that of the musical. According to animation historian Scott Curtis, this is a similar strategy engaged by early Warner Bros. cartoon shorts: "Given that the tempo of the music has already been decided upon in any given cartoon [as a result of the music selection], the 'mise-en-scène' enacts that tempo in a variety of ways . . . characters sing and . . . buildings also sway" (Curtis 1992, 200). This acknowledgement of the power of music to drive animation became a factor in the audio and visual design of subsequent shorts, specifically *Boundin'* and *One Man Band*. However, Pixar has staunchly avoided allowing music and the conventions of musical genre to become the sole driving factor of animation in its feature films; rather, it has remained committed to a more blended approach that balances the use of sound design and music in the storytelling process. In this way, the company has distinguished its house style from that of Disney's successful 2-D animated units, which produced musicals like *Beauty and the Beast* (1991) and *Aladdin* (1992). At the end of *Knick Knack*, Lasseter and Rydstrom even draw our attention to the power of a blended approach by offering a transgressive use of "score," which comically comments on the fate of the written word in light of potential computer-generated animation. While the credits roll, a voice reads along, presenting not a recitation of the written text but a rhythmic reading of the phrase "blah blah blah." The use of the phrase is both musical and satirical as it is a playful reference to both the ineffective nature of voice-over and a sense of nostalgia for what many filmgoers might recognize from Peanuts cartoons as "adult speak," which, like most film credits, is something that fails to hold anyone's attention.

This is not to say that Pixar has given up on speech and language entirely. Rather, the shorts and the Pixar house style eventually do find a place for both, but, once again, the filmmakers rethink their use in light of the particular needs of computer animation as a new film form. Just as Rydstrom's philosophy of hyperrealism borrowed the codes and emotional components of specific sound effects, so Pixar's use of the voice attempts to do the same by borrowing the credibility of celebrity voices to bring its computer-designed characters to life. With the release of the short *Mike's New Car* (2002), Pixar moved the production of their shorts in a new direction by specifically linking them to their feature-length productions through voice casting and character crossover. In this instance, the buddy team of Sully (voiced by John Goodman) and Mike Wazowski (voiced by Billy Crystal) from *Monsters, Inc.* comes together for a test drive of Mike's new vehicle, which not so surprisingly ends in madcap mayhem as the car's gadgetry gets the best of them. The voices bring a star quality that serves as a recognizable anchor for both the production design and the filmgoers. According to animator John Kahrs, "John Goodman's vocal performance was really rich and had a lot of range . . . There is a resonant warble to his voice, almost bear-like, and it fits the character so well. I would get direction from his performance and know exactly how the eyebrows are going to move and what the emotion of the scene is going to be" (*Monsters, Inc.*: Production Notes). By contrast, Billy Crystal established Mike Wazowski with vocal qualities of an East Coast origin, specifically by engaging in fast-talking New York mania, which is incorporated into the physicality of the character through frenetic arm and leg gestures. According to lead character animator Andrew Gordon, "Billy would take a line and go off on lots of tangents with ad-libs and comedy routines" (*Monsters, Inc.*: Production Notes).

The voice itself becomes a kind of special effect, which is much like the image-sound relations found in puppetry. Sound theorist Michel Chion has argued that in the 1970s the nature of the voice in Hollywood cinema changed significantly as filmgoers were made aware of its constructed nature in films like *The Exorcist* (1973) and *Star Wars* (1977), in which the voices seemed "stuck on" to various characters in makeup or masks (Chion 1999, 164–65). Chion makes the connection directly to puppetry: "We're constantly aware that voices are grafted onto bodies, only temporarily on loan" (Chion 1999, 154). Pixar's house style acknowledges this hyperawareness and in fact depends upon the filmgoer's knowledge of these actors' personas, past work, and vocal qualities to lend credibility to their virtual incarnations. Technically, these efforts are further supported by animation software that enhances vocal synchronization in relation to the characters' on-screen facial gestures. As senior animator Peter Docter (director of *Monsters, Inc.*) noted in relation to the voice and animation style on *Toy Story*, "We have a program that enables us to look at a sound wave and break it down into frames. I listened to the sound over and over again, then did an assessment of the pitch of the words" (Street 1995, 85). These cues then became the reference points for facial gestures and body movements, while "lip-synching was facilitated by a library of mouth poses that could be used to form the various sounds" (Street 195, 87).

This is not to imply that Pixar has allowed the voice to take complete priority in the design process as it might in a live-action film; rather, the vocal design supports the overall filmmaking philosophy and house style. One of the ways in which Pixar achieves this goal is to direct voice talent toward a particular kind of skewed readings. As a result, many of the vocal performances are mannered and caricaturized somewhat with an elastic quality. Billy Crystal's performance is a prime example. His vocal timing and rounded enunciations are presented in fits and starts much like someone learning to drive a stick-shift car. They beg to be animated in the form of a giant green eyeball with a large and expressive mouth, evoking the performance codes of puppetry. Once again, this approach has been formulated somewhat around the limitations of the animation software, which despite years of development still has difficulty rendering the human form.

Pixar has worked around these limitations by embracing a unique style of human character design that focuses on geometric shapes and patterns similar to those found in comic books. This style is evident in the crossover short *Jack-Jack Attack* (2005), directed by Brad Bird (*The Incredibles*). The storyline features another child-care situation, in which preteen Kari (voiced by Bret Parker) must contend with the developing superpowers of Jack-Jack, the offspring of two superheroes. Unlike the baby in *Tin Toy*, Jack-Jack's skin is smooth, and his head is round with a triangular sprig of hair on top. Kari's facial features are similarly variant, with particular emphasis on her elastic mouth and braces. As with Billy Crystal, Kari's vocal performance hinges on idiosyncratic timing from the lethargic to the manic. The comedy resides in the scale of the performance contrasts, which range from the naturalistic to the cartoony. In this way, sound takes on hard edges that match the geometric visuals. Bird notes that, in designing these characters, he strived to make them both "caricatured *and* believable," an approach that "Disney used to call . . . 'the plausible impossible'" (Corliss 2004, 80). In the end, this assessment is an apt description of the overall audiovisual design philosophy and house style of Pixar Animation Studios as it incorporates both an understanding of the animation traditions of the past and a vision of computer-generated animation in the future.

Conclusions

Pixar Animation Studio continues to use the development of short films as a training ground for new producers, directors, animators, and sound designers. Each effort continues to provide the company with valuable research and creative outcomes. Pixar's house style, therefore, is by no means fixed; rather, it continues to develop under the influence of new technologies and personnel. One glimpse of the future, however, may be found in Pixar's support of research on sound synthesis at

Cornell's Department of Computer Science, where computer graphics researchers are developing the equivalent of computer-imaging software for sound. Using physics and software modeling, researchers are simulating the sound of water, rendering noise vibrations, splashes, and even the formulation of bubbles in synchronization with computer-generated images of water provided by Pixar as if both existed in the real world (Steele 2009, 1). If widely implemented, the implications of synthesized-sound software could shift both the production of animation in film and video games radically. For instance, sounds in video games could be programmed for real-time activation and origination, coming from specific on-screen actions or environmental sources rather than from sound files for sound effects and dialogue that are recalled by the program and repeated based on activation points or predetermined gestures. Sound and image could occur in real time, just as they do in the physical world. This approach would strengthen the immersive quality of video games, which has long been one of the primary projects of game development. In animation development, sound and image constructions could be designed together by the same software and designer, a development that could once again lead to a further collapse of duties and hierarchies within the mode of production. Ultimately, the roles of the animator and the sound designer could merge, offering the potential of not only a significant shift in Pixar's house style but also a realignment of the cultural reception and our expectations related to sound and image design in the digital age.

Table 15.1 Online Resources

Cornell Harmonic Fluids Project	http://www.cs.cornell.edu/projects/HarmonicFluids
Digidesign (Pro Tools)	http://www.digidesign.com
Dolby Laboratories	http://www.dolby.com/index.html
FilmSound.org	http://filmsound.org
Pixar Animation Studios	http://www.pixar.com
Pixar's RenderMan	https://renderman.pixar.com
Skywalker Sound	http://www.skysound.com
William Whittington, PhD	http://web.me.com/williamwhittington

NOTE

1 *Foley* is the term given to those effects created on a sound stage in synchronization with the picture. *Ambiences* are layers of background noises, which often form environmental aspects such as busy street noise or waves on a beach.

REFERENCES

Amidi, Amid. *The Art of Pixar Short Films*. San Francisco: Chronicle, 2009.
Blake, Larry. *Film Sound Today*. Hollywood: Reveille, 1984.

Chion, Michel. *The Voice in Cinema*. New York: Columbia University Press, 1999.
Corliss, Richard. "All Too Superhuman." *Time* (October 25, 2004), 80.
Curtis, Scott. "The Sound of the Early Warner Bros. Cartoons." In *Sound Theory, Sound Practice*, ed. Rick Altman, 191–203. New York: Routledge, 1992.
Darley, Andrew. "Second-Order Realism and Post-Modernist Aesthetics in Computer Animation." In *A Reader in Animation Studies*, ed. Jayne Philling, 16–24. London: Libbey, 1997.
———. *Visual Digital Culture*. New York: Routledge, 2000.
Kenny, Tom. "Gary Rydstrom: Oscar-Winning Sound Designer on the Road to Pixar." *Mix—Professional Audio and Music Production* (February 2004). http://mixguides.com/consoles/tips_and_techniques/gary-rydstrom-interview-0204/ (accessed July 9, 2009).
LoBrutto, Vincent. *Sound-on-Film—Interviews with Creators of Film Sound*. Westport, Conn.: Praeger, 1994.
Milani, Matteo. "Ben Burtt Interview—Upcoming Pixar." (November 2008). http://usoproject.blogspot.com (accessed July 9, 2009).
Milsom, Erica. *A Short History of Pixar: Pixar Shorts*, vol. 1. Los Angeles: Walt Disney Home Entertainment, 2008. DVD.
"*Monster's Inc.*: Production Notes." http://www.cinema.com/articles/724/monsters-inc-production-notes.phtml (accessed July 9, 2009).
Mott, Robert L. *Sound Effects: Radio, TV, Film*. Boston: Focal, 1990.
Paik, Karen. *To Infinity and Beyond! The Story of Pixar Animation Studios*. San Francisco: Chronicle, 2007.
Pierson, Michele. *Special Effects: Still in Search of Wonder*. New York: Columbia University Press, 2002.
Place-Verghnes, Floriane. *Tex Avery: A Unique Legacy (1942–1955)*. Eastleigh, UK: Libbey, 2006.
Price, David A. *The Pixar Touch: The Making of a Company*. New York: Knopf, 2008.
Reis, Detlef. "John Lasseter's Seven Creative Principles." *Animation* 23(2) 191 (February 2009): A6.
Rydstrom, Gary. "The Brain and Creativity: A Conversation with Dr. Antonio Damasio and Gary Rydstrom." Sloan Science Seminar Series, University of Southern California–Los Angeles, January 22, 2010.
Sonnenschein, David. *Sound Design: The Expressive Power of Music, Voice, and Sound Effects in Cinema*. Studio City, Calif.: Wiese, 2001.
Steele, Bill. "Computer Graphics Researchers Simulate the Sounds of Water and Other Liquids." *Cornell University Chronicle Online* (June 1, 2009). http://www.news.cornell.edu/stories/June09/SynthSounds.ws.html (accessed July 9, 2009).
Street, Rita. "Toys Will Be Toys." *Cinefex* 64 (December 1995): 76–91.
Thompson, Kristin, and David Bordwell. *Film History: An Introduction*. New York: McGraw-Hill, 1994.
Whittington, William. *Sound Design and Science Fiction*. Austin: University of Texas Press, 2007.

CHAPTER 16

THE AVANT-GARDE IN THE FAMILY ROOM: AMERICAN ADVERTISING AND THE DOMESTICATION OF ELECTRONIC MUSIC IN THE 1960S AND 1970S

TIMOTHY D. TAYLOR

SOUND doesn't knock on your door—it comes right in.

—Raymond Scott, "Raymond Scott Sounds Off on Sound"

I would like to acknowledge Suzanne Ciani for her willingness to be interviewed, Bob Kosovsky at the New York Public Library for his help with the Eric Siday Archive, and the many participants at the Sound Studies conference in Maastricht, the Netherlands, in November 2009.

Introduction

One of the interesting aspects of music in advertising in the United States is that modernist musical techniques or idioms—atonality, widespread use of dissonance, complex rhythms, and so forth—never really found their way into the production of this music. Many writers of advertising copy who aspired to become a "serious" novelist got their start in advertising, and the same is true of many visual artists (see Bogart 1995). But nothing resembling musical modernism in the sense described here ever found a place in advertising music.

With one exception: Some new sound technologies found their way into advertising music quite quickly, following fairly closely behind their use in avant-garde music. Advertising musicians' early adoption of electronic technologies meant that many Americans, unaccustomed to listening to avant-garde music of any kind, first heard sophisticated new electronic technologies such as the Moog synthesizer through less prestigious venues such as soundtracks to science fiction films and television and radio commercials. To be sure, many had probably heard the Theremin and other electrical sounds, but the invention and widespread use of electronics in the 1960s marked a new era of the proliferation of such innovative sounds. No less a figure than Robert A. Moog wrote the following in 1977: "The listening public first became aware of the elctronic [*sic*] medium subliminally, through radio and TV commercials." Moog mentioned some of the figures I discuss later (Moog 1977, 859).

While this may seem strange, in the period when electronic technologies were appearing, advertising budgets were nearly unlimited (see Taylor n.d.). Advertising agencies spared no expense in hiring musicians, who were thus able to purchase these expensive technologies. Eric Siday, one of the first to use electronic music commercially, was paid about $5,000 per second (more than $33,000 today) in the mid-1960s ("Swurpledeewurpledeezeech!" 1966, 68).

Today, Americans are surrounded by electronic sounds of all kinds: as recorded music, as heard in films and television, as telephone hold music, or as the opening sound splash that accompanies powering up a personal computer. But such sounds were not always so ubiquitous or accepted. The early history of electronic music in the United States in the post–World War II era, whether in the realm of art or commerce, is a history of suspicion, derision, and confusion. Countless articles in the popular press critiqued or ridiculed the music and frequently greeted it with incredulity or scorn.

This chapter, then, offers a history of domestication, the domestication of sounds that were initially associated with science fiction but fairly quickly found their way into television commercial pitches for anything from coffee to beer. These sounds might have originated in public cinemas, but they became ubiquitous in family rooms across America. Helpful here is the science and technology studies (STS) work on the concept of domestication (Oudshoorn and Pinch 2003; Silverstone and Hirsch 1992). Domestication in the STS literature tends to focus on technologies, but here we are dealing with sounds, products of technologies. The focus, unlike for Silverstone and Hirsch (1992) is not the household but the computer music studio and the laboratory.

I use the domestication concept in three ways. The first is the most literal—the adoption of sounds associated with science fiction to use in selling products in everyday life. However, I also mean it to describe the complex set of processes behind the scenes by which electronic sounds were harnessed for use in selling. How did electronic musicians convince potential clients in and out of the advertising industry that such sounds could be accepted for use in advertising? Finally, I intend the term to refer to the ways in which inanimate commodities were thought to become friendly products for consumers, brought to life by electronic sounds.

Electronic Music: First Associations

It is perhaps not surprising that the first main uses of electronic technologies for making music outside of the rarefied realm of art music composition were in science-fiction films. Electronic music could represent the future as depicted in such films, a future that included "music without musicians," as the title of one article on electronic music put it in the late 1950s (Tall 1957). Electronic technologies in science-fiction films date back to the 1950s. Louis Barron (1920–1989) and Bebe Barron (1925–2008) are best known for composing the soundtrack to the film *Forbidden Planet* in 1956 (Barron and Barron 1989). As a result of this film, their prominence made them sought-after musicians for electronic music in commercials. The way their work was discussed helped contribute to popular ideas about electronic music, as this quotation from *Mademoiselle* in 1959 shows:

> Louis and Bebe Barron create film scores with electrons, activated into electronic circuits that seem to express emotional characteristics. When stimulated, the circuits respond with meaningful sounds, as if alive; the film track is composed by altering and editing them on tape. The eerie results communicate emotions without the intervention of symbols. ("Music of the Future?" 1959, 94)

For a long period after the rise of electronic music in advertising in the 1950s and into the 1960s, advertising-music composers had to contend with the public's perception of such music as strange, fit only for representing the unreal and horrific. In a speech on October 16, 1969, Eric Siday said, "the failure of electronic music to, as yet, escape the confines of the experimenters, space film sound tracks, and the world of the avant garde for much wider acceptance is that the listener has not only been expected to accept a new and strange *sound*—but to learn a whole new musical *language*."[1] In addition, Raymond Scott, a composer/inventor who was another of the earliest composers of electronic music for commercials, said in

1962 that, to the public, electronic music "sounded like nuclear war and at its tamest like 'outer space' music."[2] However, he continued, "If that was all you could do with electronic music, I'd say it had no place at all to speak of in TV background and scoring" (liner notes to *Manhattan Research Inc.* 2000, 122).

Scott's fears were reasonable. Early coverage of electronic music in the popular press in the United States frequently played up its seemingly otherworldly or horrific attributes. A 1961 article about an electronic music concert quoted a woman from the audience: "All those spiders in the sky!" (Bowers 1961, 60). An article from 1962 admitted that electronic music "*may sound like static from outer space*" and stated that sinus tones have a "spooky impersonality [that] carries an implicit connotation of space-travel, music from Mars, the machine personality" (Moor 1962, 49, 50; emphasis in original). Another article that same year noted composer Vladimir Ussachevsky's annoyance at the terms "push-button music" and "space music" ("The Sound of Hell," 1962, 86). In 1968 an article by an anonymous writer, surveying twenty years of electronic music, noted, "Electronic music had its uses. Its science fiction sounds were wonderful for TV commercials and moving-pictures scores" (Discus 1968b, 157).[3] A few months later, this author praised what he believed was a new trend in electronic composition, moving away from "abstraction and the science-fiction sound" (Discus 1968a, 165). Cartoons in the popular press throughout the 1960s showed images such as dancing robots (Moor 1962, 49) and a man in tails at a piano but looking at an electronic display instead of at the music on the stand (Berger 1969, 46).

Some writers did believe that electronic music was no passing fad, for, in the words of one critic, "too many serious people are working on it, too many rockets are shrieking up into the skies, too many children are watching TV." He lamented the cliché—even in 1961—that associated electronic music with outer space (Bowers 1961, 61).

Despite the associations that the general public had for electronic music, Scott nevertheless felt that the unfamiliar sounds could sell products in novel ways. A demo track titled "Don't Beat Your Wife Every Night!" consists of a series of Scott's audio logos (about which more later) with a voice-over of sample advertising copy, a sort of rehearsal for the use of electronic sounds in commercials. In a 1962 lecture Scott recounted how he assembled this track:

> After we put together the tape, we called in an announcer-friend of ours, "Bucky" Coslow. We said, "Listen to these electronic effects, we'll play them one at a time and whatever they make you think of—commercial-style—say it real spontaneous-like. We'll record them, then later, we'll have a mix and see what happens." We did just that, with one difference: The announce tape was edited but before we'd got a chance to sync it with the effects tape, it was run purely by accident at random against the effects tape. The effect was startling. Words and phrases that had no business showing up where they did against certain electronical effects took on a wonderfully convincing and attractive quality and seemed to indicate that electronic music for this purpose may turn out to have unusual vitality, conviction, and atmosphere plus a rather shocking flexibility. (liner notes to *Manhattan Research Inc.* 2000, 111)[4]

Scott also said, "Maybe the reason these effects are as attractive as they seem is because we are as yet not preconditioned to a species of electronic music, and consequently our ears are ready to accept a more abstract marriage between the spoken word and electronic musical effects. But regardless, we believe that possibilities are most exciting" (liner notes to *Manhattan Research Inc.* 2000, 131). Thus, even composers had to convince themselves of the utility of these new sounds for commercial purposes.

AUDIO EXAMPLE 16.1

The Rise of Electronic Music in Commercials

The decade of the 1960s was one in which the postwar rise of American consumption was reaching new heights, as is well known. Thanks to the Servicemen's Readjustment Act of 1944 (popularly known as the GI Bill, which made it possible for returning veterans to obtain an education or a mortgage), an unprecedented number of new houses were built in the 1950s, houses that were bigger than those most Americans had lived in before, marking both increased affluence and the increase in consumption of goods, which necessitated more space in which to hold them. A new and relatively prosperous middle class emerged for the first time in American history.[5]

The advertising industry grew along with the American population and its increasingly voracious consumption habits; billings to clients for advertisements increased by more than 50 percent every five years following World War II, with the first half of the 1960s increasing by nearly 78 percent, the largest increase in the couple of decades after the war.[6]

The rise of what is known in the industry as "parity products" (those that are essentially similar, such as soft drinks or fast foods) came to rely heavily on music to help create an impression of difference among similar products (see L. Cohen 2003 and Taylor n.d.). Budgets for the production of music grew, including, for the first time in any significant way, electronic music, once composers began to understand how to employ electronic music in commercials. The next step was to convince potential clients that such music could benefit advertising representations and sales of commodities.

Raymond Scott: Moments of Freshness

Raymond Scott (1908–1994) was one of the first advertising musicians to compose music for commercials using only electronics, beginning late in 1960, for major national brands such as Vicks, Lever Brothers, Alcoa, Hamm's Beer, and Parker Pen (Chusid 1999, 10).[7]

Scott (born Harry Warnow) was a Juilliard-trained musician who wrote and arranged jazz tunes for his own quintet and served as the musical director for *Your Hit Parade* on CBS radio. His long interest in electronics led to the founding of Manhattan Research, Inc., in 1946, a company that specialized in the development of electronic devices for musical use. He began producing recordings for a variety of purposes, including, beginning around 1960, advertising. At one point, the firm billed itself as Designers and Manufacturers of Electronic Music and Musique Concrète Devices and Systems. The use of the term *musique concrète* shows some cognizance of the European avant-garde of the 1940s and 1950s, which Eric Siday (discussed later) shared.

Scott believed that he had found a way to domesticate electronic music, to employ it in everyday usages such as commercials that obviated its association with "nuclear war" and "outer space music." He argued that, unlike in science-fiction films, electronic music:

> can be used in a light way, and in many mood-provoking ways, where instead of frightening the audience it will entertain them and help put over points and create memory images that are much more striking than could be done with conventional instruments or even with an unconventional handling of conventional instruments. Electronic background scoring is not intended as a replacement for conventional orchestral ensembles, but for certain moments of freshness, for certain kinds of products, and especially for an Audio Logos idea where a very special sound appears together with the appearance on screen of the product, or the sound or punctuation or some kind of audible underlining of a video moment, an announce moment or a combination of both has extremely powerful possibilities. (liner notes to *Manhattan Research Inc.* 2000, 135–36)[8]

For Scott, whose music bears a closer sonic relationship to the avant-garde than Siday's, the problem of "lightness" was a significant issue in the process of making the music approachable and domesticated:

> I do believe that we have licked the problem of lightness in electronic music and that I wouldn't be at all surprised as these months go by that the various things that we can do with it will become so wide in scope that there's a good chance that we would be hard put to find something that electronics couldn't do something for. (liner notes to *Manhattan Research Inc.* 2000, 136)

In 1957 Scott released a promotional LP titled *The Jingle Workshop*. The text accompanying the recording referred only briefly to the use of electronic technologies, showing just how carefully electronic sounds had to be presented to potential clients in this era:

> THE JINGLE WORKSHOP recognizes the growing need for a thoroughly qualified, completely staffed musical organization, equipped to apply a combination of special talents and professional skills to the production of successful TV and Radio Jingles.
>
> THE WORKSHOP feels that certain elements are vital to assure success in such a strongly competitive field:
>
> *First*—the experience and ability to analyze and satisfy the individual needs of each product.
>
> *Second*—the flair for brilliant, provocative, yet basically simple—composition of music and words.
>
> *Third*—the talent for "exact touch" orchestration.
>
> *Fourth*—the conducting skill to achieve a winning instrumental performance, perfectly balanced with vocal style and diction for proper product emphasis.
>
> *Fifth*—the electronic "know-how" to apply recording techniques and equipment with imagination.
>
> *Sixth*—the overall showmanship in taste and approach, to blend these many elements into an effective expressive performance.
>
> THE WORKSHOP sincerely believes that it has these aims in its operation, further
>
> THE WORKSHOP would welcome the opportunity to discuss your individual needs—either for new commercials or for fresh, exciting production of our existing jingles. (liner notes to *Manhattan Research Inc.* 2000, 24)

The machine Scott used for composing was nicknamed "Karloff." It drew upon two hundred different sound sources and employed several electronic tone generators. Pitch, timbre, intensity, tempo, accent, and repetition were controlled by a panel. According to a 1960 magazine article, the machine:

> can do virtually anything. It can sound like a group of bongo drums. It can give impressions which suggest common noises. It can create the mood of musical tone-poems. And it can also take the advertiser's theme music and produce limitless emotional variations on it to suit a variety of musical styles—all, of course, if Scott is at the controls. (liner notes to *Manhattan Research Inc.* 2000, 51)[9]

An article on Scott from *Advertising Age* in 1962 said that the average American was exposed to more than five hundred commercials per week (one of many such articles from this decade that worried about advertising "clutter" on the air, which clearly influenced Siday as well) and said that Raymond Scott had a way to cut through it: "Grab 'em by the ears!"—by using audio logos, "new plastic sounds," and "electronic abstractions," according to the bemused *Advertising Age* author (McMahan 1962, 119).[10]

These new electronic sounds were thought to be able to capture listeners' attention in ways that more conventional sounds could not. Some of Scott's descriptions of various advertisements illustrate his approach. About a commercial for County Fair Bread in 1962, he said:

> We were invited to do an electronic music score of an animated TV spot on behalf of County Fair Bread. The problem: to create an electronic musical impression of a calliope playing: "Where, oh where (in this case) is my County Fair Bread?" Compose a "man in a white suit"–type theme, for the central character, a slightly knocked-out magician. Put together a group of electronic impressions of typical animated cartoon sound effects. Here are the sounds we came up with:
>
> - Sound of the calliope;
> - Magician/"man in a white suit" theme;
> - Cartoon effects. (liner notes to *Manhattan Research Inc.* 2000, 130–31)

The resulting commercial sounds like a John Cageian tape experiment such as *Variations IV*, with the jingle treated to a number of effects, sometimes being warped and, to my ears, parodied. It is astonishing that the client approved these sounds, as advertising clients, then as now, are notoriously conservative and deeply afraid of offending anyone. One can only surmise that the music was heard as appropriate to accompany the visuals.[11]

AUDIO EXAMPLE 16.2

Scott's discussion of an Auto-Lite spark plug commercial in 1961 provides another justification of electronic sounds: that they could both animate and support the images in a commercial:

> Every time you saw the Auto-Lite spark plug, you saw it spark and the explosion that followed. We were able to create a distinctive sound that was somewhat realistic, somewhat impressionistic and you not only saw the spark and the explosion, but the attractive nature of the accompanying sound made a fine part of audio identification with that high-point in the video part of this commercial. Effecting an impressionistic interpretation of situations can frequently make possible a stunningly attractive effect to go with a certain moment in a TV commercial where, something that hasn't got a sound can have a sound riding "piggy back," so to speak, on the general action—which, with exposure, becomes the audio-frequency memory of the "spot" in question. (liner notes to *Manhattan Research Inc.* 2000, 135)

AUDIO EXAMPLE 16.3

Despite Scott's concerted efforts to employ electronic music in commercials in ways that were palatable to his listeners—to domesticate electronic music—he was less successful in wining national accounts than his main competitor in this era, Eric Siday. Scott, an inventor as well as a musician, seemed to be too wedded to the sounds his machines could make to become as famous as Siday, who conceptualized his electronic sounds first and foremost as acoustic ones.

Eric Siday: Identitones

London-born Eric Siday (1905–1976), a jazz musician who emigrated to the United States in 1939, began working in the advertising music world in 1949 (Cross 1968, 59). Unlike Scott, Siday became spectacularly successful. In 1966, *Time* magazine reported that about 80 percent of the U.S. population had heard at least one of Siday's compositions, making him one of the most frequently played composers in the world and one of the best remunerated, as we have seen. *Time* said that Siday's signature music work was based on the principle that a product would stick in the consumer's subconscious better if accompanied by an "instantly recognizable musical trademark" ("Swurpledeewurpledeezeech!" 1966, 68).

Perhaps the most famous commercial Siday wrote was for Maxwell House Coffee in 1959 and first broadcast on August 10 of that year.[12] This commercial re-creates the sound of percolating coffee not just as sound but as a tune almost as memorable as a good jingle. Hugely successful, the commercial was played on and off into the 1990s (" 'Perking Pot' Theme Back at Maxwell House" 1990). Siday claimed, "the Maxwell House 'perking' sound is considered one of the breakthroughs in the field of broadcast logos."[13] Elsewhere, he described it thus, in a manuscript that is undated but likely from the late 1960s:

> [T]his commercial for coffee?—No pictures of cans of coffee. No newlyweds at the breakfast table, not so far, even a mention of coffee! . . . A character was created; and starred. It was a Coffee Pot . . .
>
> As the camera slowly panned down, it revealed the coffee pot, shiny and elegant, perking away, but not with the real sound of coffee perking; each little perk had a quasi musical tonality that was a calculated characture [*sic*] of the real sound.[14]

The creation of the character of the perking pot and the use of music to serve as its soundtrack helped these sounds gain acceptance. It is likely that many viewers did not understand this music to be the product of tape manipulation, particularly since it functioned as a musical illustration of moving images.[15]

VIDEO EXAMPLE 16.1

Unlike Scott, Siday's music seems to have been conceptualized at least in part as tonal music, and he wrote at least some of his electronic compositions in conventional notation, perhaps in order to copyright them. As was the norm in the business at that time (and the present), copyrights were assigned to advertisers. Notated scores were usually just a few bars of the most memorable melodic portions of the jingle. The notated score to "The Perking Coffee Pot" was a shorthand version of melody, just twelve bars long (plus six more of slow-moving accompaniment), and none of the nuances provided by his electronic gear are notated or indicated. By contrast, the version used in the commercial lasts nearly the entire length of the sixty-second advertisement.

With the enduring success of this jingle, which was contained on many demo tapes and mentioned in many promotional materials—often as the lead example of Siday's prowess and fame—Siday began to write electronic sound logos for use in advertising to identify a product in sound. He also began writing snippets of sound to announce a news broadcast or weather repot or sports item in radio broadcasts. He called these sounds "Identitones," an updated term from "Identitunes," which was his earlier, acoustic music venture, begun in 1957 and suspended in 1964. In mid-1966 Identitunes became Identitones, Inc., and was clearly an electronic music endeavor, as Siday's standard contract indicates: "Identitones, the originator and owner of a copyrighted package of electronic sounds consisting of various sounds which may be used in radio and television as promotion aids to identify a radio and/or television stations(s) and its various program features (i.e., weather, news, time), hereby gives and grants to the LICENSEE . . ."[16] Siday provided sound logos for many major American and multinational corporations such as Westinghouse, ABC Radio, ABC Television, American Express, and Sprite. Siday wrote in a 1969 speech called "Electronic Music in Communication," "I think the most significant value I have found in my use of electronic music *is* it's [*sic*] application to the *logo*, and I consider this almost an art form in itself."[17]

Siday composed a sound logo for Westinghouse on October 17, 1963, according to the date on the score, which is only six measures (though the static part of it includes the notation "electronic gliss: [Key of F]") and for which Siday was paid $6,000 for the first six months' use alone.[18] This became another well-known Siday composition that was included on many of his demo tapes, and the copyright saga around it reveals the difficulties of claiming ownership for these new kinds of compositions. Conventionally notated but electronically realized, such compositions were a sort of hybrid of electronic and tonal acoustic music, which made Siday's compositions less difficult to accept than Scott's in this period.

AUDIO EXAMPLE 16.4

Yet Siday's electronic sounds were different enough that Siday believed they could be used to cut through the advertising clutter of his day, a concern of many in this period. "The world is satiated by conventional music. To grasp a listener today, you have to give him something new." For him, the means to this was through the "art of miniaturization—saying something that instantly stands for a corporation's personality" ("Swurpledeewurpledeezeech!" 1966, 68). Identitones produced fifty "sound images" that radio stations could use to identify themselves. Siday thought that the similarity of radio programming made it difficult for listeners to tell stations apart. Identitones consisted of station identifications that presented the station's call letters in a dozen variations, as well as various sounds that could be used to introduce weather reports, traffic alerts, and so on. Wrote *Time*:

> On rainy days, announcers give the weather forecast to the background refrains of electronic wind and rain, which comes in three intensities, drying up, drizzle and drench. Warm summer nights are depicted by impressionistic bullfrogs and nightingales, cold winter days by chilling quivers and twangs. ("Swurpledeewurpledeezeech!" 1966, 68)

Siday says, "It's all subliminal. The imagination of the listener can run riot"; he believed that "You just can't get a good drenching rain sound with an orchestra" ("Swurpledeewurpledeezeech!" 1966, 68).

Like other early electronic music composers in the commercial realm, Siday faced incredulity from some potential clients, as measured both by his promotional material, which emphasizes time and again the benefits of using electronic music in advertising and sound logos, and by some of the feedback contained in his files. Siday's promotional material frequently refers to jingle fatigue, that listeners are being "jingled to death." A document from around 1970 says, "The constant repetition of a jingle may quickly amount to boredom, creating a continuous need for costly refreshment." Siday's Identitones could come to the rescue since their electronic sound made them seem fresh: "The wide variety of ideas with uniqueness of sound provided by the Identitones technique has proved to have the lowest 'wear factor' in the industry. There are instances of these sounds still in network use after five, eight, or eleven years."[19]

Moreover, like so many electronic composers, Siday emphasized the variety of sounds available to electronic composers:

> There are literally thousands of electronic sounds available that advertisers are not aware of, some of which might be valuable. . . . Electronic music offers a new spectrum of colors in sound, simply by manipulating known sounds like the guitar, a bird call, or a machine with electronic equipment. Changing the speed of the tape, filtering, and editing all help to produce new sounds from oft-used sound sources. Pure sounds, made from electronic equipment like the oscillator, can also be used, alone or integrated.[20]

Another benefit, which appears in several promotional documents, is that Siday's electronic works (and, of course, those of others) do not require the labor of live musicians, which means that this music is cheaper to produce and residuals do not have to be paid to singers.

Yet skittish potential clients feared that the sounds were too strange. Siday received a letter from Robert B. Jones Jr., vice president and general manager at WFBR radio in Baltimore, dated November 4, 1965:

> We are very excited about this project, but at the same time realize that it offers a tremendous challenge both to you and to us. The very *technique* of your "electronic music" is so unique and different that it becomes terribly important that we not get too sophisticated, cute or off-beat in the content. In other words, our audience consists of slob housewives like Jane who must be able to both dig and enjoy what we're doing without a degree from Julliard [*sic*] or RCA. Since we must do this without pictures or lyrics, I don't think we should shy away from the obvious or the corny, provided it's good corn.[21]

Radio station WFBR became Siday's first Identitones customer.

Siday also faced the opposite reaction from those who knew a little about technology. In the mid-1960s Siday employed a salesman, Charles Barclay, who visited the major media markets in the United States with Identitones demos. A contact report by Barclay from January 1967 at station WSB in Atlanta says, "Played tape for Elmo Ellis and 2 others (Hill & Shaffer). He liked it but not overwhelmed. He's fooled around with audio oscillator and thinks we'll be quickly imitated."[22]

Still another issue that emerges from Siday's documents concerned communication. Siday would customize Identitones packages for customers (for a price, of course). However, clients were not always conversant enough with electronic technologies to know how to ask for the sounds they wanted or whether what they wanted was technologically possible. A letter from a production manager at WQUA radio, Moline, Illinois, dated January 24, 1968, requested that Siday modify an existing package to match the station's updating of its campaign:

> We'd like to give a slightly different twist to the I.D.'s . . . a longer intro to the I.D. itself and a fast stinger [a short piece of music that breaks up a radio program]. The intro to build excitement for the stinger which is the actual "taste" so to speak. The intro should be long enough for the announcer to say, "Get rid of those blues. Have a Taste of the New Life on WQUA (sig). . . . I think that the major difference in what we want and what we now have is "excitement"; we'd like a little more! Kind of like a whirling cloud of sparkling gold dust to lead into the sig.[23]

Once electronic sounds began to become normalized and catch on, Siday had clients who wanted to convert their acoustic advertising music into electronic. A query to Siday's publicist resulted in the following response in August 1972:

> Yes, it [the client's acoustic jingle] could be translated electronically, but there are several reasons why I don't think you should attempt it.
>
> First it would have to be scored (same as any musical number). Your original charts would help a lot, of course. Then, each electronic element or "instrument" would haveto [*sic*] be recorded a single track at a time and mixed through filters, spectrum shaper and other electronic devices. This is a long process of trial and error. The only simple phase is the final mix which now has a series of programmed tracks ready to blend.

> In other words, it would be as complete as an entirely original score—maybe more so because of the specific confines of the theme line. Eric's minimum fee for such a project is $5,000, plus recording charges.
>
> But even more important is the fact that due to a lack of a simple recurrent short musical phrase in the song, Eric doesn't feel that it will accomplish the memorability element that makes the Siday sounds work. The jingle arrangement is too sophisticated and subtle, unless an IDENTITONE is put in *over* the master track, and this would be entirely too busy.
>
> What might be possible, however, is to experiment with simply *tagging* [adding to] that jingle *as is* with a single IDENTITONE. Cost: $ZERO.[24]

Siday and Scott, working independently as competitors in a growing field, were well aware of each other's work. They faced many of the same problems in attempting to promote their music for use in something other than science fiction. Each had to attempt to convince both advertising agencies and advertisers that electronic music could serve their purposes without sounding freakish or off putting. Scott was the more daring of the two sonically, though with less long-lasting success. Siday's solution was to write compositions that were largely electronicized versions of music that could have been executed by acoustic instruments, though with electronic flourishes. Despite Scott's moderate and Siday's big successes, electronic sounds were still far from being ubiquitous. Later composers still had to grapple with ignorant or reluctant clients through the 1970s and into the 1980s, a period in which the use of electronic music was still the province of the technically inclined and the specialist composer, not yet mainstream.

Electronic Music in the 1970s

Siday's well-publicized successes in the world of commercial music and the fame he received from it helped recruit other musicians into the realm of advertising music. Composer/sound recordist Bernie Krause recounts the following:

> While studying electronic music at Mills College [in the mid-1960s], I read an article about a man in New York who had been paid $35,000 for seven seconds of music for an American Express commercial. Five thousand dollars a second for six notes! Impulsively, I bought an airplane ticket and found myself heading east to meet this composer.

"This composer" was Siday, who had purchased the second prototype by Robert Moog, which he had used to make that commercial. Viewing the synthesizer on the table, Krause became "absolutely convinced I was seeing the future of media music unfolding before me" (Krause 1998, 43).[25]

After meeting Siday, Krause and his partner, Paul Beaver, visited Robert Moog to purchase one of his synthesizers. Their goal was to write advertising music like Siday but on the West Coast, where they were based. Krause found it a difficult field to break in to, for most of the potential clients didn't want synthesizer music.[26] But he had some successes. Krause found that while clients were initially reluctant to hire a synthesizer composer, once they dipped a toe in the water, they all wanted the same sound that they had heard on pop records (Krause 2009).

Doubtless partly a result of the massive success of Wendy Carlos's *Switched on Bach* album of 1968 (Carlos 2001 the first classical album to generate sales of half a million, eventually going platinum) by the late 1960s, electronically generated sounds were viewed less as strange and more as novel and interesting. Employment of the Moog or other synthesizers became a selling point for advertising music composers. Thus, by the early 1970s, the use of electronic sounds was becoming increasingly common. Advertising music composers discussed the benefits of new technologies, invoking the dominant language of the industry—mood, but not forgetting the earlier associations of this music. Edd Kalehoff, a composer-performer, said the following in 1973:

> In commercials, I've found the Moog adds a strong sense of the energy that takes place when music joins picture. It's a special tool that can dramatize certain moods, something out of the ordinary, or very strange or very grotesque. It's great for electronic effect or logos, and it can replace an entire band, but it isn't a magic instrument to solve all problems and you can't save money using it to get a big band sound. (Owett 1975, 13)

A Schafer Beer commercial the same year recycles that company's jingle and features Kalehoff playing the Moog synthesizer. The ad introduces Kalehoff ("Edd Kalehoff at the MOOG synthesizer") and shows him playing the instrument, turning dials and adjusting the machine in other ways, and opening a can of beer. A 1984 ad for Kalehoff's services in a trade magazine shows him wearing sunglasses and sitting at two Moog keyboards; the top of the ad reads "There's more than one way to play it," above a logo that reads "Moog Synth King." The ad's copy sells the Moog as much as Kalehoff:

> Play it legato, ostinato, brassy, funky, highbrow or high-tech. No matter how Edd Kalehoff plays it, whether it be on this model, which he helped design with Bob Moog 15 years ago, or combined with keyboards, rhythm, voices and effects, a satisfying sound can be created to enhance any visual. Edd's classical background lends itself to orchestral arrangements which, combined with electronics, produce the richly rewarding sound that sells. Edd has composed about 2001 commercials & themes.
>
> Edd and his staff, with their extensive TV production expertise, are an asset to any size project, from an entire TV station music pack to the voice of Q-Bert . . .
>
> WARNING: THE SURGEON GENERAL HAS DETERMINED THAT EDD KALEHOFF'S MUSIC CAN BE HABIT-FORMING.[27]

VIDEO EXAMPLE 16.2]

Still, synthesizers were novelties in this era. They were used by music technology specialists and employed partly for new sounds and also as a cost-saving device for advertising agencies; an article in the late 1970s said that the upright piano was still "standard office equipment in the jingle industry" (Gorfain 1979, 51).

Suzanne Ciani: Poet of Sound

While Kalehoff might have gained some measure of fame from his use of the Moog synthesizer, the musician who became the most celebrated user of electronic technologies in advertising music after Eric Siday was Suzanne Ciani (1946–). Ciani, in contrast to Siday and many others, used the Buchla synthesizer, which, unlike the Moog, did not have a keyboard.

Ciani told me in an interview that she broke into the business by devising sounds for some fifteen-second television commercials for Macy's Department Store. She composed these at the studio at Mills College, where commercial work wasn't permitted, so "I took the scripts back with me and hid in the Mills College studio" (Pinch and Trocco 2002, 163). There, she says:

> I got to interpret what I saw poetically in sound. So if it was a commercial for a fur coat, I made [what was] to me the sound of a fur coat. If it was a key chain, I made the sound of a key chain. . . . I looked at these things and saw a different dimension in them. It wasn't just music, it was some kind of poetic interpretation of the visual that was also included. So there were notes, but there was sound also. So I loved doing that work. I just, I thought it was very original. And when I got to New York, and I was down and out and wanted to work more on my art and said, "Okay, you need money." I thought, "Well, advertising is a good resource because I will still be in the domain where I want to work." And it turned out to be a great advantage for me because . . . I got to work in the top studios with very high budgets, with the top musicians. It was . . . a very well-supported field. (Ciani 2004)

Ciani says that it was the cost of owning her own synthesizer in the early days of the instrument that impelled her to enter the world of advertising. "Here I was making $3 an hour and within a year I had an $8,5000 Buchla. That was the beginning. But it did necessitate finding a commercial outlet to pay for my habit. So that was when I started doing commercials" (Milano 1979, 33).

Trevor Pinch and Frank Trocco describe Ciani taking her Buchla into a commercial recording studio with the commercial producers present; they didn't know what to make of the machine. According to Ciani, "I'd walk into a studio without a keyboard and they'd go, you know, like they didn't know what to do, how to use it, what to write. Some of them just said, 'Do whatever you want . . . make the sound of a spaceship, make the sound of whatever' " (Pinch and Trocco 2002, 166; ellipsis in original).

In an industry that has always employed ordinary nontechnical language to describe music—since most advertisers and advertising agencies are not musicians—the difficulty of finding language to describe synthesized sounds was a problem in the early days of synthesizers:

> So you come up with a sound and if you touched one knob, suddenly everything was different. And these producers who didn't know how to talk, nobody had the vocabulary for describing sound, he'd say, "No, no, go back, go back to where you were." So I'd move he knob back and he'd say, "No, no! It's not the same" because there were so many interactions—there were maybe fifty knobs contributing to one sound. The guy used to hit my hands—whenever he liked it [and] I'd move it, he'd say, "Stop! Don't touch that, don't you touch another knob! Okay, record." (Pinch and Trocco 2002, 166-7)

Old ideas about what electronic music could illustrate persisted. According to an article in the trade press, Ciani was known for a time as a "disease composer" because she could make sounds that accompanied various illnesses in commercials. Later she was employed to make sounds that were more realistic than an actual recording, whether it was the sound of a can of Coca-Cola being opened and poured or the sounds of a jai alai match: all synthesized (R. Cohen 1977, 16). Ciani told me that for the Coca-Cola commercial:

> I went into the session, and they played the commercial, and they showed me the blank space. And I said, "All right, great, but you've got to tell me, is this going to be used only in this commercial?" "Well, what difference does that make?" "Well, because if its going to be pitched in this key of E, then I can design one thing, but if you're going to use it in every key, in every rhythmic context, in every possible place, then I have to design it differently." And they hadn't even thought of that. They thought, "Well, here we are. We've got this open space. Just put something in there." And I knew that there was a possibility that it would be used in other places. So I personally designed something that had no specific rhythm . . . that could fit in any place. The bubbles move chromatically, so they're not pitched in a key. And . . . I used a lot of the white noise because that is not specific to any key. So . . . I designed it to go in any place it could fit, and it did. They put it in every single commercial, all over the world. (Ciani 2004)

She also said the following:

> When I went in to do the Coca-Cola pop-and-pour sound, and this is typical of the desire to make a larger-than-life replica of a real sound like . . . the sound of the potato chip being crunched and the bottle opening. They would go and record the real sound, and of course it never sounded very good. And it didn't have the punch, it didn't have the life, it didn't have any of the qualities that they needed. So after they exhausted themselves recording the real thing, I would come in and be like a surgeon. I would make sounds, my domain, one of my specialties was a very small domain. You know, if I made the sound of the potato chip crunching, I might have twenty elements in there: every . . . piece of salt flying . . . was something that I designed. And then I could splice together . . . the first few milliseconds, and, you know, get a blended sound. (Ciani 2004)

AUDIO EXAMPLE 16.5

In these and other commercials, Ciani was developing what she called the "poetry of sound," mentioned earlier:

> So it wasn't so much the note music as much as it was a poetry of sound—you know, what is the sound of a fur coat? What is the sound of a key chain? What is the sound of perfume? And developing metaphors in sound. The feeling, you know, the feeling you got listening to it. Was it soft and warm? Was it hard and cold? You know, so this poetry of sound is what I really brought to the industry. (Pinch and Trocco 2002, 163)

In 1985 Ciani and associates at her company Ciani/Musica talked about a commercial for General Electric called "Beep," which illustrates her method of achieving this poetry of sound. The company wanted a talking dishwasher. Ciani synthesized a sound on one of their synthesizers that synchronized with the rhythm of the words on the screen. Later, for other beep spots, they used a Synclavier, one of the first digital synthesizers, to sample the original beep and then sequence it:

> They presented the commercial with just a close-up of this dishwasher with a lot of blinking lights. And the actual dishwasher made a sound, but the sound was just one beep. And they wanted me to re-create the beep of the dishwasher and to integrate it into this commercial. And I said, "Well, gosh, if it's thirty seconds of just listening to this one beep—people will just go crazy. Can't we do something with it?" "Well, I don't know. Let's check with the legal department because we're not sure if it would be considered false advertising if the beep were not authentic." And I said, "Well, while you're checking, I'm going to put together what I think it should be." And, of course, my Buchla synthesizer—one of the things that I loved about it was it had this feedback system of lights. So how you know what's going on in the machine is that the lights tell you. If something is triggered, the light goes on. If the envelope is decaying, the light fades. If the random voltage is fast, the light oscillates quickly. If the sequencer stage is on, the light is up that stage, and so on and so forth. So I was very used to this dialogue with a language of lights. So I took this little beeping, the lights in the commercial, and I made the machine have a personality. And it was cute. You know, when the lights went up, it went up, and when the lights did this, it did that. And it was just really cute. (Ciani 2004).[28]

In an interview I had with her, she became very emphatic: She wasn't simply adding emotion to a commercial: "I had to create a being. I had to create a personality, a character. I gave life to this thing. When they had this thing. it was just a machine with some lights on it. And by the time it was done. you wanted to hug it" (interview with Ciani 2004). Employing music to give personality to a commodity goes back to the earliest days of radio (see Taylor n.d.), but with electronic technologies employed to heighten the realism of what is depicted on-screen, a new mode of representation had been achieved.

VIDEO EXAMPLE 16.3 [29]

Conclusions

While Raymond Scott and Eric Siday struggled in various ways to convince potential clients that their electronic sounds could be successfully utilized to sell goods, effectively domesticating them, Ciani, building on their success, produced sounds that seemed to be more realistic and vivid than the actual recorded sounds, a new phase in the domestication of electronic music. With the "poetry of sound," the classically trained Ciani achieved what few in the world of advertising music had, a kind of rarefied aesthetic approach to this music. Using electronic music not as a substitute for conventional instruments, she broke from the past uses of this technology—of anthropomorphizing, adding "personality," or simply presenting electronic sounds as those of acoustic musical instruments—to participate in advertising's common use of hyperreality to sell goods.[30] Coca-Cola sounds better than it does in real life because the sound you hear in the commercial is not simply a recording of the can being opened and poured but something that sounds better and, as a result, looks better. Ciani took domestication to the next level from her electronic music forebears: Not only did she make the product seem better than it was, but she also helped create an impression with sound that life could be better with a particular product. Beep.

NOTES

1 Eric Siday, "Electronic Music in Communication," speech dated October 16, 1967, Eric Siday Archive, folder 215; emphases in original, New York Public Library, New York.

2 For more on Scott as an inventor, see Winner and Chusid, n.d.

3 Classical composers took up the science-fiction theme as well. In 1961 a reviewer in *Saturday Review* wrote approvingly of a composition by Swedish composer Karl Birger Blomdahl. His *Aniara: An Epic of Space Flight in 2038* AD, released on Columbia, is described as telling the story:

> of a spaceship loaded with refugees from our now radioactive earth. They are headed for resettlement in a colony of cold tundra on Mars. Colliding with some asteroids, the ship loses its rudder and is forced to head uncontrollably and indefinitely into outermost space. At one point the earth, by now a far distant star, a blinding sun to the eyes, explodes, and the electronic music doubles the keen heartbreak both of the incident and feelings of the galactic-bound passengers. (Bowers 1961, 61)

4 This was originally presented as Raymond Scott, "Lecture on Electronic Music in Radio & Television," *Advertising Age* convention, Chicago, July 31, 1962.

5 For histories and useful discussions of the 1950s, see Halberstam 1993, Jezer 1982, and May 1989. For a history of the era that focuses on questions of consumption, see Cohen 2003.

6 These figures were calculated from the billing figures included in Fox 1997, appendix.

7 See also McMahan 1962. For a good overview of Scott's work, see Winner 2000.

8 The problem of "lightness" persisted well into the 1980s; the associations of electronic music with science persisted quite a long time. A New York City advertising music producer said in the late 1980s that one of his spots worked because the music had a humorous edge to it. "The more contemporary you get with synthesizers, the less humor you get" (Rutter 1988, 15S).

9 This was originally published as "Commercials Go off the Beaten Sound Track," unidentified magazine article from December 1960 and reprinted. Around 1960, Scott was asked to adopt an "unusual approach" to a commercial for Vicks cough drops and tablets, and, for the first time, he employed his "Karloff" machine (liner notes to *Manhattan Research Inc* 2000, 119).

10 See also "Tip Top Jingle Money Makers" (1962).

11 Thanks are due to Raymond Knapp for this observation.

12 This latter date comes from a letter from James P. Ellis to Philip L. Tomalin at Ogilvy, Benson, and Mather, Inc., September 28, 1960, Eric Siday Archive, folder 56. However, the copyright date for the composition is May 24, 1960 (and is found in the Eric Siday Archive, folder 5); the assignment of the copyright form is dated June 26, 1959, Eric Siday Archive, folder 55.

13 Eric Siday, "For Phil Shabecoff," Eric Siday Archive, folder 185. This document was prepared for a *New York Times* article (Shabecoff 1964).

14 Eric Siday, draft article titled "Musical Identification in Contemporary Advertising," in Eric Siday Archive, folder 223; first ellipsis in original.

15 There is some confusion around the dates and genesis of this particular commercial. Despite many claims that it was electronic, its earliest version seems to have been acoustic, with temple blocks, guitar, and bass; Herbert Deutsch has written that Siday wrote the commercial on the Ondes Martenot (Deutsch 1993, 15). Robert Moog said in an interview that he didn't know how Siday made those sounds, but it wasn't with any of his equipment (Robert Moog, interview by Trevor Pinch and Frank Trocco, Analogue Music Synthesizer Oral History Project 1996–1998, Coll. 640, Series 2: Transcripts box 5, folder 7, Moog, Robert, National Museum of American History, Washington, D.C.). An advertising trade magazine that included interview comments from Siday said that the commercial was an example of *musique concrète*, the sounds produced by "tempo [*sic*] blocks and the sophisticated editing of tape" ("Audio Logos: An Image in Seven Seconds—or Less," 1964, 36). It is not clear whether there were subsequent versions that were electronic. The most common version, which circulates on the Internet (on YouTube and elsewhere), won a Clio Award in 1964 and is most likely the one that Siday describes as *musique concrète*.

16 Identitones' basic agreement form, Eric Siday Archive, folder 120. A copy of a letter from Edith Hall, ex-secretary to Austen Croom-Johnson, to the New York Department of Taxation and Finance, dated June 23, 1964, requests that Identitunes be dissolved since Croom-Johnson had died (Eric Siday Archive, folder 114).

17 Eric Siday, "Electronic Music in Communication," speech dated October 16, 1967, Eric Siday Archive, folder 215; emphasis in original.

18 According to a contract dated April 3, 1964, Eric Siday Archive, folder 38.

19 These three quotations are from an untitled and undated document c. 1970, Eric Siday Archive, folder 146.

20 Siday, "For Phil Shabecoff," Eric Siday Archive, folder 185.

21 Robert B. Jones Jr., letter to Eric Siday, November 4, 1965, Eric Siday Archive, folder 126; emphasis in original.

22 Charles Barclay, contact report, Eric Siday Archive, folder 139.

23 John Dombek, letter to Eric Siday and Charles Barclay, January 24, 1968, Eric Siday Archive, folder 146.

24 Walter Collins, letter to Richard Sonntag, August 21, 1972, Eric Siday Archive, folder 153; emphasis in original.

25 Bernie Krause, *Into a Wild Sanctuary: A Life in Music and Natural Sound* (Berkeley: Heyday, 1998), 43.

26 Bernie Krause, interview by Trevor Pinch, August 24, 1998, National Museum of American History, Analogue Music Synthesizer Oral History Project, 1996–1998, Series 2: Transcripts, box 5, folder 5, coll. 640.

27 Advertisement for Edd Kalehoff, *Backstage* (April 20, 1984), 27; uppercase passage in original.

28 For more on this beep, see Doerschuk 1985.

29 There is a YouTube video of a production session with Ciani and associates at http://www.youtube.com/watch?v=sxUTtUue5RQ (accessed Feb. 15, 2011).

30 For a useful discussion of hyperreality and music in the advertising world, see Kurpiers 2009.

REFERENCES

Archives

Analogue Music Synthesizer Oral History Project 1996–1998. National Museum of American History, Washington, D.C.

Eric Siday Archive. New York Public Library, New York.

Interviews

Ciani, Suzanne. Telephone interview by author, May 4, 2004.

Krause, Bernie. Telephone interview by author, August 4, 2009.

Unpublished Materials

Kurpiers, Joyce. Reality by Design: Advertising Image, Music, and Sound Design in the Production of Culture. PhD diss., Duke University, 2009.

Taylor, Timothy D. n.d. The Sounds of Capitalism: Advertising, Music, and the Conquest of Culture. Chicago: University of Chicago Press, forthcoming.

Discography

Barron, Louis, and Bebe Barron. *Forbidden Planet*. Small Planet PR-D-001. 1989.

Carlos, Wendy. *Switched-on Bach*. East Side Digital ESD 81602. 2001.

Scott, Raymond. *Manhattan Research Inc.* Basta 90782. 2000.

Books and Articles

"Audio Logos: An Image in Seven Seconds—or Less," *Sponsor*, June 8, 1964, 34–37.
Berger, Ivan. "The 'Switched-on Bach' Story." *Saturday Review* (January 25, 1969), 46.
Bogart, Michele H. *Artists, Advertising, and the Borders of Art*. Chicago: University of Chicago Press, 1995.
Bowers, Faubion. "Electronics as Music." *Saturday Review* (November 11, 1961), 60–61.
Cohen, Lizabeth. *A Consumer's Republic: The Politics of Mass Consumption in Postwar America*. New York: Knopf, 2003.
Cohen, Randy. "Songs in the Key of Hype: Jingles Sweeten Sales Pitch with Pop Tunes, Catchy Cliches." *More* (July/August 1977), 12–17.
Chusid, Irwin. "Beethoven-in-a-Box: Raymond Scott's Electronium."*Contemporary Music Review* 18 (1999): 9–14.
Cross, Lowell. "Electronic Music, 1948–1953." *Perspectives of New Music* 7 (autumn-winter 1968): 32–65.
Deutsch, Herbert A. 1993. *Electroacoustic Music: The First Century*. Miami: Belwin Mills.
Discus. "Brave New Worlds." *Harper's Magazine* (November 1968a), 165–66.
———. "Twenty Years of Electronic Music." *Harper's Magazine* (March 1968b,), 157–58.
Doerschuk, Bob. "Suzanne Ciani & Her Ace Apprentices Set the Pace of Commercial Synthesis." *Keyboard*, April 1985, 16.
Feehan, Eugene. "The Sound of Television Music." *Television Magazine* (February 1967), 28.
Fox, Stephen. *The Mirror Makers: A History of American Advertising and Its Creators*. 2d ed. Urbana: University of Illinois Press, 1997.
Gorfain, Louis. "Jingle Giants." *New York* (April 23, 1979), 50–53.
Halberstam, David. *The Fifties*. New York: Villard, 1993.
Jezer, Marty. *The Dark Ages: Life in the United States 1945–1970*. Boston: South End, 1982.
"The Jingle 'Hall of Fame.'" http://www.classicthemes.com/50sTVThemes/thoseOldJingles.html.
Krause, Bernie. *Into a Wild Sanctuary: A Life in Music and Natural Sound*. Berkeley: Heyday, 1998.
Liner notes to *Manhattan Research Inc*. Basta 90782. 2000.
May, Elaine Tyler. *Homeward Bound: American Families in the Cold War Era*. New York: Basic Books, 1988.
McMahan, Harry W. "Raymond Scott's 'Sounds Electronique' Accents New Emphasis on Audio." *Advertising Age* (Apr. 16, 1962), 119–22.
Milano, Dominic. "Suzanne Ciani: Supplying Synthesized Seasoning for Radio & TV Commercials, Movies, & Records." *Contemporary Keyboard* (June 1979), 32.
Moog, Robert A. "Electronic Music." *Journal of the Audio Engineering Society* 25 (November 1977): 855–61.
Moor, Paul. "Sinus Tones with Nuts and Bolts." *Harper's Magazine* (October 1962), 49–52.
"Music of the Future?" *Mademoiselle* (December 1959), 94–97.
Oudshoorn, Nelly, and Trevor J. Pinch, eds. *How Users Matter: The Co-Construction of Users and Technology*. Cambridge, Mass.: MIT Press, 2003.
Owett, Bernard. "Making Music." *ANNY* (May 30, 1975), 8.
" 'Perking Pot' Theme Back at Maxwell House." *New York Times* (Jan. 18, 1990), §D, 19.
Pinch, Trevor, and Frank Trocco. *Analog Days: The Invention and Impact of the Moog Synthesizer*. Cambridge, Mass.: Harvard University Press, 2002.
"Raymond Scott Sounds Off on Sound." *Sponsor* (Oct. 5, 1964), 42–44.
Rutter, Joyce. "Kuby Conducts." *Advertising Age* (Aug. 1, 1988), 15S.

Shabecoff, Philip. "Advertising: Use of Public Relations Rises." *New York Times* (Aug. 27, 1964), 44.
Silverstone, Roger, and Eric Hirsch, eds. *Consuming Technologies: Media and Information in Domestic Spaces*. New York: Routledge, 1992.
"The Sound of Hell." *Newsweek* (Dec. 10, 1962), 86.
"Swurpledeewurpledeezeech!" *Time* (Nov. 4, 1966), 68.
Tall, Joseph. "Music without Musicians." *Saturday Review* (Jan. 26, 1957), 56–57.
"Tip Top Jingle Money Makers." *Sponsor*, 30 April 1962, 32.
Winner, Jeff E. "Circle Machines and Sequencers: The Untold History of Raymond Scott's Pioneering Electronica." *Electronic Musician* (December 2000), 94.
———, and Irwin D. Chusid. n.d. " 'Circle Machines and Sequencers': The Untold History of Raymond Scott's Pioneering Instruments." http://RaymondScott.com/em.html (accessed Aug. 30, 2010).

SECTION VI

CONSUMING SOUND AND MUSIC: THE HOME AND BEYOND

CHAPTER 17

VISIBLY AUDIBLE: THE RADIO DIAL AS MEDIATING INTERFACE

ANDREAS FICKERS

INTRODUCTION

THE aim of this chapter is to consider the radio receiver and, more specifically, the radio dial as a mediating interface. Usually the radio dial is thought of as mediating between the operator—the listener—and the radio stations tuned in by turning the dial. Here we look at mediation in a wider sense—between a European regulatory regime of frequency allocation and the imagined European broadcasting landscape of the listener. My argument thus develops a triangular relationship between the rise of a European regime of frequency regulation, the materiality of the radio set, and the symbolic appropriation of the European broadcasting landscape. This approach requires an analysis of the material, institutional, and symbolic dimensions of a concrete technical innovation: the calibrated radio station scale. In analyzing the iconological and semantic meanings of this technical artifact, I emphasize the importance of material objects as sources for a cultural history of technology in general and of radio listening in particular.

Accordingly, it is necessary to combine history of technology approaches with media and cultural studies perspectives. This cultural history approach to sound technologies allows me to embed the act of radio listening in a broader ensemble of senses (haptic and visual) and thereby underline its effects on processes of what we might call individual mental mapping: the cognitive production of an individualized representation of experienced space (Downs and Stea 1977). The appropriation of the radio (this is my first thesis) involved the appropriation not only of a technology of communication but also of an imagined space: the ether. What I refer to as the "visualization of the hearing experience," which resulted from the introduction of the station scale as a map of an imagined voyage through the ether, downgraded the previous practice of explorative listening in search of audible radio stations (tuning) while simultaneously prioritizing our "first sense" (seeing) as a new aid in radio set operation. This deeply changed the human/machine interaction with the receiver and provided the user—who turned from a pure listener to a viewer of a topographical representation of European station names (mainly names of cities)—with a new and highly symbolic media interface. Interacting with this new interface—as suggested by the iconography of advertisements and reported in many literary reminiscences to the station scale—innervated the user's imagination of traveling the European broadcast landscape while turning the dial from one station to another.

In this process of symbolic appropriation (this is my second thesis), changes in receiver design can be interpreted as a crucial shift in the interaction with the radio, as well as in its "material meaning" (Gitelman 2004). While early radio listening required technical skills and a specific tacit knowledge, the emancipation of radio from a medium of tinkerers or hobbyists to become a mass medium was bound to the development of a user-friendly design that transformed this electrical device into a domesticated piece of furniture. Although scholars have covered this domestication of the radio receiver to some extent (Aitken 1976; Douglas 1987; Hilmes 1997; Lenk 1997; Méadel 1994), so far they have ignored the introduction of the radio dial as a crucial moment of transition in human-machine interaction. I argue that the advent of the calibrated station scale facilitated the black-boxing of radio as a complex technology and created the "user illusion" of an easily manipulable machine.

Finally, I develop a third thesis, which is that these crucial changes in the reception side of radio, in terms of both receiver design and new forms of usages, came as an unintended consequence of regulatory initiatives on the transmission side. Without the emergence of a European regulatory regime, based on the authority of the Technical Committee of the International Broadcasting Union, the development and successful implementation of calibrated station scales would not have been possible.

To understand the substantial change in radio receiver usage, it is necessary first to briefly trace the early history of amateur radio technology and to describe the advent of this new medium as a sensorial revolution. Early literary responses to radio serve as sources for a cultural contextualization of the phenomenon of "hearing through the ether" and allow me to sketch the contours of how that new "art of listening" evolved from broadcast technology. Next I discuss the crucial steps in the technological development of radio in the 1920s by focusing on the central role of radio amateurs and the tacit knowledge needed for the operation of these new electronic machines. The gradual reduction of tacit skills for radio operation involved a successful process of domestication, typical of the cultural appropriation of household technologies in the twentieth century (Silverstone 2006).

Subsequently I explore the gradual emergence of a European regulatory regime in frequency planning, mirroring the need for regulation of the fast-expanding broadcasting infrastructure of the mid-1920s. My argument concentrates on the role and importance of the Technical Committee (TC) of the International Broadcasting Union (IBU) in Brussels, which acted as a technopolitical mediator in the establishment of a European regime of frequency regulation. The successful promotion of a "technopolitics of accuracy" by the TC paved the way for the standardization of European frequencies and enabled—as an unintended consequence—the successful implementation of calibrated station scales as both technical and aesthetic innovations in radio receiver design in Europe.

In a final section I interpret these changes in receiver design as material inscriptions of these regulatory efforts. A closer look at the design of the new sets with a calibrated station scale highlights the role of design as a mediating interface between technology and user, between the hidden structure of the regulated European broadcasting landscape and processes of mental mapping and symbolic appropriation of an imagined European topography of radio stations.

The empirical evidence for my claims has been gathered by studying the Telefunken archives in Berlin and the IBU archives in Geneva, as well as a selection of major radio journals in Germany, France, Great Britain, and the Netherlands. The secondary literature used for contextualizing the phenomena discussed includes classical German accounts, as well as French, British, and American studies.

The Advent of Radio as a Sensorial Revolution

With the emergence of broadcasting, a new quality occurred in the process of the world's spatial and communicative networking (Lenk 1997). From a macrohistorical perspective, one could describe this process as a change from a visible to an invisible networking of the world. While all kinds of traffic routes and cabling (electricity, telephone, telegraph) left visible traces in private and public spaces, wireless broadcasting heralded the age of the invisible but audible networking of the world. This process resulted in an increased sensibility for the sense of hearing as a producer of cultural meaning.

Numerous literary sources report the amazement and fascination caused by the emergence of radio broadcasting. Writers, poets, columnists, and copywriters celebrated the wonders of radio. Two of the central metaphors used to describe the new sensations were "the world at home" and "the ear of the world" (Bleicher 2000). In his masterful study *Radio: An Art of Sound* from 1936, the German film and radio critic Rudolf Arnheim wrote the following: "An apparatus which technical peculiarity simply consists in enabling sounds made at a particular spot to be simultaneously reproduced in as many and as far removed places as one wishes by disrespectfully breaking through boundaries of class and country, signifies a spiritual event of primary importance" (Arnheim 1972, 226). Radio complemented the growing possibilities of travel and transportation, and it also invited people to travel mentally. As René Schickele, a German regional writer, put it in a 1931 article titled *Paneuropa der Sender*, even the farmer in the Black Forest got connected to the world: "Today, you can be lonesome without losing the connection to the world" (Schickele 1984, 141). It is worth noting that authors often used "world" as a synonym for "city." This equation of the modern world with urban life can be found in several radio poems of the 1920s and 1930s. Consider the following lines from a 1927 poem by Robert Seitz titled *Überall her aus der Welt* [From All over the World]:

> From all over the world I send you my salute
> From all over the world you were coming to me
> Alaska is on my side and China
> The city on the Seine and the town on the Newa
> Cities in decline and cities on the rise
> Rome and Sao Paulo
> I can hear your voices next to me
> And I'm going through you with your voice
> I walk on boulevards
> And pass the palaces in Granada
> I find myself amidst traffic in seaports
> And I'm all alone on the vast plains of a southern continent. (Seitz 1984, 42–43)

As shown by a 1927 essay by Lothar Band, called *Hörbilder aus dem Leben* [Acoustic Images of Life], the simple announcement of a program from a distant city could trigger a feeling of nervous tension in listeners:

> Some time ago, there was this short message full of weird tension: "In one minute Geneva will be on the air!" Hastily transmitted, these nine words are trembling with an infectious excitement. These are words of uncertainty, for we do not know if the transmission across a distance of more than one thousand kilometers will work—words feeding the hope of becoming an ear witness of an important moment in world history . . . When attending such a newsworthy event via broadcasting, we not only want to listen, no, we want our ear to see. We want to participate in this event, trying to raise it to the level of a real experience. (Band 2002, 244)

As these quotes suggest, it was not only the simple technical possibility of sending sounds through the ether that aroused people's imagination. There was also a real emphasis on the revaluation of hearing and the ear as a producer of cultural meaning.[1] In a 1925 article titled *Die Denkumschaltung durch den Rundfunk* [The Thought Switch Caused by Broadcasting], Christian Gerlich states the following:

> For thousands of years our thinking has been based on seeing. Then, suddenly, the electrical wave came upon us like lightning, the unseen. Incidents in the ether that remained secretive and mysterious because we couldn't see them suddenly became unraveled by our ears. Another of our five senses is taking over, and we have to switch from the eye to the ear. (Gerlich 2002, 210)

A similar position can be found in an article on the "aesthetics of broadcasting" by the famous German novelist Arnold Zweig. Likewise, in a piece by Richard Kolb on the "horoscope of the radio play," the microphone mutates into the listener's ear, and the "announcer and listener meet at the shared focal point of mental acoustics" (Kolb 2001, 32). All of these authors share the same enthusiasm for the positive power of imagination caused by radio listening. With and through radio, so they claim, the ear has become a serious rival to the eye. "Hear thinking" has finally challenged the modern practice of "eye thinking"! This power of radio to foster cognitive visual and spatial imaging through the internal production of "acoustic images" is a crucial aspect of radio. As Susan Douglas puts it, "The magic comes from entering a world of sound and from using that sound to make your own vision, your own dream, your own world" (Douglas 2004, 28).

From Tinkerer to Mass Medium: Functional and Symbolic Change of Radio Use

When taking into account the real acoustic output that these early beeping and screeching apparatuses produced, it is quite surprising to see such terrific fantasies and expectations surrounding the emergence of the new medium. In addition to the mediocre sound quality of early broadcasting, the search for stations was a plodding affair that needed the full attention and concentration of the radio operator and the complete silence of the attending public. This often lengthy procedure was not always successful, and early critics of radio described the "surfing" on the radio waves in the ether—the long search for an acceptable signal in the wireless frequency spectrum—as an abnormal phenomenon. The German philosopher of religion Johannes Maria Verweyen called this disease "radiotis" and lamented in 1930:

> Radiotis has caused a perturbation of the inner balance that is popularly called *Drehkrankheit* [tuning malady]. This forebodes to a certain pathological affection or even addiction to make as many turns as possible without paying any attention to the changing contents. (Verweyen 2002, 454)

Individual station adjustment (fine-tuning) was of central importance in the first decade of radio broadcasting, that is, until the emergence of calibrated station scales in the late 1920s. The fine-tuning adjustment called for a concentrated dedication of haptic and acoustic senses, or "fine motor skills" (Schmidt 1998, 316). It is impossible to underestimate the tacit knowledge needed to operate a receiver. To explore the broadcast spectrum, many circuits had to be adjusted individually by rotating the control knobs sequentially in small increments until a station was heard. Next, several precise readjustments were necessary to bring in the desired program with clarity and volume (Harrison 1998, 296). The gradual development of automatic tuning devices, then, was an indispensable condition for the transformation of the radio set from a tinkerer's or an amateur's medium to a mass medium. Central steps in this process were the embedding of previously exposed components into a cabinet, the reduction of tuning knobs, and the integration of the loudspeaker into the cabinet. From 1924 onward, there were more than seventy patent applications for single-control tuning in the United States alone (Harrison 1998, 299–300). While the radio industry sought to produce sets for everyone-they believed that radio listening should be simple, and the technological vision of the time was a receiver that could be operated with only one knob.

A 1930 advertisement by the British radio manufacturer Murphy, which read "We were not blessed with three hands," shows that simplifying the operation of a receiver was a central aim of the receiver industry in its effort to win over more of the public, especially women, as potential buyers.[2] In a *Telefunken* note dated 1927 one could read the following advice to the licensed dealer: "Please consider that most of your customers are no experts and that they cannot afford some sort of 'radio chauffeur' to operate their receiver."[3] In a 1934 issue of the French journal *Radiodiffusion*, one read this prediction: "The single-knob receiver will lead women to broadcasting, just as the electrical ignition led them to the automobile."[4] In *The Setmakers*, British broadcast historians Keith Geddes and Gordon Bussey arrive at a similar conclusion: "With the introduction of the automatic volume control, the adoption of the superhet principle, and the demise of the oscillation-provoking reaction control, the taming of the domestic receiver was complete by the early 1930s" (Geddes and Bussey 1991, 107). As Roger Silverstone and Leslie Haddon have argued on the basis of Adrian Forty's study of the English radio cabinet design between 1928 and 1993 (Forty 1972), the simplified user interface of the radio set since the mid-1920s is a fitting example of what they call "pre-domestication": "an anticipation in design itself of the artifact's likely place (in this case the home) and an attempt to offer a solution *in the design of the object itself* to the contradictions generated within the process of technical innovation" (Silverstone and Haddon 1996, 49).[5]

The gradual simplification of the operation of the radio set since the mid-1920s can be interpreted as the first successful example in the still young history of electronic consumer goods to enhance the user friendliness and create—through the intervention of engineering design and the exterior of the radio—a "user illusion." This implied that the user mastered the device without in fact having any understanding of its technical functioning.[6] The creation of this "user illusion" accompanied the development of a desktop philosophy by industrial designers who were confronted with the difficult task of creating functional simplicity in the face of structural complexity (Betts 2004). In this process of functional and symbolic transformation the apparatus became a piece of furniture whose inner mechanism was successfully hidden (Carlat 1998; Forty 1972). In the words of philosopher Norbert Bolz, the concealment of the technical interior of a machine—or, in other words, the black-boxing of technology by the development of user-friendly design—is "the rhetoric of the technology which consecrates our ignorance. And this design-specific rhetoric provides us with the user illusion of the world" (Bolz 1998).

Next to the single-knob operation ideal, the radio industry endeavored to develop and integrate the electrodynamic flat speaker into the cabinet—a process that reminds one of the domestication of the phonograph some twenty-five years earlier. In an article titled "Die Wesensform des Lautsprechers: Ein Beitrag zur Ästhetik der Technik" [The Essence of the Loudspeaker: A Contribution to the Aesthetic of Technology], Ewald Popp, a Czechoslovakian loudspeaker specialist, describes the wonder of the flat speaker and refers to the importance of the loudspeaker as "contact person" for the listener: "We hear the loudspeaker, but we see him, too. Just as we look into the face of our contact person, we involuntary direct our glance to the loudspeaker. He is the acoustic center we look at—we have to look at [him] because our senses require it" (Popp 2002, 312). In other words, the eye "listens"! The integrated electrodynamic flat speaker and the emergence of the station scales (see later) mark the beginning of what I call the return of *Schaudenken*, thinking by looking or, in other words, the increasing visualization of the hearing experience.

The Emergence of a European Radio Infrastructure

The emergence of a regulated broadcast spectrum in the mid-1920s served as the technical precondition for the development of calibrated station scales. Without a fixed frequency plan that allotted specific wavelengths as either exclusive or shared frequencies to specific stations, the calibration of receivers and the design of a radio dial with fixed station names on it would have been impossible. As I demonstrate later, however, this development actually came about as an unintended consequence. The real motive for the establishment of a European regime for frequency regulation lay in dealing with the unexpected pace of growth of broadcasting as a new medium in the 1920s. Radical innovations in the tube industry, a mass of excited radio amateurs, and the discovery of the educational and entertainment potential of broadcasting caused a genuine radio boom in both Europe and the

United States. At the same time, the great success of radio broadcasting also created a serious threat to the new medium. The uncontrolled crusade of the airwaves soon became a severe concern, and critics like Rudolf Arnheim started to complain about the problems of interference and the rising "chaos" in the ether:

> What we hear today from the loudspeaker is an artistically forceful symbol of constant war in peace, of the deficiency of central authority which we permit around us—a chaos concretized in discord and as such directly perceptible to the human ear. So the dance of music of one country comes through the funeral march of another. (Arnheim 1972, 238)

To tackle this chaos, broadcasting became the subject of intense political and diplomatic activity at both the European and the international level. Such cooperation was indeed badly needed, as Arno Huth, author of the impressive survey of international broadcast activities titled *La radiodiffusion: Puissance mondiale* (1937) convincingly argued:

> Before the international cooperation there was chaos. Each country allotted frequencies following their national needs and without taking care of the transnational nature of broadcasting signals. As a result, radio reception became an agony, and the listener suffered from constant interferences and turned into a victim of the peeping, cracking, crunching, and groaning sounds, of which he retains dreadful memories. (Huth 1937, 45)

The first European initiative to tackle this problem of increasing interference in radio broadcasting was directed by Swiss radio amateur Maurice Rambert. In October 1922 Rambert had obtained the first license for radio broadcasting in Switzerland. As one of the early radio amateurs, he personally suffered from the growing disorder in the ether, and in 1923 he and his friend Dr. Edmond Privat, president of the Bureau International de l'Espéranto in Geneva, initiated the first European meeting of broadcasters in Geneva (UIR 1944). This *conférence préliminaire pour une entente internationale en radio-téléphonie* took place on April 22–23, 1924, and was attended by some forty representatives of state administrations, private associations, radio clubs, the radio industry, and the wireless press. This preliminary conference resulted in a number of resolutions. Among other things the participants claimed an exclusive part of the frequency spectrum for radio broadcasting and for radio amateurs and also established a permanent organization for broadcasting regulation and cooperation in Geneva. Maurice Rambert, elected as chairman of the provisional executive committee, was charged with organizing a second preparatory meeting at the British Broadcasting Company (BBC) in London (March 18–19, 1925), where agreement was reached over the general lines of the future organization. On April 3–4, 1925, the Union Internationale de Radiodiffusion/International Broadcasting Union (UIR/IBU) was finally launched. The BBC's deputy managing director, Rear Admiral Charles Carpendale, became president of the First General Assembly, and his colleague Arthur R. Burrows, program director of the BBC, served as first director of the IBU's permanent office in Geneva.

From the outset, Geneva was seen as the ideal headquarters because of its reputation as host to numerous international organizations, most prominently the League of Nations, the Organisation International de Travail (OIT), the Bureau International de l'Espéranto, the International Committee of the Red Cross—and, of course, the International Telecommunication Union (ITU). Symbolically, the foundation charter of the Union International de Radiodiffusion (UIR) was signed at the offices of the League of Nations building, a ceremonial performance that attested to the vision of broadcasting as an instrument of peaceful purposes (Tomlinson 1938). All of the new UIR members—despite what their diverse intentions and interests may have been—were convinced of the exceptional importance of radio broadcasting as a political and cultural instrument. Despite the alarming tendencies toward the national or nationalistic deployment of the medium for ideological or propagandistic purposes, the central UIR actors were driven by a humanist ideology and conceived of radio broadcasting as an unprecedented instrument for the promotion of international understanding and the transcendence of economic classes, political ideologies, and territorial boundaries. Exactly one year after the founding of the UIR, all of the members signed a "gentleman's agreement" against the illegal and immoral use of radio as an instrument of propaganda. In the ensuing years, several bilateral and international conventions promoting the "moral disarmament" of the ether would be signed (Huth 1937, 386–91).

The close link between the technical and the ideological or political self-image and mission of the early European initiatives to harmonize the European ether is telling. This may be confirmed by a consideration of the development of the IBU's organizational structure. After the creation of a Technical Committee in December 1925 to prepare a document to serve as a discussion base for a frequency plan, a permanent Juridical Committee was founded in March 1926. This was followed three months later by the establishment of the Committee of Intellectual, Artistic, and Social Rapprochement and, in July 1928, by the Committee of International Relays.[7] This organizational structure mirrors the intertwinement of the material (technical), institutional (juridical), and symbolic (intellectual, artistic, and social) dimension of broadcasting as a critical infrastructure.

Planning the Spectrum: The Emergence of a European Frequency Regulation Regime

The first and most effective way to harmonize the European ether was to develop and agree on a frequency plan that allocated specific frequencies to broadcasting stations. The self-imposed task of the IBU was to organize the frequency spectrum reserved for radio broadcasting in such a way as to minimize parasitic interference and to maximize order by allotting specific wavelengths to specific national broadcasting services. In his study "Broadcasting without Frontiers," published on behalf of UNESCO in 1959, George Codding argues that the "European experiment" of assigning individual frequencies to individual broadcasting stations was characterized by three phases or approaches (Codding 1959). The first attempts that led to the frequency plans, known as Geneva (1926) and Brussels (1929), brought about voluntary cooperation between the various operators of broadcasting services (private and public) under the auspices of the IBU. The second approach, inspired by Czechoslovakia, aimed at making frequency distribution more official by convening a special European conference of broadcasting representatives of national governments. This conference, held in Prague in 1929, produced the so-called Prague Plan. Third, the IBU itself convened two conferences of administration representatives (Lucerne in 1933 and Montreux in 1939) aimed at assigning European broadcasting frequencies in conformity with frequency allocation tables established by the International Telecommunication Union in Madrid and Cairo (Codding 1952, 92).

The various initiatives regarding regulation of the frequency spectrum for radio broadcasting were based on the principle of allocating each European country at least one exclusive frequency and several shared frequencies, proportionally allotted depending on the number of existing stations, the topography and size of the country, and its economic and cultural development. These factors resulted in a formula for calculating each country's share of the spectrum. As Nina Wormbs has shown, the result—the number of allotted wavelengths per country—largely depended on the definition of Europe's size: The more countries were taken into consideration, the fewer exclusive frequencies could be allotted (Wormbs 2008, 116). This problem became evident during the negotiations for the Lucerne plan, when a large number of eastern European countries—including the Soviet Union—joined the IBU. The various plans were therefore the result of technopolitical diplomacy: trying to combine the radio engineers' rational grounds with the hidden or explicit political agendas of the national telecommunication authorities and broadcasting organizations.

The establishment of the permanent Technical Center was an important step in institutionalizing the IBU as the "ether police." The Technical Center became the headquarters of the Technical Committee and soon served as the IBU's

Table 1. The steady growth of both the number of stations and the transmitter power was a continuous challenge for the regulation of the European spectrum of broadcasting frequencies. Source: Codding (1952, 96)

	Geneva Plan	*Prague Plan*	*Lucerne Plan*	*Montreux Plan*
Year	1926	1929	1933	1939
Number of stations	123	200	257	351
Total power in kW	116	420	3.260	10.790

technoscientific conscience. In the eyes of the Technical Committee, the main remedy for the chaotic situation in Europe was the accurate control of the broadcast stations' transmitters. This was made possible by a newly designed frequency meter, calibrated to the frequency allotted to each station by the Geneva plan. One year after the creation of the IBU's permanent Technical Center in Brussels in 1926, more than fifty frequency meters had been ordered by various broadcasting organizations. The frequency meter, which became the materialization of a regulatory regime, was successfully implemented by the IBU's Technical Committee. Its politics of accuracy helped to significantly reduce the interference problem and established an effective control mechanism. The frequency meter as the materialization of a technology of transmission control found its material equivalent in the calibrated station scale on the reception side. According to Sheila Jasanoff and Brian Wynne, the emergence and implementation of the frequency meter and the calibrated station scale are examples of the coproduction of a regulatory regime and technical artifacts. Broadcasting as a new "social project" had shored up the legitimacy of both the newly produced natural knowledge (and its material manifestations) and the political order of the regulatory regime (Jasanoff and Wynne 1998, 16).

However, another—and in the eyes of the Technical Committee more problematic—difficulty was that even IBU member organizations that had ratified the plans often ignored the agreements or failed to apply them accurately. This tension between the de jure regulation and de facto violation of the European broadcast space demonstrates the critical status of the IBU as a regulatory body. As a nongovernmental organization, the IBU had no real political or juridical power to press its members to enforce the agreement, let alone to exert any kind of pressure on nonmember organizations (figure 17.1).

A positive development, however, was the increasing adjustment of the transmission characteristics at a distance by means of telegraph correspondence (and later by telephone). When the staff of the Technical Center in Brussels detected serious distortions of or variations in the power and/or frequency of the transmitted signals, they immediately contacted the station's technicians (either by telegram or a phone call). From 1926 on, this very practical and effective form of *réglage à distance* between the Technical Center and broadcast stations all over Europe became one of the main concerns of the technical staff in Brussels, reinforcing the IBU's status as the technical authority in radio broadcasting regulation. The technical advances in and the accuracy of frequency measurement in Brussels helped to establish a regime of soft regulation based on the expertise and political neutrality of the Technical Center. This was clearly the credo of Raymond Braillard, chairman of the TC and Director of the Technical Center in Brussels, who believed that the constant progress of science would help to surmount transitory difficulties if it were accompanied by strict technical application.[8] While this belief in technoscientific rationality and its inevitable practical success was widespread in the international community of radio engineers and broadcast technicians, it was constantly challenged by "irrational" interference of a political or an ideological nature.

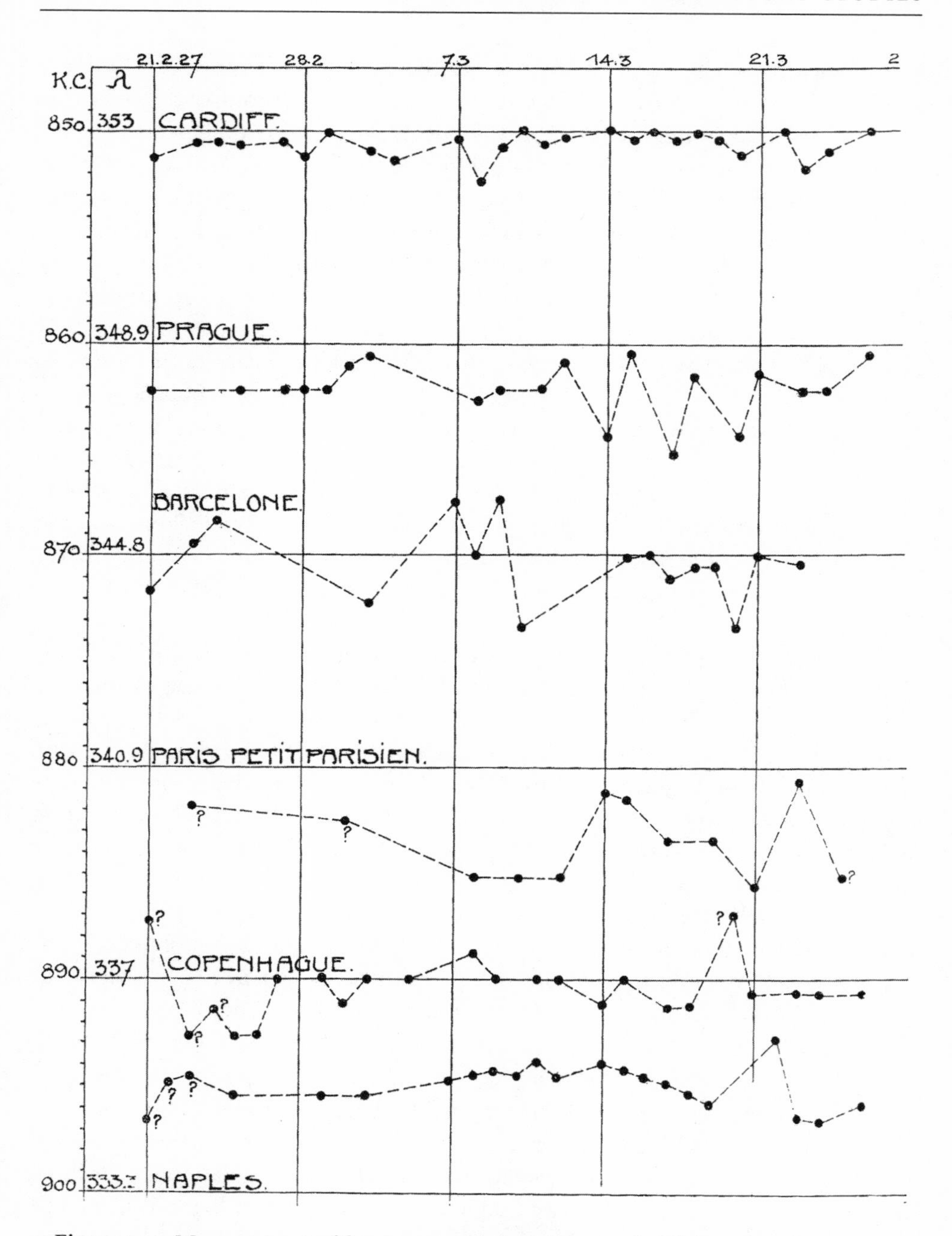

Figure 17.1. Measurement of frequency variations, International Broadcasting Union (IBU) Technical Centre, Brussels, 1927. Source: IBU Archives, Geneva.

From Technical Instruction to Scientific Understanding

From a long-term perspective, one is tempted to see in the different plans just temporary cures of symptoms, not of the actual European frequency disorder itself. The regulatory initiatives were constantly challenged by both technical and political changes. At the technical level, the gradual expansion of the frequency bandwidth used for radio broadcasting (long, middle, and, in the early 1930s, short waves) and the continuing expansion of transmitter power reduced the half-life of the plans to less than a year. Moreover, the ongoing diplomatic and political struggles on the territorial definition of the European broadcasting space foiled the TC's technocratic hopes that the frequency allocation could be managed by the application of a scientific formula. Yet from a critical historical perspective, it would not be fair to interpret or even judge the "success" or "failure" of a European regulatory regime only on the basis of the stability or life cycle of the different frequency plans (figure 17.2).

When looking at the activities and impetus of the Technical Committee specifically and of the IBU in general, one can identify a clear focus on the study and application of precise technical parameters for a smooth European and international operation of radio broadcasting. The regulation of the European ether

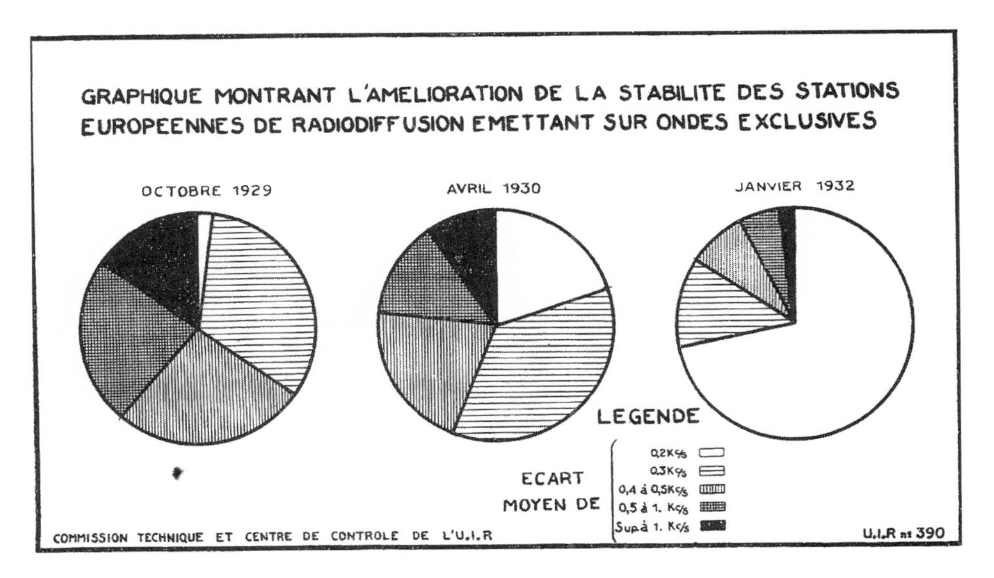

Figure 17.2. Amelioration of the frequency stability of European broadcasting stations (exclusive frequencies) between 1929 and 1932. Source: "MEMORANDUM sur la situation actuelle de la radiodiffusion européenne au point de vue des interferences et les moyens propres a y remédier." Bruxelles: Publications techniques de l'Union Internationale de Radiodiffusion No. 1, 1932, p. 26.

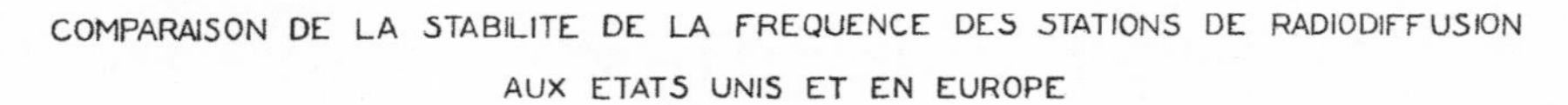

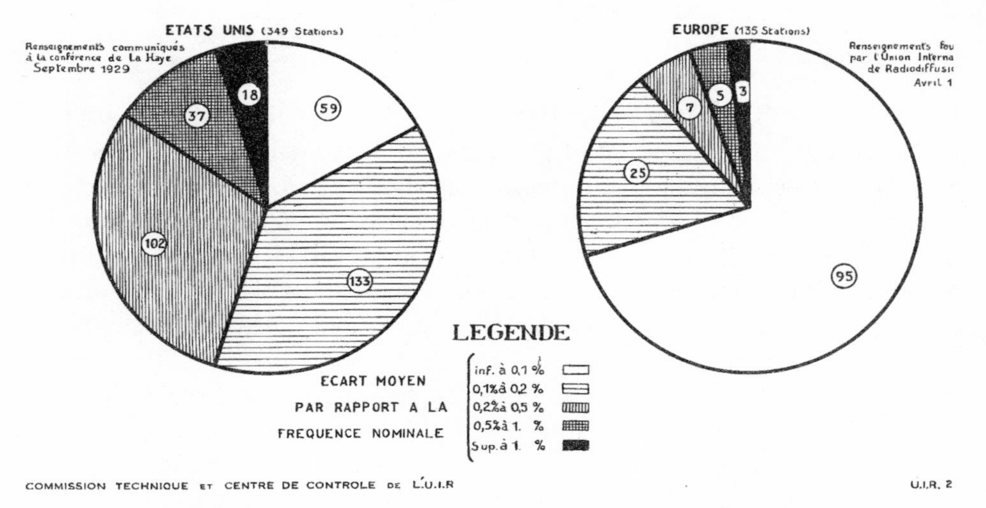

Figure 17.3. Comparison of frequency stabilization USA / Europe. The white area indicates the numbers of stations working with a very high degree of frequency stability (less than 0.1 % of deviation from the calibrated norm). Source: IBU Archives, Geneva 1932.

seemed primarily a problem of adequate description and measurement of the technical criteria of radio broadcasting (frequency meters and study of origins of interference) to be followed by a rigorous application and enforcement of these rules. As figure 17.3 shows, this *politics of accuracy*—prominently propagated by the Technical Committee and skillfully customized by the Technical Center in Brussels—has been extremely successful and undeniably produced tremendous positive effects (reduction of interference). As comparison to the situation in the United States shows, this politics of accuracy may be interpreted as a real European technopolitical success story.[9]

The negotiations during the development of the Lucerne plan demonstrate that the technical norms and standards developed by the Technical Committee had been accepted and adopted by all of the negotiating parties regardless of the governmental, private, or corporate status of the IBU member organizations.

Visualization of the Listening Act: The Invention of the Station Scale

Not only did the emergence of a real European transmission network call for rigorous regulation of the frequencies utilized by the radio stations; it also represented a technical challenge for the receiver industry. The only way to guarantee reasonably good reception in the densely populated ether was to improve receiver selectivity. A real milestone here was the development of the so-called superheterodyne circuit, which was based on the physical principle of mixing, or heterodyning, a received signal to a fixed intermediate frequency that can be more conveniently processed than the original radio carrier frequency (Witts 1935). In many countries, these receivers were called "superhets," or simply "super." Because of their excellent selectivity, they were the first sets that allowed the installation of calibrated station scales. From 1931 on, leading set manufacturers promoted their automatic adjustment circuits, which offered the "highest selectivity" and "perfect single-knob operability." The Schwarzwälder Apparate-Bau-Anstalt (SABA) dubbed its 1931 model, SABA 41 W, "the signpost to the frequency jungle!" (Ketterer 2003, 90). The first superhet with a calibrated station scale on the German market was the Mikrohet W, manufactured by Stassfurt in 1928. This model had such success on the market that other producers immediately copied the idea (figure 17.4).[10]

From 1933 on, the station scale became a standard function in mid-price-range receivers. The mechanical realization of these station scales was hardly a simple matter, however. Because of many neighboring stations in the narrow frequency bands, a complicated mechanical construction was necessary to produce the optical illusion of a scale with evenly spread stations. The conversion of a turning movement on the station selection knob into an automatic fine-tuning of the electrical circuits was a real masterpiece of electromechanical achievement. The optical solution of this problem reminds one of the invention of nonscaled maps for the London underground developed by electrical engineer Henry Beck in 1933 (Schlögel 2003, 102). The new underground map owed its success to its simplicity and clearness, constructing a symbolic nearness or connectivity of the suburbs with the city center. In other words, Beck's map brought the suburbs closer to the center. Instead of a realistic representation of the daunting bigness of greater London, it converted the real geographical distance into an invitation to the city (Vertesi 2008). Similarly, the radio station scale converted the fascinating but unfathomable "ether" into meaningful places (see figure 17.5).

The development and introduction of the calibrated station scale marked an important innovation in the technical improvement of the receiver, as well as in radio design. The technical solution of the "scaling problem" consisted in a complex counterbalancing coupling of the visual station indicator with the rotary capacitor responsible for selectivity (Rhein 1935, 97–143). Around the mid-1920s, sets that used optical elements for the indication or recovery of stations sporadically appeared, for example, numerically gauged adjusting knobs or so-called microscales, little panels that showed a small cutout of the scale (Friemert 1996, 47–49). These rapid innovations in the radio receiver industry—the pace was set by the seasonal rhythm of radio trade fairs (Riedel 1994; Bressler 2009)—sparked harsh technical and creative competition among the set manufacturers. In 1930 one of the major players in the European radio industry, the German Telefunken Society, acquired the patent and licensing rights to an invention by French inventor

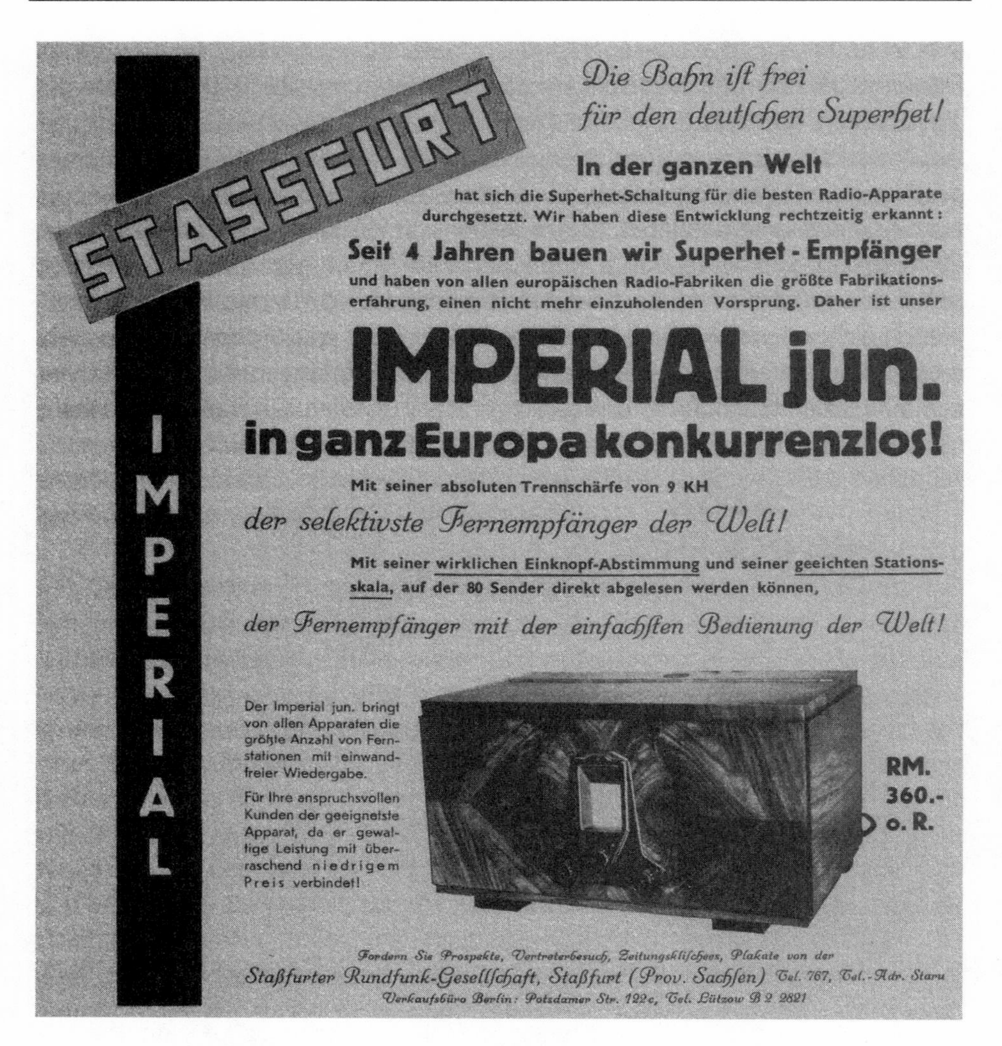

Figure 17.4. Advertisement for the radio set "IMPERIAL junior" from the German firm Stassfurt, promoting the set as "the receiver with the highest selectivity in the world," "real one-knob operability," and "calibrated station scale with 80 European stations." Source: Ketterer, Ralf. *Funken. Wellen. Radio. Zur Einführung eines technischen Konsumartikels durch die deutsche Rundfunkindustrie 1923–1939.* Berlin: VISTAS Verlag, 2003, p. 79.

Joseph-Louis Routin, who had realized a skillful solution to the "scaling problem" and distributed it under the label of "Valundia system" (figure 17.6).

Telefunken offered Routin a consultancy contract, arranging for a yearly visit to the company's laboratories in Berlin. The patent and licensing contract assigned Telefunken the rights to produce and to commercialize three of Routin's inventions: the conical scale called Nomina, the fine-tuning adjustment for capacitors, and a plug connector for an illuminated scale.[11] From 1931 on, Routin's Valundia

Figure 17.5. “No searching – just dial.” This was the slogan of the NORA ad from 1931. Source: Telefunken Archives, Deutsches Technikmuseum, Berlin.

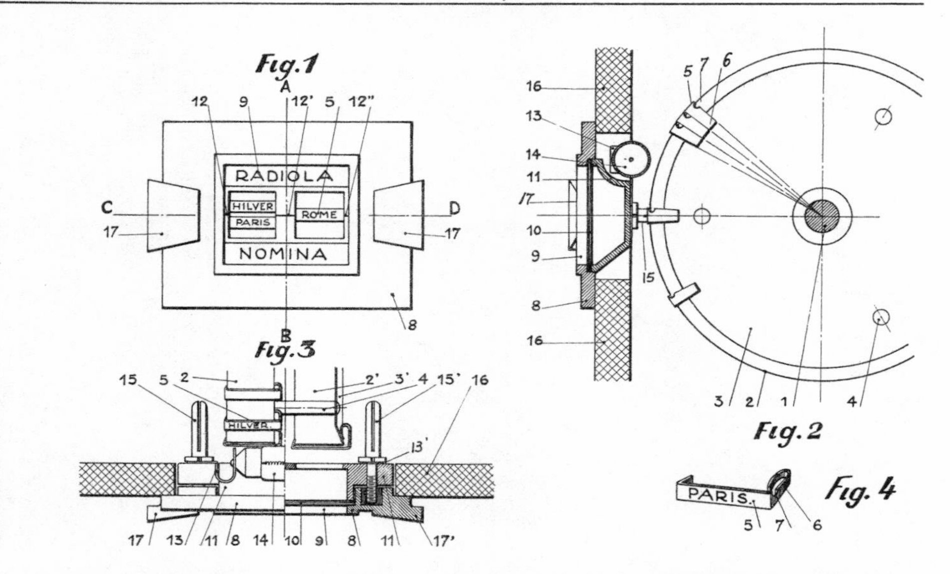

Figure 17.6. Drawing of the Valundia-System in the patent application of Joseph-Louis Routin (1930). Source: Telefunken-Archives, Deutsches Technikmuseum, Berlin.

system was commercialized by Telefunken under the name of Auto-Skala (figure 17.7).

The station scale is advertised as a signpost to or an atlas of the most important European stations. Unlike most other models on the market with fixed station scales (mainly in the form of printed scales placed on the front of the receiver), the Auto-Skala provided the listener with an opportunity to react in a flexible manner to the common changes of frequencies. In other words, the Auto-Skala was an interactive device that enabled users to customize the scale to reflect their listening habits. The booklet that came with the Auto-Skala stated the following:

> The basic newness of the Auto-Skala consists in the exchangeable station names. Hence: no dead figures, no tuning tables, no searching but easy reading of the station names. No schematical listing of all irrelevant stations in Europe and no obsolescence because of frequency changes, but comfortable adoption of your favorite stations and flexible adjustment of the terms . . . for all station names are written on changeable labels easy to replace when a station changes its frequency.[12]

The text of this advertisement draws our attention to a dimension too often neglected in the history of technology: the user (Oudshoorn and Pinch 2005).[13] Analysis of the discursive strategies detectable in both the iconographic and the semantic dimension of various advertisements promoting the new station scale gives rise to several observations. The first is that most of these texts catered to a female public. As argued earlier in this chapter, the introduction and promotion of

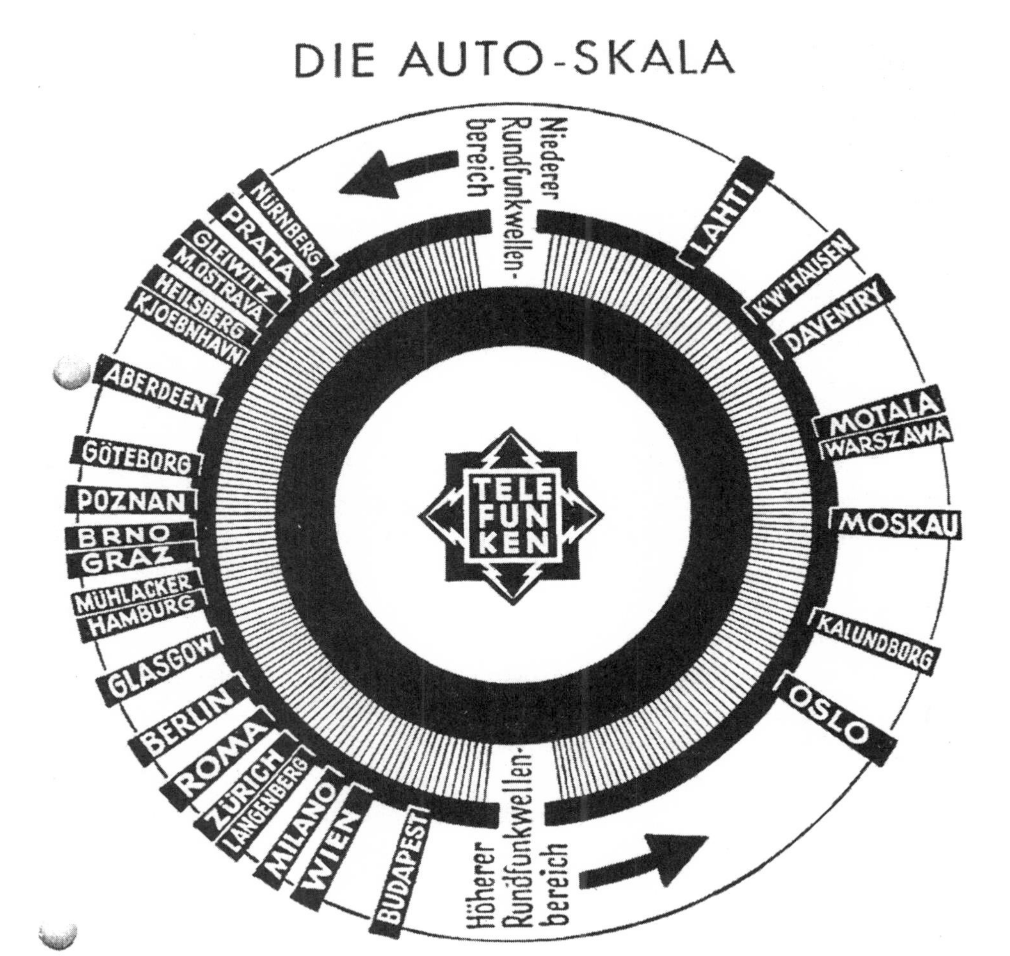

DIESES BLATT SOLL IHNEN HELFEN

die wichtigsten Sender auf Ihrer Auto-Skala erstmalig zu bestimmen. Haben Sie einen der auf dem Bild eingezeichneten Sender eingestellt, so dient das Bild als Wegweiser, an welcher Stelle ungefähr die wichtigsten europäischen Sender hörbar werden. Die Sender folgen, wie Sie sehen, in der gleichen Reihenfolge aufeinander wie auf dem beigegebenen Senderverzeichnis, nur daß das Senderverzeichnis auch die dazwischen gelegenen Stationen enthält. Sie brauchen dieses Hilfsmittel nur, bis Sie die Sendernamen erstmalig aufgesetzt haben; dann ermöglicht Ihnen die Auto-Skala das Wiederauffinden ohne jedes Hilfsmittel.

Figure 17.7. The Auto-Skala of Telefunken (1931).
Source: Telefunken Archives, Deutsches Technikmuseum, Berlin.

the station scale must be embedded in a process of the "domestication" of the radio receiver. In this process, described as a process of functional and symbolic appropriation of the apparatus, the improvement of the set's user-friendliness can be seen as the radio industry's central motivation since the mid-1920s. This shift in the identity of radio as medium (from tinkerer's medium to mass medium) was closely linked with the attraction of a massive new user potential represented by children and, in particular, women. For instance, the front page of the booklet advertising the four 1931 models introduced with the new Auto-Skala for the German market is decorated with an elegant Chinese statue: A woman who radiates self-confidence, Eastern grace, and a hint of exoticism looks down on the receiver. On the following page, a photograph shows a bourgeois living room with a chess set in front of the receiver; its caption reads: "In every elegant living room, the Telefunken receiver with Auto-Skala is an organic part of the furnishing. Its artistic shape perfectly harmonizes with every style." The portrait of a hand easily changing a label with a station name is clearly a female hand. A Dutch Telefunken advertisement of the same year promotes the company's receivers with the slogan "In your own environment" and emphasizes the fact that the station scale directly connects the listener with all European stations.[14] Despite this undeniable attention to the female listener in advertisement strategies, the construction of gender is not limited to women. In fact, many radio advertisements in this period demonstrate a well-balanced mixture of female and male elements. While aesthetic and functional adjectives addressed women as the users of the appliance and as competent managers of the household, technical information about the number of circuits, tubes, and the material basis of the chassis were meant to appeal to the juvenile or male radio connoisseur.[15]

While advertisements as a specific category of sources are perfectly suitable for a discursive analysis of gender construction (Smulyan 1994), the analysis of the appliance as material source offers additional perspectives. Here, the role of design as a mediating interface between technology and the user becomes crucial. As the British broadcast historian Paddy Scannell (2004, 86) argues, "the mediating stage in the transition from technology to domestic equipment is design. . . . Design is essential to the transformation of user-unfriendly technologies that only trained experts can use into simple user-friendly things that anyone can use."

As I argue elsewhere (Fickers 2007), we can identify at least five dimensions or functions of design when looking at the radio set. First, design comes with a "gender script." As Madeleine Akrich, Nelly Oudshoorn, Trevor Pinch, Ellen van Oost, and others have argued, technical artifacts can be analyzed as representations of gender constructions inscribed into the materiality of the technical object (Oudshoorn and Pinch 2005). Parallel to the concept of "gender script," Steven Woolgar's concept of the design process as a struggle to configure the user offers an alternative interpretation of the role of design. For Woolgar (1991), configuring is the process of identifying putative users. The writings of Joseph-Louis Routin are a striking example of how inventors and engineers have configured the potential users of their inventions. In a letter dated January 27, 1931, to Telefunken patent lawyer Levy, Routin explained:

> The radio dial with its cluster of names of stations is especially appreciated by women and children, who have an instinctive horror of numbers. It is useless to speak of lambdas, wavebands, frequencies, or kilohertz to them. But it is exactly this clientele which is of great interest to the radio industry as it represents the largest segment of the population and is most interested in domestic family distractions.[16]

Second, another function of design is that of being an agent of standardization of mass production and mass consumption. The radio as "Fordist lead product" (Wittke 1996) played an important role in the aim of industrial designers to "cultivate" and "aestheticize" mass products in the capitalist economy (König 2003). In this sense, a third function of design is of course to mediate between function and form. As philosopher Ernst Cassirer wrote in his *Form und Technik* in 1930, all intellectual examination of the world is based on a double process of *Fassen*: grasping in the sense of theoretical comprehension and grasping in the sense of physical shaping (Cassirer 1995, 52). This double process of grasping is exactly what happens in the present case: the connection between the symbolic appropriation of the world (reading of the station scale) and the physical grasping of the material world (manipulation of the radio set). Next, and closely related to this, there is the meaning of design as symbol or language. Philosophers such as Roland Barthes and Jean Baudrillard have emphasized the semantic quality of the *objet signe*, and "product semantics" is nowadays an established field in design theory.[17] Finally, the fifth function of design is the already mentioned function of concealing technology in order to create the "user illusion." As demonstrated earlier, all of these dimensions or functions of design are evident when looking at the radio set as a mediating interface.

Conclusion: On Station Scales, Mental Maps, and Imagined Topographies

One of my theses in the introduction was that the names of cities on station scales can evoke something like a mental map in the imagination of the radio listener. In *Mapping Cyberspace*, Martin Dodge and Rob Kitchin assert that the geographical space is superposed by an imagined space that allows people and organizations to react in a flexible way to the changes in real spatial geographies (2001, 14). Since the publication of *Maps in Mind: Reflections on Cognitive Mapping* by geographer Roger Downs and psychologist David Stea in 1977, the concept of "mental maps" has been widely discussed (Schenk 2000). According to Downs and Stea, a mental map is a structured image of a part of a person's spatial environment. This map does not

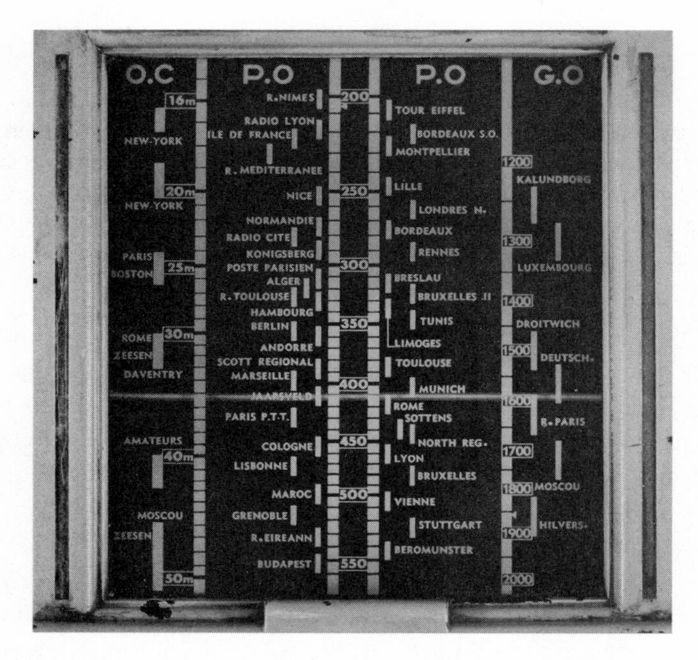

Figure 17.8. Station scale of a French receiver showing in three wavelengths: short wave (16–50 meter, left), medium wave (200–550 meter, middle) and long wave (1200–2000 meter, right). Source: Photo by Dan Talson ©

have to be correct in a geographical sense. It is mirroring the world as the person envisions the world. As Janet Vertesi argues, the iconic image of the London tube map has been incorporated into users' cognitive mapping of London despite the obvious divergence between the real topography of the city and its idealized representation on the map. Subscribing to Vertesi's plea not to separate the analysis of representations of technological systems from a community's interaction with them (Vertesi 2008, 26), I argue that the interaction with the radio station scale can be interpreted as an active process of mental mapping on the part of the users.

The dial can be read as an invitation to an imagined ether voyage, where London, Paris, Oslo, or Hilversum were just one turn away from each other. They evoked in the radio listener—who is a radio watcher, too—a mental map whose fault line could be decoded only by the listener. Station scales, then, served as early atlases of globalization, and one may interpret the use of radio in the 1920s and 1930s as a symbolic appropriation of the European broadcasting landscape. In the words of Paddy Scannell, "It is not just that radio and television compress time and space. They create new possibilities of being: of being in two places at once or two times at once" (Scannell 1996, 91). In his inspiring essay "Ästhetik der drahtlosen Telegrafie" [Aesthetic of Wireless Telegraphy], design historian Chup Friemert reinforces this thesis and links it to the functionality of the radio station scale: "The lust for the scale is the lust for the transportability of world events, as well as the lust for their disposability" (Friemert 1996, 49).

As Uta Schmidt has shown in her study on the appropriation of radio in the Weimar Republic and the Third Reich, the sense of seeing occupied a central position in the "economy of pleasure" (Norbert Elias). Listening to the radio—especially when considering the effect of illuminated station scales and "magic eyes" in the dark—evoked feelings of intimacy and aesthetic delight:

> The illuminated station scale has deeply engraved itself into the memory of many generations of radio listeners. Although many of the stations could be captured only with considerable effort on a good antenna, the stations promised a world outside the possibilities of bodily experiences and invited the listener to make fantastic imaginary travels. (Schmidt 1998, 317)

In that way, the radio dial became a virtual roadmap for the journey through the ether, the radio station mutated into a station, inviting the listener to remain there for a while. Many ego documents (e.g., novels, poems, song lyrics) describe the aesthetic fascination, juvenile excitement, and the stimulation of imaginary thought travels experienced while interacting with the radio dial with all of one's senses.[18] In the Dutch book for young people *De radio-detective: draadloze ogen* [The Radio Detective: Wireless Eyes] by Leonard Roggeveen (1930), Hans, the book's fifteen-year-old protagonist, experiences the "wonder" of traveling through Europe by simply turning the dial of his self-constructed receiver:

> Hilversum 2 was precisely in the right place. The red calibration indicator stood right in the middle of the 298-meter wavelength panel. . . . He moved along the whole dial, and most of the stations came in very clearly. Luxemburg and Rijssel played dance music, a man in Munich said, "Achtung, meine Damen und Herren." A symphony orchestra was performing in Strasbourg, "here national program," said a lady in Droitwitch. . . . Brussels followed, 1 and 2, and London, Paris, Rome—yes, Rome! Hilversum 1, Vienna, Stuttgart, Beromünster! Hans listened to an opera in Paris, to gramophone music from Rome, lovely violin music from Stuttgart. All Europe was present: He traveled in one turn on the dial from Hilversum to Vienna, from Rome to London; the countries lost their borders. The most alien peoples stood next to him—next to him in his little chamber. He heard voices from strangers hundreds of kilometers away as if they were standing next to him, and he immediately understood the meaning of what he had read in so many radio advertisements: "Make the world your neighbor." (Roggeveen 1930, 24)

The new broadcasting space was not only shaped by the technological infrastructure of sender networks around the world; it was also actively co-constructed by radio listeners, who became individual actors in the spatialization of their world (Geppert, Jensen, and Weinhold 2005). While turning the dial and tuning in a station (city name), the vast broadcast *space* was transformed into a specific, meaningful *place*. In this sense, we should interpret the acts of radio listening, viewing, and tuning as acts of symbolic appropriation of the world. To stylize the radio as the "monosensual stepchild of the century of audiovisual attractions" (Maase 2004, 47) misjudges both the aesthetic attraction and playful usages of wireless technology in a technical culture (Haring 2007) and the multisensorial experience of radio as a cultural practice.

NOTES

1 On the "supremacy" of the visual in Western culture, see Jenks (1995). For a critique of this one-dimensional concentration on the visual, see Mitchell (2005). For a sophisticated analysis of the relation between seeing and hearing in the process of modernization, see Sterne (2003).

2 "We were not blessed with three hands!" Advertisement by Murphy in *The Wireless and Gramophone Trader* (Aug. 30, 1930), 26.

3 Telefunken-Sprecher, no. 7, 1927, quoted in Ketterer (*Funken. Wellen. Radio* [2003], 167). On the disappearance of real chauffeurs in the history of the automobilist see the chapter by Stefan Krebs in this volume.

4 Quoted from Méadel (1994), 196.

5 Because of the "double articulation" of the radio as both object *and* medium—a key distinctiveness of media technologies compared to other technologies—Haddon and Silverstone speak of successful domestication only when a communication technology has been tamed as an object and socially appropriated as a medium. "With both dimensions of doubling in mind it becomes both possible, indeed essential, to see the innovation of media and information and communication technologies as a fundamentally social process, and the particular complexity of the design/domestication interface of these technologies as one which requires an understanding of the role of both producer and user in its definition." (Silverstone and Haddon 1996, 50).

6 Significantly, the notion "user illusion" was coined by the American computer pioneer Alan Kay, who first worked at Xerox PARC and later became one of the founders of Apple; he was to be instrumental in the development of human/machine interfaces. For a detailed history of the invention and development of the graphic user interface (desktop metaphor), see Friedewald (1999) and Bardini (2000). A more than inspiring introduction to the philosophical and epistemological dimensions of the "user illusion" and "desktop philosophy" is offered by the Danish science journalist Tor Nørretranders in *The User Illusion* (1998).

7 The creation of the different committees naturally affected the working methods of the whole organization. In fact, each question or problem to be discussed by the IBU was first submitted to the General Council, which subsequently relegated it to one of the committees. Next, they had to charge one of their members to produce a report, which was then examined by the General Council or—in some cases—the General Assembly. The frequent meetings of the different working committees circulated between countries, while the General Assembly met in Ouchy-Lausanne in Switzerland on a yearly basis (see UIR 1944, 16).

8 Raymond Braillard, "Rapport de la réunion de la Commission Technique à Lausanne," 31.05.1928, p. 15. In IBU/EBU archives, Geneva, box 80ter / Com. T. au Conseil.

9 A variety of factors, such as the different topographic conditions in the United States and Europe, the emergence of the commercial network system in the United States versus the predominance of the public service broadcasting in Europe, and the great diversity of languages and dialects in Europe in comparison to the United States, make the European situation rather different and complicated when compared to developments in the United States. This complexity clearly called for a more precise and accurate technical regime. On the United States, see Slotten (2000).

10 For a richly illustrated encyclopedia of early German radio sets together with short portraits of radio manufacturers, see the five volumes of Abele (1999).

11 See "Aktennotiz über de Besprechung mit Herrn Routin am 10. Dezember 1930," Deutsches Technikmuseum Berlin (DTM), Historisches Archiv, I.2.060 C, no. 1750.

12 See the booklet of the Telefunken receiver with Auto-Skala in DTM, Historisches Archiv, III Sog. 2 Firmenschriften, no. 51641.

13 For a useful sociological overview of the relationship between technology, usage, and representation see Flichy (2008).

14 "In uw eigen milieu," Telefunken advertisement in *Radio Gids, Officeel Orgaan van de Vereniging van Arbeiders-Radio-Amateurs* 5(13) (Jan. 24, 1931).

15 In her masterful study of the cultural and social appropriation of the radio set in Weimar and Nazi Germany, Inge Marßolek refers to a 1933 Telefunken ad for the Viking model to demonstrate the multifaceted character of radio ads. This Viking ad, printed as double-page leaflet, had a strict gender-related layout: The left side was headed "Wie *sie* ihn sieht" [How *she* sees it], while the right side had the headline "Wie *er* ihn sieht" [How *he* sees it]. While the left (female) side was visually dominated by a picture of a little boy, smilingly turning the knob to adjust the station scale and thereby demonstrating the simple use of the set, the right (male) side showed a photo of the interior of the set, representing its "organic and stable" assembly. The name of the model itself was a tribute to the new National Socialist power in Germany: "Viking. A proud name from the time when Germanic sailors accomplished great things in small boats." See Riedel (1994: 311–12).

16 See letter from Joseph-Louis Routin to Mr. Levy, Paris, Jan. 27, 1931, Annexe 1, p. 1, in DTM, signature I. 2. 060 C, 1750.

17 "Probably the most noteworthy development of design today is its concern for the cognitive meanings, symbolic functions, and cultural histories of form. . . . Product semantics is the study of the symbolic qualities of man-made forms in the context of their use and the application of this knowledge to industrial design. It takes into account not only the physical and physiological functions but the psychological, social, and cultural context, which we call the symbolic environment. Product semantics is an effort to understand and to take full responsibility for the symbolic environment into which industrial products are placed and where they should function by virtue of their own communicative qualities." (Krippendorf and Ritter 1984, 4)

18 At present I am collecting ego documents in order to underpin my arguments with more historical relevance on the side of the users. So far I have collected more than 150 references to the radio dial in German, French, and English novels and children's books, some 25 references in poems, and 15 references in song lyrics. In addition to that, many advertisements of the 1930s use the metaphor of the radio as an "ear to the world" as a parable for modernity and present the radio dial as a virtual atlas or road map of Europe and/or the world. A thorough narrative analysis of these ego documents and an iconographic analysis of the advertisements based on the concepts of "parables" and "visual clichés" as introduced by Roland Marchand (1985) will be the topic of a future publication.

REFERENCES

Abele, Günther. *Historische Radios: Eine Chronik in Wort und Bild.* 5 vols. Stuttgart: Füsslin, 1999.

Aitken, Hugh. *Syntony and Sparks: The Origins of Radio.* New York: Wiley, 1976.

Arnheim, Rudolf. *Radio: An Art of Sound*. New York: Da Capo, 1972.
Band, Lothar. "Hörbilder aus dem Leben." In *Medientheorie 1888–1933*, ed. A. Kümmel and P. Löffler, 244–45. Frankfurt: Suhrkamp, 2002.
Bardini, Thierry. *Bootstrapping: Douglas Engelbart, Coevolution, and the Origins of Personal Computing*. Stanford: Stanford University Press, 2000.
Betts, Paul. *The Authority of Everday Objects: A Cultural History of West-German Industrial Design*. Berkeley: University of California Press, 2004.
Bleicher, Joan Kristin. " 'Das Ohr zur Welt': Öffentlichkeitsstrategien des frühen Hörfunks." In *Öffentlichkeit im Wandel: Neue Beiträge zur Begriffserklärung*, ed. W. Faulstich and K. Hickethier, 132–43. Bardowick, Germany: Wissenschaftlicher Verlag, 2000.
Bolz, Norbert. "The User-Illusion of the World: On the Meaning of Design." In *The Context Issue: Mediametic Magazine* 9(2/3) (1998); http://www.mediamatic.net/page/5672/nl.
Bressler, Eva Susanne. *Von der Experimentierbühne zum Propagandainstrument: Die Geschichte der Funkausstellung 1924–1939*. Cologne: Böhlau, 2009.
Carlat, Louis. " 'A Cleanser for the Mind': Marketing Radio Receivers for the American Home." In *His and Hers: Gender, Consumption, and Technology*, ed. R. Horrowitz and A. Mohun, 115–38. Charlottesville: University Press of Virginia, 1998.
Cassirer, Ernst. "Form und Technik." In *Symbol, Technik, Sprache: Aufsätze aus den Jahren 1927–1933*, ed. E. Orth and J. Krois, 39–90. Hamburg: Meiner, 1995.
Codding, George A. The International Telecommunication Union: An Experiment in International Cooperation. Geneva: UNESCO 1959.
Dodge, Martin, and Kitchin, Rob. *Mapping Cyberspace*. London: Routledge, 2001.
Douglas, Susan. *Inventing American Broadcasting (1899–1922)*. Baltimore: Johns Hopkins University Press, 1987.
———. *Listening In: Radio and the American Imagination*. Minneapolis: University of Minnesota Press, 2004.
Downs, Roger, and David Stea. *Maps in Mind: Reflections on Cognitive Mapping*. New York: Harper and Row, 1977.
Elias, Nobert. *Über den Prozeß der Zivilisation : soziogenetische u. psychogenetische Untersuchungen*. Basel: Verlag Haus zum Falken, 1939.
Fickers, Andreas. "Design as Mediating Interface: Zur Zeugen- und Zeichenhaftigkeit des Radioapparates." *Berichte zur Wissenschaftsgeschichte* 30(3) (2007): 199–213.
Flichy, Patrice. "Technique, usage, et représentations." *Réseaux* 2(148–49) (2008): 147–74.
Forty, Adrian. "Wireless Style: Symbolic Design and the English Radio Cabinet 1928–1933." *Architectural Association Quarterly* 4 (1972): 22–31.
Friedewald, Michael. *Der Computer als Werkzeug und Medium: Die geistigen und technischen Wurzeln des Personal Computers*. Berlin: GNT, 1999.
Friemert, Chup. *Radiowelten: Zur Ästhetik der drahtlosen Telegrafie*. Ostfildern, Germany: Cantz, 1996.
Geddes, Keith, and Gordon Bussey. *The Setmakers: A History of the Radio and Television Industry*. London: BREMA, 1991.
Geppert, Alexander, Uffa Jensen, and Jörn Weinhold. "Verräumlichung: Kommunikative Praktiken in historischer Perspektive, 1840–1930." In *Ortsgespräche: Raum und Kommunikation im 19. und 20. Jahrhundert*, ed. A. Geppert, U. Jensen, and J. Weinhold, 15–52. Bielefeld, Germany: transcript, 2005.
Gerlich, Christian. "Die Denkumschaltung durch den Rundfunk." In *Medientheorie 1888–1933*, ed. A. Kümmel and P. Löffler, 211. Frankfurt: Suhrkamp, 2002.

Gitelman, Lisa. "Media, Materiality, and the Measure of the Digital; or, the Case of the Sheet Music and the Problem of Piano Rolls." In *Memory Bytes: History, Technology, and Digital Culture*, ed. L. Rabinovitz and A. Geil, 199–217. Durham, N.C.: Duke University Press, 2004.

Griset, Pascal. "Innovation and Radio Industry in Europe during the Interwar Period." In *Innovations in the European Economy between the Wars*, ed. F. Caron, P. Erker, and W. Fischer, 37–64. New York: de Gruyter, 1995.

Haring, Kristen. *Ham Radio's Technical Culture*. Cambridge Mass.: MIT Press, 2007.

Harrison, Arthur. "Single-Control Tuning: An Analysis of an Innovation." *Technology and Culture* 39(3) (1998): 296–302.

Hilmes, Michele. *Radio Voices: American Broadcasting 1922–1952*. Minneapolis: University of Minnesota Press, 1997.

Huth, Arno. *La radiodiffusion: Puissance mondiale*. Paris: Librairie Gallimard, 1937.

Jasanoff, Sheila, and Wynne, Brian. "Science and Decisionmaking." In *Human Choice and Climate Change*, vol.1, ed. S. Rayner and E. Malone, 1–81. Columbus, Ohio: Batelle, 1998.

Jenks, Chris. "The Centrality of the Eye in Western Culture: An Introduction." In *Visual Culture*, ed. C. Jenks, 1–25. London: Routledge, 1995.

Ketterer, Ralf. *Funken. Wellen. Radio. Zur Einführung eines technischen Konsumartikels durch die deutsche Rundfunkindustrie 1923–1939*. Berlin: VISTAS, 2003.

Kolb, Richard. *Das Horoskop des Hörspiels*. Berlin: Hesse, 1932. Quoted in Wolfgang Hagen, "Was mit Marconi begann . . . Radio und Psychoanalyse." *Neue Rundschau* 112(4) (2001): 32.

König, Getrude. "Auf dem Rücken der Dinge: Materielle Kultur und Kulturwissenschaft." In *Unterwelten der Kultur: Themen und Theorien der volkskundlichen Kulturwissenschaft*, ed. K. Maase and B. Warneken, 95–118. Cologne: Böhlau, 2003.

Krippendorff, Klaus, and Ritter. Reinhart. Product Semantics: Exploring the Symbolic Qualities of Form." *Innovation* 3 (1984): 4–9.

Lenk, Carsten. *Die Erscheinung des Rundfunks: Einführung und Nutzung eines neuen Mediums 1923–1932*. Opladen: Westdeutscher Verlag, 1997.

Maase, Kaspar. " 'Jetzt kommt Dänemark': Anmerkungen zum Gebrauchswert des frühen Rundfunks." In *Die Idee des Radios: Von den Anfängen in Europa und den USA bis 1933*, ed. E. Lersch and H. Schanze, 47–72. Constance: UKV Medien, 2004.

Marchand, Roland. *Advertising the American Dream: Making Way for Modernity 1920–1949*. Berkeley: University of California Press, 1985.

Méadel, Cécile. *Histoire de la radio des années trente: du sans-filiste à l'auditeur*. Paris: Anthropos, 1994.

Mitchell, W. J. Thomas. "There Are No Visual Media." *Journal of Visual Culture* 4(2) (2005): 257–66.

Nørretranders, Tor. *The User Illusion: Cutting Consciousness Down to Size*. New York: Viking, 1998.

Oudshoorn, Nelly, and Pinch, Trevor, eds. *How Users Matter: The Co-Construction of Users and Technology*. Cambridge, Mass.: MIT Press, 2005.

Popp, Ewald. "Die Wesensform des Lautsprechers: Ein Beitrag zur Ästhetik der Technik." In *Medientheorie 1888–1933*, ed. A. Kümmel and P. Löffler, 312–13. Frankfurt: Suhrkamp, 2002.

Rhein, Eduard. *Wunder der Wellen: Rundfunk und Fernsehen dargestellt für jedermann*. Berlin: Ullstein, 1935.

Riedel, Heide. *70 Jahre Funkausstellung Berlin: Politik, Wirtschaft, Programm*. Berlin: VISTAS, 1994.

Roggeveen, Leonard. *Draadloze oogen: de radio-detective*. Den Haag: Van Goor en Zn, 1930.
Scannell, Paddy. *Radio, Television, and Modern Life: A Phenomenological Approach*. Oxford: Blackwell, 1996.
———. "Technology and Utopia: British Radio in the 1920s." In *Die Idee des Radios: Von den Anfängen in Europa und den USA bis 1933*, ed. E. Lersch and H. Schanze, 83–94. Constance: UKV Medien, 2004.
Schenk, Frithjof Benjamin. "Mental Maps. Die Konstruktion von geographischen Räumen in Europa seit der Aufklärung." *Geschichte und Gesellschaft* 28(3) (2000): 493–514.
Schickele, René. "Paneuropa der Sender. " In *Radio-Kultur in der Weimarer Republik: Eine Dokumentation*, ed. I. Schneider, 141. Tübingen: Narr, 1984: 140–143.
Schlögel, Karl. *Im Raume lesen wir die Zeit: Über Zivilisationsgeschichte und Geopolitik*. Munich: Hanser, 2003.
Schmidt, Uta C. "Radioaneignung." In *Zuhören und Gehörtwerden*, vol. 2, ed. A. von Saldern, 259–368. Tübingen: Diskord, 1998.
Seitz, Robert. "Überall her aus der Welt. " In *Radio-Kultur in der Weimarer Republik: Eine Dokumentation*, ed. I. Schneider, 42–43. Tübingen: Narr, 1984.
Silverstone, Roger. "Domesticating Domestication: Reflections on the Life of a Concept." In *Domestication of Media and Technology*, ed. T. Berker, M. Hamann, Y. Punie, and K. Ward, 229–48. Maidenhead, UK: Open University Press, 2006.
———, and Leslie Haddon. "Design and the Domestication of Information and Communication Technologies: Technical Change and Everyday Life." In *Communication by Design: The Politics of Information and Communication Technologies*, ed. R. Mansell and R. Silverstone, 44–74. New York: Oxford University Press, 1996.
Slotten, Hugh. *Radio and Television Regulation: Broadcast Technology in the United States, 1920–1960*. Baltimore: Johns Hopkins University Press, 2000.
Smulyan, Susan. *Selling Radio: The Commercialization of American Broadcasting, 1920–1934*. Washington, D.C.: Smithsonian Institution Press, 1994.
Sterne, Jonathan. *The Audible Past: Cultural Origins of Sound Reproduction*. Durham, N.C.: Duke University Press, 2003.
Tomlinson, John. International Control of Radio Communications. Diss., Institut Universitaire des Hautes Études Internationales, vol. 41. Geneva, 1938.
UIR, ed. *Memorandum sur la situation actuelle de la radio-diffusion européenne au point de vue des interférences et les moyens propres a y remédier*. Brussels: Author, 1932.
———, ed. *L'Union Internationale de Radiodiffusion. Son histoire—son activité*. Geneva: Author, 1944.
Vertesi, Janet. "Mind the Gap: The London Underground Map and Users' Representations of Urban Space." *Social Studies of Science* 38(7) (2008): 7–33.
Verweyen, Johannes Maria. "Radiotis! Gedanken zum Radiohören." In *Medientheorie 1888–1933*, ed. A. Kümmel and P. Löffler, 454. Frankfurt: Suhrkamp, 2002.
Wittke, Volker. *Wie entstand industrielle Massenproduktion? Die diskontinuierliche Entwicklung der deutschen Elektroindustrie von den Anfängen der "großen Industrie" bis zur Entfaltung des Fordismus (1880–1975)*. Berlin: Sigma, 1996.
Witts, Alfred T. *The Superheterodyne Receiver: Its Development, Theory, and Modern Practice*. London: Pitman, 1935.
Woolgar, Steven. "Configuring the User: The Case of Usability Trials." In *Sociology of Monsters: Essays on Power, Technology, and Domination*, ed. J. Law, 58–97. London: Routledge, 1991.

Wormbs, Nina. "Standardizing Early Broadcasting in Europe: A Form of Regulation." In *Bargaining Norms—Arguing Standards: Negotiating Technical Standards*, ed. J. Schueler, A. Fickers, and A. Hommels, 112–21. The Hague: STT, 2008.

Zweig, Arnold. "Ästhetik des Rundfunks." In *Radio-Kultur in der Weimarer Republik: Eine Dokumentation*, ed. I. Schneider, 76. Tübingen: Narr, 1984.

CHAPTER 18

FROM LISTENING TO DISTRIBUTION: NONOFFICIAL MUSIC PRACTICES IN HUNGARY AND CZECHOSLOVAKIA FROM THE 1960S TO THE 1980S

TREVER HAGEN WITH TIA DENORA

Introduction

This chapter presents an empirical case study of the wide variety of nonofficial settings and reinventions of music listening, recording, and distribution technology in Hungary and Czechoslovakia from the 1960s to the 1980s.[1] As much as possible we compare the two sites, but our main purpose is to use the data to discover how individuals used sounds and to consider the question of what those sounds, coupled with their uses, enabled these individuals to do. We also use the case study to think more abstractly about sound and music as a resource for collective agency and action. Here we are interested in how collective agency takes shape in relation to what we describe as "creative constriction": the paradoxical situation whereby suppression and control generate new opportunities for creative action. We situate this creative constriction in terms of how its specific sound technologies—their construction, consumption, and appropriation—afforded alternative and socially important cultural practices that were informally learned via sonic/musical experience. These "lessons" in turn provided a springboard for nonofficial[2] modes of being that coalesced at individual and collective levels.

Forms of nonofficial listening, recording, and distribution practices in both countries during the communist era examined here involve attention to radio broadcasts, listening to and trading LPs from the West,[3] and producing *magnitizdat*—self-made recordings. Musical experience, then, is understood as the intersection of sounds, music, technologies, and places. Music, in this understanding, is a flexible medium—a liminal space—one in which all the fine shades of an actor's lifeword can be displayed. This display, we suggest, permitted music listeners to pursue—to varying degrees—alternative or independent ways of being and feeling, from dipping a toe in nonofficial waters to plunging in and never resurfacing. We use the term *official* to describe areas of society and culture that were defined and ordered by centralized powers of the state, manifested not only in institutions and agencies but also in everyday practice, as Jakubowicz puts it, "to achieve a commonality of enthusiastic commitment to building communism" (Jakubowicz 1994, 271).

Central to our argument in understanding the liminality between official and nonofficial worlds is the (social) activity of making music and musical meaning, which Small calls "musicking" (Small 1998, 8). We expand the concept of musicking to take on any number of forms, such as attending concerts, tuning in to the radio, practicing scales, humming, imagining music, singing along to an LP, making sound compilations, and bootlegging performances. The musicking we study provides and sustains a variety of moods, commitments to nonofficial culture, modes of attention to music, and emotional and knowledge structures, which we refer to as dispositions. This congruence of dispositions via musicking is the part of the collective formation of a liminal space that creates "a common shared world of time, space, gesture, and energy, which nevertheless allows diversity and unity" (Pavlicevic and Ansdell 2004, 84).

Throughout, our aim is to consider the two communist regimes in Hungary and Czechoslovakia in regard to control over music, youth policy, technology, and culture. Both "creative constriction" and "dipping into" are present in each country but take different forms. It is important to note that what follows is not intended to be a comprehensive history of sound technology and its uses in Hungary and Czechoslovakia or a fully comparative study as that is beyond the scope of this chapter. We chose to study Hungary and Czechoslovakia because the attitudes of their respective regime to popular music, the level of economic reform in each country, subsequent access to technology, and the lack of informal information exchange between music amateurs between these countries highlight some of the pathways of creative constriction that we identify in nonofficial musical life between 1960 and 1990. In what follows, we address two areas common to both regimes, nonrecordable and rerecordable sound technology. In these nonofficial cultures, two particular modes of production and distribution were held in common:

Samizdat refers to self-published textual material that ranges from manuscripts to material that is typewritten and copied by hand, typewriter, or mimeographed. The existence, form, and context of samizdat, as well as the amount produced for dissemination, varied from country to country (Machovec 2009). Self-publications were created without the permission or consultation of the authorities in either country.

Magnitizdat refers to the recording and distribution of sonic material that was not available to the public, music that was banned or censored, sound that could be seen as potentially subversive, or music that was wanted immediately. Magnitizdat, in terms of distribution in the shadow economy or black market, was not necessarily subversive but instead filled a market gap in times of shortage (Smith 1984). In the Eastern-bloc countries, a certain amount of bricolage, or "situated experimentation," was employed, of the type found in settings where resources are limited and speedy results are required (Büscher et al. 2001). For example, some of the earliest methods of magnitizdat arose from discarded X-rays in the Soviet Union in the 1950s: The emulsion on the X-ray was a material that one could engrave as one would a record (Ryback 1990, 32–33). The process of production and distribution was known as *Roentgenizdat* [playing the bones], and playable at 78 rpm on a seven-inch record player.

Hungary's "Goulash Communism" and Czechoslovakia's "Normalization"

Following the de-Stalinization process in both countries during the 1950s, economic reforms were introduced in the 1960s, which allowed for an opening and a liberalization of the market, social rights, and relaxation of creative constriction. János Kádár, who took power following the 1956 Hungarian Revolution, installed liberalizing mechanisms within Hungary's economy—which remained a socialist-planned economy but contained elements of a market system, such as private businesses and enterprises (Kaufman 1997, 31). This new era of mixed economic models was brought about primarily through Hungary's New Economic Mechanism (NEM).

The NEM also sought economic reform through deregulation, which, for example, granted more licenses to artisans and small businesses, thus reducing the state's monopoly. In this atmosphere, there was added room for amateur musicians, and rock music was permitted to reach listeners via the mass media (Szemere 1983, 123). While the NEM was supported within the Politburo, the

Central Committee, and the General Secretariat, it was eventually halted as a result of hardliners' efforts in the party, coupled with the 1968 Warsaw Pact invasion of Czechoslovakia, demonstrating that the economic business of the satellite states—even after de-Stalinization—was still of much interest to Moscow. The goals of the NEM were thus stalled in the '70s but eventually returned in practice in the early 1980s (Adair 2003).

These new reforms were characteristics of the Kádár regime's "Goulash Communism"—a form of communism particular to Hungary. It enabled the country to be one of the most reputable Eastern bloc states in which to live. Ultimately, Goulash Communism contributed to, as Adam Przeworksi (1991, 20) recalls:

> an implicit social pact in which the elites offered the prospect of material welfare in exchange for silence. And the tacit premise of this pact was that communism was no longer a model for a new future but an underdeveloped something else. . . As . . . Hungarian surveys showed, the outcome was a society that was materialistic, atomized, and cynical. It was a society in which people uttered formulae that they did not believe and that they did not expect anyone else to believe. Speech became a ritual.

Similarly, Anna Titkow (1993, 274) describes this condition as "cognitive dissonance," which emerges in the gap between reality and the representations produced in the state media and official propaganda and the gap "between the ideals of socialism as preached by the propaganda apparatus and as practiced by the system itself" (Jakubowicz 1994).

During Kádár's regime the second economy began to swell in Hungary—by 1982, more than 75 percent of the population relied on it to contribute extra income to their formal wage, not just supplementing it but in many cases amounting to a higher income in the second economy (Sampson 1987, 126). The second, or "shadow economy," manifested in a variety of ways, such as peasants on farms selling their produce, families renting out a room in their apartment, bribing a butcher for choice meat, prostitution, and so on (Sampson 1987, 121). In many parts of Eastern Europe and the Soviet Union this second economy provided a "lubricating" function, but this was even more so the case in Hungary, where it was an integral part of the planned economy and helped stem shortages and production bottlenecks (Sampson 1987, 122).

Czechoslovakia, on the other hand, continued its post-1968, Moscow-approved path of social and economic normalization under Gustáv Husák. After consolidating power in 1970,[4] the Husák-dominated socialist state began the first political process of "normalizing" Czech society, which lasted until the late 1980s. As Ulč characterized the situation in the 1970s, "[The] aim [of normalization was] the reinstitution of the *status quo ante* and expiration of the liberalizing heresies of the Prague Spring 1968. Prominent among the measures of normalization has been the introduction of thorough censorship" (Ulč 1978, 26). Although often considered a return to the Stalinist practices of Czechoslovakia during the 1950s, normalization was characterized not by overt coercion but by extrajudicial socioeconomic hardship. For example, if someone's son or daughter was caught by the police distributing banned material, that person might be prohibited by the state from attending a university (Kreidl 2004)—often such nonofficial practices endangered networks of connections between friends and neighbors.

Along with these economic and political changes in both countries came adjustments to cultural policy. In Czechoslovakia, the reconfiguration of the rock music scene was achieved in part by keeping official rock bands in line with the minister of culture's new policies, primarily by regulating hair length, limiting musical genres, instigating a "no English" policy, and editing lyrics (Ryback 1990, 143)—a dramatic turn from the thriving *bigbít* music scene in the latter part of the 1960s. At the center of this creative constriction were required "requalification exams" for musicians. These were taken every two years in order to obtain a license to play professionally or even as an amateur. As state-run institutions, the licensing agencies functioned as a "censorship mechanism." They had the authority to determine which musicians were allowed to perform based not only on exams that tested their knowledge of musical theory but also on their familiarity with Marxism-Leninism, their presentation, the lyrical content of the music, and the length of their hair (Vaniček 1997, 33–37). Saxophonist and guitarist Mikoláš Chadima describes seven points that musicians had to abide by if they wanted to pass the exam:

> First, no English band names! Second, no long hair! Third, no English texts! Fourth, be properly dressed! Fifth, don't play music which is "too wild"! Sixth, learn the rudiments of music theory! Seventh, don't argue with the adjudicators and let them inflate their ego at your expense! (Vaniček 1997, 47)

The agency tested these seven points in musical auditions, oral tests of political theory, and finally a written test of Western music theory (Vaniček 1997, 47–50). While exams to determine a musician's ability to play music still existed in the 1960s, the new requalification exams also established a musician's place in "normalized" Czech society, in that one could not pass the exam without an adequate knowledge of, for example, "the history of the worker's party . . . who the Minister of Culture was . . . or their opinions on communism" (Vaniček 1997, 49). Efforts made by the Husák government from 1970 to 1973 to curb rock's growing interest among the youth population culminated in the implementation of the exams in 1973,[5] which segregated musicians as "official" or "nonofficial" in Czechoslovakia for the remainder of the communist era.

In Hungary the state institution for the National Management of Light Music also instigated practice licenses, defined performance fees, and employed musical proficiency tests (Szemere 1983, 131). However, responses to rock music were based on strategies of a modified form of commercial inclusion rather than on a division between official and nonofficial musicians or explicit repression of the latter. At its most visible, this inclusion took the form of giving bands recording contracts but not allowing them to record some of their most popular songs. Moreover, the release of an album could be delayed for years, long after the popularity of its pieces had waned.[6] The second economy, however, took care of this bottleneck through bootlegged, copied, and exchanged audio cassette tapes.

Regardless of the Hungarian regime's restrictions, the situation was still enticing to some since being a rock musician in Hungary during the '70s and '80s permitted musicians to receive royalties and tour extensively not only throughout the Soviet bloc and the Soviet Union itself but also in other parts of Europe and the United States if sponsored by the regime. In short, Hungarian musicians still had a degree of freedom relative to their Czech counterparts. The containment was less strictly enforced, and officially there was no censor.[7] As János Kobor of the official rock star group Omega attested in 1987, "There is no strong censor, but they are, you know, careful. When we made our last album, the record company asked us about the title. They said, this 'Dark Side of Earth,' does it refer to Hungary or does it refer to the socialist world? And then after we said no, then it's all right, no problem."[8]

The Hungarian form of control by inclusion also involved a method of tactical overexposure of musicians, which diluted their potential threat. One of Hungary's first punk groups, Beatrice, was subsumed into the official music scene after a series of meetings with representatives of the official culture. This led to a TV spot to discuss punk music, competing in a song contest sponsored by Hungarian radio, and being offered a support-band role in a tour with the superstar groups Omega and Locomotive GT. Seen by fans as "selling out" and leaving their subversive message behind by playing with *the* establishment bands of Hungary, Beatrice was not helped when the state released a live concert album of the bands (Ryback 1990, 173–74). For the government it was doubly effective: overexposing a subversive band while also improving record purchases by the youth, illustrating the regime's inclination to institutionalize commercial success.

Within Goulash Communism, Hungarian official cultural life had a loose categorization of bands called "the three Ts" (*tűrt, támogatott, tiltott* [promote, permit, prohibit]): Certain bands, such as Omega, were promoted, while others were permitted to perform but not to record, and some were not allowed to play under any circumstances.

For both regimes, a key problem was not only how to address restrictions on musical practices but also how to assert policy and control in relation to sonic and technological areas. How, for example, were they to protect radio frequencies from unwanted broadcasts, and should they monitor and assess the ideological content of cultural products? These discussions over policy ebbed from the 1960s to the 1980s; however, restrictions remained in place in both countries until 1989. Moreover, technological innovation—particularly gadgets, sound, and video—themselves raised ideological questions across the bloc. Since such devices were invented and produced in the West, how could a regime import such objects without destabilizing the foundations of the communist system (Kusin 1987)? Each regime met these issues in different ways, which in turn affected access to the latest technological and cultural products such as albums, films, literature, and clothes. The regimes thus became active participants in how music was experienced by their direct involvement with and control over sonic-related matters.

Nonrecordable Sound Technology: Radio and LP Listening

Listening to the radio had always been legal in both Hungary and Czechoslovakia. In fact, it was often either encouraged or compulsory in situations such as collective listening to state propaganda in the workplace or barracks or the more passive listening found in public areas that had loudspeakers (Rév 2004). However, certain radio stations—Radio Free Europe (RFE), Voice of America, BBC—were banned, and listening to such prohibited radio content could lead to punitive measures. Nonetheless, radio, which was ubiquitous, inexpensive, and manufactured in the bloc, provided a readily available entrée into nonofficial musical experiences and did not require the listener to acquire any illegal or prohibited technology; to dip into this nonofficial practice, an individual only had to tune in at certain times to specific radio frequencies. Moments of "dipping in" were made possible simply by listening.

While this mode of listening could have been for either informative, news-related purposes or for musical enjoyment (often the two were wrapped up into one program), this activity was risky and possibly dangerous. For example, if listening in a block of apartments, neighbors could overhear the unusually unsually loud, hourly station update: "This is Radio Free Europe on the 16th, 19th, 25th, 31st, 41st, and 49th shortwave bands" (Rév 2004, 5).[9] In this sense, the radio sound—both volume and content—afforded peripheral modes of listening: overhearing and eavesdropping. Not to be confused with "bugging" or state surveillance methods, overhearing and eavesdropping emerged from the sociopolitical circumstances of living situations such as the overcrowding in urban areas and simply the dispositions of curious people. These practices were part of what can be described as the necessary "skill set" in a second economy and also included skills Grossman calls the "4 Bs of resource procurement, *bribery, bartering, black marketeering,* and '*blat,*' " a Russian term for "connections or influence" (Grossman, quoted in Sampson 1987, 128).

While we do not mean to assert that individuals were inclined to eavesdrop in these situations for malicious purposes, in some contexts information about the tenant living upstairs in a larger apartment with a balcony could be used advantageously if one wanted. On the other hand, a friendly neighbor who overheard someone listening to Western broadcasts was in the precarious situation of possessing unwanted and potentially harmful information. Underpinned by sound, these modes of listening in private spaces—overhearing and eavesdropping—heightened people's fear and at times placed them in danger.

Moreover, while listening to the radio in private, one was also subject to the noise of radio jamming from centralized locations in Budapest. Listeners in Czechoslovakia referred to Soviet jamming as "Stalin's bagpipes." As Rév (2004) argues, the noise let the listeners know that even in the confines of their own homes, the regime was still able to censor, control—and at the very least—be present in the lives of citizens (Bijsterveld 2008). Fear and anxiety could to some extent be alleviated by well-timed listening. It was possible to listen during the early hours of the day or late at night and thus avoid the jamming. Quoting K. R. M. Short (1986, 6), Rév (2004, 25) illustrates the temporal strategy of listening:

> The timing of the broadcasts is also important because the twilight hours of morning and evening are the most ineffective of Soviet-originated sky-wave jamming. This is because the western broadcasts can take advantage of the ionosphere's "solid" condition at these times, while the eastern jamming broadcasts have difficulty in achieving a reasonable reflection in their "broken" section of the ionosphere. This creates a time-related gap in the Soviet defenses.

The strategic listeners were thus able temporally to configure their sonic space and practices in order to hear more clearly and to avoid the omnipresent regime's jamming noise.

One very popular RFE program broadcast into Hungarian-speaking regions was "Teenager Party," which played Western rock to listeners and accepted postcards and letters requesting songs from the Beatles, Jimi Hendrix, the Bee Gees, Paul Simon, and other popular Anglo-Saxon bands.

These radio programs expressly provided an opportunity for informal musical learning through imitation. Nagy Feró, leader of the Hungarian punk group Beatrice, stated that budding musicians turned on the radio and learned music by "listen[ing] to 'Teenager Party' and copying the songs phonetically."[10] In this way, new informal practices were acquired through musical experience—in this case, imitation and modeling. Similarly, Canadian musician Paul Wilson, who lived and performed in the Czech Underground during the late '60s and '70s, describes radio listening in Czechoslovakia during the 1960s:

> One of the things that censorship [before the Prague Spring] did very badly was keep music out of the country. One of the things that was very marked in the 1960s was that although intellectuals found it very hard to get a hold of books it was very easy for kids to be right on top of things because records were brought in and the music was broadcast over Voice of America and other radio stations. So, there was a very current music scene [in Prague], with a lot of knock-off bands and a lot of fans of different groups just the way you'd find them in the West. (Velinger 2005)

Censorship efforts by both regimes to restrict radio listening had unintended results: They fostered a culture of exposure to new music. Each week new sounds came over the ether, and musicians developed listening strategies that in turn provided informal training. They imitated the music they heard and formed "knock-off bands." Thus, music was passed from one sound technology to another and from person to person through various practices of distribution and musical "information." These practices ultimately had implications for what actors could do and for how musical experience came to be linked in collective efforts.

Because listening privately was a common practice, it is clear that there were many actors participating *collectively yet separately* in listening; in other words, while the sonic experience was happening "in the room," it may not have fostered microsocial interaction but rather a felt, collective experience. Radio broadcasts from abroad, as well as this collective experience and interest, led both Czechoslovakia and Hungary to produce rock radio programming in the 1960s; the regime itself imitated Western forms of cultural technology. This start-up of rock radio programming as a result of popular demand reflected the growing "confusion" produced in these communist regimes by rock music and its growing significance for young people (Ryback 1990, 88). While the state frequencies expanded rock airplay in the '60s, it was inevitably halted in Czechoslovakia after the Soviet-led, five-army Warsaw pact invasion in 1968—radio programming in Czechoslovakia returned to brass-band broadcasts shortly there after.

Musical "Cues" for the Room

In addition to the collective-yet-separate radio listening in apartments and houses, people could move easily between collective listening practices afforded by LPs or recordable technology such as open-reel tape recorders, which were not transmitted through the ether. One member of the Czech Underground (František Stárek) described his radio/LP-listening practices in the late 1970s as follows:[11]

FS: While living on the commune, we used to finish meals together and then put on *Voice of America*.
TH: Where did you listen to the program?
FS: In the kitchen. It was the only room big enough for all of us to fit. We had about ten to twelve people during the week and many more [Undergrounders] at the weekend.
TH: How did you listen?
FS: Well, everybody was silent. Even the kids who normally ran around knew that they should be quiet at this time. We wanted to hear about what was going on in our country.
TH: And did you listen to music programs as well?
FS: Not really. After[ward] we'd listen to Underground [bootlegs] and drink.

The listening experience in this Underground commune illustrates the more general, fluid, and multiple listening tendencies common at the time. As listeners moved from one sound technology to another—from radio to open-reel bootleg concert recordings—the listening practices changed slightly. Listening together in a room that afforded microsocial interaction was not always possible. In such cases, technology created sound for the room: playback for private gatherings and parties in apartments, bedrooms, and weekend homes, where music provided both the backdrop for, and cues to, social interaction. Listeners, technology, and the sociopolitical context repeatedly coproduced the culture of the space—illegal, alternative, and filled with anxiety and excitement.

While the radio offered variation and the possibility of experiencing new music, this new pattern of listening behavior contrasted with LP culture in Czechoslovakia, where large collections were rare, and the same albums were often played repeatedly in domestic environments. The selection of Western LPs was quite limited in Czechoslovakia, and listeners who wanted to acquire LPs turned to the weekly black markets in Prague or asked relatives and friends abroad to send albums.

The Hungarian Record Company (HRC) had quality record-pressing technology in the 1960s and 1970s and doubled its output of LPs from 1975 to 1980 in an attempt to meet the needs of younger audiences (Szemere 1983, 130). This trend toward pop music stemmed in part from Hungary's policy of moving away from isolationism or Sovietization in the 1950s to a "windows to the West" policy in the 1970s. As Cultural Secretary Aczel addressed delegates of the Society for Dissemination of Scientific Knowledge in 1968:[12]

> There are people who demand the closing of our borders to cultural and intellectual exchange; they want to isolate our culture. It would be impossible to implement such a policy, but even if we were to succeed in doing so, we should create a situation, which would leave us ignorant and weak; a robust plant cannot be grown in a hothouse.

The culture secretary's stated position of "peaceful coexistence" with the West helped to widen the gap of rhetoric and practice in Hungary: Although there was an abundance of Hungarian LPs, foreign albums were still highly restricted, and one had to rely on methods similar to those in Czechoslovakia: obtaining LPs from relatives in the West, black markets, or by chance finding a secondhand Yugoslav-pressed album in a store.

Modes of listening are thus tied up with subsequent access to technology and musical recordings—how one would procure an object and where it was listened to all contributed to the shaping of the musical experience. According to Ruth Finnegan (2003, 183), listening as musical experience involves learning "how to feel, how to deploy particular emotions in contexts appropriate to our situation." Moreover, if emotional structures can be the seeds of knowledge production (Witkin and DeNora 1997, 5), it is possible to see musical experience molding a flexible, liminal space—a space where one can view and imagine both the nonofficial and the official lifeworlds (Eyerman 2006). Musical experience is thus at once lived, embodied, and "intertwined with culturally diverse epistemologies" (Finnegan 2003, 183) from a range of cultural resources mediated by its interaction with sound technology.

Reconfiguration and Repackaging of Music and the Actor

The listening pattern produced by playing LP collections as opposed to listening to diversified radio programming provided a further entrance point into the liminal space by learning how to hear music: Discussing the music in the physical space of listening was one way in which people were able to reconfigure themselves in an alternative manner.

In the late 1960s in Czechoslovakia, a handful of musicians who rejected the newly instigated official music standardization moved to private gatherings to listen to music, talk about music, and share the small collections of music they possessed. Here, in these private gatherings while listening to LPs, they would also experience running monologues and informal lectures on aspects of Czech cultural history, orated primarily by Ivan Jirous, art historian and band manager of the Czech Underground group, The Plastic People of the Universe.

The lectures discussed and explained current Czech rock music within the context of Czech musical revivals, art history, and political history and maintained that "even in the darkest of times, the Czechs had always been able to keep the flame of culture alive" (Jirous 2006 [1975], 7). The notion of a nonofficial lifeworld here, understood not in relation to "the Other" or based on difference, was rather the opposite: that rock music was not just a new phenomenon; instead, it was part of an experiential mode of culture that had long existed in the Czech lands—an alternative mode of being present throughout the country's history, to which the Czech Underground belonged.

These lectures and ritual listening sessions, consisting of copious amounts of beer and dumplings, were described by Paul Wilson during some of his first meetings with the Czech Underground in an apartment in central Prague:

> [Jirous] would put on his favorite records on a battered turntable jacked into an old WWII radio . . . I lay back and listened to the Velvet Underground, Captain Beefheart, the Doors and the Fugs, and as I listened, I began to feel a depth in the music I hadn't felt before, as though I were hearing it for the first time with Czech ears. (Wilson 2006 [1983], 20)

Here we begin to see how these sounds came to be empowered as exemplary of Underground life: They emerge from the intersection of sounds, sound technologies, and social rituals of listening to help construct a lifeworld; the musical experience clearly afforded Wilson a different form of comprehension or depth of listening. These rituals helped form nonofficial dispositions that came to structure the flexible medium of the Underground—a cultural space and mode of living that were particular to this community in Czechoslovakia. In this case, the musical sound was a flexible object that came to be symbolic of an alternative way of life as it was combined with other practices of specifically Czech consumption (beer, dumplings, a Czech living space).

To speak of the music's flexible affordances is to highlight how the physical space for listening provided, in part, contextualization cues. DeNora (1986, 91) describes these contextualization cues as "various conventions or ritual practices that, through experience, come to carry certain connotations which serve as the tools for the work of sense-making and meaning construction." In other words, cues help listeners to shape their interactions and appropriations of musical objects and, in the case of Czech appropriation, as they immerse themselves, to varying degrees, in transformation. In the context of Czech listening rituals, the extramusical contextualization accomplished by "running monologues and informal lectures" achieved two goals: a reconfiguration of the musical content and a reconfiguration of the actor's mode of listening and attention from "non-Czech" to "Czech." For Wilson, reconfigured as someone with "Czech ears," the music then became a soundtrack for the placement of alternative modes of being within Czech history.

Reconfiguring the space through active listening during radio and LP playback and collective listening connected actors to networks of feeling, being, and thinking and thus enabled actors to distance themselves from official society and to dip more than a toe into nonofficial culture. How far individuals dipped in and how long they stayed immersed was, of course, dependent on a web of other social and familial ties.

Recordable Sound Technology: Cassette Tapes and Open Reel

The dispositions formed through radio and LP listening served to develop practices related and translated to other sound technologies, such as the rerecordable technology of open-reel and audio cassette tapes. These sets of practices surrounding the exchange of bootlegged concerts, compilations, and LP copies became extensions of older sound technologies such as radio. These practices allowed for modes of being to shift in accordance with technological change: Cassette tape technology permitted exchange to develop into a mode of communication in its own right. Simultaneously, the new technology and the tape trading that it facilitated during the late '70s and '80s in turn enabled a wider culture of alternative sound as listeners forged new (and highly nonofficial) practices of musical exchange and recording. Tapes could now be acquired by ordering through *samizdat* magazines by contacting tape traders through postings in public places, by recording concerts, and by exchanging with friends and acquaintances. As tapes came to be associated with the practice of tape archiving/collecting, the alternative articulations between music and collective action burgeoned. The once collective-yet-separate radio listeners were thus linked via active engagement and practices to one-to-one or many-to-many interactions involved in exchange, thus generating still more network links among the actors.

This burgeoning informal exchange system—within which distribution became a cultural end in itself—in turn augmented the contextualization cues previously associated with radio and LP listening. Overall, these processes removed music from its original associations (for instance, how it was framed by artists) and from the conventional notion of creation/composition as a distinct phase, still further up the chain in arts production (Becker and Pessin 2006). Instead, tape distribution recontextualized musical works by delivering them to consumer groups that their authors were not originally intending to reach. Thus, the music distributed via tape in Czechoslovakia and Hungary during this time derived its meaning and social power from its repackaging and its novel and adaptive forms of distribution. Cassette tapes, as well as the music they contained, heightened and expanded one's personal network: Oftentimes someone owned a tape that had been received through distant connections and copied several times; the audio quality of cassettes copied many times over left a sonically anonymous trail of the network.

In other words, the music was recontextualized by its modes of distribution, which brought into being new systems of collective representations as a result of tape exchange. The sound technology of cassette tapes and open-reel tape recording thus extended the simple act of listening (e.g., to a radio broadcast or to LPs exchanged on the black market) to acts of recording, compiling, and bootlegging. New associations, images, and ideas could, in other words, be hung on tones (DeNora 1986, 93), and here the associated practice of tinkering and the participatory design feature of tape trading/bootlegging come to the fore. Members of the Hungarian punk band CPG were one such group that was persecuted for its participatory design of a compilation tape of music and RFE commentary. The trial of the "punks of Pol Pot County" took place in 1982 in the Hungarian county/town of Szeged. The members of CPG were first charged with incitement for the cassette tape they were carrying. According to Szőnyei (2005, 88), sandwiched in between songs on the tape was political commentary on the martial law in Poland, recorded from RFE. Although the musicians were tried and acquitted for disturbance of the community, the police were unsettled not only by CPG and its actions but also by the possibilities of playback in the room and dissemination to a wider audience. The broadcasts, torn from their RFE *dispositif*, were crafted and adjusted to fit CPG's own dispositions. In other words, cassette tapes, combined with radio, offered a type of *radizdat* that yielded an assertion of user control over the sonic environment.[13]

Various practices associated with exchange were in part facilitated by the regimes in each country. In Hungary in 1981, economic reforms that took the NEM into consideration were finally starting to gain momentum. The changes taking place allowed for small businesses, such as record stores, to appear and function within Hungary's liberalized/ing market. For instance, from 1970 to 1983 the number of private shops in Hungary doubled—to 19,293 (26 percent of all shops) (Sampson 1987, 125).

In part because Hungary had a more open market since the mid-1960s, the practice of taping was somewhat different from that in Czechoslovakia. Artist György Galántai was an avid taper who used a Sony tape recorder purchased through relatives in Vienna. Galántai, the founder of the alternative art archive Artpool recorded concerts for posterity and exchange abroad while receiving demo tapes from Hungarian bands.[14] In the early '80s, Galántai began to produce a series of cassette tapes to exchange as mail art, which were described as "cassette-radio,

radio work" called "Radio Artpool." The cassettes were a bricolage of interviews, music, ambient recordings, documentation of telecommunications concerts, improvisations, sound art, found sounds, spoken words, and so on. The eight-part series merged sound, posterity, distribution, and exchange into one object as an audio version of a cultural journal, thus indicating a mode of listening attention that was active, critical, and expressly disseminated via the postal system.

Sound technology and distribution in Czechoslovakia took on a different role. With a stricter regime line on popular music, there was less foreign music on the official market as there was in Hungary. Officially, there were few places to buy Western LPs, although a Czech label did release some American jazz imprints, and select albums could be purchased or ordered through the record clubs, Gramofonový Klub and HiFi Klub (Vaníček 1997, 121). Moreover, cultural centers in other countries served as places to pick up foreign music. At the Hungarian Cultural Center in Prague, for example, it was possible to listen to or lend albums by Omega and other Hungarian rock giants.

In this rather large gap between official and nonofficial musical acquisition, Czech *magnitizdat* labels and distributors came into existence. One such label, S.C.T.V (samizdat cassette tapes [and] videos) was primarily run by Petr Cibulka from his apartment in the Moravian capital of Brno, where his mother took care of the administrative side of the label (Vaníček 1997). While many people in the country were still using open-reel recordings, Cibulka in 1976 was credited with producing one of the first *samizdat* compilation tapes, which comprised many bands of the Czech Underground, who at that time had been imprisoned and set the stage for the launch of Charta 77.[15] Similar to the ways that Hungarian tapes were constituted, this first Czech compilation tape instigated the use of manipulatable, rerecordable technology as participant design and signaled the formidable quality that cassette tapes afforded as a do-it-yourself (DIY) set of practices.

A similar independent initiative was started by Mikoláš Chadima, who ran a small label, Fist Records, from his apartment.[16] Fist Records' approach differed slightly from that of S.C.T.V. in that Cibulka held that *magnitizdat* and *samizdat* should be spread widely and in quantity, echoing dissemination as an alternative mode of being. Chadima similarly used distribution as a creative mode of being; however, he focused on releases that met his idea of quality music and created cover art and liner notes for many of his releases (Vaníček 1997). Although a majority of the taping and copying was done by these two individuals, the trading and exchange operations were carried out between friends, siblings, and acquaintances and thus involved a collective of people dipping into and reconfiguring a musical experience linked not only to listening but also to *magnitizdat* creation and distribution. With tape exchange, the informal learning shifted the contextualization cues from the physical space of the room, as described earlier, to the social ritual of distribution. Actors, in turn, became reconfigured by "dipping in" while simultaneously reconfiguring the sound technology for their own use and purposes.

For example, Chadima's Fist Records often did not have sufficient personnel to complete all of the liner notes and every piece of cover art; thus, graphic designers or individuals who could help copy the art and liner notes would offer their time to Chadima, thus "dipping into" alternative practices a little bit more as tape distribution opened up new avenues.

Liner notes could sometimes be found in *samizdat* magazines such as Vokno, which was printed and distributed by the aforementioned Underground commune in Czechoslovakia. Translated roughly as "memory loss" and also a play on the word "window," Vokno's samizdat distribution web spread across the republic from 1979 to 1989 and provided, as a part of the tape exchange, the contextualization cues for listening that the physical space afforded. In addition to liner notes, Vokno often contained artwork and details of recordings, not to mention order forms and inventory for S.C.T.V and Fist Records. Those who read *samizdat* or listened to *magnitizdat* were thus drawn further into a set of consumption practices that linked listening, technology, and contextual cues provided by magazines such as Vokno and the practice of exchange.

If the creative constriction of performing, recording, and distributing music in each country helped to set up exchange and taping as novel practices, they also led to novel forms of musically mediated learning. More specifically, they paved the way for the individual and collective learning of new dispositions associated with how to live/be nonofficially. Indeed, they provided object lessons in how to communicate nonofficially.

Conclusion: The Liminality of Nonofficial Musical Experience

We have described how sets of nonofficial musicking practices took shape in relation to and in turn shaped sound technologies.[17] This mutual structuring process also gave rise to situations within which individuals could test alternative dispositions and underground culture. Within these situations, the musical object was simultaneously an aesthetic item *and* its distributive technologies. Opportunities for musical experience in Czechoslovakia and Hungary provided a liminal space, one poised between official society and nonofficial life, affording fine degrees of commitment and deliberation. Because of the heterogeneity of this space, which was neither official nor nonofficial, it was possible to gradually move into deeper, nonofficial waters and to learn about and adopt new, alternative dispositions in informal and not necessarily conscious ways. If we were to theorize about music's role in relation to collective action and collective consciousness (and so to build upon earlier work on this theme in ways that seek to delineate the actual mechanisms through which music provides "exemplars" for action and collective mobilization [Eyerman and Jamison 1998]), we would suggest that the musical experience, importantly, provides a two-way flow between individual and collective world building: (1) an aggregating conduit to the development of dispositions (learning how to) and (2) a collective distillation of dispositions (from many experiences to one collective experience). This definition of musical experience as liminal space enables individuals to take either the first steps into nonofficial life or to reaffirm and constitute their alternative disposition through control—or choice—over the sonic environment. In what we have described, the musical experience has two parts: a sensory aspect (as in listening) and an active facet (as in participating in distribution as creation).

The social changes in these nations in the late 1980s—breakdowns in different parts of government and dissident political initiatives by opposition groups—rested, we suggest, on a technologically dispersed, linked population of actors who had, at one point or another and to varying degrees, dipped into and informally acquired alternative modes of being. This "dipping in" (from toe to fully immersed) was, we suggest, a critical resource for the reconfiguration of dispositions and the increasing rejection of the creative constriction imposed by official institutions.

NOTES

1 The research involved ethnographic interviews of more than twenty individuals in both the Czech Republic and Hungary conducted 2007 and 2009. These interviews focused on musical experience related to sound technology. Interviewees were people involved in the practice of making and distributing samizdat and magnitizdat (see later explanation), as well as consumers of the material from the late 1960s to 1989. Additionally, archival resources were gathered from the Open Society Archives and Artpool in Budapest, Libri Prohibiti, Institute for Contemporary History (Ústav pro soudobé dějiny), and the Archive of Security Forces (Archiv bezpečnostních složek) in Prague.Grounded theory methods were employed for analysis of the interviews and archival data (Glaser and Strauss 1967).

2 Throughout, the term *nonofficial* is used not only to describe underground or alternative movements in each country but also to discuss the actions of any individual (as opposed to an organization or an office)—from a dissident to a merchant to a teenager.

3 The "West" is described by one Czech listener as primarily the UK and the United States.

4 Husák was both the president of the country and secretary general of the Communist Party.

5 For example, Supraphon, the national recording label, halted or canceled all music projects that contained what were considered to be too extreme Anglo-American themes. Instead, the company played and recorded brass band music.

6 Interview with Tamás Szőnyei by Trever Hagen, Budapest, Hungary, July 5, 2009.

7 Ibid.

8 Charles T. Powers, "Changing Times for Hungary's Pop Music Giant Omega," *Los Angeles Times* (Sept. 5, 1987), 1. Retrieved in HU-300-40-2 Records of Radio Free Europe/ Radio Liberty Research Institute, Hungarian Unit, Subject files.

9 As Rév (2004) notes, exiles from Hungary, who were surveyed by Radio Free Europe, consistently remarked on the volume level of the hourly updates in comparison to the volume level of the programs.

10 Agnes Sesztak, " 'Punk' Star Interviewed, Describes Trials, Tribulations," *Mozgo Vilag* 12(86–91) (1984), 48. Retrieved in HU-300-40-2 Records of Radio Free Europe/Radio Liberty Research Institute, Hungarian Unit, Subject files.

11 Interview with František Stárek by Trever Hagen, Prague, Czech Republic, Apr. 21, 2009.

12 Radio Free Europe research report, "Hungary's Windows to the West," Oct. 28, 1970; http://www.osaarchivum.org/files/holdings/300/8/3/text/34-6-85.shtml.

13 *Radizdat* means dissemination by radio, transmitting banned sonic material back to the country of its production. It also refers to the transmission of Western music and news into the bloc.

14 Interview with Julia Klaniczay by Trever Hagen, Budapest, Hungary, July 1, 2009.

15 Interview with Miloš Mueller by Trever Hagen, Prague, Czech Republic, February 3, 2007, and interview with František Stárek by Trever Hagen, Prague, Czech Republic, Apr. 21, 2009.

16 Interview with Mikoláš Chadima by Trever Hagen, Prague, Czech Republic, Jan. 28, 2009.

17 The musical experience we have described has covered tuning in to the radio and listening to vinyl or illegally recorded material; it should be noted that jam sessions, rehearsals, and live performances would also need to be included in a broader definition but are not discussed here.

REFERENCES

Adair, Bianca L. "Interest Articulation in Communist Regimes: The New Economic Mechanism in Hungary 1962–1980." *East European Quarterly* 37(1) (2003): 101–26.

Becker, Howard, and Alain Pessin. "A Dialogue on the Ideas of 'World' and 'Field.' " *Sociological Forum* 21 (2006): 275–86.

Bijsterveld, Karin. *Mechanical Sound: Technology, Culture, and Public Problems of Noise in the Twentieth Century*. Cambridge, Mass.: MIT Press, 2008.

Büscher, M., S. Gill, P. Morgensen, and D. Shapiro. "Landscapes of Practice: Bricolage as a Method for Situated Design." *Computer Supported Cooperative Work* 10(1) (2001): 1–28.

DeNora, Tia. "How Is Extra-Musical Meaning Possible? Music as a Place and Space for 'Work.' " *Sociological Theory* 4(1) (1986): 84–94.

Eyerman, Ron. "Toward a Meaningful Sociology of the Arts." In *Myth, Meaning, and Performance: Toward a New Cultural Sociology of the Arts*, ed. Ron Eyerman and Lisa McCormick, 13–35. Boulder: Paradigm, 2006.

———, and A. Jamison. *Music and Social Movements*. New York: Cambridge University Press, 1998.

Finnegan, Ruth. "Music, Experience, and the Anthropology of Emotion." In *The Cultural Study of Music: A Critical Introduction*, ed. Martin Clayton, Trevor Herbert, and Richard Middleton, 181–92. New York: Routledge, 2003.

Glaser, Barney G., and Anselm L. Strauss. 1967. *The Discovery of Ground Theory*. Chicago: Aldine.

Jakubowicz, Karol. "Equality for the Downtrodden, Freedom for the Free: Changing Perspectives on Social Communication in Central and Eastern Europe." *Media, Culture, and Society* 16(2) (1994): 271–92.

Jirous, Ivan. "Report on the Third Czech Musical Revival." In *Views from the Inside: Czech Underground Literature and Culture (1948–1989)*, ed. Martin Machovec, 7–32. Prague: Charles University Press, 2006 [1975].

Kaufman, Cathy. "Educational Decentralization in Communist and Post-Communist Hungary." *International Review of Education* 43(1) (1997): 25–41.

Kreidl, Martin. "Politics and Secondary School Tracking in Socialist Czechoslovakia, 1948–1989." *European Sociological Review* 20(2) (2004): 123–39.

Kusin, Vladimir. "Introduction: The Video Revolution in Eastern Europe." *Radio Free Europe Background Reports* (Dec. 17, 1987); http://fa.osaarchivum.org/background-reports?col=8&id=35875 (accessed Oct. 13, 2010).

Machovec, Martin. "The Types and Function of Samizdat Publications in Czechoslovakia, 1948–1989." *Poetics Today* 30(1) (2009): 1–26.

Pavlicevic, Mercédès, and Gary Ansdell, eds. *Community Music Therapy*. London: Kingsley, 2004.

Przeworski, Adam. "The 'East' Becomes the 'South'? The 'Autumn of the People' and the Future of Eastern Europe." *PS: Political Science and Politics* 24(1) (1991): 20–24.

Rév, István. "Just Noise?" In *Cold War Broadcasting Impact Conference*. 2004. http://www.osaarchivum.org/files/2004/justnoise/just_noise.html (accessed Oct. 13, 2010).

Ryback, Timothy. *Rock around the Bloc: A History of Rock Music in Eastern Europe and the Soviet Union*. New York: Oxford University Press, 1990.

Sampson, Steven L. "The Second Economy of the Soviet Union and Eastern Europe." *Annals of the American Academy of Political and Social Science: The Informal Economy* 493 (1987): 120–36.

Short, K. R. M. *The Real Masters of the Black Heavens*. (Western Broadcasts over the Iron Curtain. London: Croom Helm, 1986), 6; quoted in István Rév, "Just Noise?" 2004. http://www.osaarchivum.org/files/2004/justnoise/just_noise.html (accessed Oct. 13, 2010).

Small, Christopher. *Musicking: The Meaning of Listening and Performing*. Middletown, Conn.: Wesleyan University Press, 1998.

Smith, Gerald Stanton. *Songs to Seven Strings: Russian Guitar Poetry and Soviet Mass Song*. Bloomington: Indiana University Press, 1984.

Szemere, Anna. "Some Institutional Aspects of Pop and Rock in Hungary." *Popular Music: Producers and Market* 3 (1983): 121–42.

Szőnyei, Tamás. *Nyilvántarottak: Titkos szolgák a magyar rock körül 1960–1990*. Budapest: Magyar Narancs, Tihany Rév Kiadó, 2005.

Titkow, Anna. "Political Change in Poland: Cause, Modifier, or Barrier to Gender Equality." In *Gender, Politics, and Post-Communism: Reflections from Eastern Europe and the Former Soviet Union*, ed. Nanette Funk and Magda Mueller, 253–56. New York: Routledge, 1993.

Ulč, Otto. "Social Deviance in Czechoslovakia." In *Social Deviance in Eastern Europe*, ed. Ivan Volgyes, 23–35. Boulder: Westview, 1978.

Vaníček, Anna. Passion Play: Underground Rock Music in Czechoslovakia, 1968–1989. Master's thesis, York University, North York, Ontario, Canada, 1997.

Velinger, Jan. "The Impact of the Plastic People on a Communist Universe: An Interview with Paul Wilson." In *Radio Praha* 31 (May 2005). http://www.radio.cz/en/section/one-on-one/paul-wilson-the-impact-of-the-plastic-people-on-a-communist-universe (accessed Oct. 13, 2010).

Wilson, Paul. "What's It like Making Rock 'n' Roll in a Police State? The Same as Anywhere Else, Only Harder. Much harder." In *Views from the Inside: Czech Underground Literature and Culture (1948–1989)*, ed. Martin Machovec, 33–48. Prague: Charles University Press, 2006 [1983].

Witkin, Robert, and Tia DeNora. "Aesthetic Materials and Aesthetic Agency." *News Letter of the Sociology of Culture Section of the American Sociological Association* 12(1) (Fall 1997): 1–6.

CHAPTER 19

THE AMATEUR IN THE AGE OF MECHANICAL MUSIC

MARK KATZ

INTRODUCTION

IN 1906 the American composer and bandleader John Philip Sousa famously predicted that the rise of mechanical music—in particular, the player piano and the phonograph—would mean the end of the amateur musician. "[I]n this the twentieth century," he wrote, "come these talking and playing machines, and offer . . . to reduce the expression of music to a mathematical system of megaphones, wheels, cogs, disks, cylinders, and all manner of revolving things. . . . It will be simply a question of time," he concluded, "when the amateur disappears entirely" (Sousa 1906, 279, 280). Sousa was the most prominent and influential figure to express concern about the deleterious effect of technology on musical performance, but he was neither the first nor the last. As early as 1878, when Thomas Edison's phonograph was barely a year old, a New York journalist suggested that "with the perfecting of the phonograph to the degree Edison's industry and skill give promise, the occupation of the skilled performer will be gone" (quoted in Feaster 2007, 136). Unlike Sousa, this writer eulogized the professional musician, and in subsequent years opponents of mechanical music sometimes foretold the death of one, the other, or both. Regardless, the predictions were uniformly dire as, for example, music historian Leonid Sabaneev's 1928 prophecy that "Orchestras will perish and disappear, like the ichthyosauri" (Sabaneev 1928, 114). Some even offered specific expiration dates: American critic Winthrop Parkhurst believed it would be the year 2000, while British librarian Lionel McColvin saw it happening no later than 2031 (Parkhurst 1931, 6; McColvin 1931, 322).

Amateur music making was seen as particularly vulnerable, and it was said that few would want to expend the effort to make music for themselves when high-quality performances could be enjoyed effortlessly at home. Despite the warnings, the amateur musician never went the way of the ichthyosaur. To the contrary, from Sousa's day to the present, amateur music making not only survived but also thrived. Moreover, musical amateurism persisted, not despite the presence of the player piano, phonograph, and their descendents; in good part it flourished *in response* to the possibilities of these technologies.

This chapter explores the role of amateur music in the age of sound recording and reproduction technologies. It begins by returning to Sousa's complaint in order to evaluate concerns about the fate of the amateur in the early twentieth century. Four brief case studies then examine the complex relationship between amateurism and music technologies. The first remains in early twentieth-century America to consider the unexpected uses of the player piano and the phonograph. The next two case studies focus on musical phenomena that arose in the 1970s but remain hugely influential decades later—hip-hop and karaoke. Situated in the early twenty-first century, the final case study examines how digital music technologies—in the form of video games and mobile phone applications—challenge traditional notions of musicianship and amateurism. Most of the examples I cite come from the United States, and my claims hold most strongly for American musical life; I do, however, draw evidence from Europe and Asia as well in order to suggest the global scope of technologically mediated amateur music making.

In this chapter I take a strongly user-centered perspective. Although I grant that both design and marketing influence human engagement with technology, the case studies reveal a constant process of co-construction between users and technologies and deny any kind of determinism strong enough to predict with great specificity how a given technology may be deployed. In fact, my own refrain will be that users have exerted, much more than is typically acknowledged, a decisive influence on the development of sound technologies since the late nineteenth century. In taking this stance I am sympathetic with SCOT (the social construction of technology), given its focus on technology in its social contexts, as well as on the fields of cultural studies and media studies, which explore how users perform and construct identity through technology (e.g., Oudshoorn and Pinch 2003).

The Menace of Mechanical Music?

In the early twentieth century amateur music making existed in many forms. Amateurs often performed publicly, whether in municipal brass bands, choral societies, or community orchestras and opera troupes; they could be heard in schools, at sporting events, in parks, on the streets, and in houses of worship. In private, amateurs gave home recitals for neighbors, gathered around the parlor piano to sing popular songs, or played chamber music.

The extent of early twentieth-century amateur activity is hard to quantify with any precision, but it is possible to gauge its growth or decline indirectly by tracking the numbers of music teachers, which, at least in the United States, formed a census category. Music teachers are indeed professionals, but their clientele consists primarily of amateurs—most often children but also adults seeking to develop their musical skills. (Professional musicians tend not to take regular instruction.) Therefore, a rise or decline in the number of music teachers suggests a corresponding change in the number of amateur musicians. We can thus evaluate Sousa's claim that, because of player pianos and phonographs, "it will be simply a question of time when the amateur disappears entirely, and with him a host of vocal and instrumental teachers, who will be without field or calling" (Sousa 1906, 280). Contrary to Sousa's assertion, the years immediately following the publication of his article saw an increase in the number of music teachers in the United States. If we take the period between 1890 and 1910—one that included the introduction of commercial records and the launch of the extremely popular phonograph model, the Victrola—we see a significant increase in the number of music teachers and professional musicians. (Unfortunately for the present purposes, these two groups were counted as one category.) As table 19.1 (adapted from Harris 1915, 301) shows, moderate growth occurred between 1890 and 1900, but then a considerable jump took place between 1900 and 1910, the period in which Sousa was writing. In twenty years the number of musicians and music teachers per capita increased by 50 percent. Clearly, neither the professional musician nor the amateur suffered during that time. The same was true of later decades as well. A 1967 report claimed that the total number of amateur musicians in the United States grew by 141 percent between 1930 and 1960 (from 13,000,000 to 31,300,000), far outstripping the 46 percent population increase during the period (from 122,775,046 to 179,323,175) (Report on Amateur Instrumental Music 1967, 4).

What, then, do we make of Sousa's clearly wrongheaded claim? Rather than simply dismiss him, we should consider what motivated him and like-minded critics. Sousa may have cared about amateur musicians for many reasons; one of them, however, was that he depended on them for a good deal of his income. Every time Sousa wrote a new composition, he published it not only in its original form (usually for band) but also in a wide array of arrangements—for everything from

Table 19.1 Number of Musicians and Teachers of Music in the United States, 1890–1910

Year	*U.S. Population*	*Musicians and Music Teachers*	*Ratio*
1890	62,622,250	62,155	1 per 1008
1900	75,994,575	92,174	1 per 824
1910	91,972,266	139,310	1 per 660

banjo to piano to zither—that could be played at home by amateur musicians (Warfield 2009, 449). Sousa therefore had good reason to oppose anything he thought might discourage the amateur musician and thus slow his income stream. However, instead of proposing actions that would directly protect amateurs—if such a thing were possible—he argued that mechanical music should essentially be taxed, so that any time a phonograph disc or player piano roll was made, the appropriate copyright holder would be properly compensated. The timing of Sousa's 1906 article was no accident. The U.S. Congress was debating new copyright legislation at the time, and his essay was part of a campaign to win over public opinion for copyright reform, in particular the expansion of copyright to include mechanical reproductions of music. Sousa's efforts—which included not only his 1906 article but also his testimony before Congress—paid off, and the landmark 1909 U.S. Copyright Act included provisions that generated income for Sousa and other music copyright holders (Warfield 2009).

The amateur was therefore essentially a red herring in Sousa's article, for the copyright reform he advocated was not aimed to protect them, nor was it aimed to slow the production of rolls or records. Given Sousa's prominence in early twentieth-century American culture and the widespread discussion that his article generated, it is fair to say that he, more than anyone else, set the terms of the debate about the value and influence of sound-reproducing technologies. The terms he set were simple: The player piano and the phonograph were either bad or good for music. In surveying the literature of the time, however, it is clear that Sousa was not among the majority. Many commentators, among them prominent musicians and educators, embraced these technologies as a means to elevate musical culture in the United States (Katz 2010, 56–79). Nevertheless, whether later observers agreed or disagreed, they tended to defend one of two binaries—good or bad—and given how long the debate has been raging, it is now difficult to avoid choosing one or the other. Indeed, I offer evidence of possibly salutary effects of sound technologies on amateur music making. The purpose of this chapter, however, is not to defend mechanical music against its detractors. Rather, through the following case studies, my aim is understand how amateur musicians have responded to these technologies and thus gauge how amateurism has been transformed by—and has transformed—some of the most culturally significant and pervasive technologies of the past century.

Case Study 1: Making Music with Machines in the Early Twentieth Century

As Trevor Pinch and Karin Bijsterveld have suggested, we may fruitfully understand the influence of music technologies by observing how they breach longstanding conventions when first broadly introduced (Pinch and Bijsterveld 2003). Perhaps the central convention that the earliest sound technologies breached was that music had always been understood as a quintessentially *human* activity. Although mechanical instruments (music boxes, musical automata, and so on) had existed for centuries, these had not been seen as a serious threat to traditional musical culture or commerce. However, with player pianos and phonographs, it became possible to envision a future in which much of the music heard on a daily basis was made not by humans but by machines. Moreover, the phonograph in particular introduced something unprecedented in the history of music technology: the ability to reproduce the human voice accurately. "Encountering a machine with one attribute so singularly human," Lisa Gitelman writes in her study of the early phonograph, was for many an unsettling, even shocking experience. This shock, she explains, had "unequivocal implications for the mutual construction of the categories Man and Machine since machines became more human when the phonograph-cyborg breathed its first words" (Gitelman 2004, 332). How, then, did amateur musicians respond to this breach, this blurring of the traditional boundaries between agents and artifacts? Some might well have stopped making music while others might have continued their musical practices as before, unaffected by the availability of these player pianos and phonographs. But many others engaged with these technologies in musical ways. I see three broad categories of such engagement: treating the technologies as quasi-musical instruments, making sound recordings of amateur performances, and performing live music alongside recordings.

Owners of player pianos were not necessarily passive consumers of the music and often acted as what we might call co-performers. Many instruments either allowed or required users to control their operation as the music played. Pumping pedals or sliding levers, users could manipulate tempo, volume, and accentuation throughout the performance. As Timothy Taylor points out, the proliferation of these instruments gave rise to an "industry of how-to guides, player piano teachers, and other modes of instruction" (Taylor 2007, 285). The manufacturers touted this interactivity as an opportunity to experience music in a new way. An advertisement from the Aeolian Company proclaimed, "The Pianola thus is found to be a pleasure-giving instrument for all . . . enabling them to play on the piano with absolute correctness and with human feeling" (Roehl 1973, 7). A 1925 ad for the Gulbransen Registering Piano included this motto: "The Biggest Thrill in Music is playing it *Yourself*," while the makers of the Angelus Player Piano boasted that their instruments were equipped with the "*Melodant*, the famous *Phrasing Lever*, and other expression devices" (Roehl 1973, 35). In another form of co-performance that prefigured karaoke, many piano rolls were intended to accompany singing and included lyrics printed on the side of the rolls. These "word rolls" thus contained only half of the music and were completed only when their owners sang along.[1] Further evidence of active engagement comes in the form of the many articles and books intended to teach owners of player pianos how to play their instruments. These include *The Pianolist: A Guide for Pianola Players* by Gustav Kobbé, Ernest Newman's *The Player-Piano and its Music*, and Sidney Grew's *The Art of the*

Player-Piano: A Text-book for Student and Teacher. (Kobbé, 1907; Newman, 1920; Grew, 1922). As these books make clear, the player piano had its own idiomatic performance practice, one quite distinct from that of the traditional piano, sometimes referred to in these guides as the "hand piano."

Some phonographs allowed a modest form of co-performance as well. Aeolian's Vocalion model came with a device dubbed the Graduola. It was simply a cable with a knob attached that, when pushed or pulled, operated a shutter that muffled and unmuffled the phonograph's sound. Using the Graduola created only a mildly noticeable alteration in volume and timbre, but its modest effect did not stop Aeolian from trumpeting it as an "epoch-making achievement" in a 1915 brochure. Aeolian offered this retrospective look at the technological triumph of the graduola. "This was a device for artistically and effectually controlling tone volume. It not only added materially to the efficiency of the phonograph in its accepted field, but also greatly increased the artistic possibilities—it made of the phonograph *a practical instrument for personal musical expression—a musical instrument in the truest sense of the term!* (Aeolian Company 1915, [7]). Not to be outdone, Brunswick came up with a rival mechanism. A two-page 1916 ad in the *Saturday Evening Post* announced that a "group of experts was assigned to give human control, so that one might render any selection with his own interpretation, giving high lights or shadows to the music at will. As a result, the Brunswick phonograph has a perfected regulator which is a joy everlasting to those who wish to 'play' a phonograph" (*Saturday Evening Post*, October 16, 1916, 50–51).

One will note that much of the evidence presented here of co-performance comes from advertisements. It is difficult to find precise information on how player piano and phonograph owners used or thought of these devices, and it is reasonable to take a skeptical view of all this breathless advertising copy. Nevertheless, we can conclude from these advertisements further evidence of a technological breach. Given that traditional instruments require human interaction, we can see why the mechanical music industry would go to such lengths to assure consumers that their products were not soulless automata but real instruments that came alive only with a human touch. The industry thus had to strike a fine balance: The devices had to be easy to operate so as not to discourage users but also had to be sufficiently interactive to offer a rewarding experience. As scholar Brian Dolan writes, player pianos "helped people engage with music . . . [b]y inviting middle-class consumers to effortlessly participate in, rather than passively listen to, musical performance (Dolan 2009, xviii). Put another way, as Pinch and Bijsterveld suggest, "The blending of personal achievement (the loss of which had been feared by opponents of mechanical instruments) with democratized leisure (which had been seen as an advantage by proponents of mechanical instruments) in effect helped create a market for the player piano" (Pinch and Bijsterveld 2003, 543).

In addition to the type of co-performance described earlier, some phonograph owners treated recordings not as finished musical products but as raw materials to be combined and altered to produce novel sonic effects. Phonograph scholar Patrick Feaster has discovered a number of examples in his research. For instance, in 1906 the members of a Chicago women's club made recordings of their pet dogs barking and then, in a Dadaesque performance, played six machines simultaneously to create a kind of canine symphony. Around the same time, other home sound engineers were creating phonographic montages by recording several cylinders simultaneously or sequentially onto a single recording, while others affixed two halves of different cylinders together to create amusing juxtapositions. Such home experiments predated splicing, overdubbing, and mash-ups by decades (examples cited in Feaster 2007, 660–61).

These kinds of quasi-performative and quasi-compositional interactions were not the most pervasive ways in which users engaged with early sound technologies. More common was the second type of activity cited earlier: using the phonograph to record amateur performances. Whereas making piano rolls required a special kind of expertise and equipment, it was quite feasible for phonograph owners to make recordings at home. The earliest machines, specifically ones that used cylinders, were capable not only of playback but of recording as well. The phonograph industry, perhaps trying to close a breach, encouraged consumers to make their own recordings. Thomas Edison's company, for example, sponsored a home recording contest in 1907 (Feaster 2007, 442n66). While such gatherings might seem fanciful, countless amateur cylinders were made. One such recording was made by A. H. Mendenhall of Pomeroy, Washington, who, sometime early in the twentieth century, recorded a phonographic "letter" to send to his friend Guy Willebrand, who lived nearby. After some initial pleasantries he launches into a comic song about two Irish bricklayers, then clears his throat and proceeds with a jaunty harmonica solo; he closes by wishing his friend a Merry Christmas and a Happy New Year. Or consider the recording made by the Tindle family of New York City. Around 1910 they gathered around the family talking machine and recorded a medley of three popular songs from 1909. The singing is jovial and far from polished and is occasionally interrupted by good-natured chatter.[2]

Home recording declined with the standardization of the disc-playing machine, which was designed for reproduction, not playback. (It was not until tape recorders became available in the 1950s that home recording became popular again.) Yet live and recorded music continued to remain partners, and some phonograph owners would sing or play along with recordings. As an amateur violinist explained in response to a 1921 Thomas A. Edison, Inc., questionnaire, "I often learn how to interpret a piece by listening to Mr. [Albert] Spalding play it on the Edison—then I play it along with him (Katz 2010, 78). An amateur English cellist wrote in *Gramophone* magazine in 1925 that he found playing along with recordings preferable to the kind of "anaemic amateur orchestral society" to which he had once belonged (Luetchford 1925, 74). In Germany, Hausmusik, or amateur music making at home, may have benefited from the presence of records early in the twentieth century. "For all the complaints about its supposed displacement by records," historian Corey Ross writes, "it seems that Hausmusik was becoming more, not less, popular after the First World War. Indeed, records arguably *aided* the revival of Hausmusik" (Ross 2008, 51). One way this was possible was through specially made records (dubbed "Spiel mit" [Play with] discs) that featured professional musicians playing duets or quartets with one part missing, which would be supplied by the amateur playing along at home (Gerigk 1937; Ross 2008). (The sheet music for the missing part was included with the discs.) Similar recordings later became popular in the United States and elsewhere, produced by companies such as Music Minus One, Inc., and continued to be sold into the twenty-first century (www.musicminusone.com).

Many music teachers came out in favor of the phonograph as a means of engaging their students. In a 1916 forum in the American music journal *Etude*, J. Lawrence Erb argued that "the total effect of mechanical players has been to increase interest in music and stimulate a desire to make music on one's own account" ("Effect of Mechanical Instruments upon Musical Education" 1916, 483; see also Cecil 1903, 482). The teacher F. W. Wodell explained in the same article that he had "established a system whereby records are made by students at regular intervals, of both exercises and pieces, and reproduced for critical hearing and comparison by the pupil" ("Effect of Mechanical Instruments" 1916, 484). Oscar Saenger published a course of vocal study in which the student listened to and then imitated various exercises on several specially made discs; Hazel Kinscella published a similar method for piano (Saenger 1916; Kinscella 1924).

Clearly, the amateur did not disappear in the early days of mechanical music. For many music lovers, the player piano and the phonograph did not deaden the urge to perform; quite the opposite. Amateurs entered into a variety of partnerships with these technologies, whether they treated music machines quasi-instrumentally, used phonographs to record themselves, or played along at home with them. These new relationships helped reconfigure musical culture. Hybrid types of musical performance emerged, as did new ways of experiencing music and understanding its role in everyday life.

Case Study 2: Karaoke

In Japan, where karaoke first emerged in the early 1970s, impromptu amateur singing had long been accepted and even expected at certain kinds of social gatherings. Karaoke developed out of this tradition, and karaoke technology aided—and inevitably influenced—the practice. The earliest machines used cassette tapes but were similar to disc-playing jukeboxes in that they played music when coins were inserted. However, unlike jukeboxes, and more like the "Spiel mit" or "minus one" recordings mentioned earlier, karaoke machines were designed and intended to accompany live performance. The crucial feature of any karaoke technology—whether it uses cassettes, records, CDs, laser discs, MP3s, or some other format—is that the recordings it plays are not complete in themselves. The term *karaoke*, which can roughly be translated as "empty orchestra," points to this incompleteness.

The recordings provide instrumental backing for singers and are almost never listened to unaccompanied. Simply insofar as karaoke *demands* active music participation, it provides a counterexample to the argument that recording technology necessarily encourages a passive relationship with music. Those who interact with karaoke technology are anything but passive, and it could be said that any given group of people listening to karaoke is less an audience than a collection of performers-in-waiting. While there are professional karaoke singers and teachers, it remains an almost exclusively amateur phenomenon. When money changes hands, it is the performers who pay and the owners of the karaoke machines and venues who make money from those who sing.

Karaoke became hugely popular first in Japan and then in the rest of Asia and the world. In Japan in 1995, Hiroshi Ogawa reported, "55.4 per cent of Japanese over the age of 15 experienced karaoke [and] people who used karaoke systems did so an average of 10.5 times per year" (Mitsui and Hosokawa 1998, 47). On any given week in the late twentieth century, then, tens of millions of Japanese amateurs were singing into a microphone. Around the globe, many more millions have done the same.

The question then arises of what impact these legions of songbirds may have on musical practices. I would point to two main areas of influence: the music industry and song composition. Given the huge amounts of money spent on karaoke, the humble amateur singer wields great power in the marketplace of music. Ogawa writes, "At the start of the 1980s only a few recorded singles had registered a million sales in any given year. But since 1991 there have been more than ten million-sellers of CD singles per year. . . . The reason for this increase," he argues, "is karaoke." In other words, great demand—in the form of millions of karaoke singers gravitating toward a handful of songs—begat great supply in the form of millions of copies of certain popular songs being sold in Japan. This demand has also driven the production of specific types of songs, whether duets for women, disco songs, or *enka* (a Japanese popular music genre)—all of which witnessed a surge in the numbers of songs and hits in response to karaoke (Mitsui and Hosokawa 1998, 49–50). Of course, catering to amateur musicians is nothing new. Countless instrumental composers and songwriters of the early twentieth century and before either wrote or arranged music with the capabilities of nonprofessionals in mind. (John Philip Sousa, as cited earlier, is a perfect example.) What is different is that *this* market for amateur performance owes its very existence to the kind of sound-reproducing technology that some feared would render the amateur extinct.

Amateur karaoke singers do not merely affect supply and demand but also influence the very sound of the music they sing and the compositional practices of songwriters. As Ogawa points out, because of karaoke, "one of the important elements for hit songs changed from 'good to listen to' to 'good to sing' " (Mitsui and Hosokawa 1998, 49). That is, songwriters have to think about not only the professional singers who will record their music but also the abilities of the amateur musicians who will replace those professionals in their local bars, coffee shops, or karaoke boxes. (A karaoke box is a room rented out to private groups for karaoke singing;

it is prevalent in commercial areas throughout Japan, Korea, China, and elsewhere in Asia.) In the music genre known in Hong Kong as Cantopop, songs became less complex as karaoke versions of them became more popular. Cantopop composers started writing songs with narrower melodic ranges, fewer difficult leaps, and shorter phrases in order to accommodate the amateur singers' typically modest abilities.[3] Similarly, Shinobu Oku has suggested that karaoke has exerted a conservative influence on *enka* in Japan: "Because of the strategy of matching the songs to fans' singing ability, new singing techniques and unfamiliar musical elements are generally risky and unwelcome. Therefore, many traditional musical elements have remained in *enka*" (Mitsui and Hosokawa 1998, 71). In Hong Kong, karaoke transformed a classical genre into a popular genre. Karaoke brought Cantonese opera, once a highbrow traditional form into homes, clubs, and restaurants and made everyone a potential opera singer. As Patrick Ho, Hong Kong's secretary for home affairs, explained in 1995, "Nowadays Cantonese opera lovers are as likely to organise their own recitals at home as they are to line up for tickets to the next staged performance. From an art form designed to be watched and appreciated, Cantonese opera has evolved into one that invites participation" (Ho 2005). These are profound, even definitional changes occurring in some of the world's most popular music genres, and they have been spurred by karaoke.

In 1984 ethnomusicologist Charles Keil wrote about his encounters with what he called "mediated-and-live musical performances" in Japan, one of which was karaoke (Keil 1984). Keil made several perceptive observations that are relevant here, for they offer ways of understanding the relationship between live (especially amateur) music making and sound-reproducing technologies. Many of the performances and rituals he attended combined traditional vocal or instrumental performance with the playback of recordings; as an American with preconceived notions about how one interacts with music technologies, he was surprised that he detected "no visible friction" between those who sang or performed live and those who operated record players or cassettes. "The scenes I witnessed seemed very traditional . . . the atmosphere suggested, 'this is the way it is always done' " (Keil 1984, 93). Keil also concluded that the interactions between live and mediated music that he observed did not deaden the human musical spirit but quite the opposite: "What is striking to me in all these instances of mediated-and-live musical performances is first of all the humanizing, or better still, personalization of mechanical processes" (Keil 1984, 94). Particularly in the case of karaoke, he saw the practice as "an assertion of individualism, skill, and personal competence before others in a demanding situation where any slip or hesitation will not be compensated for by the 'empty orchestra' " (Keil 1984, 95).

Finally, Keil realized that what he witnessed was not as foreign to him as he had first thought. Turning his scholarly gaze inward he asked, "Why haven't I paid any ethnomusicological attention to the fact that since I was a teenager I have spent more time playing bass or drums to records than I have playing with other people?" (Keil 1984, 95). The answer is in part that, at least before karaoke became popular in America in the 1980s, there had been no single widespread phenomenon like karaoke in which amateurs performed alongside recordings; the answer may also be attributed to the idea, reinforced since Sousa's time, that the live and the recorded could not easily coexist. But they could and they did. In the United States, people had long made music with recording technology—they had sung, played, and conducted alongside machines and had done so frictionlessly, just like their later Japanese counterparts.

Case Study 3: Hip-Hop

In the early 1970s, not long after the first karaoke machines were being marketed in Japan, a discovery in the New York City borough of the Bronx led to the development of another globally influential, live-and-mediated—and, to a great extent, amateur—musical practice. Young disc jockeys spinning at parties noticed that dancers were especially energized by what were called breaks, the short percussion solos found on many funk, soul, and rock records. In itself, this discovery was not particularly significant; as is true of most discoveries, its real impact came from its practical application. Rather than simply let the breaks come and go in a matter of seconds, as was typical on most records, the DJs intervened. Through a great deal of trial and error, DJs found that they could extend the breaks almost indefinitely. At first, they would simply pick up the needle and return it to the beginning of the break, repeating it as desired. (This technique, developed by a DJ known as GrandWizzard Theodore, was called the needle drop.) Later, Grandmaster Flash and others developed more complex ways of repeating breaks, using two copies of a record on two turntables and employing a piece of equipment called a mixer to switch quickly and seamlessly between the two discs (Flash and Ritz 2008, 75–81).

Ironically, DJs extended the breaks by harnessing the latent powers of the very technology that demanded such brief percussion solos in the first place. It is important to understand that the short four- or eight-bar break is native to sound recordings; they are common on disc because of their often severe time limitation. It was the fundamental insight of the hip-hop DJ that the turntable itself contained the key to overcoming its procrustean nature. The response to this technological intervention was swift and profound. A new form of virtuosic solo dancing arose around the repetition of the breaks, one that has continued to evolve in the decades since. Those who danced to the breaks called themselves b-boys and b-girls, and their art later came to be known throughout the world as *break dancing*. However, DJs did more than simply extend breaks. They also transformed the sound of prerecorded discs in a more radical way—through scratching. Scratching, introduced in the mid-1970s by GrandWizzard Theodore, involved manually pushing and pulling a record back and forth underneath the stylus. The result was a percussive rasping that no longer resembled the original recorded sound.

For the most part, in the decades before hip-hop, records were simply played from beginning to end; with few exceptions, little thought had been given to any other way of interacting with recordings (for exceptions, see Katz 2010, 109–123). By the 1970s, the phonograph had existed for nearly a century; the period of its greatest interpretive flexibility would seem to have long been closed. It was a stabilized technology, one deeply embedded within established social practices. Then DJs destabilized the phonograph by violating a taboo as old as the phonograph against inappropriately touching records. For many (then and now), to touch a record is to defile it. One properly holds a disc by the edges, the fingertips making only the slightest contact with the vinyl. Yet the fundamental techniques these DJs developed—mixing and scratching—are based on a simple dictum: To best control the sound, it is necessary to touch the records. As Grandmaster Flash points out, "This was a major no-no. You *never* touched the record with your fingers." Yet he realized that the best way to manipulate a break was to put "your greasy fingertips on the record" (Flash and Ritz 2008, 76, 79). DJs like Flash and others refused to accept the "proper" use of the phonograph and made an art out of treating the turntable not as a means for the passive consumption of music but for the active creation of music. In other words, they transformed the turntable into a musical instrument.

Without the DJs' manipulation of records, hip-hop, broadly speaking, could not have flourished. We must understand that in 1973—now generally accepted as the year of hip-hop's birth—there was no separate musical category actually called hip-hop. It was not yet a distinctive genre but more a performance practice, a way of approaching *other* types of music. The performer was the DJ, and the practice was to isolate and repeat choice instrumental parts of popular songs—the breaks—at dance parties. Later, vocal parts were added over the top of these repeated breaks in the form of rapping. Eventually, what began as a practice became its own genre, one that combined dance with instrumental and vocal music.

Rapping, like DJing, is a form of live-and-mediated musical practice, though instead of manipulating recording technology to create new music, it arose as a way of accompanying turntable manipulation. Rapping, however, did not begin as a musical practice. Rappers were originally called MCs, and in fact that remains the preferred term. Short for "master of ceremonies," MC seems like an odd way to describe a vocalist, for the term usually conjures up not a musician but a host or an announcer. Standing before a microphone, an MC—in the traditional sense—might welcome an audience to an event, introduce acts at a variety show, or make announcements. At the earliest hip-hop parties, that is exactly what MCs did, and that is how they got their name. Even from the beginning, however, MCs did more than just make announcements. The more boisterous ones shouted out to the audience, trying to help their DJ by stoking the energy in the room. Some developed catchphrases: Kid Creole was known for calling out variations on "Yes yes y'all, and you don't stop, to the beat y'all and you don't stop." Cowboy was probably the first to command dancers to "Throw your hands in the air and wave 'em like you just don't care" (Fricke and Ahearn 2002, 209). And like good catchphrases, they caught on and became part of the hip-hop vocabulary. Then MCs started to do more, chanting brief rhymes to the crowd, and gradually their rhymes grew—from couplets to stanzas to nearly epic-length poems. Their numbers expanded as well: Often five rappers would stand before the crowd at a time, and crews of MCs—though still associated with a DJ—proliferated. A line had been crossed: Announcers became artists and no longer assisted the DJ but commanded equal, if not more, attention from the crowd.

The pioneering DJs and MCs began as amateurs and developed their art in the bedrooms and basements, as well as the streets and playgrounds, of the Bronx. Although some went on to generate an income from their work, most did not. More to the point, it seems unlikely that hip-hop would ever have been developed by professional disc jockeys and MCs. Mixing and scratching records or chanting over the microphone for extended periods contravened long-established practices and rules; simply put, such behavior would have been deemed unprofessional. What is also significant here is that these amateur musicians typically had not already pursued traditional types of instrumental or vocal performance. In other words, the birth and flourishing of hip-hop offers a case study of the rise of a new class of musicians, one directly tied to a technology long held to discourage, not promote, the musical amateur.

Case Study 4: The Digital Amateur—Music Video Games and Mobile Phone Music

Since the late 1990s an explosion has occurred in the amount of digitally based musical activity designed for or engaged in by amateurs. In video game arcades, on home computers, and even on phones, music is being performed and composed on a vast scale. Two examples will illuminate some of the issues that this new age of musical amateurism raises: music video games and music applications for mobile phones. I have picked these in order to illustrate two broad points: that these forms of musical activity challenge traditional notions of musical agency and creativity and that this amateur music making has become a driving force in modern musical culture.

Music Video Games

In the first decade of the twentieth century one of the hot consumer items in the United States and in parts of Europe was a modified musical instrument that allowed anyone, regardless of ability, to take part in high-level music making. This was the player piano, and the user, by altering tempo or timbre by pumping pedals or sliding levers, contributed, however modestly, to the execution of a polished musical performance. In the first decade of the twentieth-*first* century one of the hot consumer items was another modified instrument that allowed nonmusicians to participate in musical performance. The instrument was not a piano but a guitar. Actually, it was a video game controller in the form of a guitar used in the wildly popular game *Guitar Hero* (2005) and its many sequels. In contrast to the player piano, the plastic, guitar-shaped controller does not allow users to modify tempo and timbre—these are fixed. Rather, the user controls—to a certain extent—the melody; pressing buttons on the controller's fretboard in time with moving images on a television screen activates the pitches in a prerecorded popular song. If the player presses the right buttons at the right time, the full and correct melody is heard. With both the player piano and the guitar controller, users are given power over a limited number of musical parameters. In writing of the player piano Brian Dolan calls this phenomenon the *de-skilling* of musical performance. Certainly both player pianos and guitar controllers are easier to play than the instruments they are modeled on, but they also require actions not demanded of pianists and guitarists—for example, sliding levers and pressing buttons. A distinct type of expertise is thus required to master these. We can see this as a reconfiguration of skill rather than a simple subtraction of expertise; such an approach allows us to understand these technologies on their own terms rather than as mere imitations or simulations (Seaver 2010). Put another way, as Andrew Goodwin argues, "new technologies have not removed the notion of 'skill' " (Goodwin 1990 [1988], 262; see also Théberge 1997).

Games like *Guitar Hero* have excited strong opinions about their legitimacy. On one side are those who argue that these games are a positive force in modern musical life. Among them, naturally, is Alex Rigopulos, cofounder of Harmonix Music Systems, the company that created *Guitar Hero* and several related games. "Everyone comes into the world with this innate desire to make music and almost everyone tries to learn an instrument at some point. And the overwhelming majority of these people quit after a few months or a few years because it is just too damn difficult. They spend the rest of their lives loving music . . . but not having any outlet for that innate urge they feel" (quoted in Radosh 2009). Rigopulos touts both the de-skilling and democratizing effects of the games and suggests that these technologies remove the barriers that have been keeping so many from expressing their innate musicality. Once again, we see parallels with those who promoted player pianos. "Before the Pianola came," explains an early twentieth-century Aeolian ad, "how very few there were who even caught a glimpse into the grand world of harmony." However, with the Pianola, "The world of music is free to all" (Roehl 1973, 7). Furthermore, just as vehemently as Sousa and others condemned early forms of sound technologies, critics of the games dismiss their musical value. "Guitar Hero," argued professional rock guitarist John Mayer, "was devised to bring the guitar-playing experience to the masses without having to put anything into it" (quoted in Miller 2009, 405).

However, as ethnomusicologist Kiri Miller contends, neither the arguments for nor those against these games illuminate what makes them distinctive. She has coined the term *schizophonic performance* to describe the act of playing guitar, bass, or drum controllers, an act that combines "physical gestures of live musical performance with previously recorded sound" (Miller 2009, 401). Borrowing from R. Murray Schafer's term *schizophonia*, Miller highlights the split between the live and the recorded. Yet she argues that the two are not as distinct as we might think, that these music video games occupy a middle ground. On the one hand, operating game controllers can be like playing real instruments (though more so for experienced players than beginners.) For one, players sense a causal relationship between their physical actions and the sounds they hear. Second, players can experience music from the *inside* of the song and feel that they are both part of and carried along by the musical texture. Writer Helene Stapinksi describes this sense of immersion when she played the drums: "I loved the drums because they were all-consuming. They surrounded me . . . When I played the drums, I knew what those surfers felt like, inside the tube, the wave breaking over them, but at the same time carrying them, faster and faster, toward the shore" (Stapinksi 2005, 12). (For more on immersion see Grimshaw, this volume). These are the same type of sensations often described by accomplished players of *Guitar Hero* and similar games. Third, game players may develop the ability to focus on and shift attention among different parts within an ensemble, a skill required of accomplished musicians.

At the same time, there are clear differences between gameplay and instrumental play. One does not create new sounds but activates prerecorded sounds in a tempo and order not of one's choosing. Depending on the game, players have little leeway for spontaneous musical acts, whether through improvised melodies, harmonic progressions, or changes in phrasing, accentuation, timbre, or dynamics, all of which are hallmarks of live performance.

As Miller explains, "Playing *Guitar Hero* and *Rock Band* isn't just like playing a real instrument, but it's nothing at all like just listening to music." "Schizophonic performance," she concludes, "is collaborative performance: the players and audiences join the game designers and recorded musicians in stitching musical sound and performing body back together" (Miller 2009, 424). Whether it is called schizophonic performance, mediated-and-live musical performances (cf. Charles Keil), co-performance, or hybrid performance, the acts performed by the millions who play these games represent a growing norm. This type of collaborative amateur performance is nothing new, of course, and dates back to the earliest player pianos and phonographs. Nonetheless, a point may well have been reached in which the majority of amateur music making involves some sort of technological mediation.

Mobile Phone Music

In January 2007, Apple, Inc., introduced its multimedia and Internet-capable mobile phone, the iPhone. One of its most distinctive features is the App Store, a service that provides an ever-growing collection of free and paid third-party applications of every imaginable variety (www.apple.com/iphone/apps-for-iphone). Hundreds, perhaps thousands, of these apps are music related. Some of them connect to Internet radio stations or identify the lyrics to songs stored on the phone (which also doubles as a music player); others serve as electronic tuners or offer music games similar to (or in fact versions of) *Guitar Hero* and its kin. Of particular interest here, however, are those applications that allow users to create music. These cover a huge range of possibilities. Sound Shaker, for instance, allows children to create and mix sounds by tapping the touch-sensitive screen and shaking the phone. (The iPhone has an accelerometer that detects movement and spatial orientation.) PocketGuitar, iBone, and Filtatron offer guitar, trombone, and synthesizer simulations. Singing into the phone's microphone using the app called I Am T-Pain results in a digitally altered voice that emulates the rapper T-Pain's use of the robotic-sounding Auto-Tune effect. When surveying user comments (typically collected on sites dedicated to each app), it becomes clear that most engage these apps for recreational purposes and not as a means of earning a living as a performer or composer. That is to say, these apps almost exclusively serve amateurs.

The iPhone was not the first and is not the only mobile phone to offer musical applications. It has been, however, one of the most popular and influential such devices, and thus offers a valuable case study for exploring mediated amateur music making in the digital age. In particular let us consider Ocarina, an app that seemingly transforms the iPhone into a wind instrument. Ocarina was released by Smule in November 2008 and quickly caught on—in its first six months it was downloaded (at a cost of 99 cents in the United States) on more than a million devices (http://ocarina.smule.com/). Its popularity was due in part to its novelty and ease of use. The player blows into the phone's microphone and covers various combinations of four simulated airholes to produce a mellow, pleasant tone with a range of an octave and a fourth. It is also possible to switch between different scales and to play with vibrato by tilting the phone downward; it thus offers a respectably wide variety of pitches and timbres. Ocarina's popularity can also be traced to Nintendo's multimillion-selling video game *Legend of Zelda: Ocarina of Time*, released in 1998. The game is said to have spawned a craze for the real ocarina, a vessel flute typically in the shape of an elongated egg that was popular among amateurs in the early and mid-twentieth century. The residual popularity of the video game (in which an ocarina plays a significant role) and the real ocarina seems to have primed the market for Smule's version of the instrument.

Smule claims that Ocarina is the "first true musical instrument created for the iPhone" (http://ocarina.smule.com/). Whether or not it was the first, it is hard to dispute that the iPhone, when Ocarina is engaged, truly is a musical instrument. Playing Ocarina is not an act of co-performance, where users control only certain nuances while pitch, melody, and rhythm are predetermined. Because most musical parameters are under the player's control, the iPhone Ocarina is closer to a traditional ocarina than a *Guitar Hero* controller is to a guitar; it would be fair to call Smule's Ocarina simply a type of aerophone, or wind instrument.

If the iPhone is a musical instrument, it is hardly a simple one. Given that Ocarina is just one of many instrumental apps, we might call the iPhone a multi-instrument, for it can be any of a hundred types of music maker. Moreover, unlike most instruments, the iPhone is Internet capable. This means that iPhone musicians can connect to one another across the globe. Ge Wang, Ocarina's designer, touts this as a particular virtue: "Ocarina is also a type of *social instrument* that enables a different, perhaps even magical, sense of global connectivity" (Wang 2009, 303). Tap on the globe icon, and an image of the earth floating in space appears. Glowing pinpoints indicate the location of Ocarina users as if each one represents a city lighting up the night sky. (The sun never seems to rise in Ocarina's "World Listen," as it is called.) Every few seconds, snippets of music sound, each a passage played by an Ocarinist somewhere in the world, the notes ascending heavenward from the player's location (provided by the phone's GPS) in a glowing column of light.

As a sampling of the music being played around the globe on the Ocarina, consider what I heard using World Listen over the course of a few minutes one day late in 2009: "Twinkle, Twinkle, Little Star" (England), the theme from *Star Wars* (France), "The Lion Sleeps Tonight" (Portugal), "Ode to Joy" (Tokyo and Vancouver), and "Oh, Susanna" (Malaysia). The halting quality of many of the performances attests to the amateurism of the Ocarinists and also imparts a sense of authenticity—the music clearly seems to have been played by actual people. One can track users and designate them as favorites and communicate with them in one of the Ocarina Internet forums (www.smule.com/forum). One need not buy into the utopian vision that Ocarina's promoters push, however, to see at least the potential of the Ocarina and, by extension the iPhone, to connect people through music. Indeed, hundreds of videos on YouTube.com show iPhone Ocarinists playing in small ensembles, including a consort performing Led Zeppelin's "Stairway to Heaven" or Ge Wang's own Mobile Phone Orchestra (or MoPho) at Stanford University.

In addition to the Internet connectivity of the iPhone-as-instrument, another distinctive feature is its portability. The iPhone is essentially a small computer, which makes it not unlike any of countless home devices that also allow users to make music. The iPhone's portability, however, means that millions of people are carrying musical instruments in their pockets, making music wherever they please. Again, there seems to be the potential for, if not a massive democratization of music at this point, then a considerable surge in amateur music making through the iPhone.

At this early date, however, it would be unwise to issue grand pronouncements about the musical impact of the iPhone. Although Apple had sold more than seventy-three million iPhones by the end of its 2010 fiscal year (http://en.wikipedia.org/wiki/Iphone; accessed Feb. 18, 2011), relatively few users regularly employ them as musical instruments. Moreover, the device is, by nearly any standard of living, very expensive. One could easily buy a modest flute, guitar, or violin—or a few dozen plastic ocarinas—for the same price. Moreover, the polished metal and glass iPhone is not the sturdiest of music devices and certainly was not designed for long, physically intense rehearsal or performance sessions. "It doesn't feel like an instrument," objects Tod Machover, MIT computer scientist and composer. " [T]hink of a piano . . . I mean, a piano is a machine, but you've got ivory, and there's weight behind the keys, you have this really—you feel the resonance in the instrument" (Machover 2010). At this point, then, we should recognize the significance of the iPhone instrument in terms of the potential it represents. It is not hard to imagine a time when cheap, robust, and interconnected devices with the kind of appealing tactility of successful traditional instruments are in the hands of the multitudes. The unanswerable question, however, is whether the multitudes will make music with them.

Conclusion

The case studies offered here in no way cover the full spectrum of live-and-mediated amateur music making, which has flourished since the late nineteenth century. Other examples include the once-popular practice of phonograph conducting, in which one would "lead" an invisible orchestra, baton in hand, in the privacy of one's home or, in later decades, the phenomenon of the mash-up, in which one creates new music by digitally combining elements of two or more different songs on a home computer (for more on the mash-up, see Pinch and Athanasiades, this volume; see also Katz 2010, 165–174). Nevertheless, the case studies on the player piano and phonograph, karaoke, hip-hop, music video games, and musical mobile phone applications illustrate a number of important points about the development of musical amateurism in an age of mechanical music.

The first and most obvious point is that amateur music making has not disappeared. This is significant, given how deeply ingrained the idea of the mutual exclusivity of musical amateurism and musical technologies has been. Yet it is hardly the most consequential conclusion to be drawn from this study. More important is the understanding that although amateur music making continues to thrive, it has not done so unchanged. One of the clearest changes—clearly seen in the demographic of middle-class Americans—is the transformation of an amateurism dependent on sheet music to one dependent on recorded and synthesized sound or, more broadly speaking, the transformation of a written musical culture into an oral one. A related point is that amateurism changed from a live-only practice to a mediated-and-live practice. We must understand, then, that the flourishing of musical amateurism is in fact deeply tied to the rise of various types of mechanical music. Yet while amateurs have adapted many of their practices in response to these technologies, one thing has not changed: the social, communal aspects of music making. It is certainly possible to engage in mediated-and-live performance in isolation, but this has not been the prevalent practice. Both DJs and MCs developed their craft in response to the needs of dancers and partygoers; friends, family, neighbors, and colleagues gather to sing karaoke or play *Guitar Hero*, and Ocarina players around the globe connect with one another. It would be fair to say that it is the companionable interaction with others, as much as the ability to express oneself through sound, that appeals to amateur musicians. Whether through physical interaction or online social networking, users have found ways to maintain this crucial aspect of music making.

Another important conclusion is that in the twentieth and twenty-first centuries technologically mediated amateur performance became a hugely powerful creative and commercial force. To take the clearest example, hip-hop, a globally influential musical and cultural phenomenon, can trace its roots to the schizophonic performance of the turntable. Karaoke has exerted tremendous influence as well by shaping the form and function of various musical styles, while music video games have collectively sold billions of dollars' worth of merchandise and music and have influenced pop music sales (Radosh 2009). New songs and remixes are also being composed specifically for phone applications.

To note the prevalence and power of the musical amateur, however, is not to require that we celebrate it. In 1888, when the phonograph was little more than a decade old, composer Arthur Sullivan witnessed a demonstration of Edison's machine and prophesied what could happen if the technology caught on: "I am astonished and somewhat terrified at the results of this evening's experiment—astonished at the wonderful power you have developed, and terrified at the thought that so much hideous and bad music may be put on record forever" (Sullivan 1888). Sample the infinitude of amateur music preserved on records, cassette tapes, CDs, and MP3s and know that Sullivan prophesied correctly. Moreover, with the so-called Web 2.0—the second generation of Internet design, which encourages the exponential growth of user-generated content as exemplified by sites like YouTube and MySpace—all of this amateur music is accumulating fast. In his 2007 book, *The Cult of the Amateur*, Andrew Keen wrote of the development of Web 2.0: "The new Internet was about self-made music, not Bob Dylan or the Brandenburg Concertos. Audience and author had become one, and we were transforming culture into cacophony" (Keen 2007, 14). We may agree or disagree with the last part of his statement, but this is beside the point. Whether one views record scratching or the alcohol-soaked karaoke version of "My Way" as a joyous noise or just noise, we must realize that, golden or not, the turn of the twenty-first century must be reckoned as the age of the amateur.

NOTES

1 My thanks to Nicholas Seaver for pointing this out to me.

2 Both recordings discussed here have been reissued on *I'm Making You a Record: Home and Amateur Recordings on Wax Cylinder, 1902–1920*, Phonozoic compact disc 001.

3 This observation comes from an unpublished paper by composer and Hong Kong native Angel Lam, Dec. 6, 2005.

REFERENCES

Aeolian Company. *The Aeolian Vocalion: The Phonograph of Richer Tone That You Can Play*. New York: Author, 1915.

Cecil, George. "The Phonograph as an Aid to Students of Singing." *Etude* 21 (1903): 482.

Dolan, Brian. *Inventing Entertainment: The Player Piano and the Origins of an American Musical Industry*. Lanham, Md.: Rowman and Littlefield, 2009.

"Effect of Mechanical Instruments upon Musical Education, The." *Etude* 34 (1916): 483–84.

Feaster, Patrick. "The Following Record": Making Sense of Phonographic Performance, 1877–1908". PhD diss., Indiana University, 2007.

Flash, Grandmaster, and David Ritz. *The Adventures of Grandmaster Flash: My Life, My Beats*. New York: Broadway, 2008.

Fricke, Jim, and Charlie Ahearn, eds. *Yes Yes Y'all: The Experience Music Project Oral History of Hip-Hop's First Decade*. New York: Da Capo, 2002.

Gerigk, Herbert. "Hausmusik und Schallplatte: Die unsichtbaren Spielpartner für den Hausmusiker." *Die Musik* 29 (April 1937): 506–507.

Gitelman, Lisa. "Unexpected Pleasures: Phonographs and Cultural Identities in America, 1895–1915." In *Appropriating Technologies: Vernacular Science and Social Power*, ed. Ron Eglash, Jennifer L. Croissant, Giovanni Di Chiro, and Rayvon Fouché, 331–44. Minneapolis: University of Minnesota Press, 2004.

Goodman, Andrew. "Sample and Hold: Pop Music in the Digital Age of Reproduction." In *On Record: Rock, Pop, and the Written Word*, ed. Simon Frith and Andrew Goodwin, 258–73. New York: Pantheon, 1990.

Grew, Sydney. *The Art of The Player-Piano: A Text-book for Student and Teacher*. New York: E. P. Dutton, 1922.

Harris, Henry J. "The Occupation of the Musician in the United States." *Musical Quarterly* 1 (April 1915): 299–311.

Ho, Patrick. "Singing Up a Storm—in Karaoke." Hong Kong Home Affairs Bureau official website, June 2005. www.hab.gov.hk/file_manager/en/documents/whats_new/from_the_desk_of_secretary_for_home_affairs/20050604_SCMP.pdf (accessed February 19, 2011).

Katz, Mark. *Capturing Sound: How Technology Has Changed Music*. Rev. ed. Berkeley: University of California Press, 2010.

———. *Groove Music: The Art and Culture of the Hip-Hop DJ*. New York: Oxford University Press, forthcoming.

Keen, Andrew. *The Cult of the Amateur: How Today's Internet Is Killing Our Culture*. New York: Random House, 2007.

Keil, Charles. "Music Mediated and Live in Japan." *Ethnomusicology* 28 (January 1984): 91–96.

Kinscella, Hazel Gertrude. "The Subtle Lure of Duet-Playing," *Musician* 29 (January 1924): 8, 15.

Kobbé, Gustav. *The Pianolist: A Guide for Pianola Players*. New York: Moffat, Yard & Company, 1907.

Luetchford, A. L. "Playing with the Gramophone." *Gramophone* 3 (July 1925): 74.

Machover, Tod. "Tod Machover, Composer and Inventor." *Big Think* (February 3, 2010). http://bigthink.com/todmachover (video), http://bigthink.com/ideas/18530 (transcript).

McColvin, Lionel R. "Will 2031 Be without Music?" *Sackbut* (July 1931): 322–24.
Miller, Kiri. "Schizophonic Performance: *Guitar Hero*, *Rock Band*, and Virtual Virtuosity." *Journal of the Society for American Music* 3 (November 2009): 395–429.
Mitsui, Toru, and Shuhei Hosokawa, eds. *Karaoke around the World*. London: Routledge, 1998.
Newman, Ernest. *The Piano-Player and its Music*. London: Grant Richards Ltd., 1920.
Oudshoorn, Nelly, and Trevor Pinch, eds. *How Users Matter: The Co-Construction of Users and Technologies*. Cambridge, Mass.: MIT Press, 2003.
Parkhurst, Winthrop. "Exit the Interpreter." *Disques* 2 (March 1931): 6-9.
Pinch, Trevor, and Karin Bijsterveld. " 'Should One Applaud?': Breaches and Boundaries in the Reception of New Technology in Music." *Technology and Culture* 44 (July 2003): 536–59.
Radosh, Daniel. "While My Guitar Gently Beeps." *New York Times Magazine* (August 11, 2009). www.nytimes.com/2009/08/16/magazine/16beatles-t.html.
Report on *Amateur Instrumental Music in the United States 1966*. Chicago: American Music Conference, [1967].
Roehl, Harvey N. *Player Piano Treasury: The Scrapbook History of the Mechanical Piano in America*, 2nd ed. Vestal, N.Y.: Vestal, 1973.
Ross, Corey. *Media and the Making of Modern Germany: Mass Communications, Society, and Politics from the Empire to the Third Reich*. New York: Oxford University Press, 2008.
Sabaneev, Leonid. "The Process of Mechanisation in the Musical Art." *Nineteenth Century* 104 (July 1928): 108–17.
Saenger, Oscar. *The Oscar Saenger Course in Vocal Training: A Complete Course of Vocal Study for the Soprano Voice on Victor Records*. Camden, N.J.: Victor Talking Machine, 1916.
Schaefer, R. Murray. *The Tuning of the World*. New York: Random House, 1977.
Seaver, Nicholas Patrick. "A Brief History of Re-Performance". Master's thesis, Massachusetts Institute of Technology, 2010.
Sousa, John Philip. "The Menace of Mechanical Music." *Appleton's* 8 (1906): 278–84.
Stapinksi, Helene. *Baby Plays Around: A Love Affair, with Music*. New York: Villard, 2005.
Sullivan, Arthur. Toast at Little Menlo. Recorded October 5, 1888, London. http://www.gutenberg.org/ebooks/10310.
Taylor, Timothy D. "The Commodification of Music at the Dawn of the Era of 'Mechanical Music.' " *Ethnomusicology* 51 (Spring/Summer 2007): 281–305.
Théberge, Paul. *Any Sound You Can Imagine: Making Music/Consuming Technology*. Hanover, N.H.: Wesleyan University Press, 1997.
Wang, Ge. "Designing Smule's Ocarina: The iPhone's Magic Flute." In *Proceedings of the International Conference on New Interfaces for Musical Expression*, Pittsburgh, 2009, 303–307. http://www.nime2009.org/proceedings/NM090137.
Warfield, Patrick. "John Philip Sousa and 'The Menace of Mechanical Music.' " *Journal of the Society for American Music* 3 (2009): 431–63.

CHAPTER 20

ONLINE MUSIC SITES AS SONIC SOCIOTECHNICAL COMMUNITIES: IDENTITY, REPUTATION, AND TECHNOLOGY AT ACIDPLANET.COM

TREVOR PINCH AND KATHERINE ATHANASIADES

Introduction

This chapter focuses upon one form of digital music making—that created and shared among a group of Internet users. Websites dedicated to this new form of musical production and consumption are generally part of what has become known as Web 2.0. Most Web 2.0 sites, such as Facebook and MySpace, are social networking sites and may include the uploading and downloading of music as a feature.

Here we write about a website, http://www.ACIDplanet.com (henceforth known as "ACIDplanet"), which is dedicated to the production of music made by its users, who share their songs, review each other's creations, and enter special remix and mash-up competitions promoted by the website. It is one of the largest such websites, and much of the music is made by a special software program sold by Sony (who also owns the website) known as ACID. This software turns a computer into a digital recording studio and enables users to make music by downloading repetitive samples of music, known as "loops." The chapter is not an exhaustive review of online music; rather, it presents an ethnographic study (conducted 2005–2007) of the working of this one online music site.[1]

We are particularly concerned in this chapter with three themes: (1) how reputations are achieved in this online world, (2) how musical identities are negotiated, and (3) how much, if anything, there is in common between the online and offline world of musical identities and reputations. On the first theme we show the importance of a form of "transduction," whereby the sounds and music these musicians make are rendered into words and symbols that enable rankings to be carried out. On the other two themes, we show that, despite the radical possibilities that the online musical world offers for, say renegotiating musical identities, many standard (offline) practices and identities are actually reinforced at the website.

ACIDplanet's History

The independent software company Sonic Foundry of Madison, Wisconsin, initiated ACIDplanet. In 1998 Sonic Foundry launched the first version of ACID software. The program uses dropdown menus to facilitate the building up of complex, multitracked compositions by recording numerous individual tracks in real time. The user can modify or add sampled sounds or sounds from real instruments to any individual track, thus turning a computer into a complete recording studio. Based on the new opportunities that this software afforded music hobbyists, a separate Madison-based firm developed the first version of the website (named ACIDplanet for the first time), which Sonic Foundry brought in-house and launched in 1999. Dave Hollinden, who was ACIDplanet's webmaster from 1999 to 2005, explains:

> People are able to make music that didn't have any way to do it in the past . . . even if . . . they don't really have a musical, physical nature, you know as far as playing an instrument or something . . . it gives people a chance to . . . have that feeling of making music and getting it in front of people and getting a pat on the back, and patting other people on the back, and making friends, and . . . so on.[2]

The original purpose of the website (created at a time when every new business needed one) is not entirely clear. However, over time it has evolved as the primary place to post, download, and review the musical creations of people using ACID software. In 2003 Sonic Foundry was bought by Sony for a reported $18 million. From the perspective of the ACIDplanet webmasters, Sony's intentions for the site are not always clear. Direct sales of Sony software products are a small but significant part of the website's business, but the site overall is more important as a promotional vehicle for this form of music making.

Our Study

In our ethnographic work we have made a special effort to locate, observe, and interview thirty-five users of ACIDplanet in the places where they make their music and access the website. We have thus spent a lot of time in home studios, as well as the Internet cafes where some members of this new breed of musician like to operate. We mainly interviewed users who lived close to our base in the northeast region of the United States. Other interviews were conducted opportunistically in Korea and the UK, as well as with Sony personnel. Interviews were held face to face and typically lasted between one and a half and two and a half hours.

Thirty-two of our interviewees were male, twenty-five resided within the United States, and thirty were Caucasian. The age range was eighteen to fifty-two. Our interviewees had a great variety of experience in terms of musical tastes and backgrounds, personal histories, motivations for using the site, and amount of time spent on the site. Some obsessively used the site, posted all of their music there, and used it as a promotional tool for their own independent CD releases; others were casual users who had discovered the site by chance when playing with ACID software and in effect used the site as a place to store their compositions. However, even these casual users would alert their friends and family to the existence of their musical compositions, which anyone could listen to by visiting the site.

Using the Site

ACIDplanet provides for its "citizens" the options to post, listen, review, and obtain a chart position for musical compositions, participate in forums, enter remix and mash-up contests, make lists of favorite artists, and more. All of these activities, importantly, are free, and the site has no requirements of its users other than to follow basic guidelines of etiquette. It makes no promises of fame or fortune, either; Sony simply provides the web space and location, as well as the music-creation software that encourages many people to get involved in music making in the first place.

Upon visiting the site, potential users are able to perform a limited number of activities, such as listening to other users' songs (any song can be listened to immediately without downloading any software). To explore the full range of options, a user must sign up for a profile, thus becoming a "citizen" of ACIDplanet. Anyone with a valid email address can create a profile, and a user can have up to three affiliated profiles attributed to that one email address.[3] If someone has multiple email addresses, that person can create multiple groups of independent profiles. For each profile, citizens choose a name they will be known as by other users on the website. They can choose to provide information on their profile such as their identity, their musical style, influences, and equipment used. It is up to the individual citizens to decide what, if anything, they include in the ACIDplanet profile. They can, therefore, remain as anonymous, fake, honest, or informative as they wish. The users' profile on ACIDplanet becomes their main online identity in that world.

The predominant activity at the site is the users' posting of their own music (attached to a profile). Each song has its own page, and on this page the user leaves a description of the song, including a self-rating from G to R (as with movie ratings), and a genre (see figure 20.1). The genre is important because it defines the ACIDplanet charts on which the song will appear.

With regard to posting music to the site, we can see one of the major differences between ACIDplanet and other MP3 download sites: The material that users post is expected to be their original work. ACIDplanet will not condone the posting of any music that is copyright infringed. Of course, because of the sheer volume of music

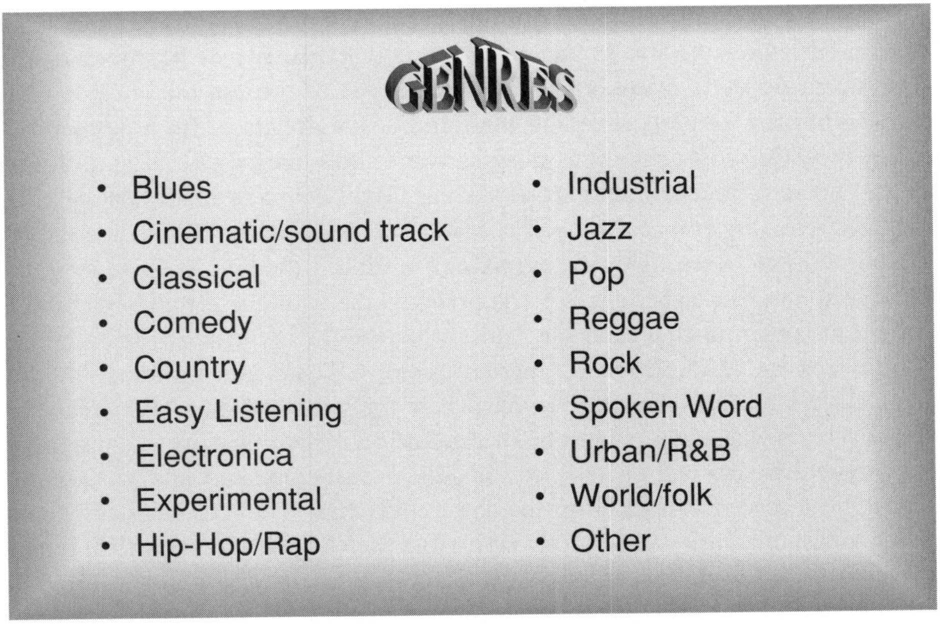

GENRES

- Blues
- Cinematic/sound track
- Classical
- Comedy
- Country
- Easy Listening
- Electronica
- Experimental
- Hip-Hop/Rap
- Industrial
- Jazz
- Pop
- Reggae
- Rock
- Spoken Word
- Urban/R&B
- World/folk
- Other

Figure 20.1 Genre of Music on ACIDplanet (as of 2007).

posted to the site, it is impossible for the webmaster to police music for pirated tunes all of the time. When citizens post songs to the website, they must agree that there has been no copyrighted music used in those songs without permission. If a citizen does post a copyright-infringing song and the posting is discovered by the webmaster or reported by another user, then that song will be deleted and that person will receive a warning. If a user continues to violate the rules and posts copyright-infringing songs, that person can be banned from the website permanently—or at least until the user reregisters at the site using a different email address.[4]

Users also participate heavily in the reviewing portion of the site. As soon as a user posts a song, other users can listen to it and leave their own comments for the original user, as well as a star rating from one to ten. The only formal parameters for reviewing are that the review cannot be malicious or obscene, and if a rating of fewer than five stars is assigned, then that review must be at least twenty-five words long. Reviewing plays a large role in the charts, as the ratings for each song determine its chart position and, consequently, the number of people who are exposed to the song. Reviewing is an interesting feature in terms of sound studies because it forms a "translation node," whereby the sonic dimension—sound—is rendered into another medium (words) (more on this later).

Another major feature of the website is the contests that it sponsors. Several of our interviewees report being first drawn to the site by discovering these contests. Remixing contests are the most common, but sometimes there are also mash-ups (where parts from separate songs are mixed together to make an entirely new musical creation), and contests for original compositions. In remix contests, well-known artists donate one of their songs to ACIDplanet to distribute in the form of separated tracks for ACIDplanet citizens to remix. Entrants can use as many of the tracks as they like and add their own interpretation of the song using their skills, instruments, and software. At any given time, ACIDplanet may host four or five contests, giving users many options regarding type of contest and style of song. Winners of these contests generally win hundreds of dollars worth of Sony products, namely the most up-to-date music software and some loop libraries for future songs. Very occasionally major artists such as David Bowie or Madonna will offer tracks for remixing or mashing up. The Madonna remix contest held in 2002 (three Madonna songs were offered for remixing) produced the biggest spike in traffic the website has ever experienced.[5] The prizes to the winners of these exceptional contests are often much greater (an Audi car for the David Bowie remix).

Citizens can customize their experiences on ACIDplanet in a number of other ways. Dave Hollinden, former webmaster of the site, explains: "ACIDplanet is somewhere . . . where you can just be an absolute hobbyist or be fairly accomplished or a power user or a real casual user and still . . . just come and upload, and, you might find a place for yourself in the community that way."[6] Citizens can create their own list of favorite ACIDplanet artists and thereby provide a way to find their preferred music at the touch of a button. A user can pay to join the ProZone (a higher tier of service), one of whose extra options is to change the color of the user's profile. While this seems a superficial benefit, citizen PhillyC says, "[I]f you're flipping through all these pages that all look the same, and you come upon something where the colors are different, that immediately catches your eye right there. And so you're gonna pay more attention to that."[7]

A big draw of ACIDplanet is the charts. Every piece of music posted is given a chart position. The all-genre chart is the most widely acclaimed. Here the top ten songs at any time are displayed on the ACIDplanet homepage, but there are charts for every genre and subgenre of music. There is also an all-time greats chart. Based on how recently a citizen posted a song and how many stars that song received from other members, that song will appear on all of the relevant charts. For example, an acoustic rock song will appear on the acoustic rock chart, the rock chart, and the all-genre chart—although it probably will not have the same ranking on all three charts. It may be number 2 on the acoustic rock chart, number 7 on the rock chart, and number 865 on the all-genre chart. The chart position (determined by a secret algorithm) depends on how many reviews a song can garner in a short period of time, as well as what the competition is like within a given genre.

Different users place different emphasis on the charts. Some acknowledge that getting into the charts is the best way to get a song heard by the largest possible group of people. Others explain that the charts are not as important to them as just getting their song heard, and still others feel that regardless of how important they say the charts are for them, they still get a tremendous ego boost when one of their songs makes number one. Says PianoPlayer of the time when his song "Baby, I'm over the Pain" went to number one for a week on the soft rock charts, "You know something? My . . . my head was like, out to here [demonstrates head being very large with arms]."[8] We have received numerous emails from users telling us when their music "charts" and giving us a link to the song. While the charts mean different things to different people, the fact remains that more people are exposed to the music at the top of the charts, so the best way for users to get their songs heard in general is to find a way to encourage more people to review them as soon as they are posted. As we will see later, this can lead to attempts to "game" the system to ensure greater chart success.

Users

Who joins ACIDplanet? Learning about ACIDplanet citizens is difficult because, although they may choose to disclose information about themselves in their profiles, there is no way of easily verifying such information. Determining which profile belongs to a person of which gender, ethnicity, or country is not easy. Although the subset of 35 from the 300,000 registered users we have interviewed tend to have accurate profiles as far as we can ascertain, they may be an atypical sample because they agreed to be interviewed. Collecting accurate demographic information from the website is almost impossible because of the possibility of fake profiles and because users may have more than one profile.[9] However, we were able to gain an understanding of users' perceptions of the demographic makeup of the site. For example, everyone we spoke with believed that there were more men than women. The typical user, according to citizen Johnnybreakbeat, is, "you know, [males from] late teens to, I'd say, probably early thirties, you know. Guys who sit in their rooms and make music. That's generally what I'm finding."[10] Indeed, our own experiences confirm the huge number of male users. Despite explicit attempts to find and interview female users and offering an interview with a female researcher who had herself worked on gender and music (Athanasiades), we were able to interview only three female users for this study.[11]

The preponderance of male users of this technology is hardly surprising, given the gender inequities found in the world of computers, synthesizers, samplers, and the like and in the electronic dance music culture so popular at the site (Wacjman 1991; Lie and Sørenson 1996; Faulkner 2000; Pinch and Trocco 2002; Rodgers 2010). The gendered aspect of ACIDplanet was apparent to the few female users we were able to interview. Women who disclosed their gender told us that they are treated differently from men by the men on the site. Citizen Hank Mohaski explains that there is not only blatantly different treatment of women but sometimes outright sexism as well:[12] "If she [a woman] puts a song up, and then you go back and read the reviews for that song, you'll see the stuff percolating up, where it's anything from like, 'Are you hot?' . . . 'What are your measurements?' All that kind of BS [bullshit] stuff that you find in a chat room."

Mohaski also noted that reviews for women artists would sometimes read like pick-up lines and that there was:

> this out-and-out sexism that, you know, "Oh, you're a girl, and you suck and . . . you know, you'll never make it," and . . . then, if those females come into the forums, it's kind of a lot of the same . . . I've known a couple female artists that have left the site, specifically because of that. Because either they weren't taken seriously as an artist or because they were basically being sexually harassed . . . I've heard that the [now defunct] chat rooms could get quite brutal at times when there was a female in there.

Another more humorous example arose when user Sonic Epiphany told of a time when he was flirting with an alternate identity of one of his male friends without even knowing it:[13]

> Sonic Epiphany: Oh, he, he creates pseudoidentities like [Spam Haters], like he has these identities like Jenny . . . McSpamHater. And then he does a song with . . . a girl singing, and her, and he just, "It's Jenny and Trixie! We're like, uh, you know, California fun and tricks."
>
> Katherine [interviewer]: Oh . . . I saw that [on the site].
>
> Sonic Epiphany: That's him! You thought it was a girl? I was flirting with him for a while, and I found out that Jenny McSpamHater was him!

In this case, Sonic Epiphany (a male user) did not know anything about the person or people behind the profile "Spam Haters" except that they said that they were girls. However, they turned out to be fictitious females created by a male friend of his. Later we return to the musical advantages that these multiple profiles sometimes give users. The Internet is an interesting place in terms of gender performances; as well as constraining what is possible, it can enable new forms of gendering to emerge (Kendall 2002).

Johnnybreakbeat's observation that most users were under the age of thirty does not seem to be true for our sample. Indeed, we have continually been surprised by the numbers of people in their thirties and forties who have become users of ACIDplanet. The reason for this seems to be the convenience of the platform and the possibility of making music at home as opposed to the tiring traveling life of a working musician. Several of our respondents work from their own home studios (often with small children around, who are instructed not to mess with the equipment), and at the end of the day or night all they have to do to distribute their latest musical composition is click to upload to the site. For older people who have spent a lifetime in the music biz, working from a laptop at home is a luxury they never dreamed of when on the road gigging.

Most users are based in the United States, where the website is located, although browsing through the site brings up users from all over the globe. Indeed, the preponderance of U.S.-based users can actually serve to attract global users. For example, one Korean user we interviewed, Chester™, made heavy metal–style songs. He derived a big kick from getting his songs positively reviewed by U.S. users because the United States is perceived to be the home of heavy metal music.[14]

To give readers a better feel for who comes to ACIDplanet, we next describe two users, stephanie K and Jim Spitznagel, in more detail.

Stefanie K

Stefanie (figure 20.2) (who spells her profile name with a lowercase s) began playing the piano at the age of five or six under the direction of her father, who was a classical piano teacher. Though he taught her classical music, she says that she "would change classical music into rock 'n' roll" whenever her father went upstairs. She stopped playing when she was fifteen, when her parents divorced and she no longer had a piano. She did not return to piano again until she was twenty-five.

Figure 20.2 Part of stefanie K's Profile on ACIDplanet (as of 2010).

Stefanie became a securities trader but gave up her career when she got married and had two children. In 2001 she was inspired to write a song about a friend who was killed in the World Trade Center attack. When she performed this song on the piano for some friends, they told her that she had some talent, and suddenly she found herself pouring out lyrics at all hours of the day. Of one song she said, "I wrote the lyrics while we were eating dinner at a restaurant—on a napkin. I write lyrics everywhere. I have pieces of paper in every bag I have with lyrics scribbled."

Stefanie began to meet other musicians online through various forums and her own website. One of the people she met was Alex Stangl, who was at that time an active member of ACIDplanet. Stefanie had never heard of the site, but as soon as she began to participate, she was completely drawn in and started actively reviewing and collaborating with different members of the website. She was so involved, in fact, that she says, "I was obsessed for a while. My husband got a little pissed off." Stefanie began spending several hours a day on the website, listening to music, working on her own music, and reviewing other people's music while her kids were at school and her husband at work. Stefanie was so caught up with the site that she had to limit herself. She says, "I used to be on the computer all night, but I cut back. I stopped. . . . I'm pretty minimal now. I have my "Favorite" list, and, you know, people that I like, [if they] put up a new song, I'll listen to it."

Most of the songs that Stefanie posts stem from collaborations. She says that many people contact her because they want her to sing on their tracks. Stefanie does not use ACID software, as she says that she simply does not have the time to learn it. However, she does have a small home studio that consists of a microphone, a keyboard, and a mixer. She usually goes into a professional studio to record her tracks because she says that her vocals come out better when they are professionally done. Generally she will bring a slew of her songs to the studio and have a producer decide which are the most promising and which need lyrical or musical work. After the session, she brings the songs back to her collaborators for reworking. Either Stefanie or one of her collaborators will post the song.

Stefanie says that her kids—especially her daughter—love her music and that all of her kids' friends listen to it (her daughter took the photograph of Stefanie for her profile, which another ACIDplanet citizen further edited). While her husband is not particularly supportive of her hobby (Stefanie subsequently divorced after our study ended), Stefanie says that he "has started to accept it more because I put my foot down. I'm like . . . this is me! It's my passion. I found it, and . . . you know this is what I wanna do.") An added bonus is that Stefanie can follow her passion from home, which means that she can still be attentive to her kids and the chores. Stefanie would love for her songs to get radio play or to be recognized by producers and be adopted by mainstream artists. She does not foresee herself actually performing the songs because she says she does not see herself becoming a woman who hangs out and performs in bars.

Stefanie has been compared to Tori Amos, Chrissie Hynde of the Pretenders, Kate Bush, and even Jewel though she disagrees with the last comparison the most. The songs she has posted on ACIDplanet that have attracted the most attention are mainly rock and pop, but she has one song she describes as pan-Asian, which she sings in Japanese. As of 2007 Stefanie had two songs in the top fifty all-time top songs chart on ACIDplanet: "Regrets and Revelations" and "Rock Me."[15]

Jim Spitznagel

Jim Spitznagel developed an interested in music when he was a teenager. When he was fourteen, his parents got him a guitar for Christmas, and since then he has been playing both as a hobby and professionally. He was in a number of bands while growing up and worked odd jobs here and there in the meantime. When he was in his midtwenties, Jim finally got settled in a cover band, Eddie and the Otters, and also owned a well-known record store in Pittsburgh called Jim's Records. Jim became a promoter and ended up working with many big-name bands before they broke out5, such as the Police, the Pretenders, the Ramones, and U2, to name a few. Jim's home studio has many photographs of him posing with famous New Wave artists. Eventually Jim retired, sold the record shop, moved to Ithaca, and started his own online music store called JIMS, or Jim's Ithaca Music Shop. He also started his own record label, Level Green Recording Company. Jim bought and became acquainted with music studio software such as ACID. Though he recorded other musicians for a while, he eventually decided to record and produce his own music exclusively. Jim has produced several of his own CDs to date and promotes them on ACIDplanet. He is a ProZone member, a perk he acquired by virtue of being a beta tester for an updated version of the site.

Jim has two profiles at ACIDplanet. His main profile, "Spitznagel" (figure 20.3), is where he posts his acid-jazz compositions. In 2005 he began posting music he had created with his recently purchased Moog Voyager synthesizer under a new name, "SMX" (which stands for Spitznagel's Moog Experiments). He created the new profile so that the reputation of his Spitznagel profile would be unaffected by how well

Figure 20.3 Part of Jim Spitznagel's Profile on ACIDplanet (as of 2010).

or how poorly his new, more experimental compositions were received in the ACIDplanet community. Jim, unlike Stefanie, does not participate in collaborations simply because he does not have the time. He writes, records, and masters all of his own music. He has a very high standard for himself and an excellent working knowledge of the software.

Jim is able to produce high-quality tracks time after time. He sees ACIDplanet, first and foremost, as a way of getting more exposure to his music—as he says, "I look at it as a marketing tool. And that's what I always saw it as." Like Stefanie, Jim's music is good enough that he regularly climbs the charts whenever he posts a new song. His goal is "getting people to hear [the song]; hopefully some of those people who hear it will wanna buy my CDs." As well as making music CDs and tracks for ACIDplanet, Jim also plays the occasional live gig using his Theremin, a recently acquired Tenori-on, and a stack of software synthesizers. He sometimes plays with another electronic artist as part of the band, the Beamis Point. In a surprising twist for both parties, Jim has recently started a musical collaboration with one of the authors of this chapter (Pinch), and they perform as the Electric Golem.

Contrasting these two citizens illustrates the different ways in which users have appropriated the website and its options. Stefanie uses ACIDplanet to test new songs, but she also sees it as a final destination for many of her songs. Jim uses ACIDplanet as a marketing tool for his CDs because it gives his music increased exposure. Stefanie mainly collaborates with other users, while Jim works mainly as a solo artist. Stefanie does not use any sort of music-creation software and has producers fine-tune her music; Jim uses music-creation software and masters all of his own work. Stefanie is a pop-rock artist, and Jim is a jazz and experimental musician.

New Practices and Terms

The existence of ACIDplanet as an online place where people can congregate, create their own musical identities, and share their music has led to several interesting developments. Citizens have created new terms to describe particular actions or events. For example, to be a "chart whore" or to partake in "chart whoring" means that a person is trying to climb the charts by using methods that are unacceptable to the ACIDplanet community.[16] One way to "chart whore" is for the original posters to create many fake profiles to review their own song in order to get more stars and to climb the charts faster. Sending mass email solicitation for reviews of a particular song is another way to chart whore. A more labor-intense way is to write meaningless reviews of other users' songs. Short positive reviews are offered along with a ten-star rating. This method (which is remarkably different from the peer reviewing practices found in the sciences) works by the implied pressure on the user to return the favor (anonymous reviewing is not allowed) and to review the reviewer's own latest song.[17] This practice is so widespread at the site that it has its own special term, R = R, or Review = Review.[18] Citizen mtkzoa explained how this worked to boost one of his songs to an even higher chart position:[19]

> Basically, this "Be Revived!" [title of song] thing is the first time—I posted it, and I got a couple of hits on it, and . . . by accident I found out I was like in third place. So I said, "Well, I'm gonna do some reviews and see if I can push it up," and I reviewed a few people, and then word gets around to other people, and they start reviewing your stuff, and the next thing you know, I was number one. It's been that way for like five days now.

Reviewing

Reviewing is one of the main activities at the site. In order to understand how it works we look in more detail at some of the first few reviews received by one of stefanie K's most popular songs, "Rock Me" (181 reviews as of February 2010), and Jim Spitznagel's most popular song, "Pharoah's Corvette" [*sic*] (90 reviews as of February 2010). The genres of both songs are very different. "Rock Me" is posted under the genre of "Rock," and "Pharoah's Corvette" under the genre of "Jazz" and the subgenre of "Acid Jazz." "Rock Me" is a guitar, drum, vocal-based song, which allows Stefanie's voice to shine. "Pharoah's Corvette" has a much more ambient sound and features what sounds like brass instruments, as well as washes of electronic sound. The reviews for these songs are typical of the site as a whole in that the authors seem to have penned them hastily—almost as part of a stream of consciousness. The reviews are frequently ungrammatical and have many spelling mistakes. They appear to have been written casually as the first reaction on listening to the song (or even while it is playing) and are rarely carefully crafted as in more formal record reviews. Nevertheless, much detailed feedback can be gleamed from them.

The very first review of "Rock Me," posted under the heading "Radio Ready," which signals the reviewers' opinion of the song's commercial potential, is as follows:

> Very Nice job by all . . . love the blues leads on the right . . . Love the placement of the backing vocals and the bull horn FX on the vocals was cool too . . . great lyrics too!...........kudos....................peace. Skip

One of the first reviews of "Pharoah's Corvette," under the heading "This is great," was actually written by stefanie K (many of the leading artists on ACIDplanet review each other's work):

> Quite intresting [*sic*] . . . you have certainly put alot of work into this mix . . . its very busy in certain places . . . I love the horn but I think you could have maybe cut it back a bit, and than [*sic*] it would be more dramatic when it appears . . . still

trying to figure out why you call it Pharoahs corvette. cool title, does it refer to King Pharoah . . . nice job overall! stefanie K

Both reviewers note the features of the tracks they "love," such as the "blues lead," "backing vocals," "great lyrics," and "horn." In the case of "Rock Me" the reviewer notes the placement of the "blues leads on the right" and also the treatment added to the vocals.

Other reviews note exact timings in the track. For example, in the case of "Rock Me," one reviewer writes, "dig it at 2:08, very cool . . . this is put together real nice . . . gutars [*sic*] sound great . . . 3:12 is cool too . . . voice sounds great . . . hot track for real . . . respect." Another reviewer is more specific: "Great guitar "solo" around 1:19. The mini-breakdown at 3:11 is very smooth—great feel." Reviewers also note in a friendly way what they think could be improved, such as in Stefanie's recommendation to Jim on his use of the horn: "Cut it back a bit" so it "would be more dramatic when it appears." Often names of well-known artists, whom the track reminds the reviewer of, are mentioned. For instance, Jim's track is described by one reviewer as being like the synthesizer-based jazz artist Adham Shaikh. Most reviews are very short, and a few are little more than positive endorsements of the artist and are explicitly of the R = R type in inviting the artist to check out the reviewer's own music. However, even those reviews that seem to imply, demand, or return reciprocation can be full of information. For instance, one of the first reviews of Pharoah's Corvette acknowledges an earlier review Jim wrote of the reviewer's own music but still offers its own detailed comments on what is good about Jim's track: "[I] like the robot voices and the cool horn playing excellent combinations and great production thnx for the review of 'Beyond the Shores of Time' ed (AE)."

Even a short tongue-in-cheek review that might verge on sexism can reveal the appeal of the music, as in this early review of Stefanie's track: "Really starting to put a rump in your hump lately (lol) Edgier and heavier . . . and we really like the direction your [*sic*] taking . . . Great work!"

These reviews are multilayered in the way they work. At the most fundamental level they allow the listener to render the sounds into the medium of words. The effect of a song on a listener is often expressed in metaphors such as "hump in your rump." However, what is striking about these reviews is how they often mimic the sort of language found in recording studios, where the studio engineers, record producers, and artists have to discuss what sounds good or bad about the unfolding recording. Many of the comments combine the technical terms that recording engineers use (e.g., FX), technical features of the music (e.g., arpeggio), references to the particular sound of well-known artists, and descriptions of the sounds themselves, such as being "heavier" or "edgier."[20] By turning sound into words the reviews make it possible for the users to communicate about sound within this community of technical practice. They enable the users to know and understand their listeners and offer guidance to all as to how to improve their music-making abilities and production skills.

Reviews are, however, double edged. Not only are they a means of translating sounds into words, thereby providing invaluable feedback and the chance to improve production techniques and so on, but they also form the basis of a user's *reputation* at the website. The number of stars the review receives and subsequent chart placement ultimately determine the artist's success at this website. As well as the reviews' shaping what we might call the "reputation economy" in this community, the reputation economy also shapes the reviews. For instance, because of the norm of reciprocation (R = R), those artists who are successful on ACIDplanet end up writing an enormous number of reviews. At the time of writing (February 2010), Stefanie had posted 2,566 reviews and Jim 3,541. In principle, reviews can be as long as the authors want them to be; the short "stream of consciousness" form most of them take follows from the need to write so many. Writing numerous reviews without repetition can itself be a problem as Jim notes: "If I feel that I've written a lot, the same thing, I'll get the dictionary out and change a few words. And I just, say, I'll change 'creative' to 'inventive.' Oh, that's good! I'll change, you know, 'cool' to 'groovy.' Oh, that's good, you know!"[21]

There is, of course, no common rubric, so to speak, for writing reviews. Different reviewers will look for different things and evaluate compositions in a variety of ways. Often, when these reviewing rubrics conflict, feelings can be hurt and users can get upset. Jim, for example, gives only positive reviews. He sees reviewing as encouragement for the reviewed to continue making music and does not feel that he is in a position to critique them and their styles. Some people, however, feel that a central element of honest reviewing is critiquing other users' music and thus rate and review people's music accordingly. Nevertheless, the reviews at the site are overwhelmingly positive. This again follows from the need for reciprocation. As in most reviewing economies, reviewers self-select and simply do not offer a review at all if they do not like the music. Even a nine-star review as opposed to a ten-star review can be seen as a negative appraisal. Very negative reviews, perhaps posted from bogus profiles, are discounted as being "drive-by shootings," written by reviewers who have ulterior motives (known as "haters"), such as to damage the reputation and chart placement of a rival artist. So again the reputation economy shapes in subtle ways the form the reviews take.

Users often get disillusioned by the reviewing process—there are regularly discussions in the forums about the lack of a standard for reviewing (encouragement or critique), the high incidence of nines and tens and the negative response when negative reviews are given. Hank Mohaski, for example, stopped reviewing a few years ago because he was so disheartened by the reviewing process:[22]

> I thought the amount of time and effort I was putting into [reviewing], uh, I wasn't getting any return on that investment, so to speak. . . . So I've become completely disillusioned with that aspect of it. Having said that, I still think it's an important and valuable mechanism to the site. You know? It . . . serves its purpose very well for the people who care about it.

For most users. reviewing allows them to gain recognition (and reputation) incrementally, both in the act of reviewing and in the act of being reviewed. With reviewing, reviewers can build a supportive, encouraging, and perhaps technically savvy persona and become more likely to have their music exposed when the R = R expectation is followed. When a user is reviewed, there is not only the information to be gleaned from a detailed review but also the boost to chart placement that a good review can give. In addition, recognition is gained since many other users can read the review. Sometimes a particular reviewer will gain a following, and this can lead to even more exposure for a song. For example, the user :SEG: says that he gained popularity among the fans of popular artist MST when MST reviewed :SEG:. This is because MST is a widely respected ACIDplanet artist; thus, people respect his tastes in other users' music.

Collaboration

As we have already noted with Stefanie, ACIDplanet offers new ways for its users to collaborate. For example, it has inspired the creation of several new online forums that exist solely for members of ACIDplanet to continue their dialogues and music creation in a smaller, more private setting. One of these forums, the Acid Exchange (known as AE), has evolved, according to our respondents, into a place where online friends hang out, chat, and create and share music. The forum is a closed site: In order to become a part of AE, users must be very active on ACIDplanet, display a high degree of technical competence in all aspects of their music, and be nominated by one or more people who are already in the forum. These forums are not part of the ACIDplanet site itself but provide a space for community building among particular groups of users.

Musical collaboration of a more straightforward nature is also catalyzed by ACIDplanet. Users state on their profiles whether they are open to collaborations. These collaborations are usually accomplished virtually, as was the case with Stefanie's collaborations. The "virtual band" aspect of ACIDplanet is enticing to some users who may recruit or advertise for particular sorts of musicians they would like to collaborate with. Sometimes virtual collaborations will turn into physically copresent collaborations in which musicians enter each other's studios and make music together or even gig together. Sometimes ACIDplanet users get together to attend gigs or trade shows. One user who lived in London told us he regularly hosted overseas users in his London apartment. ACIDplanet therefore serves as both a destination for users' music and a site where people are able to make connections and expand their music and musical identities.

The new forms of collaborations that ACIDplanet offers can sometimes lead to tensions. Stefanie described one such incident that developed in her collaboration on the song "Simple Folks." Citizen JC Kercheval wrote the instrumental for the song and then gave the song to ACIDplanet artists Mark Moffre and Richard Montefiore to develop. He stated that whoever worked the song first could have it. Mark asked Stefanie whether she would collaborate with him on it, and they produced a new song using JC's original tracks. However, as soon as they finished it, they discovered that Richard Montefiore had already recorded his version and posted it on ACIDplanet. Usually, unless the song is for a remix contest, it is against unwritten policy to post a song that someone else has already performed. However, because JC had given his tracks to both Richard and Mark and because the two songs were somewhat different, Mark and Stefanie decided to post their version—with an explanation of what had happened and encouraging people to listen to Richard's version as well. In this spontaneous competition, both songs climbed the charts—and although Richard told people that it was a friendly competition, Stefanie says that he was really angry about the situation.

Musical Identity

Online identities and their flexibility have long been seen as key components of the Internet (e.g., Turkle 1997). One of the fascinating things about making online music is that it allows the musician's identity to be detached from the physical creation of the music in a new way. It also enables new forms of multiple musical identities not easily realizable in any other way. The separation of the original work of art from its reproduction (Benjamin 1968) is of course a long-standing feature of recorded music (Mowitt 1987). Advances in recording techniques and technologies of synthesizers and samplers, including software synthesizers, have further distanced the musician from the work of art, but, clearly, there are still musicians who are identified as somehow having enacted or realized works of art (Pinch and Bijsterveld 2003; Katz 2004). For instance, although not realizable in live performance because of its complex studio techniques, the music of *Sergeant Pepper* is still identified with the agency of the Beatles (although people might refer to the "fifth Beatle," Sir George Martin, in recognition of his production skills). With online music, such as that posted at ACIDplanet, musical identity and agency are much more complicated things: Often all there is is a profile attached to the music. The identity of the musicians (if any) and their agency in creating the music is now often solely revealed by the profile, and because musicians on ACIDplanet can have more than one profile they can, as we saw with the case of Jim and his experimental Moog Voyager music, maintain *more than one musical identity*. Furthermore, because a profile can be easily abandoned and a new one created, this form of online music making gives musicians an opportunity endlessly to re-create themselves.[23]

Of course, in that online identities have real-world counterparts, the affordance ACIDplanet offers might seem of limited value. However, what must be borne in mind is that many musicians on ACIDplanet release their music only on the website and never perform or meet up, as it were, "live." The links between profiles (if they exist) may of course indicate that it is one and the same person behind the profile, but the formal separation of the reviews and the reputations they garner means that a separate online music identity can be viably pursued.

One of our respondents, Hank Mohaski, claims to have many different online identities. Although he generally operates under what he deems his "main identity" of Hank Mohaski, he has at least twenty other profiles. Each of these was created to represent a different type of music. For example, his profile "Tape Jazz Massacre" is devoted entirely to music that comprises samples of songs and sound bytes he has collected over a period of years. Another profile is for the "Zanzibar Cheap Band," which plays music for elevators and supermarkets. Another example of a user with multiple profiles is SkySwim. He was the creator of "Spam Haters," the female hard-rock group who gained the attention and love of hundreds of users on ACIDplanet.

ACIDplanet as a Community

The "community" aspect of ACIDplanet, which makes it unique and differentiates it from other file-sharing and music-downloading sites, is still evolving. Furthermore, the crucial role the users play was realized very early on by the site's developers. The original webmaster told us the following:

> [I]n the end, what we create is a shell that's filled by the users, you know. . . . [Y]ou could launch this same site today empty, and people would come and look around like [it was] a big empty house and leave. And the users feel a lot of ownership in the site, and sometimes when we talk about bringing in a policy they don't like, they'll be very quick to say, "You know, we provide all the content for this site, you know. Don't forget that."

The webmasters, who are also musicians and users of the site, are very aware of the role that users play in shaping the website.[24] They pride themselves on listening to users and providing features that allow them to offer feedback and suggestions for what they like or do not like and what they would like to see developed in the future. Julian Bain, the webmaster (as of 2007), explains how this interaction with users has been successful in the past and compares ACIDplanet to a different, failed Sony site:[25]

> There was another site that Sony had, where they wanted a place for consumers to put what they made with Sony products. And that was Screenblast. But they went at it from the point of view of what they, Sony, thought [it] should be. And they had a huge staff, and they put together a very nice-looking site. But it didn't work out. And I think a lot of the reason for that, compared to us, who [are] just two people basically working on a daily basis, is that it's very user-driven, for our site, and Screenblast didn't turn out to be as [much of a] user-feedback-oriented site.

The users at ACIDplanet not only provide most of the content for the site but also form a devoted community who regularly return to the site to post songs, enter competitions, post reviews, and in many cases actually care about what is happening at the site. Users who quit the site are sometimes bombarded by emails from other users to return (and often do return). This community appears to be more organic than regular file-sharing systems.

Part of this process of community formation lies in the special form of mediated interaction that the website offers. Most social networking websites offer users a chance to interact by posting content, having friends, posting links, making lists of favorites, and engaging in certain forms of mediated interaction (Boyd 2006). The problem all such websites face is that every interaction with the website is in actuality discrete. The community can only form around the discreet users interacting with the traces (in the case of ACIDplanet, the songs and reviews) left at the website by other discrete users (Welser 2008). The more users that are involved, the greater the possibility that others can monitor these traces and the interactions users have with them (in turn producing new traces)—forming in a way an "audience" for monitoring the interactions (e.g., a review posted to a piece of music). All of this "sociality" takes place even though in principle there may be only one user at the website at any one time. The sociality of such a community comes from the norms and obligations built up incrementally among discrete users over the course of many such interactions. The more such interactions take place and are witnessable by many users, the more likely it is that any emerging community norms will be reinforced. In short, almost every interaction at the site is a public interaction visible (and often audible) to all users. This is unlike most social networking sites, where users can, via privacy settings, restrict content to a selected group of users.

ACIDplanet seems to go beyond most purely social networking sites not only by encouraging free access to most parts of the site but also by offering mechanisms whereby potentially *large numbers of users* are directed to particular parts of the content that users in turn produce. Special sociotechnical instruments, scopic focusing technologies (Knorr Cetina and Grimpe 2008)[26]—such as the charts and contests, which generate hierarchically ordered outcomes—allow users to focus on certain selected pieces of content (and thereby boost the reputation of the profile associated with that content). Part of this process, as we have seen, involves "translation nodes" or "transductions," whereby the sonic dimensions of the site are rendered into other symbolic forms (mainly words and numbers).

In the wider field of popular music, hit charts (not to forget also the gaming of such charts with the payola scandal, in which radio DJs were paid by record companies to promote certain records) and contests are very old ways of building reputations and directing consumers of music to particular songs. What the website does is to take these conventional resources and rework them in a new way. The key thing is that the working of these mechanisms in the online world becomes heavily dependent on the activities of users. Activities such as reviewing (which determines chart positions) and the judging of song contests by users (some contests award prizes for successful judges as well)[27] do not make these mechanisms any more transparent (the exact formula for chart placement, for instance, is a well-guarded secret) but do give users involvement and empowerment. This means that reputation is social not only in the sense that the outcomes of charts, contests, and the like are socially available to all users but also in the way the outcome is reached, which is itself dependent upon the users' social activities. This "double sociality" helps turn discrete users into an active and organic community. Because everyone is potentially both a producer and a consumer, everyone shares the norms and obligations that develop, and there are fewer divisions between "fans" and "artists" than in other musical communities.

Conclusion: A New Sonic Sociotechnical Economy of Reputation

Underpinning key features of the activities at the ACIDplanet website, such as reviewing and building musical identities, are the new economies of reputation that online technology enables (Masum and Zhang 2004; Resnick et al. 2006; David and Pinch 2006). The musical profiles and identities that musicians establish on ACIDplanet are subject to a complicated set of sociotechnical processes whereby some identities succeed more than others. Some of these features are obvious: Profiles that chart a lot are more successful than those that do not; competition winners gain recognition, and those profiles that never attract anything other than negative reviews for their songs or, perhaps worse still, no reviews at all are unlikely to gain recognition in the community. As with any community, however, the mores and norms are nuanced and, as in the case for the fast-changing technology of the Internet, somewhat in flux.

Ultimately the reputation of users at ACIDplanet depends upon the sounds they produce and how others judge these sounds. Interestingly, although sound is at the core of the site—after all, it is a music site—most of the interaction users experience is via written and symbolic traces such as profiles, reviews, charts, and the framing of contests and their outcomes, not to mention the numerous forums. The translation, or what Stefan Helmreich and others in this volume call the "transduction" from sound to another medium, is at the core of the reputation system. In short, the importance of the new technology is that it allows music and sound to be not only more easily produced but also turned into a symbolic form that a community of users can appraise and rank. The point we wish to stress is that the website, ancillary technology, and users together form what we call a *sonic sociotechnical community*.

Sound, as always, thus gets mediated by technology and community. What is ultimately unique about this community is that producers are consumers and vice versa, and, as we have seen, this directly affects the norms, identities, and reputations by means of which the community operates. Online music of this sort is capable of not only *radically reconfiguring* the ways musicians can form identities and reputations but also *building upon and reinforcing* rather conventional forms of musical identity and ways of allocating status (such as charts and contests).

In this chapter we have shown not only the new possibilities that this new form of digital music making offers but also some of the problems and contradictions that arise. Although new musical identities are possible, old gendered identities are still very much manifest. Although new systems of building reputation emerge, they are based on the "charts" and "star" potential of older systems. The worlds of online and digital music are still in flux as the recording industry responds to a computer company, Apple, which is becoming the major distributor of music via its online store, iTunes.

Contradictions abound as to the future of online music production and consumption in general. Some see it as the way of the future, and others as a blind alley that will lead to little change and domination by a few major labels allied to the new distributors of online music such as iTunes. When diffracted through our users' stories and their clear attachment to ACIDplanet and the user-generated community around it, the brave new world of online music appears perhaps a little less brave.

NOTES

1 For accounts of online music in general see Beer (2005) and Baym (2007).

2 Interview with Dave Hollinden, Aug. 15, 2005.

3 It is possible for a group of musicians to post their music from one profile, but there must be one registered user who "owns" the profile.

4 For discussion of the complicated issues surrounding online musical copyright see Howard-Spink (2005), McCleod (2005), and Gillespie (2007).

5 As determined by Daily Reach; www.alexa.com (accessed Oct. 19, 2006).

6 Interview with Dave Hollinden, Aug. 15, 2005.

7 Interview with PhillyC, July 7, 2005.

8 Interview with PianoPlayer, July 22, 2005.

9 Of course, it is likely that Sony has compiled more accurate information as the company has access to web logs and individual IP addresses.

10 Interview with Johnnybreakbeat, July 16, 2005.

11 Katherine Athanasiades wrote her undergraduate thesis at Swarthmore College on women Czech rock musicians (*Rockerka* and Gender: Exploring Women in Rock in the Czech Republic, May 2005). Of the three women users we interviewed, one said she agreed to be interviewed only because the interview was to be conducted by a woman researcher.

12 Interview with Hank Mohaski, Aug. 19, 2005.

13 Interview with Sonic Epiphany, July 14, 2005.

14 Interview with Chester™, Aug. 8, 2005.

15 As of 2010 these songs no longer appeared in the all-time top fifty.

16 The use of the term *whore* further reflects the gendering of the site.

17 As of 2009 reviews can be posted anonymously. This change has been made almost certainly to try to prevent R = R from occurring. More recently, star ratings are no longer visible in each review, though what can be seen is an "artist score" and a "song score" on a scale of one to one hundred. Additionally, users can post a simple or an advanced review (the advanced review has separate components to comment on and rate, such as "production values") in order to try to fulfill the requests of many users to have a more meaningful reviewing system.

18 There was even a song posted at the ACIDplanet website titled "R = R."

19 Interview with mtkzoa, June 27, 2005.

20 For a detailed analysis of this sort of talk see Tom Porcello, "Speaking of Sound: Language and the Professionalization of Recording Engineers," *Social Studies of Science* 34 (2004): 733–58.

21 One suspects many reviews on ACIDplanet are copied. In our study of reviews at Amazon.com we found about 1 percent of copied material (David and Pinch 2006).

22 Interview with Hank Mohaski, Aug. 19, 2005.

23 This possibility of having more than one musical identity has been practiced before in the offline world, where, for instance, bands such as Devo and XTC recorded under different names. ACIDplanet, by making this option democratically available to all, shows again the tensions between change and continuity at the website.

24 The users' role in the development of technology has recently gained much attention (e.g., Oudshoorn and Pinch 2003).

25 Interview with Julian Bain, Aug. 18, 2005.

26 Karin Knorr Cetina (e.g., Knorr Cetina and Grimpe 2008), in her work on financial markets, suggests that computer screens and other sociotechnical devices that bring the whole market to the trader can be thought of as "scopic" systems. The sorts of devices we have in mind do the opposite because, rather than bringing the whole market (actually a representation of it) to a user, they bring a particular piece of content to a large number of users. Most voting technologies popular on the Internet for rank ordering items are of this type.

27 Contests used to be judged by the artists themselves or "their people" at the record label. However, under pressure from users who wanted more say in determining the outcome of these competitions, in 2007 ACIDplanet introduced a new form of contest called "the mosh" (now renamed the "duel"), in which users can vote for the winning entry and in the process also be rewarded for their voting behavior (see Pinch and Athanasiades 2009).

REFERENCES

Bakker, J. I. (Hans), and T. R. A. (Theo) Bakker. "The Club DJ: A Semiotic and Interactionist Analysis." *Symbolic Interaction* 29 (2006): 71–83.

Baym, Nancy K. "The New Shape of Online Community: The Example of Swedish Independent Music Fandom." *First Monday* 12(8) (August 6, 2007). http://www.firstmonday.org/issues/issue12_8/baym/index.html (accessed January 4, 2008).

Beer, David, ed. "Music and the Internet." *First Monday*, Special Issue 1 (July 2005). http://www.firstmonday.org/issues/special10_7/ (accessed January 4, 2008).

Benjamin, Walter. "The Work of Art in the Age of Mechanical Reproduction." In *Illuminations*, ed. H. Arendt, 219–53. New York: Schocken, 1968.

Boyd, Danah. "Friends, Friendsters, and Top 8: Writing Community into Being on Social Network Sites." *First Monday* 11(12) (December 4, 2006). http://www.firstmonday.org/issues/issue11_12/boyd/ (accessed January 4, 2008).

David, Shay, and Trevor Pinch. "Six Degrees of Reputation: The Uses and Abuses of On-Line Reputation Systems." *First Monday* 11(3) (March 2006). http://www.firstmonday.org/issues/issue11_3/david/ (accessed January 4, 2008).

Faulkner, W. "The Power and the Pleasure? A Research Agenda for 'Making Gender Stick' to Engineers." *Science, Technology, and Human Values* 25(1) (2000): 87–119.

Friedman, Eric, and Paul Resnick. "The Social Cost of Cheap Pseudonyms." *Journal of Economics and Management Strategy* 10 (2001): 173–99.

Gillespie, Tarleton. *Wired Shut: Copyright and the Shape of Digital Culture*. Cambridge, Mass.: MIT Press, 2007.

Gomart, E., and A. Hennion. "A Sociology of Attachment: Music Amateurs, Drug Users." In *Actor Network Theory and After*, ed. J. Law and J. Hassard, 220–47. Oxford: Blackwell, 1999.

Howard-Spink, Sam. "Gray Tuesday, Online Cultural Activism, and the Mash-Up of Music and Politics." *First Monday*, Special Issue 1 (July 4, 2005). http://www.firstmonday.org/issues/issue9_10/howard/ (accessed January 4, 2008).

Katz, Mark. *Capturing Sound: How Technology Has Changed Music*. Los Angeles: University of California Press, 2004.

Kendall, L. *Hanging Out in the Virtual Pub: Masculinities and Relationships Online*. Berkeley: University of California Press, 2002.

Knorr Cetina, Karin, and Barbara Grimpe. "Global Financial Technologies: Scoping Systems That Raise the World." In *Living in a Material World: Economic Sociology Meets Science and Technology Studies*, ed. Trevor Pinch and Richard Swedberg, 161–90. Cambridge, Mass.: MIT Press, 2008.

Lie, M., and K. H. Sørenson. *Making Technology Our Own? Domesticating Technology into Everyday Life*. Oslo: Scandinavian University Press, 1996.

Masum, Hassan, and Yi-Cheng Zhang. "Manifesto for the Reputation Society." *First Monday* 9(7) (July 5, 2004). http://firstmonday.org/issues/issue9_7/masum/index.html (accessed January 4, 2008).

McCleod, Kembrew. "Confessions of an Intellectual (Property): Danger Mouse, Mickey Mouse, Sonny Bono, and My Long and Winding Path as a Copyright Activist–Academic." *Popular Music and Society* 28(1) (2005): 79–93.

Mowitt, John. "The Sound of Music in the Era of Its Electronic Reproducibility." In *Music and Society: The Politics of Composition, Performance, and Reception*, ed. Richard Leppert and Susan McClary, 173–97. New York: Cambridge University Press, 1987.

Oudshoorn, Nelly, and Trevor Pinch, eds. *How Users Matter: The Co-Construction of Users and Technology*. Cambridge, Mass.: MIT Press, 2003.

Pinch, Trevor, and Katherine Athanasiades. "Performing Online Interaction: From Market Pitchers to Mashups and Moshes." Unpublished paper, 2009.

Pinch,Trevor and Karin Bijsterveld. "Should One Applaud? Breaches and Boundaries in the Reception of New Technology in Music." *Technology and Culture* 44 (2003): 536–59.

Pinch, Trevor, and Frank Trocco. *Analog Days: The Invention and Impact of the Moog Synthesizer*. Cambridge, Mass.: Harvard University Press, 2002.

Resnick, Paul, Richard Zeckhauser, John Swanson, and Kate Lockwood. "The Value of Reputation on eBay: A Controlled Experiment." *Experimental Economics* 9(2) (2006): 79–101.

Rodgers, Tara. *Pink Noises: Women on Electronic Music and Sound*. Durham, N.C.: Duke University Press, 2010.

Shiga, John. "Copy-and-Persist: The Logic of Mash-Up Culture." *Critical Studies in Media Communication* 24(2) (2007): 93–114.

Sterne, Jonathan. *The Audible Past: Cultural Origins of Sound Reproduction*. Durham, N.C.: Duke University Press, 2003.

Théberge, Paul. *Any Sound You Can Imagine: Making Music/Consuming Technology*. Hanover, N.H.: Wesleyan University, 1997.

Turkle, Sherry. *Life on the Screen: Identity in the Age of the Internet*. New York: Simon and Schuster, 1997.

Wajcman, J. *Feminism Confronts Technology*. Cambridge, UK: Polity, 1991.

Welser, H. T., M. A. Smith, E. Gleave, and D. Fisher. "Distilling Digital Traces: Computational Social Sciences Approaches to Studying the Internet." In *Handbook of Online Research Methods*, ed. N. Fielding, R. M. Lee, and Grant Blank, 116–40. London: Sage, 2008.

SECTION VII

MOVING SOUND AND MUSIC: DIGITAL STORAGE

CHAPTER 21

ANALOG TURNS DIGITAL: HIP-HOP, TECHNOLOGY, AND THE MAINTENANCE OF RACIAL AUTHENTICITY

RAYVON FOUCHÉ

THE history of the black diaspora is replete with examples of the ways music has enabled various black cultural communities to cope with racial oppression, carve out livable physical and psychological niches within an adversarially racialized world, and create supportive environments for creative expression (Floyd Jr. 1995). From the spirituals sung by slaves during work in the nineteenth century to the protest songs that motivated black freedom movements around the world in the mid-twentieth century, music has been a constitutive part of the historical fabric that defined black social, political, and cultural agency (Ward 1998). With the global unraveling of de jure segregation, the direct link between music and black empowerment has waned. By the early 1970s in the United States, black music was beginning to lose some of its visible and audible political urgency. Creative energies were redirected to foment a new movement supporting urban cultural expression. This movement would eventually be known as hip-hop (George 1998). Though hip-hop is not singularly black or only a genre of music, some of the most dominant representations of hip-hop center upon two sets of artifacts: turntables and vinyl records. These objects have been positioned to represent the nostalgic pasts, the vibrant presents, and the emerging futures of hip-hop.

For most black social, political, and cultural movements, words and iconic figures have been the dominant representations of black life (Green 2005). Empowering and polarizing figures like Frederick Douglass, Booker T. Washington, Frantz Fanon, Martin Luther King Jr., and most recently Barack Obama have orated black consciousness. For hip-hop, the lyrics, performances, and significant figures are critical to the historical construction of hip-hop. However, turntables and vinyl records demand similar historical and cultural recognition that in certain instances trump the power of a human actor. A head-bobbing and headphone-clad DJ bent over a set of turntables while spinning, mixing, and scratching has become an iconic representation of hip-hop (Goldberg 2004). Yet, within the last decade a new type of DJ has challenged this image: the digital DJ. These DJs are still seen grooving with headphones perched on their head, but they are now hunched over a new set of musical tools that digitally process inputs, signals, and sounds. The digital revolution came late to hip-hop turntables but had profound implications for the culture nonetheless. On the surface it would seem that this technological evolution would be welcomed, but by some it was not. For those wedded to analog turntables, needles, and vinyl records, these new devices were undermining a valorized historical representation of hip-hop. When forward-thinking or opportunistic companies (depending on your subject position) designed new technologies to take advantage of digital music, fissures within this community of hip-hop aficionados emerged. The crisis was not only about the technology,production, and manipulation of sound but also about one's access to, membership in, and identification with hip-hop. For some of the most ardent hip-hop heads, these new technologies could change the community forever. Yet in our current technological age, such change should not be seen only as an either/or option.

For hip-hop this tension, which had social, cultural, acoustic, and racial underpinnings, was resolved through a new technology aptly named a "digital vinyl system," which allowed participants to stay true to their version of the roots of hip-hop while simultaneously embracing the sonic and technical opportunities available with digital instruments. This controversy over technological choice and use that defines a community was resolved by embracing both sides of the contested space through a new technology created by an interested third party. Thus, a new technology provided a solution to a technocultural controversy. One of the first globally available technologies to fill this space was a hardware/software digital vinyl system named FinalScratch, which enabled DJs to map digital media onto a specially encoded vinyl record using an analog turntable. Surprisingly, this device was not created by hip-hop fans or turntablists but by a small Dutch software company, N2IT. The development, reception, and integration of digital vinyl systems illustrates how technological tension between conceptions of hip-hop culture, identity, and authenticity can be resolved by a mediating technology.

Within the large body of scholarship on black cultural expression and community formation, technology and the ways that black people all over the world have used, created, and redefined technology are somewhat overlooked. Arguably, the role of technology in these spaces is most often viewed as either an adversarial or a benign affordance that human actors overcome or instrumentally use (Eglash et al. 2004). This chapter explains how sound-producing technology, in the form of vinyl records and turntables, functions within communities that endow these devices with cultural value. Hip-hop is used to center the discussion on the ways in which turntables and vinyl records are ascribed a racial authenticity not seen in other music communities where DJs exist. The racialized and nostalgic value of these artifacts for those inside, outside, and on the fringes of hip-hop is linked to familiar narratives of the destruction of black cultures, histories, and pasts that can be symbolized by events like the transatlantic slave trade or the multiple appropriations of black music and culture (Gilroy 1993). In this historical context, what marks hip-hop as unique is that it is a product of a cultural moment in which black people have had the power and authority to construct and defend hip-hop as a cultural community (Dyson 1993).

I begin with the premise that the amorphous object of inquiry known as "hip-hop culture," similar to other music cultures, is a deeply technological way of life (Braun 2002; Chanan 1995; Frederickson 1989; Katz 2004; Lysloff 2003; Manuel 1993; Miller 2004; Morton 2000; Pinch and Trocco 2002; Taylor 2001; Théberge 1997). As hip-hop has evolved from a style, form, and sound of music associated with a multiracial community of artists, musicians, and partygoers in the boroughs of New York City to a global movement that melds, bends, and intertwines disparate aesthetic, cultural, social, and sonic knowledge and experiences, a few elements have remained relatively constant. Some of the most constant have been technology and sounds or, more specifically, the artifacts and the beats. What is interesting about this consistency is how little the technology changed over the first several decades of hip-hop. Turntables and vinyl records, for hip-hop's short life, have been the dominant technological tools of cultural expression supporting musical creativity and exchange. Without the tools and the desires for cultural expression, hip-hop would not exist. Jeff Chang (2005), Murray Forman and Mark Anthony Neal (2004), Nelson George (1998), Patrick Neate (2004), and Tricia Rose (2008) are a few of the many who have written eloquently about the local and global circulations of hip-hop. However, the literature on technology and hip-hop has not been as broad. In particular, the works of Joe Schloss (2004), Mark Katz (2004), Alexander Weheliye (2005), and Kodwo Eshun (1999), as well as facets of the Afrofuturist movement, have been instrumental in building these links. This work is excellent, but it has not fully explored the ways in which turntables and vinyl records acquired the cultural values and social power to render them critical elements in the story of hip-hop.

Undoubtedly hip-hop is larger than turntables and vinyl records, but I do not address its limits here. Many scholars, fans, and artists have done that already and will continue to fight this never-ending battle (Bennett 1999; Condry 2006; Forman 2002; George 2001; Haskins 2000; Keyes 2002; Kitwana 2003; Maxwell 2003; Mitchell 2001; Potter 1995; Rose 2008). Historically, hip-hop has been composed of four elements: DJing, MCing (rapping), b-boying (break dancing), and graffiti writing (Austin 2001; Forman and Neal 2004; George 1998; Rahn 2002). There are multiple overlapping forms of DJing: club DJs (who spin records for a dancing/party audience), production DJs (who create beats for a variety of performances), and turntablists (for whom the turntable is a performance instrument) (Bennett 1999; Kitwana 2003; Brewster and Broughton 2000; Poshardt 2000; Reighley 2000). As hip-hop has expanded and elements of it have waned, the DJ has been viewed as the keeper of its history (Souvignier 2003; Broughton 2003). With the responsibility of maintaining hip-hop comes the added pressure of remaining true and, most important, authentic to its roots. Not surprisingly, the term *authenticity* has engendered conflict. As a cultural historian of hip-hop, an authentic DJ is expected to possess the highest level of turntablist skill, a massive collection of vinyl records, and a voluminous mental archive of beats, breaks, and sounds. As hip-hop has become more global, the demands of this essentialized form of authenticity become harder and harder to uphold. What is interesting about the authenticity required of hip-hop is that it is technological in nature (Maces 2008).

In *Faking It: The Quest for Authenticity in Popular Music*, Barker and Taylor (2007) contend that authenticity by performers and audiences of all musical genres has been a never-ending quest in every corner of the art form. The problem has been that authenticity means something a bit different to everyone. Some artists and audiences are more interested in a representational authenticity that is displayed by "really" playing or singing while performing. Others are moved by personal authenticity, in which artists express themselves on the basis of their experiences. Performing from a location of cultural authenticity speaks to another constituency. Baker and Taylor argue that the need for authenticity has inspired groundbreaking music, as when Jimmie Rodgers pioneered country music, but it has also done damage: "White blues fans . . . defined the genre in the name of authenticity to exclude anything too jazzy or upbeat, thus enforcing a snobbish and racist exclusion of certain blues artists from the cannon because they were too sophisticated." This fantasized form of racialized music authenticity "lauded the most primitive blues artists they could find, such as John Lee Hooker, from whom blacks turned away. In this way, the quest for authenticity did tremendous damage to the blues by codifying certain traditions and limiting innovation" (Baker and Taylor 2007, xi). What is dissimilar about hip-hop is that it was born in a different racial historical moment. In this moment black artists, fans, and consumers led the processes by which authenticity was defined, and it is a definition wrapped in turntables and vinyl records (Harrison 2009). Musicians and scholars have attempted to manage the issue of how to integrate new technology for centuries. Simon Frith's (1986) work on popular music and technology has thoughtfully destabilized the assumption that technology undermines or falsifies the authentic nature of music. Paul Théberge (1997) has eloquently explored the ways in which music and electronic tools have been coproduced. Similarly, Bernstein and Rockwell (2008) trace the development of the community of electronic artists, musicians, and performers that for a short period of time called the San Francisco Tape Music Center home. In order to learn how hip-hop aficionados responded to new technologies that challenged what many understood as a taproot of hip-hop, this chapter explore the ways in which turntables and vinyl records as conduits of a nostalgic and racialized technological authenticity required hip-hop to integrate new devices that did not undermine the community.

Hip-Hop and Race in the Digital Age

This is a story about the complexities of musical and technological evolution in the digital age (Goodwin 1988; Greene and Procello 2004). Questions concerning change and transformation have been central to the critical study of technology. In the past century thinkers like Lewis Mumford (1934), Siegfried Giedion (1948), Leo Marx (1964), Thomas Kuhn (1962), Langdon Winner (1978), Bruno Latour (1988), Wiebe Bijker and Trevor Pinch (1984) and many others have theorized about how and why different groups of people—whether a nation-state or cultural community—migrate from one technological era, period, or domain to another. If this is a story about technological change, how does hip-hop fit into these historical narratives? Generally, it does not. Examining race is vital to understanding hip-hop as a technologically inflected movement. The sound of hip-hop is not just about the music but equally about the sonic disruptions, stylistic reconfigurations, and aesthetic beauty that hip-hop released from and into urban and eventually suburban landscapes. Music has always emanated from a variety of nooks and crannies within cities, and the machinery that inhabit these urban environments has profoundly shaped black music (Dinerstein 2003). In the United States, this changed in the mid-1970s. The industrial cacophony of machinery still bathed the city with familiar sounds, but new technologies of listening allowed for many more public performances of music. The compact nature of urban dwellings has always pushed activity to the street. Black music consumers led the way not only by filling the streets with black music but also by creating a new aesthetics of sounds connected to a visual arts movement (Smethurst 2005). Hip-hop is about the music and the cultural expression of a sense, a feeling, or a style and about performing these sensibilities with technological artifacts. Historically familiar images of Kangol hat and shell-toed, Adidas-clad performers are balanced by massive-speakered sound systems and shoulder-mounted portable boom boxes that dotted the urban street with black-informed sound, music, and technological use. This confluence of sound, music, technology and blackness, precipitated by acts of black vernacular technological creativity, brought the world hip-hop (Fouché 2006). Hip-hop and turntables allow one to glimpse the ways contemporary technology is interwoven into evolving perceptions of culture, identity, and authenticity, specifically within the context of sound. They also allow one to consider how technological identity and the need for a racialized perception of authenticity functioned within the hybrid culture of hip-hop.

The participants in hip-hop have been amazingly technologically innovative, but how have race and racial identity contributed to and inspired this work? From transforming the turntable into a musical instrument and developing scratching, practitioners of hip-hop have embraced technologies as a means to reshape music. Though Tricia Rose's (1994) seminal work, *Black Noise*, focused primarily on rap music, some of the most valuable insights were about race, music, and technology. In arguing for a fuller understanding of the relationships between black music culture and technology, she concluded:

> Rap technicians employ digital technology as instruments, revising black musical styles and priorities through the manipulation of technology. In this process of techno-black cultural syncretism, technological instruments and black cultural priorities are revised and expanded. In a simultaneous exchange rap music has made its mark on advanced technology and technology has profoundly changed the sound of black music. (Rose 1994, 96)

If Rose were writing the same section today, she would undoubtedly extend this prescient statement beyond rap music to the broader category of hip-hop.

Beyond the music practices, artists have collaborated with engineers and technologists to rethink and redesign the tools of the trade. Artists like Grandmaster Flash and Pete Rock and companies like Rane, Vestax, and Akai have participated in and promoted the collaborations in equipment design. Specifically, DJ legend Grandmaster Flash worked with Rane to design the innovative Empath mixer to be true to his hip-hop sound sensibilities. Flash has spoken about the contentious design negotiations that were necessary in order to have a mixer that satisfied his technological and cultural needs and wants (Fouché 2006). The short-lived magazine *Scratch* dedicated its pages to the beats and the technology of making those beats and wonderfully chronicled this symbiotic relationship. *Scratch* regularly ended interviews with a "toolbox" byline indicating technologies of choice. In an interview, hip-hop producer and DJ Just Blaze had an overflowing toolbox that included Akai MPC4000, Vestax QFO, Roland V-Synth, Roland JV-2080, Roland JV-3080, Roland XV-5080, Korg Triton Rack, Minimoog Voyager, and Alesis Andromeda (Sharp 2005, 74). Just Blaze gladly used any and all tools that were available. The back sections of the magazines were divided into different "lab" spaces for readers to enter and learn about new techniques, artifacts, and practices. By highlighting the technological craft and skill of hip-hop, *Scratch* emphatically exclaimed that hip-hop is as much about the sound and the music as it is about the technology and the artifacts. More important, it confirmed its deep technological underpinnings.

In most cases hip-hop has embraced new and emerging technologies of sound recording and reproduction and has smoothly made the transition from its analog origin to its digital present. Yet, the turntable and vinyl have been somewhat of a sticking point. The art of DJing is an indispensible skill that provides entrée into the constantly moving inner sanctum of hip-hop and is also a way to display a level of credibility and commitment to the art form. Even for technophiles, turntables and vinyl records represent an indispensable hip-hop past, without which the present and future could not exist. Just Blaze, one of these avowed technophiles, is also a vinyl aficionado. He uses his voluminous vinyl collection for inspiration and as a material display of his musical knowledge, his commitment to hip-hop, and his reverence for the history of hip-hop. Just Blaze explains:

> I'm a collector, so a lot of times I'll buy records that I won't even make a beat off of, but it will be like a classic breakbeat where there have only been a few thousand pressed . . . To find a classic hip-hop record from over the years, to find a mint 12-inch is like having a part of history, so I'll go pay $200 for that. I'll never use it in a beat. I'll never spin it when I'm DJing because I wouldn't want to mess up the record. (Sharp 2005, 72)

Just Blaze's respect for the history of hip-hop is expressed in his affection for vinyl. His comments also touch upon the meaning of consuming and possessing artifacts of hip-hop. For him, as well as others, it is important to possess a part of hip-hop history because he has taken on part of the collective responsibility of archiving the music. Currently the desire to experience or possess the artifacts of hip-hop is reflected in strong markets for certain vinyl records, Technics 1200 turntables, and classic hip-hop mixtapes.

Hip-hop has been as much about the distribution of sound as it has about the creation and manipulation of sound. Throughout the history of hip-hop, artists have managed the reproduction and exchange of their music through the mixtape (Bell 2007). The term *mixtape* refers to any form of recorded music/sound on a distributable medium. The development and commercial release of the compact audiocassette by Philips Electronics in the early 1960s was a technological innovation that hip-hop eventually exploited. Though the medium has migrated from inscribed music/sound onto magnetic tape within a plastic enclosure to encoded data on compact discs (and more recently the 1s and 0s of online digital files), the importance of the mixtape has not changed. Historically mixtapes allowed hip-hop artists to sell their spontaneous creative work to interested fans and enabled fans to make the experience transportable and replayable. More recently, the mixtape has become a promotional tool of artists and the music industry; however, for hip-hop, mixtapes have always been an important way to extend the party by conveying the experience of a live performance, presenting new artistic creativity, and spreading the sound of hip-hop. Thus, mixtapes, by distributing hip-hop rather than creating and manipulating sound, did not challenge the roots of hip-hop. This distinction marks an important difference in the ways that hip-hop has embraced technology. Those technologies that were perceived to augment or extend hip-hop's reach without undermining it were readily accepted.

Most of the technological incursions that spread within hip-hop, as with mixtapes, were acceptable, but something happened when the technologies changed from analog to digital. The mixtape, for all intents and purposes, was perceived as an analog technology in that the music was inscribed and visually recognizable on magnetic tape. If one looks closely, one can discern the topographic contours of the music recorded on an analog technology like a cassette tape or a vinyl record.

However, once one enters the digital realms, the music/sound loses its visible and tactile qualities. The sonic information is now housed in a series of black boxes represented by CDs, computers drives, and lines of code that viscerally disconnect the creator and the user from a level of musical materiality. The transition from analog to digital reframed the understood materiality of sound, music, and hip-hop itself. The reframing challenged prior assumptions of what hip-hop was. In this instance, the change from analog to digital precipitated a small but meaningful rift within certain subcommunities of hip-hop. These fissures produced two loosely defined camps that fought over the cultural past, present, and future of hip-hop. The factions constructed boundaries of authenticity around the appropriate or inappropriate use of technological artifacts. Historical argumentation that contended that analog was true to the roots of hip-hop clashed with futurist visions of the next step in hip-hop. What it meant to be a member of the hip-hop community could hinge upon one's position on analog and digital technologies in relation to the use of turntables and vinyl records. Of course, there was much slippage between these two communities. Many hip-hop artists use the latest and greatest production tools and equipment, but the place of the analog turntable and vinyl records is central to a favored, constructed history of hip-hop.

A racial subtext also exists. Black history outside of the African continent is replete with examples of exploitation through the application of technology. From shackles and ships that extracted African slaves from their homeland to the processes by which their bodies were reduced to commodities, black people have had an adversarial relationship with technology (Walton 1999). Hip-hop has endowed a specific black community with the power to define and create an empowering past. Interwoven into this past are vinyl records and turntables. It is in this context that digital technologies, which have the ability to overshadow or even erase this past, have caused many to hesitate. The tension between analog and digital, or old and new, has been a popular trope within twentieth-century studies of technology (Edgerton 2010). In the case of turntables, one can see a spinning record, the stylus tracking a groove, and the spatial orientation of how much music has been played. The turntable is seen and heard in a familiar machine-age way, whereas digital technology is not seen or heard in the same way. The sonic feedback from a digital turntable comes primarily from the quietly rotating motor, which is perceived as analog but in actuality is at the limits of analog. The hidden laser that extracts information from a CD makes very little appreciable noise. The ocular assault of a digital turntable is a powerful juxtaposition of the bland visuals of an analog turntable. The digital turntable is marked by its silence. One can conceive of the stream of 1s and 0s passing through the virtually silent circuits, but the process by which these bits and bytes are converted into music is a mystery to most. It is the lack of mechanical age feedback that separates analog from digital in the minds of many (Mindell 2002).

The power of this dominant historical narrative, in concert with a racial subtext, required most hip-hop followers to take a position, tacitly or explicitly, on analog and digital. This tension raises interesting questions about the recent past and how transitions from analog to digital can disrupt the way in which a community defines itself. In specific, it can be disruptive if this community defines itself through the use of certain tools. When these tools change, does the community change as well? This is a familiar question within science and technology studies, but it raises questions about how to understand the history of hip-hop through the lens of technology. This technocultural rift reveals the tension within hip-hop between technology and sound. If hip-hop is about the technology, then it matters deeply which technologies are used to produce the music. If it is about the sound, then it does not matter which tools are deployed to create hip-hop. Thus, this technological evolution affects the intertwined perception of the impact of technology on the sound of hip-hop. Many artists, producers, and hip-hop fans did not follow or participate in these debates. If pushed to the extreme, they may have produced a paradigmatic shift or technological closure that would have left analog technology (and those who use it) in the "loser" categories of outmoded technologies or discarded artifacts. However, together with music consumers, organizations and individuals invested in the survival of hip-hop as an evolving culture of technology developed technologies that enabled a balance between old and new, past and future.

The language of the digital revolution speaks to a disappearing analog past. Digital rhetoric is put into direct opposition to the analog. Supporters of a digital future contend that a reliance and connection to the analog past is an anchor that prevents one from embracing the unbridled potential of the digital age. Some would even argue that we have "crossed over," as did a 2002 *Wired Magazine* infographic that illustrated that, from that year on, most consumers would purchase cameras, music, and televisions in a digital rather than an analog format ("The Great Crossover" 2002). The mounting pressure to cross over obscures the actual experience of existing on both sides of this illusory divide. If we think about this analog/digital divide as also a machine age/computer age, visible/veiled, sound/silence divide, it becomes apparent that most technological objects we encounter incorporate aspects of each polarity. However, the proliferation of the digital has overshadowed this liminal space between the two realms.

Researchers like Henry Jenkins, who call our current moment a "convergence culture," have put forth some of the most popular language about the digital age. In his words convergence culture "represents a cultural shift as consumers are encouraged to seek out new information and make connections among dispersed media content" (Jenkins 2006, 3). Jenkins and others are interested in the convergence of digital media and information, both now and in the future. As catchy as convergence is, it is important to take a step back and study the moments before we merge into digital convergence to understand how digital technologies are not just replacing analog technologies or converging but are also very accurately simulating analog devices. We have not reached a point of convergent homeostasis. What does it currently mean that, even though we have a plethora of digital technologies, older, nostalgia-laden analog technologies from handmade shoes to classic automobiles resonate deeply with many people around the world? The perception is that they are more human or humane because they lack some robotlike efficiencies (Georges 2003).

It is important to understand the extremes of analog and digital, as well as the synergistic space in between, if we are to grasp the contemporary relationships between technology and hip-hop. For hip-hop, technologies that exploit the synergies between analog and digital have the potentially preserved collective and individual identities while maintaining user tactile and mechanical control. One current example of this synergy can be seen in software music-production tools. Propellerhead's *Reason*, a commercially available, software-based ensemble of music-production tools, has waded into this space with its graphical interface. When working with the software, the user first sees the front of a graphical interface, which resembles familiar production equipment with the requisite collection of knobs, buttons, and switches and the familiar software dropdown menus and folders. However, it is the "back" of the interface that is most interesting. The rack system rotates virtually to display the back of the system with interconnecting cables. A user can click and drag to manipulate the cables to simulate cable patching that would be necessary to use the physical studio hardware. Why would a company model the back end for a piece of software? There is no need to patch cables or even look at the rear of the device in a virtual space. Propellerhead's promotional material provides a rationale for simulating the analog experience of moving cables:

> Since Reason is made to look and work just like hardware equipment, we thought that the best and easiest way to handle the audio routing and such would be to do it like you do it in the studio: Flip the rack around! By hitting the [tab] key, the back of the rack is displayed and here all the cables and connector jacks are visible. . . . This form of patching means a fantastic freedom for the producer since you can create the most crazy noises by combining machines in the rack.[1]

This synergy between the analog practice of patching cables and the digital activity of clicking nicely embraces both forms of actions. Arguably the success of *Reason* is directly connected to this synergy. The designer and programmers chose to consciously embrace the synergy of both the analog and the digital to produce a piece of software that did not alienate the potential users by eliminating familiar actions within a virtual space. Yet the case of Reason is distinctly different from that of hip-hop. The racial history of the technologies of hip-hop and what they mean for an origins story of hip-hop provides an additional layer.

Hip-Hop and the Emergence of Digital Vinyl Systems

Supported by an explosion in music and engineering, the 1960s saw a new moment in relationships between sound and technology. In the 1950s the transistor radio made music portable. In the 1960s, the compact audiocassette enabled portability and the added benefit of recording. The audiocassette profoundly restructured the processes by which everyday people listened, captured, and preserved sound. Although the audiocassette was important to the history of hip-hop, the development of new turntables in the late 1960s and early 1970s was even more influential. In 1965 the Matsushita Corporation, under its Panasonic division, launched the Technics brand to compete in the unfolding high-end, consumer hi-fi audio market in Japan. Loudspeakers were the first product it produced, and in 1969 Technics introduced a turntable, the SP-10. This was one of the first consumer turntables to eliminate belts and employ a motor to directly drive the platter on which a vinyl record rests. Technics named the next iteration of this turntable the SL-1100. Early hip-hop artists found this turntable to their liking due to its strong motor, durability, and fidelity—two things Technics turntables have been known for since then. In 1972 Technics released the SL-1200, which over the next few decades became the turntable of choice for multiple musical styles. It became the most popular visual representation of hip-hop and has now reached cult status. Other turntables are more powerful and more flexible, but the 1200s have maintained their tradition and authenticity. Part of this history is connected to the early adoption of Technics turntables by some of the first hip-hop artists, like DJ Kool Herc during hip-hop's early days (Chang 2005). The turntable's partner is the vinyl record, and the 1970s vinyl phonographic recordings proved their ability to reproduce sound with a decent level of accuracy. For hip-hop DJs, this combination was unbeatable. The power to manipulate sounds, beats, and musical expressions through the use of two turntables and a mixer in between, with a seemingly endless array of vinyl records, solidified the connection of hip-hop to these artifacts.

The symbiotic relationship between hip-hop's cultural output and the technological artifacts significantly dictated the public representation of hip-hop. Anyone could enjoy the culture of hip-hop, but to be a real hip-hop head one not only had to know beats, breaks, and rhymes but also use Technics 1200s and vinyl records (Neate 2003). Some did stray from this tacit directive, but most supported this vision. Similar to the way material culture scholar Ian Woodward (2009) writes about the way in which individuals and families narrate their lives through the objects in their homes, the history of hip-hop is partially narrated through analog turntables and vinyl records. This has much to do with the consistency and stability of the technology. To stay relevant, hip-hop DJs had to possess the newest, latest, and hottest beats or mixes, which often required entering uncharted musical terrain. As the sonic compilations increased in complexity, the familiar became less familiar. However, as DJs, artists, and producers extended the boundaries of hip-hop, the technologies remained fundamentally the same and subsequently became foundational objects. As hip-hop narrativized itself and manufactured multiple origins stories, some of the only things that could be viewed as original, authentic, and immutable were the turntable and vinyl records.

By the late twentieth century, the compact disc had led to the demise of the audiocassette and relegated vinyl records to a minor commercial space inhabited by audiophiles and collectors, of which some were DJs (Millard 1995). The illusion of the authentic analog hip-hop DJ weathering the digital storm, abetted by the narrative of hip-hop, empowered the DJ to keep the culture, craft, skill, and sound out of digital reach. This would all change in 2001, when Pioneer released the CDJ-1000. Until then, digital DJing had not been considered a viable possibility. Before the introduction of the CDJ-1000, rumors had been swirling about Pioneer's efforts, and it did not disappoint. This digital turntable allowed for the manipulation of CDs. With a jog wheel users could slow down, speed up, or scratch audio files similar to the way they handled vinyl. What set this device apart from prior digital controllers was that it was the first device to actually emulate vinyl and enable a user to manually control a CD within the device. Many smaller companies had tried to design and manufacture compact disc controllers that functioned similar to an analog turntable, but most did not succeed. They in part failed technically, but they also failed culturally by not fully understanding the users' needs and the resistance to technological change (Oudshoorn and Pinch 2003). However, Pioneer succeeded from a technical standpoint, which opened a cultural point of entry. This device changed the game. When the CDJ-1000 appeared, debates began to rage among DJs and hip-hop fans and in clubs but mostly on multiple forums about the efficacy of the technology (e.g., its feel, latency, acoustic properties). The arguments regularly centered on authenticity and how to preserve it.

A few African American DJs also felt that the digital turn was about erasing the roots of the culture and practice or, more important, its blackness. One New York DJ, DJ Soos, commented, "Who are all the vinyl DJs, the brothers? The digital DJs are rich white kids from the suburbs who can afford to spend $2,000 plus on that digital shit!"[2] As much as he did not want to talk about it, his comments illustrate how the technological divide reinforced race and class difference. He was familiar with the digital-divide rhetoric of African American technological illiteracy and was concerned that this perception would be reproduced within black musical and cultural realms. This sentiment had everything to do with the history of hip-hop and how the DJ community identified with the artifacts its members used. Self-identified as a hardcore vinyl DJ, DJ Soos felt he was protecting the history, the culture, and his own identity from an outside technological, cultural, and racial onslaught. He was well versed in the unproven claim that young white men from the United States are the dominant consumers of hip-hop. He was equally knowledgeable about the wholesale subjugation of African American people, and he saw the CDJ-1000 as merely an extension of this tradition (Tate 2003). In this regard he commented, "Once again the white boy is coming to take our culture or, should I say, borrowing my blackness to make a buck? The reason they get into the digital shit is that they don't have the skills for vinyl. They can borrow black culture but cannot feel it."[3] Clearly what this DJ was getting at was that he felt that DJing was part of his culture, and he identified with the culture and the artifacts of this cultural community. His identity as a DJ was bound up in black technological aesthetics represented by vinyl records and turntables. Interspersed within his comments was a claim that white people could not culturally access hip-hop. He also contended that there was something essential about hip-hop that can be understood only by black people. The response from a white DJ who appreciated vinyl but was primarily a digital DJ was "Psst. They just want to keep hip-hop black and for themselves. Hip-hop is for everybody!" Nonetheless, he acknowledged that a link exists between blackness, DJing, and hip-hop. He concluded: "All my heroes are black, and I do think of DJ and turntablism being a black-folks art form, but the game has changed. The technology has changed the game. You can't deny a set of CDJs."[4] These two commentaries illustrate that the tensions between analog/digital and Technics 1200s/CDJ-1000s had strong black/white racial overtures. In the context of the United States, it is nearly impossible to disentangle these issues and debates from this country's troubling history of institutionalized racism.

Race is relevant to black people around the globe; however, the politics of race vary greatly depending on local history, identity, and culture (Marable and Agard-Jones 2008). This American-centric black/white discussion is one of many that took place around the globe about the authenticity of analog and digital. However, what was at stake for everyone regardless of geographical location was the culture and practice of DJing and consequently hip-hop. The members of this ever-changing community agreed that the controversy was much less about the technological switch from analog to digital and more about the effect of this change on their collective identity and about who had control over the sound and means of hip-hop. If the technology on which the DJ community bases its collective identity changes, then the community changes, the culture changes, the historical underpinnings fade, and the community's collective identity might be destroyed. But no one wanted this to happen. No one wanted to undermine the art form and cultural practices of the community they loved. Even the most ardent vinyl DJ I interviewed would comment on how amazing the digital turntables were even though the DJ personally could not embrace them. The DJ's identity and location within the larger hip-hop world were as a vinyl DJ, and that precluded embracing the nonanalog.

As these events unfolded, small groups around the world were working on a technical solution to this problem. One avenue under development was custom-cut vinyl records. Smaller companies, like the now defunct Kingston, designed devices that straddled a turntable and received a music signal from an analog or a digital input source to drive a cutting blade to inscribe sound into a black vinyl disk. This process, if carefully done, allowed for a simpler means of producing individualized custom vinyl. The dream was to produce a readily available, low-cost, easy-to-use device for everyday DJs. The Japanese company Vestax was the only company to succeed in creating one of these devices with a price point in a consumer's reach (US$10,000). Announced in the summer of 2001, the Vestax VRX-2000 is a stunning and efficient machine, but by its own admission Vestax has not sold many in the last few years.[5] As enticing as this expensive device was for some, it was not a hip-hop community-wide solution to the analog/digital problem. The community was at a crossroads, with no apparent resolution in sight. However, two outside parties, one first in the Netherlands, and the next in New Zealand, designed and invented solutions that eventually were amenable to most DJs, turntablists, and the community and culture of hip-hop.

On November 16, 1998, a small Dutch company, N2IT, distributed the following press release of their new device, FinalScratch (initially called Scratch):

> FinalScratch is the first computer tool that lets DJs map digital media to spinning vinyl, the DJ's musical instrument. The hardware/software combination gives DJs the ability to use turntables to synchronize and interact with various media types. FinalScratch's architecture allows the user to synchronize multiple tracks of audio and video and manipulate them the way a DJ would. It lets users "needle drop" to set song position, change the RPM of the record for pitch and tempo manipulation, "scratch" by changing direction backwards and forwards, and synchronize each track's tempo based on beats per minute. DJs no longer need multiple hardware players to accomplish this task. Using FinalScratch, DJs no longer have to interact with digital media using a computer mouse when mixing. (FinalScratch, 1998)

On January 10–11, 1998, N2IT first exhibited this hardware/software combination at the BeOS Operating System Developer Conference. This might seem to be a strange place for such technology to debut but not when one understands that the developers of FinalScratch were programmers rather than DJs in any traditional sense. The lead developer, Marc-Jan Bastian, began working on FinalScratch as a technical problem presented at a hacker conference in the Netherlands in 1996. A discussion about digital music, vinyl records, and controllers resulted in a technical challenge, and Bastian led the way to find a solution.

FinalScratch and subsequent digital vinyl systems allowed a user to control digital music on a laptop using specially encoded vinyl records and turntables as controllers. Proprietary software and an analog/digital signal converter facilitated the communication between the laptop and the turntables. On the screen a user saw a moving graphical representation of the music/sound encoded on the laptop. Users could manipulate this data stream similar to the way they mixed, cut, and reassembled the information inscribed in vinyl records (Blum 2002). With a digital vinyl system, a DJ could create the illusion that no digital music was in use because it relied on familiar analog turntable techniques. Early promotional materials for FinalScratch indicate that the company saw its product as a bridge between an analog and digital divide. Specifically, one infographic shows the FinalScratch hardware device, or "hub," centrally positioned on a bright red line. This line divides the image in half, with digital above and analog below. Above the red line and in the digital realm, a laptop is shown running the FinalScratch software. Below the red line and in the analog realm one sees two turntables and a mixer. Thus, the device, metaphorically and physically, enables DJs to maintain the historical, nostalgic, and authentic elements of the analog realm while simultaneously embracing the technological innovations presented by the digital without undermining human control over artistry and skill.

Initially viewed skeptically by hip-hop DJs, techno/electronic DJs embraced the device. Clearly these different groups of DJs had only partially similar needs and expectations. Techno/electronic DJs thrived with new technologies that enabled the creative manipulation of sound. The two most central to its development were Richie Hawtin and John Acquaviva, who did not have the same racial and historic connection to the turntables as hip-hop DJs. For this community of electronic enthusiasts, it was a natural fit. Acquaviva would even say, "We were techno DJs, but for all intents and purposes we were pretty conventional. FinalScratch made us practice what we preached" (Acquaviva 2002). The commercial version of FinalScratch emerged in early 2002 with much enthusiasm but received mixed reviews. The main problem was that it was finicky, to say the least, with regular software crashes and connectivity problems of all types, as well as the need for significant amounts of processing power to make it run smoothly. The risk that it would fail during a performance was much too high for hip-hop DJs, but Acquaviva's and Hawtin's audiences were more forgiving with their technological experimentation. The amount of online and offline chatter about this device was extensive, but the DJs that found themselves acting as translators between hip-hop and other musical genres often had the most interesting things to say. For instance, DJ Spooky wrote the following in response to a post about the troubling potential of the digital in 2001:

> The whole situation is what I've been talking about for a while. . . . [T]his will bring digital music into a relationship with not only how we create sound but also how we think about file systems, media for data storage, and almost all other aspects of how contemporary culture now is a culture of the "operating system" of networks of discourse. . . . [I]t always makes me chuckle to see kids in Japan and Finland, and even like remote Pakistan have a whole generation of budding cell phone composers and graphic designers. . . . I call it the Bin Laden effect— it makes people improvise their graphic design impulse in response to the social environment they find themselves enmeshed in, but of course, they also want sounds to accompany the process so they use cell phones to compose music for greetings and to select what kind of people they feel like hanging out with. . . . I tend to think that this will create some kind of compact multi-media platform that can handle almost all aspects of digital creativity within a couple of years . . . think about how much the use of home computers and consumer audio electronics brought what was usually limited to academics and large corporations into the grasp of the average consumer . . . the kind of software that's described in this article will pretty much do the same thing for digital sound production and all that that implies. (Miller 2001)

Although DJ Spooky was clearly excited about its potential, many others were lukewarm. FinalScratch stumbled along, but soon hip-hop DJs like Jazzy Jeff began looking more favorably upon the technology in the months after the September 11, 2001, attack on the World Trade Center in the United States and the explosion of digital file sharing. Traveling with crates of records was bulky but doable. However, after 9/11, Jazzy Jeff indicated that it had become extremely difficult and prohibitively expensive (Erenberg 2005). Extra baggage charges were cutting into his profits. With the rise of Napster and its file-sharing competitors, DJs could access vast troves of music quickly and, most important, easily. Digital file systems allowed quick sorts through large numbers of records. Thus, DJs went from a few crates of carefully selected music to gigabit hard drives of digitally encoded music. A developing network of technical and social events began to make this sketchy technology look much better. Once DJs realized that they could still perform with turntables and vinyl and maintain their authenticity, it became all the more attractive. Early adopters were often apologetic about using the technologies and defensively stated that they used the devices only because it was the only way to make money while traveling internationally. Some were concerned about being construed as a digital DJ. FinalScratch eventually succumbed to technical problems, but, in 2004, collaboration between the highly respected Rane Corporation and Serato Audio Research, founded by Steve West and A. J. Bertenshaw, produced Scratch Live, which quickly gained wide acceptance among hip-hop DJs and became the market leader.

This is not a case of winners or losers; the dynamics around analog turntables, vinyl records, and authenticity still persists. I contend that hip-hop is far from being in homeostasis but is tentatively lodged in an uncomfortable balance. In an interview about his use of Serato Scratch Live, DJ Jazzy Jeff, stated, "I still go out and buy records, but now I immediately record them digitally for Serato. I record direct from my DJ mixer and play the record straight. I don't use any plug-ins or maximizers on my files before or after I record from vinyl because I want it to sound like a record." Jeff finished this interview by repositioning himself as authentic by appealing to the art, skill, and craft of DJing:

> I'm a purist—this [Serato] wasn't supposed to work. I never in my life thought there would be a program like this. And I've been waiting for somebody in the crowd to get mad that I'm playing songs off my computer. But more than anything, I've gotten all the purists saying "Yo, that's the most amazing shit in the world" 'cause I'm doing the exact same stuff. It does not change the integrity of DJing. It doesn't make you DJ any better—the needle can still jump. If you are a bad DJ, with Serato, you still are a bad DJ. (Erenberg 2005, 96)

Other DJs like Marley Marl welcome the technology and haven't looked back. In response to the question about his affinity for technology he stated, "Technology today, you got to work with it because this enables you to do things quicker. Time is money . . . Come on, man, with technologies today I could do [beats] in my car on the way to the city. I could get on a flight to Japan and make two records and four beats before we land[ed]" (Creekmur 2006, 90). Nevertheless, there are still holdouts. One is Hank Shocklee, who is best known for producing the revolutionary Public Enemy's *Fear of a Black Planet*. He laments the loss of the analog, and his thoughts about "feeling the music" resonate with the young African American DJ interviewed previously, DJ Soos. Shocklee argued as follows:

> Your enjoyment of hip-hop is lessened because you're not listening to hip-hop in its true analog form. You're listening to it in its digital form. Digital is a re-creation of hip-hop but it is not perfect. It's like a photocopy of analog. Everybody says the golden era was the greatest moment in hip-hop. It was also the moment where everything was analog. You went from wax to drum machine

> to tape . . . As opposed to you going vinyl to drum machine to some sort of digital CD. That's a whole different mechanism. This is what killed rock 'n' roll. Distortion does not register the same in the digital realm as it does in the analog realm. Because now distortion becomes a sound as opposed to having a physical presence. (Matthews 2006, 88)

The tension is still as much about the technology as it is about the sound. By focusing on the disagreements emerging as turntable signal processing moved from analog to digital, this chapter presents a case for the ways in which new technologies that were designed to mediate between analog and digital enabled hip-hop to repair the growing rift associated with the choice of tools one used. These technological synergies found in analog and digital—now known as digital vinyl systems—preserved collective and individual identities and maintained a user's tactile and mechanical control while simultaneously embracing all that the digital had to offer. As a result of these new technologies the larger hip-hop DJ community was able to maintain itself and its collective and community identities.

The intersections of hip-hop, technology, and sound help us to understand the ways the materiality of sound is embedded and circulated within society. In a most basic sense, as Trevor Pinch and Karin Bijsterveld suggest: " 'Follow the instruments' in the same way that in the early days of S&TS we learned to 'follow the actors' "(Pinch and Bijsterveld 2004, 639). An examination of hip-hop technologies in use, as well as the cultural arenas in which individuals, artifacts, and belief systems interact, links together the equally interesting but still under-developed work on race and technology.

NOTES

1 "The Reason Rack." http://www.propellerheads.se/products/reason/index.cfm?article=devices_rack&fuseaction=get_article (accessed August 1, 2009).

2 DJ Soos, interview by author, Jan. 15, 2006.

3 Ibid.

4 DJ Light Flash, interview by author, Feb. 1, 2006.

5 Hisao Kaneko, chief design engineer, Vestax Corporation, interview by author, May 31, 2007.

REFERENCES

Acquaviva, John. "FinalScratch." Transcript. Red Bull Music Academy, London, 2002.

Austin, Joe, *Taking the Train*. New York: Columbia University Press, 2001.

Barker, Hugh, and Yuval Taylor. *Faking It: The Quest for Authenticity in Popular Music*. New York: Norton, 2007.

Bell, Walter, dir. *Mixtape, Inc.* DVD. University City, C.A.: Codeblack Entertainment, 2007.

Bennett, Andy. "Hip Hop am Main: The Localization of Rap Music and Hip Hop Culture." *Media, Culture, and Society* 21 (1999): 77–91.

Bernstein, David W., and John Rockwell. *The San Francisco Tape Music Center: 1960s' Counterculture and the Avant-Garde*. Berkeley: University of California Press, 2008.

Bijker, Wiebe E., and Trevor Pinch. "The Social Construction of Facts and Artefacts: or How the Sociology of Science and the Sociology of Technology Might Benefit from Each Other." *Social Studies of Science* 14 (1984): 399–441.

Blum, Jason. "Stanton Final Scratch: Digital Vinyl for PC." *Remix* (December 2002): 72–74.

Braun, Hans-Joachim, ed. *Music and Technology in the Twentieth Century*. Baltimore: Johns Hopkins University Press, 2002.

Brewster, Bill, and Frank Broughton. *Last Night a DJ Saved My Life: The History of the Disc Jockey*. New York: Grove/Atlantic, 2000.

Broughton, Frank, and Bill Brewster. *How to DJ Right: The Art and Science of Playing Records*. New York: Grove, 2003.

Chanan, Michael. *Repeated Takes: A Short History of Recording and Its Effects on Music*. London: Verso, 1995.

Chang, Jeff. *Can't Stop Won't Stop: A History of the Hip-Hop Generation*. New York: St. Martin's, 2005.

Condry, Ian. *Hip-Hop Japan: Rap and the Paths of Cultural Globalization*. Durham, N.C.: Duke University Press, 2006.

Creekmur, Chuck. "Touch the Sky." *Scratch* (January/February 2006): 89–92.

Dinerstein, Joel. *Swinging the Machine: Modernity, Technology, and African American Culture between the World Wars*. Amherst: University of Massachusetts Press, 2003.

Dyson, Michael Eric. *Reflecting Black: African-American Cultural Criticism*. Minneapolis: University of Minnesota Press, 1993.

Edgerton, David. "Innovation, Technology, or History: What Is the Historiography of Technology About?" *Technology and Culture* 51 (2010): 680–97.

Eglash, Ron, Jennifer L. Croissant, Giovanna Di Chiro, and Rayvon Fouché, eds. *Appropriating Technology: Vernacular Science and Cultural Invention*. Minneapolis: University of Minnesota Press, 2004.

Erenberg, Jesse. "Jazzy Jeff, a Cut above with Rane/Serato Scratch Live." *Scratch* (September/October 2005): 96–97.

Eshun, Kodwo. *More Brilliant than the Sun: Adventures in Sonic Fiction*. London: Quartet, 1999.

"FinalScratch." Press release, November 12, 1998.

Floyd, Samuel A., Jr. *The Power of Black Music: Interpreting Its History from African to the United States*. New York: Oxford University Press, 1995.

Forman, Murray. *The 'Hood Comes First: Race, Space, and Place in Rap and Hip-Hop*. Middletown, Conn.: Wesleyan University Press, 2002.

———, and Mark Anthony Neal. *That's the Joint! The Hip-Hop Studies Reader*. New York: Routledge, 2004.

Fouché, Rayvon. "Say It Loud, I'm Black and I'm Proud: African Americans, American Artifactual Culture, and Black Vernacular Technological Creativity." *American Quarterly* 59 (2006): 639–59.

Frederickson, Jon. "Technology and Music Performance in the Age of Mechanical Reproduction." *International Review of the Aesthetics and Sociology of Music* 20 (1989): 193–220.

Frith, Simon. "Art versus Technology: The Strange Case of Popular Music." *Media, Culture, and Society* 8 (1986): 263–279.
George, Nelson. *Buppies, B-Boys, Baps, and Bohos: Notes on Post-Soul Black Culture.* Cambridge, Mass.: Da Capo, 2001.
———. *Hip-Hop America.* New York: Viking, 1998.
Georges, Thomas Martin. *Digital Soul: Intelligent Machines and Human Values.* Boulder, Colo.: Westview, 2003.
Giedion, Sigfried. *Mechanization Takes Command: A Contribution to Anonymous History.* New York: Oxford University Press, 1948.
Gilroy, Paul. *The Black Atlantic: Modernity and Double Consciousness.* Cambridge, Mass.: Harvard University Press, 1993.
Goldberg, David Albert Mhadi. "The Scratch Is Hip-Hop: Appropriating the Phonographic Medium." In *Appropriating Technology: Vernacular Science and Cultural Invention*, ed. Ron Eglash, Jennifer L. Croissant, Giovanna Di Chiro, and Rayvon Fouché, 107-144. Minneapolis: University of Minnesota Press, 2004.
Goodwin, Andrew. "Sample and Hold: Pop Music in the Digital Age of Reproduction." *Critical Quarterly* 30 (1988): 34–49.
"The Great Crossover: Digital Technology Finally Surpasses Analog." *Wired* (November 2002): 58–59.
Green, Ricky. *Voices in Black Political Thought.* New York: Lang, 2005.
Greene, Paul D., and Thomas Porcello, eds. *Wired for Sound, Engineering, and Technologies in Sonic Cultures.* Middletown, Conn.: Wesleyan University Press, 2004.
Haskins, Jim. *The Story of Hip-Hop.* London: Penguin, 2000.
Harrison, Anthony Kwame. *Hip Hop Underground: The Integrity and Ethics of Racial Identification.* Philadelphia, Penn: Temple University Press, 2009.
Jeffs, Rick. "Evolution of the DJ Crossfader." *RaneNote* 146. Rane Corp., 1999.
Jenkins, Henry. *Convergence Culture: Where Old and New Media Collide.* New York: New York University Press, 2006.
Katz, Mark. *Capturing Sound: How Technology Has Changed Music.* Berkeley: University of California Press, 2004.
Keyes, Cheryl L. *Rap Music and Street Consciousness.* Urbana: University of Illinois Press, 2002.
Kitwana, Bakari. *The Hip-Hop Generation: Young Blacks and the Crisis in African American Culture.* New York: Basic Civitas, 2003.
Kuhn, Thomas. *The Structure of Scientific Revolutions.* Chicago: University of Chicago Press, 1962.
Latour, Bruno. *Science in Action: How to Follow Scientists and Engineers through Society.* Cambridge, Mass.: Harvard University Press, 1988.
Lysloff, René T. A., and Leslie Gay, eds. *Technoculture and Music.* Middletown, Conn.: Wesleyan University Press, 2003.
Maces, Bruno. "Technology and Authenticity." *New Atlantis* (Spring 2008): 63–78.
Manuel, Peter. *Cassette Culture Popular Music and Technology in North India.* Chicago: University of Chicago Press, 1993.
Marable, Manning, and Vanessa Agard-Jones, eds. *Transnational Blackness: Navigating the Global Color Line.* New York: Palgrave Macmillan, 2008.
Marx, Leo. *The Machine in the Garden.* New York: Oxford University Press, 1964.
Matthews, Adam. "Distortion to Static." *Scratch* (July/August 2006): 86–89.
Maxwell, Ian. *Phat Beats, Dope Rhymes: Hip Hop Down Under Comin' Upper.* Middletown, Conn.: Wesleyan University Press, 2003.

Millard, Andre. *America on Record: A History of Recorded Sound*. New York: Cambridge University Press, 1995.

Miller, Paul D. "Re: Analog to Digital Dj Mixes Coded Language." November 4, 2001, http://www.nettime.org/Lists-Archives/nettime-l-0111/msg00017.html (accessed February 21, 2011).

———. *Rhythm Science*. Cambridge, Mass.: MIT Press, 2004.

Mindell, David A. *Between Human and Machine: Feedback, Control, and Computing before Cybernetics*. Baltimore: Johns Hopkins University Press, 2002.

Mitchell, Tony. *Global Noise: Rap and Hip-Hop outside the USA*. Middletown, Conn.: Wesleyan University Press, 2001.

Morton, David. *Off the Record: The Technology and Culture of Sound Recording in America*. Newark, N.J.: Rutgers University Press, 2000.

Mumford, Lewis. *Technics and Civilization*. New York: Harcourt, Brace, 1934.

Neate, Patrick. *Where You're at: Notes from the Frontline of a Hip-Hop Planet*. New York: Riverhead, 2003.

Oudshoorn, Nelly, and Trevor Pinch, eds. *How Users Matter: The Co-Construction of Users and Technologies*. Cambridge, Mass.: MIT Press, 2003.

Pinch, Trevor, and Karin Bijsterveld. "Sound Studies: New Technologies and Music." *Social Studies of Science* 34 (October 2004): 635–48.

Pinch, Trevor, and Frank Trocco. *Analog Days: The Invention and Impact of the Moog Synthesizer*. Cambridge, Mass.: Harvard University Press, 2002.

Poshardt, Ulf. *DJ-Culture*. Trans. Shaun Whiteside. London: Quartet, 2000.

Potter, Russell A. *Spectacular Vernaculars: Hip-Hop and the Politics of Postmodernism*. Albany: State University of New York Press, 1995.

Rahn, Janice. *Painting without Permission: Hip-Hop Graffiti Subculture*. Westport, Conn.: Bergin and Garvey, 2002.

Reighley, Kurt B. *Looking for the Perfect Beat: The Art and Culture of the DJ*. New York: Pocket, 2000.

Rose, Trisha. *Black Noise: Rap Music and Black Culture in Contemporary America*. Hanover, N.H.: University Press of New England, 1994.

———. *The Hip-Hop Wars: What We Talk about When We Talk about Hip-Hop—and Why It Matters*. New York: Basic Civitas, 2008.

Schloss, Joseph. *Making Beats: The Art of Sample-Based Hip-Hop*. Middletown, Conn.: Wesleyan University Press, 2004.

Sharp, Shawn Lewis. "Confessions of Fire." *Scratch* (September/October 2005): 68–74.

Smethurst, James. *The Black Arts Movement: Literary Nationalism in the 1960s and 1970s*. Chapel Hill: University of North Carolina Press, 2005.

Souvignier, Todd. *The World of DJs and the Turntable Culture*. Milwaukee: Leonard, 2003.

Tate, Greg, ed. *Everything but the Burden: What White People Are Taking from Black Culture*. New York: Harlem Moon, 2003.

Taylor, Timothy D. *Strange Sounds: Music, Technology, and Culture*. New York: Routledge, 2001.

Théberge, Paul. *Any Sound You Can Imagine: Making Music/Consuming Technology*. Hanover, N.H.: Wesleyan University Press, 1997.

———. "The Network Studio: Historical and Technological Paths to a New Ideal in Music Making Social." *Studies of Science* 34 (2004): 759–81.

Tremayne, Jim. "With a Hot New Mixer on the Market & a Revitalized Career in Motion, DJ Pioneer Grandmaster Flash Finds That Necessity Is Still the Mother of Invention." *DJ Times* 16(3) (March 2003): 57–62.

Walton, Anthony. "Technology versus African-Americans." *Atlantic Monthly* (January 1999): 14–18.
Ward, Brian. *Just My Soul Responding: Rhythm and Blues, Black Consciousness, and Race Relations*. Berkeley: University of California Press, 1998.
Weheliye, Alexander. *Phonographies: Grooves in Sonic Afro-Modernity*. Durham, N.C.: Duke University Press, 2005.
Winner, Langdon. *Autonomous Technology: Technics-out-of-Control as a Theme in Political Thought*. Cambridge, Mass.: MIT Press, 1978.
Woodward, Ian. "Material Culture and Narrative: Fusing Myth, Materiality, and Meaning." In *Material Culture and Technology in Everyday Life*, ed. Phillip Vannini. 59–72. New York: Lang, 2009.

CHAPTER 22

IPOD CULTURE: THE TOXIC PLEASURES OF AUDIOTOPIA

MICHAEL BULL

INTRODUCTION

> I have never cherished anything I bought as much as this little device. When I was a child, I used to watch a kids [*sic*] show called "the music machine" and I always dreamed of having something like that. A device that plays any song there is. The iPod comes pretty close to the fulfillment of this childhood fantasy. (iPod user)
>
> It has dramatically changed the way I listen to music. I use my iPod every day, generally for four to six hours a day. I listen to it at work, at home, in my car, on the subway etc. Whilst I frequently carried a personal CD player before, the iPod has become a necessity. When I leave the house, I now check my pockets for four things: My wallet, my keys, my mobile phone, and my iPod. I never go out without all four on my person. (iPod user)

In July of 2007 the *Guardian* newspaper in the UK reported the case of a Muslim woman juror who had been discharged from a murder trial after she was caught listening to her iPod, concealed under her hijab, during important prosecution evidence. The judge had heard traces of "tinny music" throughout the trial but thought that it must have been either his imagination or a defect of his own hearing. The woman juror was subsequently charged with contempt of court. Perhaps the female juror was picturing herself in a fictionalized courtroom, imagining the identities of those around her, or perhaps she was merely lost in the soundtrack of her day—chosen from the multiple playlists of her iPod. Given that the judge had heard her "tinny music" on several occasions it is unlikely that the juror had heard any of the court proceedings at all. The case reminds me of my own iPod research, in which a respondent who was a professor at an American university had stated that he had listened to his iPod while sitting on stage during his university's graduation ceremony, passing the time in pleasurable auditory reverie.[1]

Both of these examples are very twenty-first-century examples of the privatized nature of much music listening—the iPod is a miniature device that fits discreetly into one's pocket, and, once operational, it needs no further attention, unlike its predecessor, the Sony Walkman, where the user had to clumsily replace cassette tapes, CDs, or minidisks. The seamless and relatively invisible nature of use—the earpieces are designed to be discreet enables iPod users to create lengthy periods of potentially uninterrupted use during a wide variety of everyday activities. These examples involve risk—the woman juror was caught and prosecuted, and the professor would have been embarrassed and maybe disciplined if caught. They also demonstrate the pull between the desire for individual auditory pleasure, which technologies such as the Apple iPod provide, and certain public prohibitions against that use. Both of the people mentioned are using the quintessentially mobile technology of the Apple iPod while sitting stationary in places they don't particularly want to be in. These examples demonstrate that today any space whatsoever can potentially be transformed into a private auditory space of listening, thus empowering the users as they transcend the geography of the space inhabited, endowing it with their own auditory significance. It is precisely this ability to sonically transform increasing amounts of everyday experience that poses new questions in relation to users' underlying dispositions toward managing their daily experience.

The study of iPod use can cast new light on users' attitudes toward public places, others, and their own cognitive management of experience, thereby redefining the power dynamic between the user and others and their own cognitive processes. MP3 technology has transformed music reception and users' ability to pursue their pleasurable and privatized auditory interests. This capacity to sonically privatize space is possessed by the majority of those living in the industrialized world through the use of dedicated MP3 players like the Apple iPod or through mobile phones, which possess MP3 capability. Use occurs in any area of everyday life, from the domestic environment of the home, to the impersonality of the street, to the working spaces of the office, to those engaged in sports, and to the exceptional spaces of the theaters of war in Iraq and Afghanistan (Pieslak 2009).

The untrammeled pleasure of taking your own soundworld with you resonates through urban and cultural theory. It poses a set of theoretical problems relating to the nature of public and private existence and to how urban dwellers experience urban space. On the one hand, the continuous use of these technologies might be interpreted as an act of liberation—giving users increased control over their environments and themselves (de Certeau 1988). Alternatively, as in the Marcusian analysis, use is interpreted as an act of colonization, whereby users become increasingly dependent upon the use of these technologies in order to satisfactorily survive and manage their daily routines. In the following pages I investigate iPod use through the lens of forms of toxic audiotopias. The epistemological starting point is the observation that the history of cities is in some sense a history of how we come to share social space. If this is the case, we are entitled to ask, What type of culture is the audiotopia of iPod use, and what does it tell us about the contemporary nature of mediated public and private urban space?

The privatizing and colonizing impulse, which I associate with iPod use, has a long prehistory. We have throughout the twentieth and twenty-first centuries increasingly moved to music both through the provision of Fordist technologies, such as those created by the now defunct Muzac Corporation, which created sonic environments of uniformity for consumers to move through, or by the Hyper-Post-Fordist technologies of MP3 players such as the Apple iPod, through which, with the aid of a pair of headphones, users create individualized and mobile soundscapes.

The multiple provision and use of these technologies in the age of mechanical reproduction has produced an age of sonic saturation and colonization in which urban spaces—both private and public—are transformed, resulting in the continual redefinition of and contestation of the meanings attached to the way in which individuals inhabit space and place. Early users of the gramophone and radio frequently sequestered social space collectively by listening on public beaches or during family picnics. From the 1950s on, the mass use of portable radios redefined the meaning of public space (Douglas 1999). More recently, ghetto blasters and automobile sound systems have privatized public space (Gilroy 2003). In its fluidity, sound has seeped into the spaces of everyday life like no other sense.[2] This "colonization" extends to the cognition and desire of users themselves. Importantly, this is frequently a "desired," "active," and pleasurable colonization, whereby users reclaim experience, time, and place (Bull 2007). It is precisely in this area of the intoxicating immersion of the sonic that the twin concepts of "audiotopia" and "toxicity" are situated in the following pages. The chapter focuses primarily upon the use of Apple iPods as the most common, dedicated MP3 player and because my primary research is located in their use. However, the following analysis is equally relevant for other MP3 players and users.

Audiotopia, Toxicity, and Pleasure

"Audiotopia" refers to both the intense pleasure described by iPod users as they listen to music on their iPods and to their desire for continuous, uninterrupted use.

Audiotopia comprises the desire for immersive auditory experience whereby the users' chosen soundworld eradicates the preexisting soundworld that users inhabit. Frances Dyson comments upon the intimate relationship between sound technologies and immersion: "Sound is the immersive medium par excellence. Three dimensional, interactive, and synesthetic, perceived in the here and now of an embodied space, sound returns to the listener the very same qualities that media mediates: that feeling of being here now, of experiencing oneself as engulfed, enveloped, enmeshed, in short, immersed in an environment. Sound surrounds" (Dyson 2009, 4) (see also Grimshaw, this volume). The mediated re-creation of the users' sense of the "here and now" is central to the following analysis of toxic audiotopia. Instrumental in creating this immersive state is the use of headphones, which transforms the users' relationship to the environment and creates sonic privacy. The science fiction writer William Gibson captures the power of sonic privatization in his description of using his Walkman, the iPods' precursor: "The Sony Walkman has done more to change human perception than any virtual reality gadget. I can't remember any technological experience since that was quite so wonderful as being able to take music and move it through landscapes and architecture" (Gibson 1993).

Intrinsic to auditory privatization is headphone use. The habitual use of headphones began with radio reception in the 1920s. Siegfried Kracauer commented as early as 1924 on the transformative and seductive power of radio listening through headphones:

> [W]ho would want to resist the invitation of those dainty headphones? They gleam in living rooms and entwine themselves around heads all by themselves . . . Silent and lifeless, people sit side by side as if their souls were wandering about far away. But these souls are not wandering according to their own preference; they are badgered by the news hounds, and soon no one can tell anymore who is the hunter and who is the hunted. (Kracauer 1995, 333)

Contemporary iPod users, in contrast to Kracauer's radio listeners, claim to be masters of their sonic world as a result of the personalized choice of music embedded in their iPods. The impulse to construct a privatized auditory environment is itself located in Western cultural values. Jonathan Sterne notes that:

> as a bourgeois form of listening, audile technique was rooted in a practice of individuation: listeners could own their own acoustic spaces through owning the material component of a technique of producing that auditory space—the "medium" that stands in for a whole set of framed practices. The space of the auditory field became a form of private property, a space for the individual to inhabit alone. (Sterne 2003, 160)

Headphones reempower the ear against the contingency of sound in the world, bolstering the individualizing practices of sound reception. The transformative nature of using headphones is illustrated by the following iPod user:

> Although being alone listening to music through conventional speakers can make listening pleasurable, having pod headphones in can make that sensation

> infinitely more pleasurable. The feeling of blocking out other sounds and by implication other sources of interference only heightens the pleasure. You may also be made aware of subtle sounds that you can only hear when wearing headphones. This "new discovery" enhances it still further. You may have heard a track dozens of times via speakers and never heard a particular piece of music before just because you were wearing earphones. (Hillary)

Embodied in these practices is the desire to control the sonic environment. Many iPod users claim to prefer to live in a totally mediated, privatized auditory world with their music collection at their fingertips. They are not merely enjoying their own privatized auditory reveries but are also taking pleasure in being able to sonically control the very nature of their everyday life. It is here that the sonic pleasure of audiotopia resonates socially. Music reception for iPod users is no longer the straightforward "mutual tuning relationship" that Ernst Bloch commented upon. He distinguished music from the visual by arguing that "a note of music comes with us and is 'we' unlike the visual, which is primarily an 'I' divorced from the other" (Bloch 1986, 287). The relational aspect of iPod use differs insomuch as experience becomes primarily an audiovisual one in which states of mediated "we-ness" eclipse direct forms of experience by technologically mediated forms of experience (Adorno 1976; Dant 2008). The mutuality of music reception becomes increasingly asymmetrical in its privatizing, yet mediated nature because users are immersed in the chosen sounds of the culture industry contained in their iPod. Sounds fed directly into the users' ears through headphones often placed directly into the ears—directly into the experiencing subject—act so as to reduce the outside world to silence. In its immersive qualities, mediated sonic experience often appears to be more "immediate" to users of iPods than nonmediated experience.[3]

The mediated immediacy of iPod use may lead to a form of social toxicity. The term *toxicity* should, however, be handled with caution. Western consumer relationships to a whole host of consumer practices and technologies from shopping, gambling, and watching television to playing video games has frequently been described as compulsive—yet compulsive behavior need not be equated with any detrimental social attribute or harmful effect. The inability to stop reading a thriller novel, for example, because one is engrossed in it might be considered in positive terms. In the main, however, "toxicity" appears largely as a dystopian state whereby users' engagement with pervasive new technologies potentially decreases their capacity to disconnect from their use of these technologies (Rheingold and Kluitenberg 2006, 29). Toxicity refers primarily to the negative moment embodied in mediated technological connectivity. This has a long cultural and theoretical history stretching from Heidegger to Marcuse. Marcuse commented as early as the 1960s that "solitude, the very condition which sustained the individual against and beyond society, has become technically impossible" (Marcuse 1964, 68). For Marcuse this referred to both the social construction of mediated subjective desire and the nature of social spaces that were a consequence of such a disposition. The introjection of a range of social dispositions to consume was to be understood through the notion of "objective alienation," whereby users had fully normalized alienated forms of experience. In this image of "toxicity," consumers become victims of the marketing strategies of companies and the design and form of their technologies. Implied in Rheingold and Kluitenberg's understanding of "toxicity" is the users' inability to experience forms of nonmediated experience they sometimes desire. Chatfield points to the desire for "toxic" immersion among video-game users:

> People are perilously drawn by temptation to withdraw from real life's complexities into a solipsistic, simpler world. In the case of video games, in particular, it's hard to deny this kind of escape isn't a large part of their appeal. Escape, simplification and control play their part in all games and in electronic games has reached a remarkable pitch of sophistication." (Chatfield 2010, 73)

In the following pages I problematize, develop, and reevaluate the notion of "toxicity" in relation to the specifically auditory dynamic of iPod use. Before doing so I wish to clarify the position taken in the following pages. I do not claim that all iPod use is toxic in nature but merely that it is one, albeit common, possibility of use. Neither do I claim that users who embody potentially "toxic" modes of usage are fundamentally different from users who do not. Toxicity is merely a structural possibility of use that becomes socially significant with mass usage as described earlier. If "toxic pleasures" relate to a continuous withdrawal from the physical immediacy of experience and its re-creation engendered through the continuous creation of privatized sonic environments, then we require a nuanced analysis of both the pleasures and the consequences of such practices and a normative understanding of users' rationale, which underpins such practices.

Toxicity is divided into three categories in the following analysis: In the first instance toxicity refers to users' inability to disconnect from use despite their potential realization of the problematic nature of use. Central to this form of toxicity is the seductive nature of an empowering, anesthetizing potential embodied in iPod use. Second, it refers to modes of habitual use whereby users do not experience this "tethering" as at all negative but merely pleasurable. "Toxicity" in this case refers to the negative appraisal of the shared nature of social space, which is to be replaced with the audiotopia embodied in iPod use. Third, "toxicity" refers to the actual physical nature of damage that can occur through loud, continuous use of the iPod and need not require any cognitive recognition by the user of this possibility. If there is recognition of this potential, then toxicity might refer to "state one," where users find it difficult to change their use.

Toxic Pleasures 1: Seductive Audiotopia

Digitalization has enabled iPod users to stream music for every conceivable situation through the creation of playlists, which can be created and changed while on the move. Users possess unparalleled control over their daily sonic life both externally, via the aestheticization of their experience, and cognitively, through their ability to regulate their moods and volition through the micromanagement of their music listening. Perpetual sonic connectivity becomes a seductive audiotopia in the daily lives of many iPod users, transforming their relationship to the environment they inhabit and move through. Embodied in iPod use is a range of sonic rituals that structure users' daily life, and as such the iPod might be understood as an urban "Sherpa," enabling users to successfully maneuver their way through daily urban life. However, recognition of this pleasurable, hermetically sealed activity sometimes produces an ambivalent response:

> I walked around with a Sony Walkman attached to me whilst walking or roller-skating my way around many major cities in the world during the 1980s. In retrospect I wish I had not been so stuck in my own head, so disconnected from the natural symphony of place. My community members feel so alien to me now. They stand in line in front of me, dancing ever so slightly to their tune, often oblivious to what's happening around them and completely closed off from the niceties of the neighbourly "hello." (Alison, hppt://www.radioopensource/the-age-of-shuffle/)

Alison, an ex-Walkman user reminds us of both the historical nature of auditory privatization and the desire for a shared public space in the city. The writer Gabriel Sherman describes his own ubiquitous iPod use in the following terms:

> Almost anywhere I went, I plugged in and tuned out. Need cash from the ATM? The Shins' melodic New Slang would accompany me. Picking up my laundry at the Wash and Fold? How about Rachael Yamagata's sultry swooning? My music even joined me in the bathroom each morning before work. With more than 1,000 songs at my thumb tip, I could satisfy any desire, any time. My iPod was like a drug. I live in my own self-imagined movie, instantly tailoring the soundtrack to fit, or inspire, my emotions . . . I even acquired the telltale signs of an addict. Just before leaving places, I fidgeted nervously while contemplating what song I would queue up. (Sherman)

Sherman describes the ambiguous pleasures of living in a pleasurable audiotopia; his use highlights both the utopian and dystopian elements of iPod use. This dystopian moment of use I refer to as the toxic pleasures of use. "With my earphones in, I became deaf to the urban orchestra playing around me. Even worse, my iPod had sapped the energy that makes New York more exhilarating than the places we all escaped from. I had traded one kind of suburban isolation for another." Sherman, like many other users of a range of mobile technologies from the mobile phone to the iPod, finds himself increasingly dependent upon these technologies to maintain his daily life, of being "tethered" to a wide range of communication technologies, of being available 24/7 (Ito, Okabe, and Matsuda 2005). Sherman struggles with the ambiguous nature of his iPod use, realizing that he is caught between the pleasures of a privatized aesthetization of the city and his compulsive use pattern. His description highlights the multifaceted pleasures associated with continual or habitual use. The transcendence of the "here and now" is embedded in the seductive pleasures of sonic toxicity, and this transcendence is frequently aesthetic in nature.

The aesthetic colonization of urban space described by many iPod users is in part a technological tale whereby urban experience becomes synonymous with technological experience. This technological experience is both pervasive and increasingly taken for granted in wide areas of daily life. It is simultaneously empowering and dependent.

IPod use seamlessly joins together disparate experience by unifying the complex, contradictory, and contingent nature of the world beyond the user. The success of these strategies depends upon the creation of an all-enveloping wall of sound through which the user looks. Users report that iPod experience is at its most satisfying when no external sound seeps into their world to distract them from their dominant and dominating vision.

IPod users invariably aim to create a privatized sound world that is in harmony with their mood, orientation, and surroundings, enabling them to respatialize urban experience through a process of solipsistic aestheticization. IPod users aim to habitually create an aesthetically pleasing urban world for themselves as a constituent part of their everyday life. In doing so, they create an illusion of omnipotence through mediated proximity and "connectedness" engendered by their iPod use.

For many iPod users the street is orchestrated to the sounds of their favorite playlists:

> The world looks friendlier, happier, and sunnier when I walk down the street with my iPod on. It feels as if I'm in a movie at times. Like my life has a soundtrack now. It also takes away some of the noise of the streets, so that everything around me becomes calmer somewhat. It detaches me from my environment, like I'm an invisible, floating observer. (Berklee)

Susan, a manager from Toronto, describes the iPod's transformative power over her urban environment:

> I find when listening to some music choices I feel like I'm not really there. Like I'm watching everything around me happening in a movie. I start to feel the environment in the sense of the mood of the song and can find that I can start to love a street that I usually hate or feel scared for no reason. (Susan)
>
> I see people like I do when I watch a movie . . . there is a soundtrack to my encounters . . . music to accompany my thoughts about others. It dramatizes things a bit. It fills the silent void. (June)

Streets perceived as silent are in reality a complex of sounds—June's observation that her iPod fills the "silent void" is indicative of users' experiencing the world solely as a function of mediated sound. Mundane, yet nevertheless unmanageable urban life is transformed through iPod use, which creates movement and energy in the user where there was none before. The use of the iPod provides a "buffer" between the user and the recognized reality of the city street.

Users often describe the world experienced through iPod use as a movie script in which they play the central role. The selection of "sad" music, for example, is used to match the users' mood, transposing those feelings to the streets passed through. The world and the users' experience within it gain significance precisely through the creation of an enveloping and privatized soundworld. In the users' world of aesthetic euphoria, experience is simplified, clarified—the aesthetic impulse provides an unambiguous sense of purpose and meaning for users, who are creating a "space" within which to unwind and unravel their emotions. When attended to, the street becomes a function of their mood and imagination mediated through their iPod.

IPods are both interactive and noninteractive in the sense that users construct fantasies and maintain feelings of security precisely by not interacting with others or their environment. Through the power of a privatized soundworld the world becomes intimate, known, and possessed. Imagination is mediated by the sounds of the iPod, which become an essential component in the users' ability to imagine at all. Users are often unable to aestheticize experience without the existence of their own individual soundtrack acting as a spur to the imagination.

In this ordering of cognition the user surpasses the disjunction that exists between their own soundtrack, the movement of others, and the environment they pass through. Without the iPod, they experience the world out of sync. The polyrhythmic nature of the city relativizes their own place within the world and makes them just one more piece of an anonymous urban world.

Toxicity and aestheticization can be interpreted as having utopian implications for users—to aestheticize is to transcend the mundane world as it is experienced—aestheticization remains an active mode of appropriating the urban, transforming that which exists, and making it the user's own (Marcuse 1964). In this instance iPod use reempowers the subject in relation to urban space. The French anthropologist Marc Augé has described cities in terms of "non-spaces": spaces semiotically void of interest such as shopping centers and parking lots. Now, users can re-create meaning for themselves. Through the privatized experience of music, they subjectively endow all space with meaning. The use of Apple iPods permits urban theorists to increasingly understand the city as a privatized audiovisual creation of the user and to reprioritize the unique role that sound plays in the construction of daily life (Bull 2000, 2007). Now any site of use can be refashioned to mimic the auditory desire of the user—in essence all spaces are "nonspaces" to be refashioned by iPod users.

In addition, iPod users transform the world into conformity with their predispositions—the world becomes part of a mimetic fantasy in which the "otherness" of the world in its various guises is negated. This is an important strategy for iPod users who subjectivize space—consume it as if it were a commodity. In the process, immediate experience is fetishized. Technologized experience can be understood as fetishized experience—experience becomes real or hyper-real precisely through its technologization—through technological appropriation. The utopian impulse to transform the world occurs only in the imaginary—in its technologized instrumentality, the world remains untouched. The use of the iPod provides a buffer between the user and the recognized reality of the city street, invoking Kracauer's observation that "the world's ugliness goes unnoticed" (1995). Users prefer to live in this technological space whereby experience is brought under control—aesthetically managed and embodied—while the contingent nature of urban space and the "other" is denied.

Yet, in this denial of contingency lies a liberating moment understood as a form of reenchantment. I suggest that in this positive moment iPod practices can be understood as the coming together of a reenchantment of the city, understood through Walter Benjamin's concept of "aura" and Michel de Certeau's notion of street "tactics." IPod users are, in effect, individualizing each journey while simultaneously making urban space their own. In their auditory empowerment they create their own, unrepeatable, individual journey. In doing so they create, contra Walter Benjamin, their own auratic experience of the city—nonreducible to any other as it is framed—indeed, made through the mediated sounds emanating from their iPod.

Toxic Pleasures 2: The Spatial Secession of iPod Culture

Just as Sherman becomes aware of the "secessionist" potential of the iPod, so for other users it is precisely this secessionism that is attractive. The negative moment of the auditory pleasures of privatized listening lies precisely in the fictionalization or negation of the "other," whereby forms of urban reciprocity and of urban recognition are denied. In enacting these strategies, the empowerment of the subject within iPod use problematizes the very way in which users might "recognize" others (Honneth 1995).

In the following pages I appropriate Henderson's notion of "spatial secession" in order to view this "negative" moment of iPod use. Henderson, who was primarily concerned with the division of space through automobile use, argued that automobility was often embraced as a tool of "spatial secession" rather than as a pleasure in itself: "[M]obility is not just movement but also an extension of ideologies and normative values about how the city should be configured and by whom" (Henderson 2006, 295). Secessionist automobility refers to the use of a car "as a means of physically separating oneself from spatial configurations like higher urban density, public space, or from the city altogether" (Henderson 2006, 294). Secessionist values equally play a central role in the world of the iPod user, from the street to the automobile—users enact isolationist strategies with their headphones on or encased within the shell of the automobile, with the iPod docked into its sound system.

IPod use embodies forms of urban retreat, which has become a dominant metaphor in the dystopian image of urban life, whereby urban citizens attempt to maintain their sense of "self" through a range of distancing mechanisms from the "other." The iPod is the latest technological addition to the urban citizen's ability to neutralize urban space.

Benjamin understood the city as reducing the subject's capacity for thought due to the need for constant response to the contingency and plenitude of urban life. Urban "technology [had itself] subjected the human sensorium to a complex kind of training" (Benjamin 1973, 171). This training has enabled the subject to retreat from urban space by neutralizing it. Technologies of separation such as the iPod have progressively empowered urban citizens precisely by removing them from the physicality of urban relations. Urban experience undergoes a "ghettoization" through these secessionist strategies, furthering the technological intervention of subjects intent on urban retreat, transforming the polyrhythmic nature of urban space into the interiorized, monorhythmic sounds of users: "It's as though I can part the seas like Moses. It gives me and what's around me a literal rhythm, I feel literally in my own world, as an observer. It helps to regulate my space so I can feel how I want to feel, without external causes changing that" (Susanna).

IPod users invariably never willingly interact with others while engaged in solitary listening; interruption destroys the seamless reverie of use. It also indirectly relates to the silencing of experience in the city, which predates iPod use (Sennett 1994). The iPod in this instance acts as both a "gating" and a "tethering" technology. Indeed, gating assumes a metaphysical stance in iPod use: The drawing of circumscribed circles around the subject is both physical and metaphorical. This is one of the historical legacies of the values of individualism and the concurrent demand for privacy—an overriding desire to be left alone while in urban culture. IPod use increasingly becomes a habitual "mode of being in the world" (Geurts 2002, 235), in which users choose to live in an increasingly privatized and "perpetual sound matrix" through which they "inhabit different sensory worlds" (Howes 2004, 14) while sharing the same social space.

The public placing of and consumer use of a range of technologies enables the urban citizen to carry out most traditionally public tasks with little or no interpersonal contact, which furthers the architecture of isolation; exchanges are increasingly taking place between subjects and machines in urban culture, making interpersonal exchange obsolete. Cognitively, consumers increasingly expect, feel comfortable with, and desire no communication while out in public:

> Tracy expects to have wordless interactions with store attendants. When given a choice, Tracy will also use the "You-Check" line recently available in her neighbourhood's Fred Mayer, which allows her to scan her own purchases and credit card: "I love it," she says. In everyday life, she wants to "get in and out." (Jain 2002, 394)

The normative foundation of the "nonplaces" of urban culture becomes etched into consumers' social expectations as they partake of "public culture."

Physical and cognitive zoning becomes second nature in urban culture: "[T]he basic idea of zoning is that every activity demands a separate zone of its own" (Kunster 1998, 120). Urban infrastructures increasingly complement a range of technologies from the automobile to the iPod. These technologies emphasize connectivity within circumscribed circles and spaces while simultaneously alienating subjects from the copresence of one another in public space, as the following iPod users describe:

> I rarely even speak—I just hand them my credit card and say thank-you. (Mark)
>
> I tend not to notice people when I'm plugged in. I'm usually too preoccupied with myself to look at others. (Elizabeth.)
>
> It removes an external layer. I see people and things as inanimate or not-fully-connected. It seems that I have an external connection they lack. It's quite odd, actually . . . Yes. With the iPod and news talk radio files, I am having an interactive session with the anchor. When I look at the people around me, they appear to be two-dimensional and without significance. (Jonathan)

Progressive withdrawal from the cosmopolitan city both motivates iPod use and furthers it:

> I then started wearing it [the iPod] while shopping. I did it to control my environment and desensitize myself to everything around me. What I found interesting was that the more I wear my iPod, the less I want to interact with strangers. I've gotten to the point where I don't make eye contact. I feel almost encased in a bubble . . . I view people more like choices when I'm wearing my iPod. Instead of being forced to interact with them, I get to decide. It's almost liberating to realize you don't have to be polite or do anything. I get to move through time and space at my speed. (Zuni)

To be separated from others is simultaneously a mark of distinction (Bourdieu 1986) and a mark of alienation. City spaces become enacted spaces that are enjoyed and modified (de Certeau 1988; Lefebvre 1991) and also structural spaces that contextualize all behavior.

IPod users frequently glide through the urban street silently, and silence is imposed upon others as they passivize the looks and remarks of others. The silencing of the "other" is a strategy of control that represents a refusal to communicate with others in public. The users' privatized sonic landscape permits them to control the terms and condition of whatever interaction might take place, producing a web of asymmetrical urban relations in which users are invariably in control:

> A person with headphones on gives off an appearance of not wanting to be disturbed. There are times, mostly at work or walking to and from work, when I just want to be left alone. Wearing the iPod insulates me from other people in my surroundings. (Amy)

Users are aware of the symbolic meaning of the white wires of the iPod dangling from their ears: a combination of distinction and power. Empowerment is a product of withdrawal, and withdrawal creates an empowering sense of anonymity: "I use my iPod in public as a 'privacy bubble' against other people. It allows me to stay in my own head" (John).

IPod use permits users to redraw or redefine "personal space." Goffman considered personal spaces to be "the space surrounding an individual anywhere within which an entering causes the individual to feel encroached upon" (Goffman 1969, 54). Goffman's own definition represented an unrecognized historical moment of bourgeois urban sensibility—the entitlement to personal space. IPod use enhances this sensibility toward the ownership of space whereby the encroachment of others loses its physicality as users fail to notice or respond to the physical touch of others (Pinch 2010).

Forced interruption is invariably experienced negatively: "Sometimes I feel violated if I have to turn it off for an unplanned reason." The breaking of the auditory bubble represents recognition of the fragility or contingency of auditory empowerment. An involuntary and sudden return to the world, as others experience it, is invariably experienced as unpleasant. Maintenance of control frequently implies a denial of difference:

> In America, people are often loud and rude and it's sometimes hard to concentrate effectively. In Phoenix, we have a lot of Mexican immigrants. They don't learn English and they have no control over their children. I believe in mutual respect when in public places. It was becoming increasingly difficult for me to shop without encountering a bombardment of Spanish or screaming kids. The iPod lets me filter them all out. I'm much calmer now when I shop. The iPod lets me overlook the lack of courtesy. Using the iPod helps control my concentration. Since I'm familiar with the music, I can let it float to the back of my consciousness. (Tracy)

Tracy achieves a state of equilibrium precisely by withdrawing into herself. Sennett has described this form of behavior as representing "an early sign of the duality of modern culture: flight from others for the sake of self-mastery" (Sennett 1990, 44). The secessionist practices of iPod users embedded in this second form of toxic audiotopia reflects a negative moment of urban experience in which the urban subject uses a range of communication technologies to partition and remake experience. In doing so the very nature of collective urban space is thrown into question.

Toxic Audiotopia 3: Damaging Silence?

Toxic audiotopia also has potential physical consequences for users. The primary aim of many, as we have seen in this chapter, is to immerse themselves in the sonic environment created by the iPod. To achieve this, the sounds of the outside world are canceled out. This results in high volumes of listening:

> I usually turn it on as loud as it will go or until it sounds crappy. I don't like background noise when I am listening to my iPod. (Mary)
>
> Always loud—I can never hear anything around me. (Andy)

The seductive nature of the iPod—the proximity of loud music is also potentially toxic to the users' own sense of hearing. This third form of toxicity deals with the actual physical damage to users' hearing that results from loud and continuous usage. In addition to listening at loud volumes, users increase their music listening on average twofold with the purchase of an iPod. Music also does not appear to suffer from the routinization effect of other mobile technologies such as the mobile phone. After two years in follow-up interviews, respondents claimed to have maintained their level of use over the previous two years.

A recent study of sonic impairment published in the United States found that 16 percent of American adults have some degree of hearing loss. It was estimated that iPod users who turn up the volume to about 90 percent for on average two hours a day, five days a week will develop significant hearing loss. One author of the report stated that "one patient I had used his headphones instead of earplugs when he was on his construction job. He thought as long as he could hear his music over the sound of his saws, he was protecting his ears—because he liked the sound of his music but didn't like the sound of the construction noise. He has a good 50 dB to 55 dB of noise-induced hearing loss at 28" (Portnuff 2009).

IPod users vary the volume in accordance with their surroundings—the more the background noise, the greater the volume. For example, the use of an iPod on the underground system, where the ambient sound is high, encourages users to listen at near maximum volume; also, the volume used steadily increases without the users' conscious awareness as they continually attempt to compensate for the potential intrusion of ambient noise.

Portnuff (2009) found that teenaged iPod users not only played their music louder than older users but also were also unaware of how loud they were playing it. The playing of music at maximum volume for a mere five minutes a day could impair hearing permanently. The European Union's Scientific Committee on Emerging and Newly Identified Health Risks (SCENIHR) published its research findings in 2008 and found that that the numbers of young people with dangerous levels of noise exposure has tripled in the last twenty years while occupational noise levels had decreased in the same period. Although SCENIHR found that the majority of MP3 users were not at risk, it also found that 5–10 percent of users were at high risk due to their patterns and duration of use; these users listened to music for more than one hour a day at a high-volume control setting. The report concluded that the numbers of EU nationals subject to both temporary and permanent hearing loss was on the order of anything up to ten million.

Rawool and Colligon-Wayne (2008) found that general patterns of music reception among American youth tended to favor an intense auditory experience—that part of the pleasure was precisely the physicality of sonic intoxication. They found that "behavioural patterns consistent with a subjective sense of being addicted to loud music existed in 9% of their participants" (Rawool and Collington-Wayne 2008, 5). The intoxicating sound of music pumped directly into the ears constitutes a central element of pleasure for many users. The toxic damage to the ears is often not recognized or is traded in terms of immediate pleasures verses long-term potential damage that for many users remains an abstract possibility.

Conclusion

In this chapter I have analyzed the nature of the pleasures of auditory toxicity, which goes beyond the proprioceptive into the nature of our social world and the communication technologies that citizens habitually use. In doing so I recognize that iPod use should not be divorced from a range of other media and communication technologies habitually made use of. For example, automobile use, with its combination of secessionist cocooning and pleasures of mobility, are incorporated and developed through iPod use, which in turn is embedded in historical modes of portable radio reception. The nature of use is also embedded in how urban theory has understood and explained what living in the city consists of. Ever since Simmel argued that individualism in the city was premised upon the neutralization of the other through the construction of a "blazé" attitude, in which the "other" was effectively neutralized, the trope of urban retreat has featured heavily in urban literature (Sennett 1990, 1994). In contrast to this, writers from de Certeau onward have offered a more positive evaluation of an open sensory sensibility to urban experience that equates that experience with an openness to a wide array of experience that focuses upon the recognition of "difference," upon which self-realization is to be understood. Yet, following Walter Benjamin's assertion that technology has trained the human sensorium, the chapter has asked what type of training the use of technologies like the iPod signifies. The answer appears to be a complex mix of secessionism, creativity, and toxicity. The intense sonic immersion embodied in iPod use itself contains elements of both toxicity and creativity. This duality of use produces its own paradoxes as evidenced in the following iPod user's comment: "I didn't realize how much I yearn for control and probably peace and quiet. Strange, since I'm blasting music in my ears. I think I'm really tired of living on someone else's schedule. The iPod has given me some control back" (Janet).

Janet's audiotopia is based upon a defensive understanding of the lack of power embodied in nonmediated experience. The noise of uncontrolled culture is managed through the immersive sounds of her iPod, producing a sense of cognitive ease and silence. Thus lies the paradox of toxic audiotopias: Sound produces silence, connectivity produces separation, and mediated toxicity produces control. IPods are one element of the changing sound matrix of contemporary culture. The paradoxical use of Janet and other users needs to be put into a wider social context of urban separation and control in which the technology of the iPod is merely one more timely technology.

NOTES

1 The following empirical examples derive from primary research undertaken in 2005 and 2006. More than a thousand iPod users filled out a thirty-four-question questionnaire over the Internet. The respondents answered requests posted in the *New York Times*,

BBC News Online, Guardian Online, Wired News, and *MacWorld*. These requests were then syndicated and replicated in a wide a variety of newspapers and magazines worldwide. Respondents came mainly from the United States, the UK, Canada, Australia, and Switzerland but also included fewer responses from France, Italy, Spain, Denmark, Finland, and Norway. Twenty percent of respondents were then asked follow-up questions in response to their initial answers. In addition to the Internet sample, a few UK users have also been interviewed face to face. Selected follow-up interviews were conducted in 2007.

2 Teenaged users of mobile phones with MP3 capacity sometimes use their phones as mini–ghetto blasters while riding on trains and buses, holding the phone out toward others, thus claiming public space in alternative, if not particularly convivial, ways. The use of mobile phones in this way is called "sodcasting" in the UK and has produced a fervent debate in the pages of the *Guardian* newspaper concerning the "antisocial" nature of sequestering public space with such "treble-heavy" sound (*Guardian*, Aug. 13, 2010, P3).

3 I take a critical theory perspective on the way in which new technologies such as the Apple iPod integrate the user in new ways into what I refer to as "commodity culture." Dant (2008) takes a "circuit of culture" approach to reach similar conclusions as to the structurally integrating nature of use. From this point of view, "individualism" is merely a structured ideological response embodied in the cognitive practices of users. Dant also points to the power of the Apple brand in integrating the user into a form of "soft capitalism." For more on this see Bull (2008).

REFERENCES

Adorno, Theodore. *Introduction to the Sociology of Music.* New York: Continuum, 1976.

Augé, Marc. *Non-Places: Introduction to Anthropology of Supermodernity.* London: Verso, 1995.

Benjamin, Walter. *Illuminations.* London: Penguin, 1973.

Bloch, Ernst. *The Principle of Hope.* Oxford: Blackwell, 1986.

Bourdieu, Pierre. *Distinction: A Social Critique of Taste.* London: Routledge, 1986.

Bull, Michael. *Sound Moves: iPod Culture and Urban Experience.* London: Routledge, 2007.

———. *Sounding Out the City: Personal Stereos and the Management of Everyday Life.* Oxford: Berg, 2000.

———. "Technology as Fashion." In *The International Encylopedia of Communication*, vol. 11, ed. W. Donsbach, 5023–27. Oxford: Blackwell, 2008.

de Certeau, Michel de. *The Practice of Everyday Life.* Berkeley: University of California Press, 1988.

Chatfield, Tom. *Fun INC.: Why Games Are the 21st Century's Most Serious Business.* London: Virgin, 2010.

Dant, Tim. "iPod . . . iCon." *Studi Culturali* 5(3) 2008: 335–73.

DeNora, Tia. *Music in Everyday Life.* New York: Cambridge University Press, 2000.

Douglas, Susan J. *Listening In: Radio and the American Imagination, from Amos 'n' Andy and Edward R. Murrow to Wolfman Jack and Howard Stern.* New York: Times Books, 1999.

Dyson, Frances. *Sounding New Media: Immersion and Embodiment in the Arts and Culture.* Berkeley: University of California Press, 2009.

Geurts, Kathleen. *Culture and the Senses: Bodily Ways of Knowing in an African Community*. Berkeley: University of California Press, 2002.

Gibson, William. *Time Out*. (October 6, 1993), 49.

Gilroy, Paul. "Between the Blues and the Blues Dance: Some Soudscapes of the Black Atlantic." In *The Auditory Culture Reader*, ed. Michael Bull and Les Back. Oxford: Berg, 2003.

Goffman, Ervin. *The Presentation of Self in Everyday Life*. London: Penguin, 1969.

Henderson, J. "Secessionist Automobility: Racism, Anti-Urbanism, and the Politics of Automobility in Atlanta, Georgia." *International Journal of Urban and Regional Research* 30(2) (2006): 293–307.

Honneth, Axel. *The Fragmented World of the Social: Essays in Social and Political Philosophy*. New York: SUNY Press, 1995.

Horkheimer, Max, and Theodore Adorno. *The Dialectic of Enlightenment*. London: Penguin, 1973.

Howes, David. *Sensual Relations. Engaging the Senses in Cultural and Social Theory*. Ann Arbor. University of Michegan Press, 2004.

Ito, M., D. Okabe, and M. Matsuda. *Personal, Portable, Pedestrian: Mobile Phones in Japanese Life*. Cambridge, Mass.: MIT Press, 2005.

Jain, S. "Urban Errands: The Means of Mobility." *Journal of Consumer Culture* 2(3) (2002): 419–38.

Katz, Mark. *Capturing Sound: How Technology Has Changed Music*. Berkeley: University of California Press, 2005.

Klein, Naomi. *No Logo: Solutions for a Sold Planet*. London: Flamingo, 2000.

Kracauer, Sigfried. *The Mass Ornament: Weimar Essays*. Cambridge, Mass.: Harvard University Press, 1995.

Kun, Josh. *Audiotopia: Music, Race, and America*. Berkeley: University of California Press, 2005.

Kunster, Harold. *Home from Nowhere: Remaking Our Everyday World for the Twenty-First Century*. New York: Simon and Schuster, 1998.

Lanza, Joseph. *Elevator Music: A Surreal History of Muzak, Easy Listening, and Other Moodsong*. New York: Picador, 1994.

Lefebvre, Henri. *The Production of Space*. Cambridge, Mass.: Blackwell, 1991.

Marcuse, Herbert. *One-Dimensional Man*. London: Routledge, 1964.

Morley, David. *Home Territories: Media, Mobility, and Identity*. London: Routledge, 2000.

Pieslak, Jonathan. *Sound Targets: American Soldiers and Music in the Iraq War*. Bloomington: Indiana University Press, 2009.

Pinch, Trevor. "The Invisible Technologies of Goffman's Sociology." *Technology and Culture*: 51(2) (2010): 409–24.

Putnam, R. *Bowling Alone: The Collapse and Revival of American Community*. New York: Simon and Schuster, 2000.

Rawool, Vishakha, and Lynda Collington-Wayne. "Auditory Lifestyles and Beliefs relating to Hearing Loss among College Students in the USA." *Noise and Health* 10 (2008): 1–10.

Rheingold, H., and J. Kluitenberg. "Mindful Disconnection: Counterpowering the Panopticon from Inside." *Open* 11 (2006): 29–36.

Schiffer, Michael B. *The Portable Radio in American Life*. Tucson: University of Arizona Press, 1991.

Sennett, Richard. *The Conscience of the Eye*. London: Faber, 1990.

———. *Flesh and Stone*. New York: Norton, 1994.

Sterne, Jonathan. *The Audible Past: Cultural Origins of Sound Reproduction*. Durham, N.C.: Duke University Press, 2003.

WEBSITES

http://www.colorado.edu/news/r/f9b28bae908df1b6e7fc9fa33e90f99d.html (Portnuff 2009)
http://www.eurekalert.org/pub_releases/2009.../uoca-nil021809.php.(Portnuff 2009)
http://ec.europa.eu/health/ph-risk/risk-en.htm (SCENIHR).

CHAPTER 23

THE RECORDING THAT NEVER WANTED TO BE HEARD AND OTHER STORIES OF SONIFICATION

JONATHAN STERNE AND
MITCHELL AKIYAMA

From the Postsonic to the Sonified

In the spring of 2008, the incunabula of sound recording briefly broke into headline news.[1] A group called First Sounds had played a recording that dated back to 1860 (Rosen 2008; Maugh 2008). Anyone with an Internet connection could now hear "the world's oldest recording." This accomplishment raised a number of epistemological and historical questions. What does it mean to speak of a sound recording made in 1860? Thomas Edison's phonograph, first demonstrated in 1877 and patented in 1878, is still widely considered to be the first device capable of recording sound and playing it back (Read and Welch 1976; Chanan 1994, 1995). However, First Sounds had succeeded in playing back a recording from a device that predates Edison's: the phonautograph. Invented in 1857 by

Édouard-Léon Scott de Martinville, the phonautograph was conceived as a "sound writer." It channeled sound through a conic funnel to vibrate a small membrane, which in turn vibrated a stylus, thereby creating visible tracings on a recording surface—first a piece of smoked glass and later a sheet of paper rolled around a cylinder. Scott called the images it produced "phonautograms," literally, "speech self-writings." The mechanical similarities between Scott's phonautograph and Thomas Edison's cylinder phonograph are striking, a point that was not lost on nineteenth-century inventors. In the early 1870s Alexander Graham Bell had experimented with the phonautograph as a means of teaching deaf children. It was a modified version of the device, employing a human ear as a diaphragm that, according to Bell, "gave him the idea for the telephone." Edison also cited the phonautograph as an antecedent to his machines (Gitelman 2006; Sterne 2003; Bell 1878; Edison 1888). Yet, for all the similarities between the phonograph and the phonautograph, there existed no technology to play back phonautograph recordings as sound, at least until sometime shortly before 2008 (Chanan 1994, 1995; Hankins and Silverman 1995; Lastra 2000; Read and Welch 1976; Sterne 2003).

In this chapter we use the story of the phonautograph as a way in and out of investigating the development of a cluster of practices called "sonification," or the transformation of nonsonic data into audible sound. The story of the phonautograph is remarkable because it was conceived as a technology for visualization. Its recordings were never supposed to be heard, but now, thanks to the availability of tools for transforming the nonsonic into the audible, we are able to play back and listen to a recording that was intended to be seen, not heard. To call this technological process "playback" is not quite accurate since, as we will see, a great deal of reconstruction went into First Sounds' efforts. Rather, we consider the possibility of listening to a phonautograph recording as marking an aesthetic and epistemic shift in the history of sound. This innovation is more than just a technological feat; it also speaks to a significant change in the ways in which we construct distinctions among the senses. After further introducing the phonautograph, we review a range of practices of sonification, all of which are instructive insofar as they move data and experience between the sonic and nonsonic registers. Our conclusion then returns to the case of the phonoautograph in order to advance some speculative propositions regarding the present conjuncture in the history of sound. The relationship to sensation central to sonification (and the assemblages of media and practices on which it depends) is most distinctively characterized by the ability to transform data destined for one sense into data destined for another. This is not a claim about synesthesia or, worse, a desire to restore a so-called balance of the senses in an idealized sensorium. Rather, we argue that this extreme plasticity lays bare the degree to which the senses themselves are articulated into different cultural, technological, and epistemic formations. The implication for sound studies is simple to grasp but difficult to handle. If we acknowledge the plasticity of sound, we must also acknowledge the limits of a commonsensical definition of the field's object as "what can be heard" or "sound as an effect in the world."

Scholars in the humanities and social sciences still too often treat particular technologies or cultural forms as if they are predestined for or determined by a single sense. If we take the proposition of the plasticity of sound seriously, it is no longer possible to maintain such assumptions. Recent decades of sensory history, anthropology, and cultural studies have rendered banal the argument that the senses are constructed. However, as yet, sound scholars have only begun to reckon with the implications for the dissolution of our object of study as a given prior to our work of analysis.

Why anchor a chapter on sonification to a visualization technology like the phonautograph? Scott's phonautograph is often accorded a place in the history of sound technologies, but it was neither a sound-reproduction technology nor a sonification technology. The phonautograph's relationship to sound could best be described as *postsonic*. It produced an image after the occurrence of a sound—a record of a past sound event that it did not seek to reconstruct sonically. According to Scott, the phonautograph was never intended to play back its recordings. Scott even derided Edison's phonograph on the grounds that it "merely reproduced sound—it was not a *sound-writer*."[2] The phonautograph was part of a boom in nineteenth-century devices that rendered visible otherwise invisible aspects of the natural world. Through a variety of technical feats, these technologies rendered the processes of nature—sound, electricity, biological processes and rhythms—as visual data that adhered to an orderly pattern. Scott and many of his contemporaries believed that to see sound was to better know it. The point of the phonautograph was to render sound available to careful visual examination, not to make recordings for playback (Hankins and Silverman 1995; Levin 1990). First Sounds' intervention was to ignore the inventor's intent and instead consciously and deliberately reconstruct phonautograph recordings *as if* they had been made for playback.

It is a cliché of technological history that inventors' intentions have a way of not working out and that whatever purpose or intention is built into a technology can change over time. The usual story is one of invention, innovation, engagement with users, and reconstruction ad infinitum as technologies achieve moments of temporary closure, only to be opened up again and reconfigured from a new, unexpected angle (Bijker 1995; Jenkins 2006; Pinch and Bijker 1984; Pinch and Oudshoorn 2003). The story of the phonautograph, however, is a little different. A minor device in the history of sound recording, it never really left relative obscurity. Even its most ardent users generally moved on to other options as the nineteenth and twentieth centuries rolled on (Scripture 1906). The phonautograph appeared only as a curiosity in histories of sound recording (when it was mentioned at all) until it was resurrected by a few media historians at the turn of the twenty-first century (Hankins and Silverman 1995; Lastra 2000; Levin 1990; Sterne 2003). Its brief reemergence and short-lived posthumous fame signal a shift that occurred in sonic epistemology in the intervening years. While Scott and his contemporaries sought to transform sounds into images, today groups of scientists, artists, and technologists work to convert nonsonic phenomena so that they may be heard and therefore known as and through sound. In its time, the phonautograph helped construct sound as an effect that could be reproduced without reproducing its cause. First Sounds' accomplishment of sonifying a phonautogram reveals the plasticity of sound, its ability to convey data across different sensory registers. In addition, their work reveals the degree to which data are fluid and are not necessarily tethered to any one sense.

In the spirit of this handbook's handbookness, we would like to call attention to the usefulness of the concept of articulation for our argument. Drawn from the field of cultural studies, articulation theory might best be described as antiessentialism in action (Grossberg 1992; Slack 1996; Hall 1986). Jennifer Daryl Slack and J. Macgregor Wise define articulation as "the contingent connection of different elements that, when connected in a particular way, form a specific unity." For them, elements can be "made of words, concepts, institutions, practices, and effects, as well as material things." Articulation theory is meant to draw attention to "the movement and flows of relationships" (Slack and Wise 2006, 127). So far, one might note similarities to Latour's actor-network theory (Latour 1996) or the social construction of technology tradition cited earlier. Inasmuch as all three approaches are constructivist, there is a natural affinity. However, articulation theory and the cultural studies tradition from which it draws are more particularly attentive to questions of power and, for a lack of a better term, the density or cogency of articulations. Articulations:

> can and do change over time. But here, too, the speed and direction of change is contingent. Some articulations remain relatively tenacious; they are rather firmly forged and difficult to disarticulate. . . . Others might be more easily broken and thus subject to disarticulation and rearticulation. It all depends on the particulars of the nature of articulations at any particular historical moment. (Slack and Wise 2006, 128)

In this chapter we use articulation in two ways: We show how sonification articulates a range of practices that render data for the ear, and we explore the ideas and practices articulated to hearing within sonification. However, our larger historical point is that, at the present moment, the articulations between particular senses and kinds of sense data are incredibly weak. If this is the case, it suggests that sound scholarship must be ever more vigilant about that shifting border between the sonic and the nonsonic.

Our chapter therefore advances three nested methodological propositions. We argue for attending to the modularity of sensory technologies; for the modularity of the relations between senses, subjects, and technologies; and, ultimately, for the modularity of the senses themselves.

Parameters of Sonification

Alternately known as *sonification* and *auditory display*,[3] the practice of turning data into sound as yet has no obvious epistemic center, and its overall impact on the arts and sciences is as yet unclear. Simply put, sonification is the use of nonspeech sound to convey information. This involves the translation or transposition of data from a given sensory mode into sound—a rearticulation of sorts. Techniques for analyzing or presenting data have tended toward the visual at least since the Renaissance (Ihde 2009). While there are early examples of attempts to use sound as a primary means of monitoring the world—Leonardo da Vinci is credited with inventing a form of sonar as early as 1490—sonification is a relatively recent approach to rendering information sensible and has been promoted variously as a corrective or a supplement to visual display.

In her chapter in this volume, Alexandra Supper elegantly renders the shape of current debates about the definition of sonification and the policing of the term's borders among people who work in the field. For our chapter we have chosen a broad definition of sonification to highlight both the range of practices that might be enclosed in its big tent and the connections between sonification and other sonic practices. According to Gregory Kramer and Bruce Walker (2004), two pioneers of auditory display, sonification falls into three basic categories: alerts and notifications, auditory icons, and audification. Alerts are the simplest instruments in the sonification toolbox. Fire alarms, sirens, the beeping of a microwave, and so on generally signify one thing: There's a fire; get out of the way, an emergency vehicle is trying to pass; your soup is now hot. They are neither able nor meant to communicate anything more subtle than a binary possibility: "The fire alarm signals that there is or is not a fire; not where or how hot it is" (Kramer and Walker 2004, 151).

Auditory icons—or "earcons"—are capable of more subtle types of communication. Associated mainly with computing (Robare and Forlizzi 2009; Gaver 1989; Blattner, Sumikawa, and Greenberg 1989), auditory icons might give feedback on a process—scrolling through a window or moving a folder to the trash, for example. In the late 1980s, William Gaver (1989) developed a sonic interface for Apple called SonicFinder. The interface was an ambitious attempt to sonify computing but was incorporated into mainstream operating systems only in a limited fashion. Not only did Gaver's system alert the user that a process was occurring, it also employed a level of sonic nuance that could communicate subtle features of these operations. For example, selecting a file would trigger a "hitting" sound that would vary in pitch and timbre depending on its contents or size. Remnants of this approach remain buried in the user interfaces of common software applications; in many applications, brief blips and sound effects, for example, signal the arrival of an email or its successful departure. Microsoft has programmed a veritable cacophony of noises into its Office applications, which users can enable or disable as they please. We know of no writer who actually uses the many earcons available in MS Word, but they are there and available should you wish to be the first.

While alerts and auditory icons communicate symbolically, audification functions isomorphically. Through the modification or the transduction of other vibrational or wave phenomena, audification renders data more immediately meaningful to a listener. In the early 1960s seismologist Sheridan Speeth (1961) proposed that amplifying and speeding up recordings of seismic movements could allow listeners to differentiate between earthquakes and explosions caused by bombs. Sonocytology—the sonification of cells—operates similarly by amplifying the inaudible vibrations of cellular life. As Sophia Roosth (2009) notes, yeast cells vibrate at frequencies within audible range. The volume, however, is far below the threshold of human hearing. Using scanning probe microscopes to record the vibration of these cells, sonocytologists have managed to sonify yeast cells and listen in on experiments. These scientists have found that adding alcohol to the cells' medium causes them to vibrate more rapidly, a phenomenon that, when sonified, results in higher-pitched sound, giving the impression that they are screaming.

Many recent developments in sonification have happened thanks to computers or, more specifically, digitization. Here we confront one of the defining features of "new" media as described by Lev Manovich (2001, 45-48): transcoding. Because digital files exist as both computer data and as aesthetic representations of the sensible world (in the form of an interface), a wide range of possibilities exists for both the manipulation of data and the interface itself. When an audio editor displays the waveform of a recording, it offers a simple and fairly straightforward kind of transcoding. Given audio data, it renders a picture that represents some aspects of the audio recording's waveforms. More creative options are also possible; for instance, an application called Metasynth allows users to actually generate sounds from the data found in pictures (http://www.uisoftware.com/MetaSynth/index.php). When lasers map the curves of old phonographic cylinders or trace the vibrations of old phonautograph tracings, they are similarly performing a kind of transcoding. *Pace* Manovich, transcoding is not necessarily a digital process. The study of pulsars in radio astronomy, for example, has for a long time rendered the rotation of distant stars as sonic data because the regularity of their rotations can be better heard than seen. However, transcoding becomes considerably more common in the digital realm because software allows for greater ease and facility in the manipulation of data sets. The pulsar example is instructive because it reveals one of the central features of audified knowledge, a feature that works because of transcoding—let us call it "effective indexicality."

Audification requires an assumption about the indexicality of the reproduction or manipulation of a phenomenon. The listener must believe that the sound produced is, for all intents and purposes, made of the same stuff as the object it is meant to represent. While a microwave's beep is symbolic and arbitrary, audification is indexical in that patterns of sound are directly caused by patterns in the data. Here we borrow the distinction between indexes and symbols from Charles Sanders Peirce's semiotic theory. Peirce defined a symbol as a sign that has an arbitrary relationship to its signifier, such as words in language. He defined an index as a sign with a causal relationship to its signifier—a weathervane, for example (Peirce 1955, 107; Turino 1999). In audification, patterns in data are made to "cause" particular sounds to occur, which in turn render those patterns sensible in ways that may be harder to discern visually. Although there is an arbitrary dimension to the relationship—someone must choose which sounds to associate with which data—the resulting relationship is meant to be experienced as one of cause and effect.

Human hands also construct Peirce's weathervane, and its shape as a rooster or an arrow, while arbitrary, does not affect its indexical relationship with the wind. Because the indexicality here is a deliberate effect of a human-made representational system, we call it "effective"—it is more important for the audifier to believe that the relationship between sound and object is indexical than it is for the relationship to actually be indexical. (Some forms of indexicality in Peirce are less "human made," as in the pain caused by a touching a hot stove, though there, too, we would note that a certain level of belief in pain as an accurate index of danger is required.)

Whether it apprehends information through symbolic or indexical representation, proponents of audification argue that the ear is particularly good at certain things, superior even to the eye. The ear's ability to perform monitoring tasks has been exploited for decades with devices like the Geiger counter, sonar, and heart monitors (Kramer 1994). We can listen, recognizing patterns or anomalies, without tiring or even necessarily devoting our full attention to the task at hand. The sonification of patterned data—stock market trends, network traffic, vital signs, and so on—capitalizes on the ear's ability to detect anomalies that might "pop out" of a continuous stream of information (Kramer 1994; Hermann, Niehus, and Ritter 2003). The eye isn't as proficient at picking out discontinuities while the ear is highly attuned to "wrong notes." In contrast to hearing, vision is often described as a focused sense, capable of parsing only one thing at a time. The ear, however, can discern and parse multiple sounds simultaneously. Proponents of sonifications characterize hearing as a faculty that is more sensitive to direction, a useful strength in analyzing spatial data or receiving and acting upon directional cues. Sonification discourse also suggests that the ear is more sensitive to affect than the eye (Barrass and Kramer 1999; Polli 2005), and, as such, talk about and around sonification often partakes of a long-standing romantic ideology that is attached to hearing in many cultural fields. The idea that the ear and the sonic arts are somehow inherently closer to the seat of emotion and affect is strongly rooted in Western traditions that shape the way we think about music, speech, and other forms of auditory expression (Flinn 1992, 7, 9; Sterne 2003, 15). We do not need to believe in the truth of these assertions, however, to appreciate their effectiveness in shaping how practitioners relate to, make use of, and construct arguments about sonification.

Aesthetics of Sonification

Perhaps it is the romantic idea of sound's affectivity that has led to growing interest in sonification in the art world. While it is true that in most cases the making audible of information is largely utilitarian, in many others the lines between scientific and artistic production are blurred. As Alexandra Supper notes (in chapter 10), this is a major point of concern among some advocates of sonification since the field is itself seeking legitimacy among scientists. This attempt at disciplinary respectability is complicated by the relative impermeability of the boundaries between art and science. As Supper shows, sonification researchers who are inclined toward interdisciplinary projects often have no choice but to align themselves with the terms, language, and expectations of one field or the other. For practitioners whose work might be more clearly aligned with the empirical, presenting a project in an artistic mode might undermine its credibility.

There are precedents for the hybridization of the empirical and the expressive. Georgina Born and Andrew Barry (2010) develop a genealogy of the emerging field they call "art-science." They identify three "logics" operating in these fields: accountability, innovation, and ontology (105). First, the aestheticization of data often brings a work to the attention of a public that would not necessarily normally have access to it. Insofar as an artistic mode of presentation brings researchers' work into a wider public, it is submitted to scrutiny, and researchers are forced to account for their findings. Second, the art-science pairing is often touted as producing innovations through synergy and collaboration. Third, artistic interpretations and presentations of research often work to call the very nature of scientific inquiry into question. Here we find art and science seemingly offering each other a chance at mutual legitimacy. Science stands to benefit from art insofar as the aesthetic putatively performs the function of making the empirical accountable (109). On the other hand, art practices gain epistemological credibility thanks to their supposedly creative contributions to knowledge (110).

Since its beginnings as a coherent field, this push-pull between art and science has played out in sonification discourse (Supper 2010). Despite its appeals to scientific rigor, sonification is an interpretive and aesthetic process. This is true for the actual sounds it produces, the reactions of listeners, the interfaces of sonification technologies used, and the relationships between sonic and nonsonic data in specific sonifications.

Projects by several individuals—and in many cases teams—present sonified data sets as works of art. Andrea Polli's (2005) "Atmospherics/Weather Works" employs a fifteen-channel sound system to re-create significant storms in the New York/Long Island area. Polli and her team mapped several variables—atmospheric pressure, relative humidity, wind speed, and so on—to an assigned set of recordings that included vocals sounds and wind instruments, as well as environmental and insect sounds. As such, "Atmospherics" doesn't simply present recordings of the storms; rather, it uses arbitrarily assigned sounds to create an affective representation of weather. Polli's stated intentions make it clear that her work is meant to straddle the line between art and science:

> As an artist sonifying atmospheric data, I am interested in the creation of new forms of data interpretation. As individuals and groups are faced with the need to interpret more and more large data sets, a language or series of languages for

> communicating this mass of data needs to evolve. In my artwork, I have tried to develop strategies for the interpretation of data through sound that has both narrative and emotional content because I believe that an emotional connection with data can increase the human understanding and appreciation of the forces at work behind the data. (Polli 2005, 33)

Other similar examples include "Life Music," a collaboration between composer John Dunn and biologist Mary Anne Clark (1999) that sonifies DNA sequences by assigning pitches to amino acid sequences; Thierry Delatour's "Molecular Music" (2000) makes audible the vibrational spectra of molecules; *Navegar é preciso, viver não é preciso* by composer Alberto de Campo and sociologist Christian Dayé (2006) sonically represents a variety of socioeconomic data using the time/space coordinates of Magellan's 1519–1522 circumnavigation of the globe.

The practice of using data as raw material for art has a pedigree that reaches back at least as far as the conceptual movements of the 1960s. Sound artists, including John Cage, Alvin Lucier, and Charles Dodge, along with installation artists like Hans Haacke, Sol Lewitt, and Dan Graham, broke away from the prior conception of artists as individuals that realize works according to their own uniquely personal vision. One of Cage's many contributions to the dismantling of the romantic notion of the artist as inspired genius was the work *Atlas Eclipticalis*, in which he superimposed music paper on top of star charts and plotted musical compositions as though they were constellations. In *I Am Sitting in a Room* Alvin Lucier sonified the acoustic characteristics of a room by recording his speech and then playing the recording back into the room, rerecording, playing it back, rerecording, and so on until his vocal signature was thoroughly eroded by successive generations of mediation.

Works like these brought natural and artificial systems to the fore by putting a process or phenomenon ahead of the artist's ostensibly personal "vision." What these works also implicitly proposed was that phenomena beyond the perception of human senses might somehow be represented or reified. Lev Manovich, writing about contemporary visual art that has followed a similar line, suggests:

> If Romantic artists thought of certain phenomena and effects as un-represantable [*sic*], as something which goes beyond the limits of human senses and reason, data visualization artists aim at precisely the opposite: to map such phenomena into a representation whose scale is comparable to the scales of human perception and cognition. (Manovich 2002)

As we've seen, this move to render data sensible is at the core of sonification. In this sense artists and researchers not only are allied in seeking a common objective but often also work together and occasionally even take on both roles. These artist/researchers, who often publish in journals like *Leonardo* or *Computer Music Journal*, have tended to present their work as a fusion of art and science. The justification is that sonification can provide real insight into physical or informational processes, as well as serve to create or reinforce a sense of immediacy or relevance. The artistic presentation of sonified data is the verso of the phonautogram's recto. Resolutely experimental in its approach and outcomes, the phonautograph occupied an ambiguous place between art and science. Perhaps this is a central feature, more generally, of the process of transcoding from one sense to another. Not quite synesthetic, not quite scientific in a classic sense, the conversion of data between the senses nevertheless presents itself as having a certain empirical weight and is conditioned by the aesthetic sensibilities of inventors, experimenters, and users.

Accessibility

Sonification discourse has predominantly been concerned with the interpretation and display of data, but it has also been widely touted as an important tool for creating accessibility for people who are blind. The benefits of sonic alerts, earcons, and audified information for those with limited or no vision are obvious, but there is one sonification technique that is arguably even more useful: human echolocation. Through rigorous training, some blind individuals have learned to navigate the world with remarkable aptitude. Daniel Kish, the most notable proponent of human echolocation, lost his sight in his infancy but has gradually learned how to negotiate the world acoustically (Kish 2009). By emitting vocal clicks, Kish manages to negotiate the world with a grace that is incomprehensible to sighted people. Even more astonishing, he is able and has taught others to ride bicycles. He is also able to sense and describe objects around him in detail; with training and practice echolocation is not only able to reveal the presence and placement of objects but can also help a listener to discern size, texture, and density.

In that echolocation is a technique that uses sound to convey or relay information, it is consistent with the most basic definition of sonification. However, unlike other forms, it does not operate in the symbolic realm, creating representations of data or phenomena. Rather, it is a technique for emplacement, for an immediate experience of a space. The sounds that the echolocator produces and then receives do not represent or tabulate the world; they present it in its immediate plenitude. In the same way that the reflection of light off an object constitutes the unmediated experience of seeing, the reflection of sound off a thing constitutes an unmediated experience of hearing. This is arguably true of a recent technological extension of human echolocation, FlashSonar (Kish 2009). Developed in part by Kish, the device emits clicks at a rate, volume, and endurance beyond the capacity of a human. While the sound that the FlashSonar employs has been arbitrarily chosen, the "image" of the world that the listener receives is not represented; it is immediate.

In all of these examples of sonification and the discussions it has precipitated, one might detect a slight corrective. The argument is that data have historically been largely represented by visual means. This should come as no surprise given the inherent visual bias of Western science (Ihde 2007, 2009). There is a long history in both scientific thought and the philosophy of science of privileging vision as the sense most likely to yield scientific knowledge. Vision has not only been favored epistemologically but has also been granted metaphorical dominance. The prevalence of light and sight metaphors have been with us at least since the ancient Greeks; Don Idhe and others have brought awareness to this visualist legacy in scientific thought (Peirce 1955; Foucault and Gordon 1980; Virilio 1989; Jay 1993; Hillis 1999; Turino 1999). To be more precise, this bias is not toward *vision* as such but a particular construction of sight as a sense that objectifies and separates subject from object. One result is that while there exists a long history of rendering sound as visual data—a history coterminous with the rise of modern acoustics and hearing science (Kittler 1999; Ferguson and Cabrera 2008), sonification has only recently gained traction as an epistemological innovation in itself.

Reanimation as Sonification

Here we return to the First Sounds' work, for in a way their project inverts the old visualist bias. If sound could be transcoded into image, it stood to reason that image could be transcoded into sound. The phonautograph was designed to turn sound into image, not to play it back. However, First Sounds were able to reverse this process and thereby render the images on the phonautograms as sound. The group's 2008 success in playing back one of Scott's phonautograms was accomplished using technology inspired by new practices in the preservation of old phonographic cylinders. Because some cylinders are close to one hundred years old, laying a needle on them might potentially damage (if not destroy) them in the act of playback. Instead of laying a needle on the grooves of a wax cylinder, newer apparatus use a laser to scan the media; computer software then reconstructs the audio. In the case of the phonautogram, a sheet of paper blackened by the smoke of an oil lamp and then scratched with a stylus, the process is similar; a "virtual stylus" scans the document and converts the markings into digital information that can be rendered as sound (Hennessey and Giovannoni 2008).

Again, unlike phonographic cylinders, phonautograms were never supposed to produce sound: Scott's phonautograms were meant to be "read." What is perhaps most significant about First Sounds' achievement is that they succeeded in contravening Scott's intentions by reading these objects so that we can hear them. This step has been problematic since the process was invented: The patterns that phonautograms display do not have any apparent meaning to the untrained viewer. This opacity prevented early would-be users from finding much use for the objects. Bell, for example, had hoped to use the device to train deaf students to speak by teaching them to modulate their voices until the phonautogram looked like one from his own speech. While elegant, the plan was unsuccessful because the phonautograph did not provide the necessary visual detail (Bell 1878). This remained the case until it became possible for lasers to scan cylinders. Following this innovation, the members of First Sounds hypothesized that it might also be possible to extract sound from phonautograph recordings (though the idea of hearing phonautograms probably extends all the way back to their first public presentation). They retrieved Scott's original patent deposits from France and then digitally encoded them at Berkeley Labs. Their first few attempts were largely unintelligible; because Scott's phonautograph lacked a governor and had to be cranked by hand, the recording speed was erratic and had to be calibrated to be intelligible. First Sounds had the arduous task of manually reworking the speed of the recording so that it played at a consistent pitch and rate (http://firstsounds.org; Giovannoni to Sterne 2008). Sound machines always need a little help from their human auditors, but in this case the phonautograph required at least three kinds of assistance—from its inventor, from First Sounds, and from us, its listeners.

First Sounds' first success was an 1860 recording of "Au Claire de la Lune." The original phonautogram had notations that signaled which part of the line was made by which word. First Sounds had known what they were looking for all along; what these clues offered was a map. Phonautographic recording couldn't take advantage of modern engineering practice: The machine was not capable of time stamping and did not use reference tones or pitches. We have already seen that, from Scott's perspective, phonautograms, though sound recordings in the sense that they were sonic vibrations inscribed on paper, were not sound recordings in the sense we think of them today. Rather, they were something closer to sound inscriptions, fossils, or traces. Scott did not intend for them to be heard or understood in sonic terms; rather, they were meant to help observers understand sound through visualization. Perhaps the great irony of the device is that it failed as a sound write[r] but, with a good deal of help, ended up being involved in the historical reconstru[c]tion of sound from 1860.

LAUGHING WITH THE DEAD

The admittedly minor career of Scott's phonautograph and its pho[...] illustrates the articulated nature of all sound technologies and so[...] that is to say, the phonogram is necessarily understood differently [...] mations. In the formations of nineteenth-century science—in an[...] paradigm where seeing and knowing were closely coupled—th[...] was a part of a long line of machines that rendered physical p[...] this sense, it was almost incidental that the phonautograph w[...] transduction of sound since it shared an epistemic relatio[...] devices that traced a whole range of physical processes[...] nineteenth-century sound reproduction, the phonautogra[...]

important shift toward machines that treated sound as a reproducible effect. In this alternative formation, the phonautograph's ability to transduce sound is essential for its historical and cultural legibility. Today, neither of those formations holds, and the phonautograph's means shifts again as a result. The possibility of playing back a phonautogram was unimportant to Scott; the device was a means of making sound more comprehensible in a paradigm where comprehension often meant visibility. However, the possibility of playing back a phonautograph is now important to people for a range of reasons. Sonification offers a utopian proposal similar to Scott's hopes for phonautography—but in reverse. It promises to turn a world of information into sound. First Sounds' work therefore lies somewhere in between the audification of nonsonic data and the digital transcoding and reconstruction of analog sonic data. In either scenario, Scott's intentions for his machine have become irrelevant. A phonautogram's indexicality makes it recognizable when sonified or, more accurately, *almost* recognizable if one is instructed in what to listen for.

There is an obvious novelty factor in hearing an 1860 recording, but there is more to it than that. By way of conclusion, let us step beyond that to consider some possible historical, aesthetic, and epistemological implications of First Sound's sonification of phonautograms. As we have argued throughout, sonification is a particular articulation of the sonic and the nonsonic, one that points to an increas- vagueness of the borders around the audible world. Sound studies of the present nt must therefore wrestle in new ways with the boundaries of their objects. d must let go of its axiomatic assumptions regarding the givenness of a r domain called "sound," a process called "hearing," or a listening subject. a call for a kind of everything-goes postmodernism but rather a reminder latedness of sensory technologies, sense data, and the senses themselves. sonification, we note an increasingly forceful articulation of the able and susceptible to transcoding. Thus, we end by wrestling ctions to the phonautogram playback. Here are three possible

closer to Scott" interpretation. Most writers on the phonauto- t there is an undeniable morphological similarity between Edison's phonograph. Nonetheless, there is an epistemic ne side and Edison on the other because of the latter's s is the case, one could conceivably read our present ott.

ter software, we live in an era of unprecedented ult not only of sonification practices but also rs have chosen to display sonic data and to process through the design of computer sion and not a necessary consequence of casions to "look at sound" in a digital . Those of us who have mixed audio re tactile mixing board know well the umbers and to draw straight curves in our edits not because they sound right but because they look right. First Sounds' digital sonification of Scott's phonautogram was similarly calibrated to the medium's visual appearance.

2. Despite the prevalence of visualization, we are also living in an age of unprecedented sonification, which is potentially *a reversal of Scott's process.* In some important ways, First Sounds' enterprise demonstrates the degree to which contemporary thinking inverts Scott's proposition—that the purpose of manipulating sound was to transduce it into visible data. Today, anything can be turned into sound, and in some cases it is quite useful to do so. Sound reproduction has extended into an ever-increasing number of spaces of everyday life (becoming mundane in the process), and more and more nonsonic phenomena have come to be comprehensible through sound. Friedrich Kittler has suggested that the sound-reproduction technologies of the late nineteenth century deprivileged the voice. Instead, it treated all sounds the same; it submitted them to a logic that dictates that "frequencies are frequencies" regardless of their source (Kittler 1999). Sonification takes this one step further by converting the frequencies of any given data set into an audible spectrum. There is no better example of this than our ability to now listen to sound recordings that were never supposed to be reconstituted as sound. With the phonautogram, as with the yeast cell and the pulsar, we can now listen to that which was never meant to be heard.

3. There is also a broader conclusion to be made here about *the plasticity of data in digital schemes and the dissolution of old knowledge about the senses.* In our age, the very idea of audition is so different from that of Scott's era that one must wonder whether the sound represented by his phonautograms is really the same sound that we hear coming out of the speakers when we play it back today. In Scott's time, sound was viewed as a substance, and different sounds were given different values. Bell and Edison probably still understood sound as having a certain "thisness" to it out in the world even though they also understood it as reproducible. For them, speech or music (or birdsong or wind) had a certain ontological coherence and essential character. In current use, sound-reproduction technology constructs sound as nothing more than a stratification of vibration by the body. Those vibrations within the audible range are considered sound; those outside the audible range are interpreted haptically or are not perceived at all. But even the unity of "sound" as a perceptual category is an illusion of language. We think of the ear as a transducer, as something that changes "sound" into vibrations and then into electrical and neural signals that are experienced as sound. While this integration exists at the perceptual level, it does not necessarily exist at the biological level. Cognitive scientists currently believe that different regions of the brain process different kinds of sounds. If there was a gap between Bell and Edison's understanding of sound and Scott's, one could posit another gap between our sonic world and the sonic world in which all three inventors operated.

If there is discontinuity on one stratum, there is continuity on another. In a visceral way, the work of First Sounds pushes the audible history of recording back seventeen years from 1877 to 1860. If Scott's recording is intelligible to listeners *as a recording*, albeit a bad one, then the old Victorian obsession with the voices of the dead serenading the living sticks with us even now when the dead sing to us every day on classic rock radio stations. We might also tip our hats to Freud's comments on the uncanny nature of recording (Freud 1989). When the "Au Claire de la Lune" recording was played back on BBC, the news reader couldn't control her laughter as her script moved from technological marvel to obituary (Richards and Sherwin 2009). If she could not help laughing upon hearing a dead voice (barely) reanimated, perhaps it is because, in the age of sonification, we have the power to make anything laugh back.

NOTES

1 We would like to thank the editors, Douglas Kahn, and each other for helpful suggestions on the chapter.

2 Édouard-Léon Scott de Martinville, *Le Problème de la parole s'écrivant elle-même: La France, l'Amérique* (Paris, 1878), cited in Hankins and Silverman (1995, 137) and in Levin (1990). Levin cites Scott in support of his thesis that, just as early cinema was heralded as a transparent reproduction of images that would supersede national languages, so the prehistory of sound recording articulated an "analogous discourse of democratization and univocal, natural signs." page 36

3 While the two terms have generally been used interchangeably, Thomas Hermann (2008) has recently suggested that "sonification" be used to denote the principle of rendering data as sound while "auditory display" should refer to the apparatus or technical system. Acknowledging this distinction, we use the term *sonification* as a catch-all.

REFERENCES

Barrass, Stephen, and Gregory Kramer. "Using Sonification." *Multimedia Systems* 7 (1999): 23–31.

Bell, Alexander Graham. "The Telephone: A Lecture Entitled Researches in Electric Telephony by Professor Alexander Graham Bell Delivered before the Society of Telegraph Engineers." October 31, 1877. New York: Society of Telegraph Engineers, 1878.

Bijker, Wiebe. *Of Bicycles, Bakelites, and Bulbs: Toward a Theory of Sociotechnical Change.* Cambridge, Mass.: MIT Press, 1995.

Blattner, Meera M., Denise A. Sumikawa, and Robert M. Greenberg. "Earcons and Icons: Their Structure and Common Design Principles." *Human-Computer Interaction* 4 (1989): 11–44.

Born, Georgina, and Andrew Barry. "Art-Science." *Journal of Cultural Economy* 3(1) (2010): 103–19.

Chanan, Michael. *Musica Practica: The Social Practice of Western Music from Gregorian Chant to Postmodernism.* New York: Verso, 1994.

———. *Repeated Takes: A Short History of Recording and Its Effects on Music.* London: Verso, 1995.

De Campo, Alberto, and Christian Dayé. "Sounds Sequential: Sonification in the Social Sciences." *Interdisciplinary Science Reviews* 31(4) (2006): 349–64.

Delatour, Thierry. "Molecular Music: The Acoustic Conversion of Molecular Vibrational Spectra." *Computer Music Journal* 24(3) (2000): 48–68.

Dunn, John, and Mary Anne Clark. " 'Life Music': The Sonification of Proteins." *Leonardo* 32(1) (1999): 25–32.

Edison, Thomas Alva. "The Perfected Phonograph." *North American Review* (1888): 647.

Ferguson, Sam, and Densil Cabrera. "Exploratory Sound Analysis: Sonifying Data about Sound." Paper presented at the International Conference on Auditory Display, Paris, 2008.

Flinn, Caryl. *Strains of Utopia: Gender, Nostalgia, and Hollywood Film Music*. Princeton, N.J.: Princeton University Press, 1992.

Foucault, Michel, and Colin Gordon. *Power/Knowledge: Selected Interviews and Other Writings, 1972–1977*, 1st American ed. New York: Pantheon, 1980.

Freud, Sigmund. *Civilization and Its Discontents*. New York: Norton, 1989.

Gaver, William W. "The SonicFinder: An Interface That Uses Auditory Icons." *Human-Computer Interaction* 4 (1989): 67–94.

Gitelman, Lisa. *Always Already New: Media, History, and the Data of Culture*. Cambridge, Mass.: MIT Press, 2006.

Grossberg, Lawrence. *We Gotta Get Out of This Place: Popular Conservatism and Postmodern Culture*. New York: Routledge, 1992.

Hall, Stuart. "On Postmodernism and Articulation: An Interview with Stuart Hall." *Journal of Communication Inquiry* 10(2) (1986): 45–60.

Hankins, Thomas L., and Robert J. Silverman. *Instruments and the Imagination*. Princeton, N.J.: Princeton University Press, 1995.

Hennessey, Meagan, and David Giovannoni. *The World's Oldest Sound Recordings Played for the First Time*. 2008. http://www.firstsounds.org/press/032708/release_2008-0327.pdf (accessed February 22, 2011).

Hermann, Thomas. "Taxonomy and Definitions for Sonification and Auditory Display." Paper presented at the International Conference on Auditory Display, Paris, 2008.

———, Christian Niehus, and Helge Ritter. "Interactive Visualization and Sonification for Monitoring Complex Processes." Paper presented at the International Conference on Auditory Display, Boston, 2003.

Hillis, Ken. *Digital Sensations: Space, Identity, and Embodiment in Virtual Reality*. Minneapolis: University of Minnesota Press, 1999.

Ihde, Don. "From da Vinci to CAD and Beyond." *Synthèse* 168 (2009): 453–68.

———. *Listening and Voice*. Albany: State University of New York Press, 2007.

Jay, Martin. *Downcast Eyes: The Denigration of Vision in Twentieth-Century French Thought*. Berkeley: University of California Press, 1993.

Jenkins, Henry. *Convergence Culture: Where Old and New Media Collide*. New York: New York University Press, 2006.

Kish, Daniel. "Seeing with Sound: What Is It Like to 'See' the World Using Sonar? Daniel Kish, Who Lost His Sight in Infancy, Reveals All." *New Scientist* (April 11, 2009).

Kittler, Friedrich. *Gramophone-Film-Typewriter*. Trans. G. Winthrop-Young and M. Wutz. Stanford: Stanford University Press, 1999.

Kramer, Gregory. *Auditory Display: Sonification, Audification, and Auditory Interfaces*. Reading, Mass.: Addison-Wesley, 1994.

———, and Bruce Walker. "Ecological Psychoacoustics and Auditory Displays: Hearing, Grouping, and Meaning Making." In *Ecological Psychoacoustics*, ed. J. G. Neuhoff. San Diego: Elsevier Academic, 2004.

Lastra, James. *Sound Technology and American Cinema: Perception, Representation, Modernity*. New York: Columbia University Press, 2000.

Latour, Bruno. *Aramis, or the Love of Technology*. Cambridge, Mass.: Harvard University Press, 1996.

Levin, Thomas Y. "For the Record: Adorno on Music in the Age of Its Technological Reproducibility." *October* 55 (Winter 1990): 23–48.

Manovich, Lev. *The Anti-Sublime Ideal in Data Art*. 2002.

———. *The Language of New Media*. Cambridge, Mass.: MIT Press, 2001.

Peirce, Charles S. *Philosophical Writings of Peirce*. New York: Dover, 1955.

Pinch, Trevor, and Wiebe Bijker. "The Social Construction of Facts and Artefacts: Or How the Sociology of Science and the Sociology of Technology Might Benefit Each Other." *Social Studies of Science* 14(3) (1984): 399–441.

Pinch, Trevor, and Nelly Oudshoorn, eds. *How Users Matter: The Co-Construction of Users and Technologies*. Cambridge, Mass.: MIT Press, 2003.

Polli, Andrea. "Atmospherics/Weather Works: A Spatialized Meteorological Data Sonification Project." *Leonardo* 28(1) (2005): 31–36.

Read, Oliver, and Walter L. Welch. *From Tinfoil to Stereo: Evolution of the Phonograph*. New York: Sams, 1976.

Richards, Jonathan, and Adam Sherwin. "Why BBC Radio 4 Newsreader Charlotte Green Got the Giggles and Couldn't Stop." *Times Online* March 28, 2009).

Robare, Paul, and Jodi Forlizzi. "Sound in Computer: A Short History." *Interactions* (January/February 2009): 62–65.

Roosth, Sophia. "Screaming Yeast: Sonocytology, Cytoplasmic Milieus, and Cellular Subjectivities." *Critical Inquiry* (Winter 2009): 332–50.

Scripture, Edward Wheeler. *Researches in Experimental Phonetics: The Study of Speech Curves*. Washington, D.C.: Carnegie Institute of Washington, 1906.

Slack, Jennifer Daryl. "The Theory and Method of Articulation in Cultural Studies." In *Stuart Hall: Critical Dialogues*, ed. D. Morley and K.-H. Chen. New York: Routledge, 1996.

———, and J. Macgregor Wise. *Culture + Technology: A Primer*. New York: Lang, 2006.

Sterne, Jonathan. *The Audible Past: Cultural Origins of Sound Reproduction*. Durham, N.C.: Duke University Press, 2003.

Turino, Thomas. "Signs of Imagination, Identity, and Experience: A Peircean Semiotics Theory for Music." *Ethnomusicology* 43(2) (1999): 221–55.

Virilio, Paul. *War and Cinema: The Logistics of Perception*. New York: Verso, 1989.

Index